Titles in This Series

Volume

1 **Markov random fields and their applications,** Ross Kindermann and J. Laurie Snell

2 **Proceedings of the conference on integration, topology, and geometry in linear spaces,** William H. Graves, Editor

3 **The closed graph and P-closed graph properties in general topology,** T. R. Hamlett and L. L. Herrington

4 **Problems of elastic stability and vibrations,** Vadim Komkov, Editor

5 **Rational constructions of modules for simple Lie algebras,** George B. Seligman

6 **Umbral calculus and Hopf algebras,** Robert Morris, Editor

7 **Complex contour integral representation of cardinal spline functions,** Walter Schempp

8 **Ordered fields and real algebraic geometry,** D. W. Dubois and T. Recio, Editors

9 **Papers in algebra, analysis and statistics,** R. Lidl, Editor

10 **Operator algebras and K-theory,** Ronald G. Douglas and Claude Schochet, Editors

11 **Plane ellipticity and related problems,** Robert P. Gilbert, Editor

12 **Symposium on algebraic topology in honor of José Adem,** Samuel Gitler, Editor

13 **Algebraists' homage: Papers in ring theory and related topics,** S. A. Amitsur, D. J. Saltman, and G. B. Seligman, Editors

14 **Lectures on Nielsen fixed point theory,** Boju Jiang

15 **Advanced analytic number theory. Part I: Ramification theoretic methods,** Carlos J. Moreno

16 **Complex representations of GL(2, K) for finite fields K,** Ilya Piatetski-Shapiro

17 **Nonlinear partial differential equations,** Joel A. Smoller, Editor

18 **Fixed points and nonexpansive mappings,** Robert C. Sine, Editor

19 **Proceedings of the Northwestern homotopy theory conference,** Haynes R. Miller and Stewart B. Priddy, Editors

20 **Low dimensional topology,** Samuel J. Lomonaco, Jr., Editor

21 **Topological methods in nonlinear functional analysis,** S. P. Singh, S. Thomeier, and B. Watson, Editors

22 **Factorizations of $b^n \pm 1$, $b = 2, 3, 5, 6, 7, 10, 11, 12$ up to high powers,** John Brillhart, D. H. Lehmer, J. L. Selfridge, Bryant Tuckerman, and S. S. Wagstaff, Jr.

23 **Chapter 9 of Ramanujan's second notebook—Infinite series identities, transformations, and evaluations,** Bruce C. Berndt and Padmini T. Joshi

24 **Central extensions, Galois groups, and ideal class groups of number fields,** A. Fröhlich

25 **Value distribution theory and its applications,** Chung-Chun Yang, Editor

26 **Conference in modern analysis and probability,** Richard Beals, Anatole Beck, Alexandra Bellow, and Arshag Hajian, Editors

27 **Microlocal analysis,** M. Salah Baouendi, Richard Beals, and Linda Preiss Rothschild, Editors

28 **Fluids and plasmas: geometry and dynamics,** Jerrold E. Marsden, Editor

29 **Automated theorem proving,** W. W. Bledsoe and Donald Loveland, Editors

30 **Mathematical applications of category theory,** J. W. Gray, Editor

31 **Axiomatic set theory,** James E. Baumgartner, Donald A. Martin, and Saharon Shelah, Editors

32 **Proceedings of the conference on Banach algebras and several complex variables,** F. Greenleaf and D. Gulick, Editors

33 **Contributions to group theory,** Kenneth I. Appel, John G. Ratcliffe, and Paul E. Schupp, Editors

34 **Combinatorics and algebra,** Curtis Greene, Editor

Titles in This Series

Volume

Titles in This Series

Volume

Dynamics and Control of Multibody Systems

CONTEMPORARY MATHEMATICS

Volume 97

Dynamics and Control of Multibody Systems

Proceedings of the AMS-IMS-SIAM Joint Summer Research Conference held July 30–August 5, 1988 with support from the National Science Foundation

J. E. Marsden, P. S. Krishnaprasad, and J. C. Simo, Editors

AMERICAN MATHEMATICAL SOCIETY

Providence • Rhode Island

The AMS-IMS-SIAM Joint Summer Research Conference in the Mathematical Sciences on Control Theory and Multibody Systems was held at Bowdoin College, Brunswick, Maine from July 30 to August 5, 1988 with support from the National Science Foundation, Grant DMS-8613199.

1980 *Mathematics Subject Classification* (1985 *Revision*). Primary 58FXX, 70HXX, 70QXX.

Library of Congress Cataloging-in-Publication Data

AMS-IMS-SIAM Joint Summer Research Conference in the Mathematical Sciences on Control Theory and Multibody Systems (1988: Bowdoin College)

Dynamics and control of multibody systems: proceedings of the AMS-IMS-SIAM joint summer conference held July 30–August 5, 1988, with support from the National Science Foundation/J. E. Marsden, P. S. Krishnaprasad, and J. C. Simo.

p. cm.—(Contemporary Mathematics, ISSN 0271-432; v. 97)

"The AMS-IMS-SIAM Joint Summer Research Conference in the Mathematical Sciences on Control Theory and Multibody Systems was held at Bowdoin College, Brunswick, Maine from July 30 to August 5, 1988...": T.p. verso.

Includes bibliographies.

ISBN 0-8218-5104-7 (alk. paper)

1. Control theory—Congresses. 2. Dynamics—Congresses. I. Marsden, Jerrold E. II. Krishnaprasad, P. S. (Perinkulam Sambamurthy),1949– . III. Simo, J. C. (Juan C.), 1952– . IV. Title. V. Title: Multibody systems. VI. Series: Contemporary Mathematics (American Mathematical Society); v. 97.

QA402.3.A4546 1988 89-15019
629.8′312–dc20 CIP

---

*This volume is dedicated to Roger Brockett
on the occasion of his 50th birthday.
Happy Birthday, Roger!!*

Contents

Foreword

by Roger Brockett

The conference on control theory and multibody dynamics as conceived and developed by the organizers was a happy event, covering a range of topics all of which could be viewed as applications of geometrical methods to problems arising in dynamics and control. I am pleased to have been honored by its organizers and participants. At several points during the conference I was reminded of a meeting on dynamics and control held at NASA/Ames in 1976. As compared with that earlier meeting, new areas of applications discussed at Bowdoin included work on the mathematics behind numerical methods in mechanics, the discussion of the dynamics of complex kinematic chains such as arise in automotive engineering and robotics, and various applications of strongly nonholonomic systems as in Berry's phase. By the time we shared the Down East conference banquet, those who listened to the lectures closely were able to say something mathematical about the deformation of lobster shells and the kinematic chain model of a lobster claw. For the most part papers in this volume are faithful to the idea that by using mathematics one can say a lot with a few words. This leaves room in the foreword for a few anecdotes and impressions written in a different style.

At breakfast one morning at Bowdoin several of us were sharing a table with a distinguished conferee from another AMS conference. In the same spirit as I might ask "What is elliptic cohomology?", he asked "What is control theory?" Of course television newscasters are able to dispense with much more difficult questions in an arbitrarily small amount of time, but we were reduced to giving examples. We could have done better. There have been researchers such as Rayleigh and Nyquist whose work has had enormous influence in part because they adopted a control systems point of view. The 1943 Gibbs lecture of H. Bateman, "On the Control of an Elastic Fluid," has admirable scope and impressive depth. It is exemplary in that it discusses important control-theoretic matters, feedback stability, the effect of delays, etc., and is able to capture the flavor of the systems point of view. That is, Bateman discusses dynamical systems as filters, passes freely between problems involving the transmission of sound, the transmission of electrical signals, the control of fluids, etc. He is a candidate for the pantheon of control theory because he uses input/output thinking in a way which is mathematically natural and succeeds in bringing out previously unobserved mathematical structures in real problems.

Suppose our breakfast companion has asked, "Why control theory and dynamics?" This is a more precise question and it requires a more detailed answer. I was fortunate to be starting my graduate studies in control in 1960. It was an exciting time; a flurry of new results were completely changing the nature of the subject. Bellman's work on dynamic programming, Pontryagin's maximum principle, recursive estimation, the blending of Fourier methods with Liapunov stability, and the use of Ito calculus were all introduced within a few years. In this period there was a great interest in dynamics because of Sputnik; aerospace problems associated with trajectory optimization and orbit transfer were common fare. (In contrast to the indifference one sees today, I recall that on the day of the first manned Mercury flight my probability class was cut short because the professor said "I can't think with that thing flying around up there.") Moreover, because of what seemed at the time to ban an all-consuming interest in optimal control, the calculus of variations in its various forms, including the variational formulations of dynamics, was never far from the center of the stage. Insofar as mathematics was concerned, the principal mode of thought was analysis. Since the early 1950's Solomon Lefschetz, first at Princeton and later at RIAS, had been attempting to resuscitate the subject of differential equations. Partly because of ongoing work in the Soviet Union, his interests included control theory viewed as an offshoot of differential equations. The style in vogue with respect to dynamics was that of Whittaker and Goldstein, and control theory and dynamics were closely linked.

This history has had a significant effect on the situation we find today even though, to a large extent, geometry has replaced analysis as the *lingua franca* in the common ground between control and dynamics. Although, as chronicled in the pioneering book of Abraham and Marsden, once started the geometrization of dynamics proceeded quite rapidly, control theory changed rather more slowly after the halcyon days of the early 60's. However, in time it became clear that differential geometry had the potential to express key control theoretic ideas in a natural way. A number of people took part in these developments. In my case, I had been fascinated for some time with the idea of nonholonomic constraints, partly because of the role they play in the planar integration scheme found in Vannevar Bush's differential analyzer and partly because they seemed mysterious. As I attempted to find a comfortable way of thinking about them, and the related matter of the Caratheodory statement of the second law of thermodynamics, I eventually realized that there was a wide variety of ways in which geometrical ideas could be helpful in forming a unified view of control.

What does the future hold? Developments in computer engineering have given us microprocessors and digital processing chips which make possible the implementation of a new level of mathematical sophistication in control systems. Developments in material science have put at our disposal magnetostrictive, piezoelectric, electro-rheological, etc., materials whose use in control mechanisms calls for a fresh look at modeling issues. There is significant progress in synthesizing controllable mechanisms on the scale of tens of microns. Since neither science nor technology stands still we can be sure that in the future, as in the past, there will be interesting challenges and a continuing need to meld what we know with ongoing developments.

Introduction

The study of complex, interconnected mechanical systems with rigid and flexible articulated components is of growing interest to engineers and mathematicians. The rich history of the subject derives primarily from the work of mechanism designers and the work of aerospace engineers interested in the modeling and control of complex multibody spacecraft. A more recent source of inspiration is in the area of robotics (eg. the control of multifingered hands, contact problems, etc.).

Recent work in this area reveals a rich geometry underlying the mathematical models that appear in this context. In particular, Lie groups of symmetries, reduction, and Poisson structures play a significant role in explicating the qualitative properties of multibody systems. A fresh look at covariant formulations of elasticity has also proved to be very useful. Geometric ideas should play an even more crucial role in the design of reliable numerical schemes for computer simulation of models of multibody systems, and the underlying control theory of Hamiltonian systems with symmetry is beginning to be worked out in a systematic manner.

In engineering applications, the question of exploiting the special structures of mechanical systems is important. For mechanical systems with symmetry (such as rigid bodies with internal rotors) one wants to take into account conservation of angular momentum and the special Poisson bracket structures on the associated reduced phase space. Certain mechanical problems involving control of interconnected rigid bodies can be formulated as Lie-Poisson systems, possibly with forcing and damping. It is likely that dynamic models of robotic manipulators are also amenable to such a formulation. The utility of this formalism lies in the natural way it describes certain controllability and stability questions.

The dynamics and control of interconnected rigid and flexible structures such as robotic, aeronautic, and space structures, involve difficulties in modeling, in mathematical analysis, and in numerical implementation. First of all, the formulation of the basic dynamical models is a nontrivial step. In modeling rotating systems with continuum-mechanical components such as plates, shells, or beams, nonlinear models (e.g., the so-called geometrically exact models) display behavior which is in certain instances qualitatively quite different from what is observed in linear and semi-linearized models. For example, if one views a plate or beam equation of say Euler-Bernoulli type as an approximation of a geometrically exact model, then the processes of *attachment* to a rapidly rotating rigid body (think of helicopter blades) and *approximation* do not commute. Done in the wrong order, the procedure can lead to spurious softening and hence to completely erroneous results even for small deflections. Thus, proper attention to dynamic modeling is a first crucial step. This is probably important even for "robust" techniques—all too often practical problems are "fixed up" in unsatisfactory ad hoc ways to balance bad modeling.

Some painful modeling lessons of this sort were learned in the early days of the U.S. space exploration program. In particular, they brought home the point that dynamic interactions between rigid and flexible components of a spacecraft are important. A famous example in this regard was the Explorer I mission in 1958. Energy dissipation in whip antennas attached to a passive spinning body caused instability and an end-over-end tumble. The key lesson here was that passive spin stabilization about the minor axis, while feasible in rigid bodies, is not possible in the presence of dissipative flexible components.

Experience with Explorer I and later with missions such as the Orbiting Geophysical Observatory III in 1966 (in this case excessive oscillations were induced by control system interactions with flexible beams) led to a vigorous program of research in multibody systems with flexible components. The approximate analytic and numerical techniques developed in the course of this research were quite successful in suggesting good designs for spacecraft of modest size and flexibility.

A new generation of spacecraft with large flexible components (radar arrays, solar collectors, truss structures) are presenting new challenges to our abilities to model and accurately predict the dynamic behavior of such structures in space. It is necessary to have the proper tools to do this since the requirements on the performance of these new spacecraft are quite unprecedented. The Hubble Space Telescope is expected to have a pointing accuracy of 0.01 arc second on rms jitter of less than 0.007 arc seconds.

Refined mathematical models and analyses will be necessary to attack such problems with a degree of confidence. New control methodologies will be necessary to maintain the effects of dynamics interactions in such large space structures within prescribed limits.

Recent developments in Hamiltonian dynamics and coupling of systems with symmetries (such as invariance under Euclidean motions) has shed new light on some of these issues. Likewise, engineering questions have suggested new mathematical structures. This commonality leads to new and interesting applications of Hamiltonian methods, such as the energy-momentum-Casimir method (for determining nonlinear stability) and bifurcation of Hamiltonian systems with symmetry (for uncovering nontrivial branches of new solutions when system parameters are varied). Other tools borrowed from Hamiltonian and Lagrangian systems theory have led to new results on periodic solutions of systems near equilibria, including information on their spatial symmetry, and to the proof of nonintegrability in certain regimes of phase space. These results also suggest how to construct numerical integration schemes that preserve energy and angular momentum, avoiding systematic biases or oscillations. When dissipation is added, all of these techniques appear to be naturally compatible with Lyapunov and invariance principle methods.

Further information on the above ideas and the larger context of control theory, dynamics, and its applications can be found in the panel report *Future Directions in Control Theory,* chaired by Wendell Fleming, and available through SIAM.

It was these sorts of developments that motivated us to organize this conference and to assemble this volume. We hope it will be a useful addition to the literature and to the cooperation between engineering and mathematics. It is a pleasure to thank all the participants for their enthusiasm and their contribution to the conference and its proceedings. Finally, it is our sincere pleasure to dedicate this volume to Roger Brockett on the occasion of his 50th birthday. He has been a great source of inspiration to many of us.

Contemporary Mathematics
Volume **97**, 1989

AN ENUMERATIVE THEORY OF EQUILIBRIUM ROTATIONS FOR PLANAR KINEMATIC CHAINS

J. Baillieul*

Abstract

In the growing literature on the dynamics of multibody systems, partial results have recently been given enumerating the relative equilibrium configurations of freely rotating planar kinematic chains. In this note, we show that the equilibria in question may be found by solving a set of algebraic equations, and thus the enumerative problem may be treated using the language and methods of algebraic geometry and intersection theory. We define a system of algebraic equilibrium equations, real solutions of which correspond to the configurations in question. It is shown that the equations are "degenerate" in the sense that (for chains longer than three links) they have solutions of positive dimension "at infinity." Although Bezout's theorem may not be applied to such a system, we are nevertheless able to derive a recursive formula for the exact number of solutions (some of which may be complex). Since real solutions correspond to actual equilibria, our number is an upper bound on the number of equilibrium configurations.

1. Introduction: The study of gravitational/centrifugal equilibria in uniformly rotating systems of both distributed and lumped masses continues to attract the interest of researchers in theoretical mechanics. A recent sense of excitement has emerged as the focus of inquiry has expanded from problems rooted purely in celestial mechanics to include the study of models of artificial systems such as complex elastic and articulated space structures. Work by Baillieul and Levi (1987), Baillieul (1987, 1988), Bloch (1987), Posbergh (1988), and Sreenath *et al.* (1988), for example, is directed at understanding the dynamics of rotating mechanical systems consisting of rigid bodies with elastic appendages and kinematic chains of rigid bodies interconnected by various types of joints. In particular, Sreenath *et al.* have studied the rotational mechanics of planar kinematic chains wherein the rotation of the system occurs in the plane in which the links of the chain are constrained to move. This work has motivated the work reported in the present paper where we take up the problem of enumerating relative equilibria for such rotating chains. (A relative equilibrium consists of a constant velocity rotation of the entire chain about its center of mass with no net translation of that center of mass and with no relative motion between any two points in the chain.)

It will be observed that the equations defining the relative equilibria have only trigonometric nonlinearities of a certain type, which makes it possible to rewrite the n equations for

* The author gratefully acknowledges support from the Air Force Office of Scientific Research under Grant No. AFOSR 85-0144

an n-jointed chain as 2n algebraic (=polynomial) equations in 2n unknowns. The advantage of characterizing the equilibria in terms of solutions to algebraic equations is that enumerative results in algebraic geometry and intersection theory may be applied. Pursuing this idea allowed us to enumerate the relative equilibria in a similar problem involving rotating chains in a gravitational field wherein the rotation was about an axis aligned with the field but perpendicular to the joint axes in the planar chain (Baillieul, 1987). It will be shown that the algebraic equilibrium equations for the problem of Sreenath *et al.* are formally degenerate in that the hypothesis of Bezout's theorem does not hold. In fact, these equations will be shown to be similar to the system of algebraic equations studied by Baillieul and Byrnes (1982) in carrying out a critical point analysis of an electric power system model. In both cases, the equations may be shown to have solution components "at infinity," and when the chain in question has more than three links, these components have positive dimension. Baillieul and Byrnes (1982) have indicated that the power system problem is similar to the classical problem of five conics which was solved heuristically by Chasles' in the 19-th century, and with greater rigor and more completeness much more recently. (See Kleiman, 1980.) The techniques applied by Baillieul and Byrnes to bound the number of solutions to their power system equations are essentially the same techniques that had been applied by others to enumerate solutions to the five conics problem. To establish the enumerative significance of these bounds (i.e. to show the bounds are tight), one must appeal to the genericity of the equation parameters. Unfortunately, the relative equilibrium equations for the rotating chain problems of Sreenath *et al.* are not "generic"; the coefficients in these equations satisfy a set of relations. Hence a naive attempt to solve the enumerative problem by the method used in Baillieul and Byrnes (1982) ("blowing up the components at infinity") does not yield the correct number of solutions to the equations. This is illustrated in Section 2 below by appealing to a direct calculation for a chain with 4 links (3 joints). For a chain with 2 links, the number of equilibrium solutions will be seen to be (obviously) 2, while for a chain with 3 links, the number of equilibrium solutions is either 4 or 6, depending on the parameters in the problem. These results for two-joint chains have been reported earlier by Sreenath *et al.*, and they are consistent with the Baillieul and Byrnes estimate.

In Section 3, we state and prove our main theorem regarding the number of solutions to the algebraic equilibrium problem. The proof involves an inductive argument and the explicit type of calculation used in elimination theory. (See Van der Waerden, 1973.) It has the advantage of yielding the exact general number of solutions to the algebraic equations. This number is an upper bound on the number of solutions to the actual equilibrium equations since only real solutions to the algebraic equations define actual relative equilibria.

2. Formulation and Low Order Examples: In Baillieul (1988), we consider a rotating system as depicted below in the Figure consisting of a single strand planar kinematic chain of rigid bodies attached at one end to a distinguished rigid body base. We suppose that this system is free to move in the plane; that is there are no forces impeding the movement of any part of the system with respect to the ambient two dimensional space. The links or bodies which comprise the structure are connected to one another in a single strand chain by frictionless revolute joints. To facilitate our study of the qualitative dynamics of such a system, we have made certain modeling assumptions: (i) Each link except for the base is massless. (ii) There is a distinguished coordinate frame (the "body frame") fixed in the base with its origin coinciding with the joint between the base and the first link. The location of each particle in the system relative to a frame of reference which is fixed with respect to the ambient space is given by prescribing the particle's coordinates with respect to the body frame together with the position and orientation of the body frame with respect to the fixed frame. (iii) The i-th link is characterized by its length r_i and the angle ψ_i which it forms with respect to the body-frame axis, and (iv) the mass of the j+1-st joint is m_j. We denote the mass of the base link by $\mathbf{m}_b$ and assume that the center of mass of the base link has coordinates $(c, 0)$ with respect to the body-frame. Choosing to study a kinematic chain of point masses interconnected by massless links avoids inessential complexity. Note, however, that we have retained a distributed mass description of the base link. This is, in part, to make contact with our earlier work (Baillieul, 1988) and in part to facilitate the inductive reduction in proving our main theorem in Section **3**.

With this notation, and following the exposition of Baillieul (1988), the configuration space is seen to be $SE(3) \times S^1 \times \cdots \times S^1$, where for each configuration, the $SE(3)$-component prescribes the position and orientation of the body frame with respect to a fixed frame, and the n S^1-components are the joint angles ψ_k prescribing the link configurations relative to the body frame.

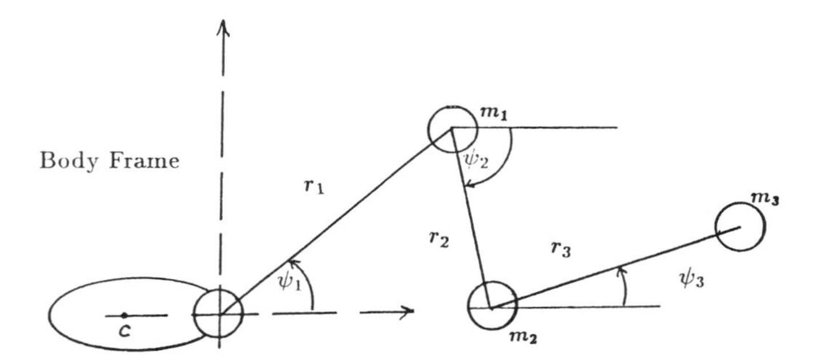

Figure 1: *A rigid body with planar kinematic chain attachment*

As depicted in the figure, the j+1-st joint is located (with respect to the body frame) at

$$\begin{pmatrix} u_j^1 \\ u_j^2 \end{pmatrix} = \begin{pmatrix} r_1 \cos\psi_1 + r_2 \cos\psi_2 + \cdots + r_j \cos\psi_j \\ r_1 \sin\psi_1 + r_2 \sin\psi_2 + \cdots + r_j \sin\psi_j \end{pmatrix}.$$

The complete dynamical model of this system is given in Baillieul, (1988). In the present paper, we shall only be concerned with equilibrium rotations in which each joint angle ψ_k is constant and where the entire system rotates at a constant angular velocity about its center of mass which is motionless relative to the fixed frame. A straightforward calculation using the dynamical model presented in Baillieul (1988) shows that these equilibrium rotations are characterized by the equations

$$\sum_{j=1}^{n} \alpha_{ij} \sin(\psi_i - \psi_j) + c\beta_i \sin\psi_i = 0 \quad \text{for } i = 1, \ldots, n \tag{2.1}$$

where

$$\begin{aligned} \alpha_{ij} &= (\mathbf{m}_b + m_1 + \cdots + m_{i-1})(m_j + \cdots + m_n) r_i r_j \quad \text{if } i < j \\ \alpha_{ij} &= \alpha_{ji} \\ \beta_i &= \mathbf{m}_b (m_i + \cdots + m_n) r_i \end{aligned}$$

Systems of equations with trigonometric nonlinearities of this type have been studied (see e.g. Baillieul and Byrnes, 1982) by noting that with the change of variables $x_i = \cos\psi_i$, $y_i = \sin\psi_i$, the system (2.1) is transformed into a system of algebraic equations

$$\sum_i \beta_i y_i = 0 \tag{2.2}$$

$$\sum_j \alpha_{ij}(x_j y_i - x_i y_j) + \beta_i y_i = 0 \quad i = 1, \ldots, n-1 \tag{2.3}$$

$$x_i^2 + y_i^2 - 1 = 0 \qquad i = 1, \ldots, n. \tag{2.4}$$

The connection between these systems of equations is given by the following:

Lemma 1: *There is a one-to-one correspondence between solutions to (2.1) in which $-\pi < \psi_i \leq \pi$ for $i = 1, \ldots, n$ and real solutions to the system (2.2)-(2.4).*

Although the above transformation doubles the number of unknowns as well as the number of equations, it permits us to exploit the rich classical theory of equations as well as some modern results in algebraic geometry and intersection theory. Baillieul and Byrnes (1982) took this approach in studying the critical point equations arising in certain electric power systems models. It was found that the classical Bezout theorem bounding the number of solutions was inapplicable because the system admitted solutions of positive dimension at infinity. Because of the obvious similarity between the power system equations and the above equations describing equilibrium rotations, it is not surprising that the system (2.2)-(2.4) also has solutions of positive dimension "at infinity". In order to pursue this remark, we introduce an additional variable z and write down a system of homogeneous equations corresponding to the affine system (2.2)-(2.4).

$$\sum \beta_i y_i = 0 \tag{2.5}$$

$$\sum \alpha_{ij}(x_j y_i - x_i y_j) + c\beta_i y_i z = 0 \quad i = 1, \ldots, n-1 \tag{2.6}$$

$$x_i^2 + y_i^2 - z^2 = 0 \qquad i = 1, \ldots, n. \tag{2.7}$$

It is standard to view solutions to the system (2.5)-(2.7) as points in $\mathbf{CP}^{2n}$. There is a one-to-one correspondence between solutions to (2.2)-(2.4) and solutions to (2.5)-(2.7) in which $z \neq 0$. (Cf. Shafarevich, 1974, p. 30.) Points in $\mathbf{CP}^{2n}$ with $z \neq 0$ are called *finite* while points with $z = 0$ are said to be ***infinite*** or ***at infinity***. Enumerative results from algebraic geometry (such as Bezout's theorem) count both finite and infinite solutions. Thus although only finite solutions are of interest in studying (2.1), we must investigate solutions to (2.5)-(2.7) at infinity if we wish to correctly enumerate (the finite) solutions to (2.2)-(2.4). Solutions at infinity satisfy (2.5) together with

$$\sum_j \alpha_{kj}(x_j y_k - x_k y_j) = 0 \quad k = 1, \ldots, n-1 \tag{2.8}$$

$$x_k = \pm i y_k \quad k = 1, \ldots, n \tag{2.9}$$

(where $i = \sqrt{-1}$). We prove the following

Lemma 2: *The hypersurfaces in $\mathbf{CP}^{2n}$ defined by the homogenized equilibrium equations (2.5)-(2.7) intersect at infinity in an algebraic set containing two $n-2$ dimensional planes.*

Proof: Clearly the solution set to (2.5),(2.8),(2.9) contains

$$\mathbf{CP}_{+}^{n-2} : \ x_j = i y_j, \ z = 0, \ \sum \beta_j y_j = 0$$

and

$$\mathbf{CP}_{-}^{n-2} : \ x_j = -i y_j, \ z = 0, \ \sum \beta_j y_j = 0$$

proving the lemma.

From this lemma, it follows that Bezout's theorem does not provide a meaningful count of solutions to either (2.2)-(2.4) or (2.5)-(2.7). It also suggests that there might be parallels between the enumeration problem for the equilibrium equations (2.1) and the power system problem treated in Baillieul and Byrnes (1982). In fact, using the modified Bezout theory developed in this reference to account for the planes at infinity, we find the number of (possibly complex) solutions to the algebraic equilibrium equations (2.2)-(2.4) to be bounded above by $\binom{2n}{n}$. This is essentially the same bound that applied to the problem treated in Baillieul and Byrnes (1982), but we shall show that the number of solutions to the system (2.2)-(2.4) does not achieve this bound for $n > 2$. Before stating the general result in the next section, we treat three examples: the cases $n = 1, 2, 3$.

Example 1: In the case n=1, (1.1) reduces to $\sin \psi_1 = 0$ which obviously has the two solutions $\psi_1 = 0, \pi$ (and only these two solutions) in the interval $-\pi < \psi_1 \leq \pi$.

Example 2: When n=2, the system (2.2)-(2.4) may be rendered

$$(m_1 + m_2) r_1 y_1 + m_2 r_2 y_2 = 0 \tag{2.10}$$

$$m_2 r_2 (x_2 y_1 - x_1 y_2) - (m_1 + m_2) c y_1 = 0 \tag{2.11}$$

$$x_1^2 + y_1^2 - 1 = 0; \qquad x_2^2 + y_2^2 - 1 = 0. \tag{2.12}$$

This set of equations is easily solved using successive elimination. (See Van der Waerden, 1973.) Solving (2.10) for y_2 and substituting this into (2.11) yields

$$[m_2 r_2 x_2 + (m_1 + m_2) r_1 x_1 - (m_1 + m_2) c] y_1 = 0. \tag{2.13}$$

Two cases must be treated separately: When $y_1 = 0$, we must also have $y_2 = 0$. The solutions to (2.1) corresponding to this case are what we shall call the "elementary solutions"

$$\begin{array}{ll} (\psi_1, \psi_2) = (0, 0), & (\psi_1, \psi_2) = (0, \pi) \\ (\psi_1, \psi_2) = (\pi, 0), & (\psi_1, \psi_2) = (\pi, \pi) \end{array}.$$

These are the only solutions in the case $y_1 = 0$, and they exist for all values of the parameters.

If $y_1 \neq 0$, we have the linear equation

$$(m_1 + m_2)r_1x_1 + m_2r_2x_2 = (m_1 + m_2)c \tag{2.14}$$

which together with (2.10) and (2.12) determines another set of (possibly complex) solutions. Indeed, it is not difficult to show that this system admits the explicit solution

$$x_1 = \frac{(m_1 + m_2)^2(r_1^2 + c^2) - m_2^2r_2^2}{2(m_1 + m_2)^2r_1c} \qquad x_2 = \frac{m_2^2r_2^2 + (m_1 + m_2)^2(c^2 - r_1^2)}{2(m_1 + m_2)m_2r_2c}$$

with corresponding values of y_1, y_2 chosen to satisfy (2.10) and (2.12). Because the signs of y_1 and y_2 must differ, it is clear that there are precisely two solutions to (2.10), (2.14), (2.12). These solutions will be real precisely when $|x_i| \leq 1$, $i = 1, 2$. To interpret this condition parametrically, there is simplification and no loss of generality in scaling the mass and link length parameters so that $c = -1$, $m_1 = 1$, and $m_2 = m$. Then we find the solutions to (2.10),(2.14),(2.12) will be real precisely when

$$|r_1 - 1|(1 + m) \leq mr_2 \leq (1 + m)(r_1 + 1). \tag{2.15}$$

We may summarize these calculations as follows:

Proposition 1: *For $n = 2$, the equilibrium equations (2.1) admit either 4 or 6 solutions in the region $-\pi < \psi_i \leq \pi$, $i = 1, 2$. Counting multiplicities, there are 6 solutions precisely when the inequality (2.15) is satisfied.*

Remark 1: The product of the degrees of equations (2.2)-(2.4) is 8, and, as Bezout's theorem predicts, this is the number of solutions to the homogenized system (2.5)-(2.7). It of course includes the two solutions at infinity identified in Lemma 2, and the upper bound of $\binom{2n}{n}$ on finite solutions is tight in this case.

Remark 2: The pair of equations (2.14), (2.10) has the interesting physical interpretation that the centers of mass of the base link and the last link both coincide with the centers of mass of the entire system. Indeed, the parameter ranges in which there are 6 solutions to (2.1) may be interpreted as prescribing precisely those physical dimensions such that it is possible for the centers of mass to coincide.

Remark 3: The content of Proposition 1 has in fact previously been reported by Sreenath *et al.* (1988). The figures they present describing the equilibrium configurations, however, fail to respect the physical interpretation given above.

Example 3: When n=3 we again subdivide consideration of the system (2.2)-(2.4) into the cases $y_1 = 0$ and $y_1 \neq 0$. When $y_1 = 0$, it is clear that there will be 6 (possibly complex) solutions to (2.2)-(2.4) corresponding to $x_1 = 1$ together with 6 other solutions corresponding to $x_1 = -1$. When $y_1 \neq 0$, by using (2.2) to eliminate y_3 from the equations in (2.3) we obtain the equivalent system

$$(m_1 + m_2 + m_3)r_1y_1 + (m_2 + m_3)r_2y_2 + m_3r_3y_3 = 0 \tag{2.16}$$

$$(m_1 + m_2 + m_3)r_1x_1 + (m_2 + m_3)r_2x_2 + m_3r_3x_3 = (m_1 + m_2 + m_3)c \tag{2.17}$$

$$y_2\left[(\mathbf{m}_b(m_2+m_3)r_1x_1+(\mathbf{m}_b+m_1)(m_2+m_3)r_2x_2+(\mathbf{m}_b+m_1)m_3r_3x_3-\mathbf{m}_b(m_2+m_3)c\right] \\ +(\mathbf{m}_b+m_1+m_2+m_3)m_1r_1x_2y_1=0 \tag{2.18}$$

$$x_i^2+y_i^2-1=0 \quad i=1,2,3. \tag{2.19}$$

If (2.17) is used to eliminate x_3 from (2.18), (2.18) may be replaced by

$$r_1(x_2y_1-x_1y_2)+cy_2=0. \tag{2.20}$$

Proceeding to eliminate variables, it is again possible to explicitly write down solutions to the system prescribed by (2.16), (2.17), (2.20), and (2.19). There are precisely four solutions (counted with multiplicity), and while it is somewhat tedious to write out the explicit formulas in general, under the simplifying assumption that $c=-1$, $m_1=m_2=m_3=m$, and $r_1=r_2=r_3=r$ they are given in Table 1. In general, our calculations may be summarized as follows.

Proposition 2: *For n=3 the algebraic equilibrium equations (2.2)-(2.4) admit 16 solutions (counting multiplicities).*

Remark 4: When n=3, $\binom{2n}{n}=20$. Thus Proposition 2 calls into question the enumerative significance of this bound. The problem, of course, is that $\binom{2n}{n}$ is only an upper bound on the number of solutions, and a precise enumeration requires a more detailed analysis. (Cf. Kleiman's (1980) remarks on the five conics problem.) It is perhaps interesting to note that despite the apparent similarity between the equilibrium equations (2.1) and the critical point equations of Baillieul and Byrnes (1982), it is only in the latter case that the bound we compute by "blowing up" the lines at infinity appears to be tight. This is because the coefficients in the equations (2.1) are very far from being generic.

$$\begin{pmatrix} x_1\\ x_2\\ x_3\\ y_1\\ y_2\\ y_3\end{pmatrix}=\begin{pmatrix} -\frac{1}{2r}\\ -\frac{1}{2r}\\ -\frac{1}{2r}\\ \mp\frac{1}{2r}\sqrt{4r^2-1}\\ \pm\frac{1}{2r}\sqrt{4r^2-1}\\ \pm\frac{1}{2r}\sqrt{4r^2-1}\end{pmatrix} \qquad \begin{pmatrix} x_1\\ x_2\\ x_3\\ y_1\\ y_2\\ y_3\end{pmatrix}=\begin{pmatrix} -\frac{8r^2+9}{18r}\\ \frac{8r^2-9}{6r}\\ -\frac{8r^2-9}{6r}\\ \mp\frac{4}{9r}\sqrt{(\frac{9}{4}-r^2)(r^2-\frac{9}{16})}\\ \pm\frac{4}{3r}\sqrt{(\frac{9}{4}-r^2)(r^2-\frac{9}{16})}\\ \mp\frac{4}{3r}\sqrt{(\frac{9}{4}-r^2)(r^2-\frac{9}{16})}\end{pmatrix}$$

Table 1: *Showing the 4 solutions to (2.2)-(2.4) in which* $y_1\neq 0$.

3. Enumeration of Solutions to the Algebraic Equilibrium Equations: Rather than attempting to refine the modified Bezout theory of Baillieul and Byrnes (1982) to enumerate the solutions to (2.1), we shall use an inductive argument based on elimination theory (see Van der Waerden, 1973) to show that the solution set to (2.2)-(2.4) for $n>1$ may be written as the disjoint union of four sets of points, two of which correspond to solution sets for "virtual" n-1 joint kinematic chains and two of which correspond to solution sets for "virtual" n-2 joint chains.

Theorem: *The number of finite solutions to the algebraic equilibrium equations (2.2)-(2.4) (counting multiplicities) is* $\mathcal{R}(n)$ *where*

$$\mathcal{R}(0) = 1,\ \mathcal{R}(1) = 2 \text{ and for } n \geq 2$$
$$\mathcal{R}(n) = 2\mathcal{R}(n-1) + 2\mathcal{R}(n-2).$$

Proof: The proof consists of separately treating the cases $y_1 = 0$ and $y_1 \neq 0$. When $y_1 = 0$, we know that either $x_1 = 1$ or $x_1 = -1$. When $x_1 = 1$, the equations (2.2)-(2.4) become the equations of an n-1-joint chain with parameters $\bar{\mathbf{m}}_b = \mathbf{m}_b + m_1$, $\bar{m}_k = m_{k+1}$ $(k = 1,\ldots,n-1)$, $\bar{c} = \frac{1}{\bar{\mathbf{m}}_b}(\mathbf{m}_b c + m_1 r_1)$, and $\bar{r}_k = r_{k+1}$ $(k = 1,\ldots,n-1)$. Similarly, if we substitute $x_1 = -1$ and $y_1 = 0$ into the equations (2.2)-(2.4), we also obtain the equations of an n-1-joint kinematic chain. Hence, if we count possible multiplicities, the induction hypothesis implies that there are $2\mathcal{R}(n-1)$ solutions to (2.2)-(2.4) in which $y_1 = 0$.

If $y_1 \neq 0$, then using (2.2) we may eliminate y_n from the $k = 1$ equation in (2.3). This yields

$$\beta_1 x_1 + \beta_2 x_2 + \cdots + \beta_n x_n - (m_1 + \cdots + m_n)c = 0. \tag{3.1}$$

Together with (2.2), this implies that the center of mass of the base link coincides with the centers of mass of the rest of the system. We claim that the n-1 links numbered 2 through n assume an equilibrium configuration as if the base and first links were removed. It is useful to rewrite the general equation (2.3) as follows:

$$\begin{aligned}
&\mathbf{m}_b(m_k + \cdots + m_n)r_1(x_1 y_k - x_k y_1) + (\mathbf{m}_b + m_1)(m_k + \cdots + m_n)r_2(x_2 y_k - x_k y_2) \\
&\quad +(\mathbf{m}_b + m_1 + m_2)(m_k + \cdots + m_n)r_3(x_3 y_k - x_k y_3) + \cdots \\
&+(\mathbf{m}_b + m_1 + m_2 + \cdots + m_{k-2})(m_k + \cdots + m_n)r_{k-1}(x_{k-1}y_k - x_k y_{k-1}) \\
&\quad +(\mathbf{m}_b + m_1 + m_2 + \cdots + m_{k-1}))(m_{k+1} + \cdots + m_n)r_{k+1}(x_{k+1}y_k - x_k y_{k+1}) \\
&\quad + \cdots + \\
&\quad +(\mathbf{m}_b + m_1 + \cdots + m_{k-1})m_n r_n(x_n y_k - x_k y_n) - \mathbf{m}_b(m_k + \cdots + m_n)c y_k = 0
\end{aligned} \tag{3.2}$$

Using (3.1) and (2.2), we may eliminate the variables x_n and y_n from this equation to obtain

$$\begin{aligned}
(m_1 + \cdots + m_{k-1})c y_k &- (m_1 + \cdots + m_{k-1})r_1(x_1 y_k - x_k y_1) \\
&- (m_2 + \cdots + m_{k-1})r_2(x_2 y_k - x_k y_2) \\
&- \cdots - \\
&- m_{k-1}r_{k-1}(x_{k-1}y_k - x_k y_{k-1}) = 0
\end{aligned} \tag{3.3}$$

Now multiply through by $\mathbf{m}_b(m_k + \cdots + m_n)/(m_1 + \cdots + m_{k-1})$ and add the result to (3.2) to obtain

$$\begin{aligned}
(\mathbf{m}_b + m_1 + \cdots + m_{k-1})\Big[&\frac{m_1(m_k + \cdots + m_n)}{m_1 + \cdots + m_{k-1}} r_2(x_2 y_k - x_k y_2) \\
&+ \frac{(m_1 + m_2)(m_k + \cdots + m_n)}{m_1 + \cdots + m_{k-1}} r_3(x_3 y_k - x_k y_3) \\
&+ \cdots + \frac{m_1 + \cdots + m_{k-2})(m_k + \cdots + m_n)}{m_1 + \cdots + m_{k-1}} r_{k-1}(x_{k-1}y_k - x_k y_{k-1}) \\
&+ (m_{k+1} + \cdots m_n)r_{k+1}(x_{k+1}y_k - x_k y_{k+1}) \\
&+ \cdots + m_n r_n(x_n y_k - x_k y_n)\Big] = 0.
\end{aligned}$$

This may be rewritten as

$$\begin{aligned}
&\bar{\mathbf{m}}_b(\bar{m}_{k-2}+\cdots+\bar{m}_{n-2})\bar{r}_1(\bar{x}_1\bar{y}_{k-2}-\bar{x}_{k-2}\bar{y}_1)\\
&\quad+(\bar{\mathbf{m}}_b+\bar{m}_1)(\bar{m}_{k-2}+\cdots+\bar{m}_{n-2})\bar{r}_2(\bar{x}_2\bar{y}_{k-2}-\bar{x}_{k-2}\bar{y}_2)+\cdots\\
&+(\bar{\mathbf{m}}_b+\bar{m}_1+\cdots+\bar{m}_{k-4})(\bar{m}_{k-2}+\cdots+\bar{m}_{n-2})\bar{r}_{k-3}(\bar{x}_{k-3}\bar{y}_{k-2}-\bar{x}_{k-2}\bar{y}_{k-3})\\
&\quad+(\bar{\mathbf{m}}_b+\bar{m}_1+\cdots+\bar{m}_{k-3})(\bar{m}_{k-1}+\cdots+\bar{m}_{n-2})\bar{r}_{k-1}(\bar{x}_{k-1}\bar{y}_{k-2}-\bar{x}_{k-2}\bar{y}_{k-1})\\
&\quad+\cdots+(\bar{\mathbf{m}}_b+\bar{m}_1+\cdots+\bar{m}_{k-3})\bar{m}_{n-2}\bar{r}_{n-2}(\bar{x}_{n-2}\bar{y}_{k-2}-\bar{x}_{k-2}\bar{y}_{n-2})\\
&\quad-\bar{\mathbf{m}}_b\bar{c}(\bar{m}_{k-2}+\cdots+\bar{m}_{n-2})\bar{y}_{k-2}=0 \qquad (3.4)
\end{aligned}$$

where

$$\bar{\mathbf{m}}_b = m_1+m_2$$

$$\bar{m}_k = m_{k+2} \quad (k=1,2,\ldots,n-2)$$

$$\bar{r}_k = r_{k+2} \quad (k=1,2,\ldots,n-2)$$

$$\bar{c} = -\frac{m_1 r_2}{m_1+m_2}$$

and the variables $\bar{x}$, $\bar{y}$ are defined by the affine isomorphism

$$\begin{pmatrix}\bar{x}_k\\ \bar{y}_k\end{pmatrix}=\begin{pmatrix}x_2 & y_2\\ -y_2 & x_2\end{pmatrix}\begin{pmatrix}x_{k+2}\\ y_{k+2}\end{pmatrix} \quad (k=1,\cdots,n-2).$$

By the induction hypothesis, this system of equations has (counting multiplicities) $\mathcal{R}(n-2)$ solutions. For $k=3,\ldots,n$, we write

$$\begin{pmatrix}x_k\\ y_k\end{pmatrix}=\begin{pmatrix}x_2 & -y_2\\ y_2 & x_2\end{pmatrix}\begin{pmatrix}\bar{x}_{k-2}\\ \bar{y}_{k-2}\end{pmatrix} \qquad (3.5)$$

and these expressions may be substituted into (2.2) and (3.1) to yield the system

$$\begin{aligned}
\beta_1 x_1+\bar{\beta}_2 x_2-\bar{\beta}_3 y_2&=\gamma\\
\beta_1 y_1+\bar{\beta}_3 x_2+\bar{\beta}_2 y_2&=0
\end{aligned} \qquad (3.6)$$

where $\gamma=(m_1+\cdots+m_n)c$, $\bar{\beta}_2=\beta_2+\beta_3\bar{x}_1+\cdots+\beta_n\bar{x}_{n-2}$, and $\bar{\beta}_3=\beta_2+\beta_3\bar{y}_1+\cdots+\beta_n\bar{y}_{n-2}$. Rewrite (3.6) as

$$\beta_1 x_1=\gamma-\bar{\beta}_2 x_2+\bar{\beta}_3 y_2$$

$$\beta_1 y_1=-\bar{\beta}_3 x_2-\bar{\beta}_2 y_2.$$

Squaring both sides, adding, and taking note of the relations (2.4), we obtain

$$\beta_1^2=\gamma^2-2\bar{\beta}_2\gamma x_2+2\bar{\beta}_3\gamma y_2+\bar{\beta}_2^2+\bar{\beta}_3^2.$$

This, together with the equation $x_2^2+y_2^2=1$ yields two solutions for (x_2,y_2), each of which may be substituted into (3.6) and (3.5) to yield corresponding values for x_1, y_1, and x_k, y_k for $k=3,\ldots,n$. Hence, we have shown that for each of the $\mathcal{R}(n-2)$ solutions $(\bar{x}_k,\bar{y}_k)$ $(k=1,\ldots,n-2)$ to the reduced system (3.4), there corresponds two solutions to (2.2)-(2.4). There are no other solutions to (2.2)-(2.4) with $y_1\neq 0$.

Taking the $2\mathcal{R}(n-1)$ solutions where $y_1=0$ together with the $2\mathcal{R}(n-2)$ solutions in which $y_1\neq 0$, we have a total of $\mathcal{R}(n)=2\mathcal{R}(n-1)+2\mathcal{R}(n-2)$ solutions, proving the theorem.

We conclude with a short table values of $\mathcal{R}(n)$, and we note that this sequence does not appear in Sloane (1973).

n	0	1	2	3	4	5	6	7	8	9
$\mathcal{R}(n)$	1	2	6	16	44	120	328	896	2448	6688

Table 2: *The number of solutions to the algebraic equilibrium equations.*

REFERENCES

1. Baillieul, J. (1987) "Equilibrium Mechanics in Rotating Systems," Proc. IEEE Conf. on Decision and Control, Los Angeles, Dec. 9-11, 1987, pp. 1429-1434.
2. Baillieul, J. (1988) "Linearized Models for the Control of Rotating Beams," Proc. IEEE Conference on Decision and Control, Austin, Texas, Dec. 7-9, 1988, pp. 1726-173.
3. Baillieul, J. and Byrnes, C.I. (1982) "Geometric Critical Point Analysis of Lossless Power System Models," IEEE Trans. Circuits and Systems, CAS-29 (November), pp. 724-737.
4. Baillieul, J. and Levi, M. (1987) "Rotational Elastic Dynamics," *PHYSICA D*, v. 27D, pp. 43-62.
5. Bloch, A.M. (1987) "Stability and Equilibria of Deformable Systems," Proc. 26-th IEEE Conference on Decision and Control, Los Angeles, Dec. 9-11, 1987, p. 1443.
6. Kleiman, S.L., (1980) "Chasles' Enumerative Theory of Conics: A Historical Introduction," in *Studies in Algebraic Geometry*, vol. 20, MAA Studies in Mathematics, pp. 117-138.
7. Posbergh, T.A., (1988) "Modeling and Control of Mixed and Flexible Structures," PhD Thesis, Systems Research Center, University of Maryland.
8. Shafarevich, I.R., (1974) *Basic Algebraic Geometry*, Grund. der math. Wissenschaften 213. Springer-Verlag, New York.
9. Sloane, N.J.A. (1973) *A Handbook of Integer Sequences*, Academic Press, New York.
10. Sreenath, N., Oh, Y.G., Krishnaprasad, P.S., and Marsden, J.E. (1988) "Dynamics of Coupled Planar Rigid Bodies," *Dynamics and Stability of Systems*, vol. 3, no. 1-2, pp. 25-49.
11. Van der Waerden, B.L., (1973) *Einführung in die algebraische Geometrie*, in Grund. der math. Wissenschaften 51. Springer-Verlag, Heidelberg.

Aerospace/Mechanical Engineering
Boston University
Boston, MA 02215

Contemporary Mathematics
Volume **97**, 1989

Stability and Stiffening of Driven and Free Planar Rotating Beams

*Anthony M. Bloch** and *Robert R. Ryan*

Dedicated to Roger Brockett on the occasion of his 50th birthday

Abstract

In this paper we analyze the stability of a beam rotating about its longitudinal axis and rotating about an axis perpendicular to the plane of rotation. We present firstly a general model for the beam which can be consistently linearized to include all important coriolis and centrifugal effects, and then we abstract a model containing the essential features. We compare the case where the beam is driven at constant angular velocity with the case where it rotates freely. In the former case we get linear models that we can solve explicitly, while in the latter case we have an essentially nonlinear model the stability of which may be analyzed by the Energy-Momentum method. An essential part of our stability analysis is an eigenvalue inequality due to Boyce, Di Prima and Handelman.

Introduction

The analysis of the dynamics and stability of interconnected rigid and flexible bodies is of great interest from both a theoretical and a practical point of view. Theoretically it presents us with the problem of unifying the classical theories of rigid body motion and elasticity, while there are many practical applications in the analysis of space structures and robotic mechanisms.

Two recent analyses of the stability of systems consisting of a rigid body with flexible appendage are those of Baillieul and Levi [1] and Krishnaprasad and Marsden [9]. The former analysis was from a Lagrangian and the latter from a Hamiltonian point of view, utilizing the so-called Energy-Casimir method [6]. These papers give a general analysis of the rotational behavior

---

* Supported in part by NSF Grant DMS-87 01576 and AFOSR Grant AFOSR-ISSA-87-0077 and by the U.S. Army Research Office through the Mathematical Sciences Institute of Cornell University.

of the systems in question. In this paper we give a detailed analysis of motion-induced stiffening and softening effects on the beam model which are critical to the analysis of stability in certain configurations. We compare the effect of centrifugal stiffening in the cases where we have an inextensible unshearable beam attached to a rigid hub rotating about its longitudinal axis and where the beam is rotating in a plane perpendicular to the axis of rotation. The axial case was examined in Baillieul and Levi [1] and the planar case without stiffening in Bloch [3] and [4].

The importance of stiffening has been discussed by Simo and Vu-Quoc in [11] and [12] and by Kane, Ryan and Bannerjee in [8]. In these papers the importance of the stiffening terms is verified by simulation. It is our aim in the present paper to give rigorous stability results for the axial and planar systems, taking into account stiffening effects and to explain the difference in the bifurcation behavior of the two systems.

Our results for stability in the planar case depend on an eigenvalue inequality developed by Boyce, Di Prima and Handelman.

We begin by presenting a very general consistently linearized model based on that developed by Kane, Ryan and Bannerjee, and then we abstract the essential features needed for our analysis.

We also compare here the cases where the beams are driven at constant angular velocity and where they are free to rotate. In the former case the models are linear and we can solve the equations explicitly by separation of variables. In the latter case, due to angular momentum conservation, the models are essentially nonlinear but we can derive results on the stability of equilibria by the Energy-Momentum method. It turns out that, for the planar system, the precise inequality result needed to prove nonlinear stability of the trivial equilibrium is that of Boyce, Di Prima and Handelman.

The outline of the paper is as follows. In Section 1 we present a very general model of the rotating beam. In Section 2 we specialize to the cases needed here, deriving the equations from first principles to illustrate the relative importance and physical meaning of the various terms. In Section 3 we first present the stability theorems for the various cases discussed above and then give the proofs. A small extension of the Boyce, Di Prima, Handelman inequality is given as a key to the proofs in the planar case. Finally, in Section 6 we discuss some finite-dimensional models which reflect rather accurately the behavior of the infinite-dimensional systems.

§1 *General System Model*

In this section we consider a very general model of a beam rotating about a rigid hub with prescribed angular and translational hub velocities. We then show how to reduce the model to the case of interest in this paper. We consider a uniform, homogeneous, flexible beam B fixed at one end (point 0) to a free-flying rigid base body A . The beam is characterized by a mass per unit length ρ , cross-sectional area A_0 , shear area A_S, principle second area moments I_2 and I_3 , effective torsional stiffness χ' , Young's Modulus E, and shear modulus G . Denote its unstretched length by ℓ .

The motion of the base is characterized by the inertial translational velocity v^0 of point 0 and the inertial angular velocity ω^A of A given by

$$v^0(t) = v_1\mathbf{a}_1 + v_2\mathbf{a}_2 + v_3\mathbf{a}_3 \tag{1.1}$$

$$\omega^0(t) = \omega_1\mathbf{a}_1 + \omega_2\mathbf{a}_2 + \omega_3\mathbf{a}_3 \tag{1.2}$$

where $\mathbf{a}_1, \mathbf{a}_2, \mathbf{a}_3$ form a dextral set of mutually perpendicular unit vectors fixed in A, while v_i and ω_i $(i = 1, 2, 3)$ are scalar temporal functions.

The deformation of B, on the other hand, is described in terms of six displacement quantities $u_1, u_2, u_3, \theta_1, \theta_2,$ and θ_3 which are functions of both space and time. The first three of these quantities, namely $u_1, u_2,$ and u_3, represent the $\mathbf{a}_1, \mathbf{a}_2,$ and $\mathbf{a}_3$ measure numbers of the displacement vector u of a generic point P on the beam. Thus, if P is located at a distance x from point 0 when B is undeformed, then the position vector P^{op} from 0 to P in any deformed or undeformed state can be written as

$$P^{op} = x\mathbf{a}_1 + u = (x + u_1)\mathbf{a}_1 + u_2\mathbf{a}_2 + u_3\mathbf{a}_3 \quad . \tag{1.3}$$

The other three displacement quantities, namely $\theta_1, \theta_2,$ and θ_3, represent a set of successive rotation angles used to describe the orientation of a cross-section of B relative to A.

In terms of these previously introduced displacement quantities, a set of six consistently linearized partial differential equations of motion of B can be formulated.

$$\begin{aligned} &\rho u_{1,tt} + 2\rho(\omega_2 u_{3,t} - \omega_3 u_{2,t}) - [(EA_0, u_{1,x})_{,x} + \left(\omega_2^2 + \omega_3^2\right)u_1 - (\omega_1\omega_2 - \omega_{3,t})u_2 \\ &\quad - (\omega_1\omega_3 + \omega_{2,t})u_3] = -\rho(v_{1,t} + \omega_2 v_3 - \omega_3 v_2) + \rho x\left(\omega_2^2 + \omega_3^2\right) \end{aligned} \tag{1.4}$$

$$\begin{aligned} &\rho u_{2,tt} + 2\rho(\omega_3 u_{1,t} - \omega_1 u_{3,t}) - \{[GA_S(u_{2,x} - \theta_3)]_{,x} + \rho\left(\omega_1^2 + \omega_3^2\right)u_2 \\ &\quad + \left[\frac{\rho}{2}\left(\omega_2^2 - \omega_3^2\right)(\ell^2 - x^2)\theta_3 u_{1,t}\right] \\ &\quad - [\rho(v_{1,t} + \omega_2 v_3 - \omega_3 v_2)\theta_3]_{,x} - \rho(\omega_1\omega_2 + \omega_{2,t})u_1 - \rho(\omega_2\omega_3 - \omega_{1,t})u_3\} \\ &\quad = -\rho(v_{2,t} - \omega_3 v_1 - \omega_1 v_3 - \rho x(\omega_1\omega_2 + \omega_{3,t}) \end{aligned} \tag{1.5}$$

$$\begin{aligned} &\rho u_{3,tt} + 2\rho(\omega_1 u_{2,t} - \omega_2 u_{1,t}) - \{[GA_S(u_{3,x} + \theta_2)]_{,x} + \rho\left(\omega_1^2 + \omega_2^2\right)u_3 \\ &\quad - \left[\frac{\rho}{2}\left(\omega_2^2 - \omega_3^2\right)(\ell^2 - x^2)\theta_2\right]_{,x} \\ &\quad + [\rho(v_{1,t} + \omega_2 v_3 - \omega_3 v_2)\theta_2]_{,x} - \rho(\omega_1\omega_3 - \omega_{2,t})u_1 - \rho(\omega_2\omega_3 + \omega_{1,t})u_2\} \\ &\quad = -\rho(v_{3,t} + \omega_1 v_2 - \omega_2 v_1) - \rho x(\omega_1\omega_3 - \omega_{2,t}) \end{aligned} \tag{1.6}$$

$$\frac{\rho}{A_0}(I_{22}+I_{33})\theta_{1,tt}-\frac{2\rho}{A_0}(\omega_2 I_{33}\theta_{3,t}-\omega_3 I_{22}\theta_{2,t})-[GK'\theta_{1,x}]_{,x}$$
$$+\frac{\rho}{A_0}(I_{33}-I_{22})[-(\omega_2^2+\omega_3^2)\theta_1+(\omega_1\omega_2-\dot{\omega}_3)\theta_2-(\omega_1\omega_3+\dot{\omega}_2)\theta_3]$$
$$=-\frac{\rho}{A_0}(I_{33}-I_{22})\omega_2\omega_3-\frac{\rho}{A_0}(I_{22}+I_{33})\omega_1 \tag{1.7}$$

$$\frac{\rho I_{22}}{A_0}\theta_{2,tt}-\frac{2\rho}{A_0}\omega_3 I_{22}\theta_{1,t}-[EI_{22}\theta_{2,x}]_{,x}+[GA_S(u_{3,x}+\theta_2)]$$
$$+\frac{\rho}{A_0}\left\{[\omega_1\omega_2(I_{33}-I_{22})-\dot{\omega}_3(I_{33}+I_{22})]\theta_1+(\omega_1^2-\omega_3^2)I_{22}\theta_2+\omega_2\omega_3 I_{22}\theta_3\right\}$$
$$=-\frac{\rho}{A_0}I_{22}(\dot{\omega}_2+\omega_3\omega_1) \tag{1.8}$$

$$\frac{\rho I_{33}}{A_0}\theta_{3,tt}+\frac{2\rho}{A_0}\omega_2 I_{33}\theta_{1,t}-[EI_{33}\theta_{3,x}]_{,x}-[GA_S(u_{2,x}-\theta_3)]$$
$$+\frac{\rho}{A_0}\left\{[\omega_1\omega_3(I_{22}-I_{33})+\omega_2(I_{22}+I_{33})]\theta_1+\omega_2\omega_3 I_{22}\theta_2+(\omega_1^2-\omega_2^2)I_{33}\theta_3\right\}$$
$$=-\frac{\rho}{A_0}I_{33}(\dot{\omega}_3-\omega_1\omega_2)\ . \tag{1.9}$$

These equations account for extension, torsion, shear and bending in two transverse directions and translational inertial effects.

Now we restrict to the special cases of interest here. (More general models are easily considered -- for a remark on extensibility see Section 4.) We firstly set translational motion to zero and neglect shear effects. Then we consider firstly the case where the beam is rotating in the a_1 - a_2 plane with angular velocity ω_3, and secondly the case where it is rotating about the a_1-axis with angular velocity ω_1. We also assume inextensibility. This yields the equations

$$\rho u_{2,tt}+\omega_{3,t}x-\rho\omega_3^2 u_2+EIu_{2,xxxx}-\omega_3^2\rho\frac{d}{dx}\left[\frac{\ell^2-\lambda^2}{2}u_{2,x}\right]=0 \tag{1.10}$$

and

$$\rho u_{2,tt}+EIu_{2,xxxx}-\omega_1^2 u_2=0 \tag{1.11}$$

respectively.

These are the equations for the driven systems. In the following section we shall see these derived again from first principles, together with angular momentum conservation equations and a term for the displacement of the beam from the center of mass of the rigid hub.

§2 *Lagrangian for the Axial and Planar Beams*

We now derive directly the Lagrangian for the axial and planar systems, in each case taking into account spin induced stiffening. We shall see that this effect is significant in the planar case, but is a high order effect in the axial case.

We suppose that our inextensible thin beam of length ℓ is attached to a hub of moment of inertia I which is spinning about its center of mass. We suppose for convenience that the beam has constant mass density $\rho = 1$.

Consider firstly the planar case. Here the beam is clamped to the hub at a distance R from the center of mass with its undeformed configuration in the direction of the vector from the center of mass to the point of attachment. We denote the transverse displacement of the beam as $y(x)$. Also we denote by u the (negative) extension of the beam along the x-axis due to bending.

Then we have the geometrical (kinematic) constraint

$$x = \int_0^{x+u} \left(1 + \left(\frac{\partial y}{\partial \xi}\right)^2\right)^{\frac{1}{2}} d\xi \quad . \tag{2.1}$$

Now assuming $\frac{\partial y}{\partial \xi}$ is small and using Taylor's theorem, we have

$$x \int_0^{x+v} \left(1 + \frac{1}{2}\left(\frac{\partial y}{\partial \xi}\right)^2\right) d\xi \tag{2.2}$$

or

$$u \approx -\frac{1}{2} \int_0^x \left(\frac{\partial y}{\partial \xi}\right)^2 d\xi \tag{2.3}$$

and

$$u_t \approx -\int_0^x y_\xi Y_{\xi t} d\xi \quad . \tag{2.4}$$

The kinetic energy of the rod in the planar case is then given by

$$\int_0^\ell \frac{d\mathbf{r}}{dt} \cdot \frac{d\mathbf{r}}{dt} \, dx$$

where $\frac{d\mathbf{r}}{dt} = \mathbf{r}_t - \hat{\omega}\mathbf{r}$, $\frac{d\mathbf{r}}{dt}$ denoting the derivative with respect to the space fixed axes, $\mathbf{r}_t$ denoting the derivative with respect to the body fixed axes and $\hat{\omega}$ being given by $\hat{\omega} = \begin{bmatrix} 0 & -\omega \\ \omega & 0 \end{bmatrix}$.

Neglecting the translational motion, we have

$$\mathbf{r} = [R + x + u, y]^T \approx \left[R + x - \frac{1}{2}\int_0^x \left(\frac{\partial y}{\partial \xi}\right)^2 dx, \ y\right]^T \quad .$$

Hence, with the approximation (2.2) we have

Lemma 2.1 *The total kinetic energy of the rod for the planar system is*

$$\frac{1}{2}\int_0^{\ell}\left(\int_0^x y_\xi y_{\xi_t} d\xi\right)^2 - 2\omega y\int_0^x y_\xi y_{\xi_t} d\xi + \omega^2 y^2 + y_t^2 - 2\omega y_t\left(R + x - \frac{1}{2}\int_0^x y_\xi{}^2 d\xi\right)$$

$$+ \omega^2\left(R + x - \frac{1}{2}\int_0^x y_\xi{}^2 d\xi\right)^2 dx \ . \tag{2.5}$$

Now if we integrate by parts and ignore terms of orders 3 or higher, we obtain

Lemma 2.2 *The Lagrangian for the planar system is*

$$L = \frac{1}{2} I\omega^2 + \frac{1}{2}\int_0^{\ell}\left[(y_t - \omega(R + x))^2 + \omega^2 y^2 - \omega^2\left(R(\ell - x) + \frac{\ell^2 - x^2}{2}\right)y_x{}^2\right]dx$$

$$- \int_0^{\ell} EI(y_{xx})^2 dx \ . \tag{2.6}$$

This yields

Lemma 2.3 *The equations of motion for the free planar system are*

$$y_{tt} - \omega_t(R + x) - \omega^2 y + EIy_{xxxx} - \omega^2\frac{d}{dx}\left[\left(R(\ell - x) + \frac{\ell^2 - x^2}{2}\right)y_x\right] = 0 \tag{2.7}$$

$$\frac{d}{dt}\left[I\omega + \left[\int_0^{\ell}\left[B(x + R)^2 + y^2 - \left(R(\ell - x) + \frac{\ell^2 - x^2}{2}\right)y_x{}^2\right]dx\right]\omega - \int_0^{\ell} y_t(R + x)dx\right] \tag{2.8}$$

with boundary conditions

$$y(0) = y'(0) = y''(\ell) = y'''(\ell) = 0 \ . \tag{2.9}$$

Consider now the free axial system. Here the beam is clamped to the hub at its center of mass and perpendicular to the plane of the hub. Denote the transverse displacement of the beam in this case by $y(z)$. Similar calculations to those above yield the following results.

Lemma 2.4 *Under approximation (2.2) the kinetic energy of the free axial beam is*

$$\frac{1}{2}\int_0^{\ell}\left[(y_t{}^2 + \omega^2 y^2) + \left(\int_0^z (y_\xi y_{\xi_t} d\xi)\right)^2\right]dz \ . \tag{2.10}$$

Neglecting terms of order 3 or higher in the Lagrangian we have

Lemma 2.5 *The Lagrangian for the free axial beam is*

$$L = \frac{1}{2} I\omega^2 + \frac{1}{2}\int_0^\ell \left(y_t^2 + \omega^2 y^2\right)dz - \frac{1}{2}\int_0^\ell EIy_{zz}^2 dz \ . \tag{2.11}$$

with corresponding equations of motion

$$y_{tt} + EIy_{zzzz} - \omega^2 y = 0 \tag{2.12}$$

$$\frac{d}{dt}\left[\omega\left(\int_0^\ell EIy^2 dz + I\right)\right] = 0 \ . \tag{2.13}$$

Equations (2.10) - (2.12) are the equations found in Baillieul and Levi [1].

Remark 1 We see from equations (2.5) and (2.10) that the stiffening term induced from the axial contraction is a low order (significant) effect in the planar case but is of high order in the axial case. This is crucial for the stability results in the following sections.

Remark 2 For the beam being driven at constant angular velocity, the relevant equations are simply (2.7) and (2.12) with $\omega_t = 0$. An important point is that the driven systems are linear, while the free systems are nonlinear due to the angular momentum conservation laws given by (2.8) and (2.13). Note also that the equations for the driven systems are precisely those derived in Section 1, modulo the hub displacement R and the sign of ω_t , the latter simply being due to our sign convention for the matrix $\hat{\omega}$.

§3 *Stability*

In this section we consider and compare the time evolution and stability of the equilibria of the driven axial and planar systems and the stability of the equilibrium of the free axial and planar systems. In the driven case the linearity of the equations permits a direct analysis of the solutions by separation of variables, while stability of equilibrium may be analyzed by Lyapunov methods. In the free cases (which are nonlinear) we can analyze nonlinear stability of the equilibrium by the Energy-Momentum method (see [6]). Our main point here is to contrast the stability results in the axial and planar cases (or, equivalently, in the planar case with and without stiffening -- see Bloch [3]). The results in the planar case depend on an eigenvalue inequality developed by Boyce, Di Prima and Handelman [5], which in turn is based on a method of Lamb and Southwell [8].

We first state the results. In the following let $\omega_i^2(0)$ denote the eigenvalues of the operators EIy_{xxxx} with boundary conditions (2.9).

Theorem 1 (Baillieul and Levi). *The free axial beam (2.12) - (2.13) has stable trivial equilibrium state if* $M < I\omega_1(0)$. *The first stationary state is always stable and all higher stationary states are unstable.*

Theorem 2 *The driven axial beam has bounded time evolution if* $\omega < \omega_1(0)$. *The trivial equilibrium state is stable if* $\omega < \omega_1(0)$. *The first mode is a stable equilibrium state and all higher modes are unstable equilibria.*

Theorem 3 *The free planar system (2.7) - (2.9) has stable trivial equilibrium states for all* ω .

Theorem 4 *The driven planar system has bounded time evolution and stable trivial equilibrium state for all* ω .

Remarks on Theorem 1 This is proved (in the dissipative case) in Baillieul and Levi [1]. The proof relies essentially on linearization and a spectral analysis. In the Hamiltonian (nondissipative) case, nonlinear stability of the first stationary state and of the trivial state for $M < I\omega_1(0)$ may be proven by the Energy-Momentum method. The proof of this is a variation of that used to prove the corresponding result for the planar unstiffened system in Bloch [3].

Proof of Theorem 2 The equation of motion for the axial driven system is

$$y_{tt} + EIy_{zzzz} - \omega^2 y = 0 \quad . \tag{3.1}$$

Consider firstly the time evolution of the system. We solve by separation of variables. Letting $y(z, t) = T(t)Z(z)$ we get the boundary value problem

$$EI\, Z''''(z) = (\omega^2 + k)Z(z) \tag{3.2}$$

$$Z(0) = Z'(0) = Z''(\ell) = Z'''(\ell) = 0 \quad , \tag{3.3}$$

and the corresponding initial value problem

$$T''(t) + kT(t) = 0 \quad . \tag{3.4}$$

Now the eigenvalue problem (3.2) and (3.3) has the sequence of distinct eigenvalues $0 < \omega_1{}^2(0) < \omega_2{}^2(0) < \omega_3{}^2(0), \ldots,$ by self-adjointness and compactness of the inverse operator. Hence we have $\omega^2 + k_i = \omega_i{}^2(0),\ 0 < \omega_1{}^2(0) < \omega_2{}^2(0), \ldots$. Hence by (3.4) the time evolution is bounded if $\omega^2 < \omega_1{}^2(0)$.

Now consider the equilibria. These are given by the solutions of

$$EIy_{zzzz} - \omega^2 y = 0 \quad .$$

Hence there is a trivial equilibrium $y = 0$ for any ω^2, and a sequence of equilibria, the eigenvectors $y_i(0, z)$ corresponding to each $\omega_i(0)$.

First consider the trivial solution. The Hamiltonian

$$\int_0^{\ell} \left(EIy_{zz}^2 - \omega^2 y^2 + y_t{}^2\right) dz \tag{3.5}$$

is conserved and has a critical value at $y = 0$. Further, by a Poincaré inequality argument, $\int_0^{\ell} EIy_{zz}^2 \geq \int_0^{\ell} \omega_i{}^2(0)y^2$. Hence (3.5) is a Lyapunov function for the trivial solution if $\omega < \omega_1(0)$. Hence the zero solution is stable if $\omega < \omega_1(0)$.

Now consider the other equilibria. We have

$$y_{tt} = -EIy_{xxxx} + \omega_i{}^2(0)y \tag{3.6}$$

with the eigenvalues of the operators on the right hand side of (3.6) being

$$\omega_i{}^2(0) - \omega_1{}^2(0)\,,\ \omega_i{}^2(0) - \omega_2{}^2(0)\,, \ldots .$$

Hence we have stability in the first stationary state but not otherwise. ■

Now, for the proof of Theorems 3 and 4 we need the following lemma which is a minor extension of a result of Boyce, Di Prima and Handelman [5].

Lemma 3.1 *Let* $\omega_1{}^2(0)$ *denote the first eigenvalue of the operator* EIy_{xxxx} *and let* $\omega_1{}^2(\omega^2)$ *denote the first eigenvalue of the operator* $EIy_{xxxx} - \omega^2[\phi(x)y_x]_x$ *where* $\phi(x) = R(\ell - x) + (\ell^2 - x^2)/2$. *Then*

$$\omega_1{}^2(\omega^2) - \omega^2 \geq \omega_1{}^2(0) . \tag{3.7}$$

Proof For the case $R = 0$, the proof may be found in Boyce, Di Prima and Handelman [5] which, in turn, uses the principle due to Lamb and Southwell [8]. We extend their argument to our case as follows.

Let $\ell = 1$ for convenience and let

$$H(y) = \int_0^1 EI(y'')^2 + \omega^2\left[R(1 - x) + \frac{1 - x^2}{2}\right](y')^2\, dx . \tag{3.8}$$

This is the energy corresponding to our eigenvalue problem.

Now let Y be the class of all functions having continuous fourth derivative and satisfying the boundary conditions $y(0) = y'(0) = 0$ and let Y' be the class of all functions which have continuous fourth derivatives and satisfy the boundary conditions $y(0) = 0$.

Then, by the minimum principle (see [5]),

$$\omega_1^2(\omega^2) = \min_Y \frac{H(y)}{\int_0^1 y^2 dx} .$$

(Note that the full boundary conditions are automatically satisfied by the minimizing function.)

Further,

$$\min_Y \frac{H(y)}{\int_0^1 y^2 dx} \geq \min_Y \frac{\int_0^1 EI(y'')^2 dx}{\int_0^1 y^2 dx} + \min_{Y'} \frac{\int_0^1 \omega^2 R(1-x)(y')^2 dx}{\int_0^1 y^2 dx} + \min_{Y'} \frac{\int_0^1 \omega^2 (1-x)^2 dx}{\int_0^1 y^2 dx} . \quad (3.9)$$

Corresponding to the nonrotating beam we have

$$\omega_1^2(0) = \min_Y \frac{\int_0^1 EI(y'')^2 dx}{\int_0^1 y^2 dx} .$$

Note the differential equation corresponding to the minimization problem which is the third term on the right hand side of (3.1) is

$$[(1 - x^2)y']' + \frac{2\lambda_1^2}{\omega^2} y = 0$$

where $\lambda_1^2 = \min_{Y'} \frac{\int_0^1 \frac{\omega}{2}(1 - x^2)}{\int_0^1 y^2 dx} y'^2 dx$.

This is, of course, Legendre's equation, and taking into account the boundary conditions we find the first eigenvalue to be 2, corresponding to the first Legendre polynomial. Hence $\frac{2\lambda_1}{\omega^2} = 2$ or $\lambda_1 = \omega^2$.

Now, corresponding to the second term in (3.1), we have an eigenfunction which is a (positive) linear function of R . Hence we have

$$\omega_1^2(\omega^2) \geq \omega^2 + \omega_1^2(0) + RK^2$$

which gives the result. ∎

Proof of Theorem 4 First we consider time evolution. We have the equation

$$y_{tt} - \omega^2 y + EIy_{xxxx} - \omega^2[\phi(x)y_x]_x = 0$$

with ω a constant parameter.

Again we use separation of variables, letting $y(x, t) = X(x)T(t)$. This yields

$$T'' + kT = 0 \tag{3.10}$$

$$X'''' - \Omega^2[\phi X']' = (\Omega^2 + k)X \ , \tag{3.11}$$

where k is a constant, subject to the boundary conditions

$$X(0) = X'(0) = X''(\ell) = X'''(\ell) = 0 \ . \tag{3.12}$$

Solving the eigenvalue problem gives

$$k_i = \omega_i^2(\omega^2) - \omega^2 \ .$$

Now from the lemma we know $\omega_1^2(\omega^2) - \omega^2 \geq \omega_1^2(0)$ where $\omega_1^2(0) > 0$. Hence $k_1 > 0$ and the first mode is stable. But $\omega_2(\omega^2) \geq \omega_1(\omega^2)$ from self-adjointness. Hence all modes are stable and the time evolution is bounded.

Now we consider stability of the (trivial) equilibrium $y = 0$. We have Hamiltonian

$$H = \frac{1}{2}\int_0^\ell \left[y_t^{\,2} - \omega^2 y^2 + \omega^2 \phi y_x^{\,2}\right] dx + \frac{1}{2}\int_0^\ell EI(y_{xx})^2 \ dx \ .$$

This is critical at 0, conserved and definite if $\omega^2 \leq \omega_1^2(\omega^2)$, which we know is true from the lemma. Hence the trivial equilibrium is stable.

To prove Theorem **3** we use the Energy-Momentum method (which is equivalent to the Energy-Casimir method in the case where momentum is a Casimir -- see [6]). This was introduced into the analysis of connected rigid and flexible bodies by Krishnaprasad and Marsden in [9]. The Energy-Casimir method has been used in a wide variety of problems in fluid and plasma physics (see [6]). For an application in a Lagrangian context see Bloch [3]. The idea is essentially to extend the method of Lyapunov by using energy plus other conserved quantities.

Here we find a λ such that $H + \lambda M$, where M is the conserved momentum, has a critical value at the equilibrium of interest. Then we show that the second variation of $H + \lambda M$ is definite at the equilibrium. This proves "formal" stability. To prove full nonlinear stability (in the sense of Lyapunov) we need to show that the second variation dominates a suitable norm on the phase space.

Proof of Theorem 3 We have

$$H+\lambda M = \frac{1}{2}I\omega^2 + \frac{1}{2}\int_0^\ell \left[\left(y_t - \omega(R+x)\right)^2 + \omega^2 y^2 - \omega^2\phi(x)y_x{}^2\right]ax + \frac{1}{2}\int_0^\ell EI(y_{xx})^2dx$$

$$+\lambda\left\{\left[I + \int_0^\ell [(x+R)^2 + y^2 - \phi(x)y_x{}^2]dx\right]\omega - \int_0^\ell y_t(R+x)dx\right\} \tag{3.13}$$

where $\phi(x) = R(\ell - x) + (\ell^2 - x^2)/2$.

Now calculating $\delta(H+\lambda M)$, we find the conditions for $\delta(H+\lambda M)$ to be zero at equilibrium $(\omega, y, y_t) = (\omega_e, y_e, 0)$ are:

$$\omega_e\left[\varphi + \int_0^\ell y_e{}^2 + (R+x)^2dx\right] - \omega_e\int_0^\ell \phi(x)y_x{}^2dx$$

$$+\lambda\left[\varphi + \int_0^\ell [y_e{}^2 + (R+x)^2]dx - \int_0^\ell \phi(x)y_x{}^2dx\right] = 0 . \tag{3.14}$$

$$\omega_e{}^2\int_0^\ell y_e dx + EI\int_0^\ell \frac{\partial^4 y_e}{\partial x^4}dx + \int_0^\ell \omega_e{}^2\frac{d}{dx}(\phi y_x)dx$$

$$+\lambda\left[\omega_e\int_0^\ell 2y_e dx + \lambda 2\omega_e\int_0^\ell \frac{d}{dx}(\phi(x)y_x)dx\right] = 0 \tag{3.15}$$

$$-\omega_e\int_0^\ell (R+x)dx - \lambda\int_0^\ell (R+x)dx = 0 . \tag{3.16}$$

Now if $\lambda = -\omega_e$ (3.14) and (3.16) are satisfied and (3.15) becomes

$$EI\frac{\partial^4 y_e}{\partial x^4} - \omega_e{}^2 y_e - \omega_e{}^2\frac{d}{dx}(\phi(x)y_x) = 0$$

which is the equilibrium condition.

Now we find the second variation to be given by

$$\delta^2(H+\lambda M)_e = I(\delta\omega)^2 - (\delta\omega)^2\int_0^\ell \phi(x)y_x{}^2dx + \int_0^\ell \left(\delta y_t - \delta\omega(R+x)^2\right)dx$$

$$+\int_0^\ell \left[(\delta\omega)^2 y_e{}^2 - \omega_e{}^2(\delta y)^2 + \omega_e{}^2\phi(x)(\delta y_x)^2 + EI\left(\frac{\partial^2\delta y}{\partial x^2}\right)^2\right]dx . \tag{3.17}$$

We can see this is definite as follows. The first part of the expression

$$I(\delta\omega^2) - (\delta\omega^2)\int_0^\ell (\phi(x)y_x{}^2dx + \int_0^\ell \left(\delta y_t - \delta\omega(R+x)\right)^2dx + \int_0^\ell (\delta\omega)^2 y_e{}^2 ,$$

is just equivalent to our kinetic energy expression. However, we know this is definite to the requisite order since this is just $\int_0^\ell \frac{d\mathbf{r}}{dt} \cdot \frac{d\mathbf{r}}{dt}\, dx$ modulo higher order terms.

On the other hand the remainder of the expression is, by integration by parts,

$$\int_0^\ell - \omega_e{}^2(\delta y)^2 - \omega_e{}^2 \frac{d}{dx}[\phi(x)\delta y_x]\delta y + EI\left(\frac{\partial^4 \delta y}{\partial \lambda^4}\right)\delta y dx \ .$$

Now by a Poincaré inequality argument for the operator $EI\frac{\partial^4 \delta y}{\partial x^4} - \omega^2 \frac{\partial}{\partial x}[\phi(x)\delta yx]$, this is stable if $\omega_e{}^2 < \omega_1{}^2(\omega_e{}^2)$. But we know this is satisfied by the lemma. This proves (formal) stability.

Remark

Stability results for an extensible rod may be found in a similar fashion. In this case one finds that there is a nonzero extension at which equilibrium is reached, reflecting the balance between the centrifugal forces and the tension in the beam. We remark also that in the presence of shear, the planar beam can go unstable; see the paper of Simo, Posburgh and Marsden in these proceedings.

§4 *Discrete Models and Concluding Remarks*

It is interesting that the dynamics and bifurcation behavior of the planar beam system can be well represented by a simple discrete planar system. This discrete system is described in detail in [4]. Here we briefly describe some of the relevant results.

Consider the following "model" of an inextensible beam attached to a rigid hub. We have a rigid body with moment inertia I rotating with angular velocity $\Omega(t) = \dot{\phi}(t)$ about its center of mass. The beam is modelled by a mass attached to a light rod of length L which, in turn, is attached to the body by a hinge at a distance d from the center of mass. The deflection of the rod is given by an angle θ and the motion is constrained by a torsional spring of spring constant k_r.

The Lagrangian for this system is

$$L = \frac{1}{2} I\Omega^2 + \frac{1}{2} m\Big((d^2 + L^2 + 2d \cos \theta)\Omega^2 + L^2\theta^2 + (2L^2 + 2Ld \cos \theta)\dot{\theta}\Omega\Big) - k_r \frac{\theta^2}{2} . \qquad (4.1)$$

Here of course we can analyze stability of the full nonlinear system by the Energy-Casimir method in the case where the system rotates freely, or by Lyapunov's method in the case where the system is driven at constant angular velocity. (We remark that there is a natural Poisson structure for these systems -- see [4].) We find

Theorem 4.1 *A sufficient condition for the stability of the equilibrium* $(\theta, \theta_t, \Omega) = (\theta_e, 0, \Omega_e)$ *for the driven or free system (5.1) is*

$$mdL \cos \theta_e \Omega_e^2 + k_r > 0 \ . \tag{4.2}$$

In particular, the trivial solution is always stable as in the full beam model.

We can also consider a model of an extensible beam, supposing now that our mass is able to move in a longitudinal direction with stretch s and that there is a longitudinal spring with spring constant k_T.

Then the Lagrangian is

$$L = \frac{1}{2} I\Omega^2 + \frac{1}{2} m\{ (d^2 + (L+s)^2 + 2(L+s)d \cos \theta)\Omega^2 + (L+s)^2 \dot{\theta}^2 + \dot{s}^2 + (2(L+s)^2$$

$$+ 2(L+s)d \cos \theta)\, \dot{\theta}\Omega + 2d \sin \theta\Omega\, \dot{s}\} - k_r \frac{\theta^2}{2} - k_T \frac{s^2}{2} \ . \tag{4.3}$$

We can then prove

Theorem 4.2 *The equilibrium state* $(\Omega, \theta, \theta_t, s, s_t) = (\Omega_e, \theta_e, 0, s_e, 0)$ *of the driven or free system (4.3) is stable if* $k_T > m\Omega_e^2$ *and*

$$(k_T - m\Omega_e^2)(\Omega_e^2 m(L+s_e)d \cos \theta_e + k_r) - \Omega_e^4 m^2 d^2 \sin^2 \theta_e > 0 \ .$$

In particular, if $k_T > m\Omega_e^2$ the undeflected $(\theta = 0)$ state is again always stable.

In conclusion, we remark that the models analyzed here have stable behavior which accurately reflects that of real rotating beams and of simulations (see eg. [7] or [11]). The behavior of the discrete models just discussed is an additional check on the behavior of the full beam model.

References

[1] J. Baillieul and M. Levi, "Rotational Elastic Dynamics," Physica 27D (1987) 43-62.

[2] A. M. Bloch, "The Dynamics of a Free Flexible Body," in C. I. Byrnes, C. F. Martin and R. E. Saeks (eds.), Analysis and Control of Nonlinear Systems, Elsevier (North Holland), 1988, 281-288.

[3] A. M. Bloch, "Stability and Equilibria of Deformable Systems," Proceedings 26th IEEE Conference on Decision and Control, IEEE, 1987, 1443-1444.

[4] A. M. Bloch and R. R. Ryan, "Approximate Models of Rotating Beams," Proceedings 27th IEEE Conference on Decision and Control, IEEE, 1988, 1230-1235.

[5] W. F. Boyce, R. C. Di Prima and G. H. Handelman, "Vibrations of Rotating Beams of Constant Section," in Proceedings Second U. S. National Congress of Applied Mechanics, Ann Arbor, Michigan (1954) 165-173.

[6] D. Holm, J. E. Marsden, T. Ratiu and A. Weinstein, "Nonlinear Stability of Fluid and Plasma Equilibria," Physics Reports 123 (1987) 1-116.

[7] T. R. Kane, R. R. Ryan and A. K. Bannerjee, "Comprehensive Theory for the Dynamics of a General Beam Attached to a Moving Rigid Base," J. Guidance, Control and Dynamics 10, No. 2 (1987) 139-151.

[8] H. Lamb and R. V. Southwell, "The Vibration of a Spinning Disc," Proc. Roy. Soc. (A) 99 (1921) 272-280.

[9] P. S. Krishnaprasad and J. E. Marsden, "Hamiltonian Structures and Stability for Rigid Bodies with Flexible Attachments," Archive for Rational Mechanics and Analysis 98, No. 1 (1987) 73-93.

[10] R. R. Ryan, "Use of Discrete Examples to Study Dynamic Modeling Methods for Flexible Bodies Undergoing Large Overall Motion and Small Deformation I," preprint.

[11] J. C. Simo and L. Vu-Quoc, "On the Dynamics of Flexible Beams Under Large Overall Motion -- The Plane Case: Part 1," J. Applied Mechanics 53 (1986) 863-869.

[12] J. C. Simo and L. Vu-Quoc, "The Role of Non-linear Theories in Transient Dynamic Analysis of Flexible Structures," J. Sound and Vibration 119 (3) (1987) 487-508.

[13] N. Sreenath, Y. G. Oh, P. S. Krishnaprasad and J. E. Marsden, "The Dynamics of Coupled Planar Rigid Bodies," to appear.

Mathematical Sciences Institute
and Department of Theoretical & Applied
Mechanics, Cornell University
and Department of Mathematics
The Ohio State University

Department of Mechanical Engineering
and Applied Mechanics
University of Michigan

Contemporary Mathematics
Volume **97**, 1989

CHARACTERIZATION OF HAMILTONIAN INPUT-OUTPUT SYSTEMS

P.E. Crouch[1] **A.J. Van der Schaft**

ABSTRACT. This paper reviews results obtained by the authors and others on the characterization of Hamiltonian input-output systems. Input-output systems are modelled by systems of vector fields on a manifold, parameterized by the inputs, together with a set of output functions, or observations, on the manifold. Hamiltonian input-output systems represent a subclass of input-output systems in which a symplectic structure is imposed in various ways. In the characterization of these systems one wishes to identify the presence of this additional structure, through the resulting structure on the spaces of input and output trajectories. This structure manifests itself in various distinct ways.

1. INTRODUCTION.

In this paper we survey a series of results obtained by the authors [5] and others on the characterization of Hamiltonian input-output systems. As a preliminary example consider a simple mechanical system consisting of a mass suspended by a nonlinear spring, and the whole system subject to an external force applied to one end of the spring. The equations of motion may be written in the form

$$m\ddot{x} = mg - k(x - h)^3 + F$$

where m, g, k and h are constants and x represents the displacement of the mass. Setting $u = F + mg$, $q = x - h$, $p = m\dot{q}$, we obtain the equations

$$\dot{q} = \frac{p}{m}$$

$$\dot{p} = -kq^3 + u\,.$$

If $H_o = \frac{p^2}{2m} + \frac{kq^4}{4}$, $H_1 = q$ and we set $y = q$ we obtain a system of equations in following form

$$\dot{q} = \frac{\partial H_o}{\partial p} - u\frac{\partial H_1}{\partial p}$$

$$\dot{p} = -\frac{\partial H_o}{\partial q} + u\frac{\partial H_1}{\partial q}$$

1980 Mathematics Subject Classification (1985 Revision). 43C10,58F05

[1] Partially supported by N.S.F. Grant No. ECS 8703615.

$$y = H_1.$$

Viewing u as an input variable, and y as an output variable enables us to consider the system of equations above as an input-output system, with a special structure, namely that of a Hamiltonian system, which we define formally later in the section. The question which this paper addresses can be stated in the following way:

Which input-output experiments need to be performed to ensure that an input-output system does have a Hamiltonian structure?

We presume that the internal structure is not available to the person performing the experiments, but rather that person sees the system as a "black box" in which he/she has access only to the input functions $u(\bullet)$ and corresponding output functions $y(\bullet)$.

For the current purposes we define an input-output system as one which has a finite dimensional state space representation in the following form

$$\Sigma \qquad \dot{x} = f(x,u),\ x \in M,\ u \in \Omega \subset \Re^m$$

$$y = h(x,u), \quad y \in \Re^p$$

where M is an analytic manifold and for each $u \in \Omega$, $x \to f(x,u)$ is a complete analytic vector field on M. We assume also that the maps $(x,u) \to f(x,u)$, $(x,u) \to h(x,u)$ are analytic. Let Σ^L denote the subclass defined by

$$\Sigma^L \qquad \dot{x} = g_0(x) + \sum_{i=1}^{m} u_i g_i(x)\,,\ x \in M,\ u \in \Omega \subset \Re^m$$

$$y_i = H_j(x)\,,\ j = 1 \ldots p,\ y \in \Re^p.$$

Before we define the subclass of Hamiltonian input-output system we discuss some important external (state independent) representations of input-output systems. The input-output map of an initialized system Σ, $x(0) = x_0$, consists of the causal mapping $u(\bullet) \to y(\bullet)$ generated by solving the differential equations defining Σ. We view the input functions as belonging to space $U([0,\infty))$ where for any interval $I \subset \Re$ we let $U(I)$ denote the space of right-continuous piecewise constant Ω-valued functions on I. We view the output function as belonging to the space $Y([0,\infty))$ where for any interval $I \subset \Re$ we let $Y(I)$ denote the space of continuous $\Re^p$ valued functions on I.

Conversely given a causal mapping $F:U([0,\infty)) \to Y([0,\infty))$, we say that F has a realization by an input-output system Σ, if there exists an initial state $x_0 \in M$ such that the input-output mapping defined by the initialized system Σ coincides with F.

The input-output behavior of a system Σ consists of the set of all pairs (u,y) where $u \in U((-\infty,\infty))$ and $y \in Y((-\infty,\infty))$, such that there exists an absolutely continuous M valued function x on $(-\infty,\infty)$, with the property that the mapping $t \to (\, u(t), y(t), x(t)\,)$ satisfies the equations Σ a.e. on $(-\infty,\infty)$. Denote this set by Σ_e, and denote by Σ_i the set of all 3-tuples (u,y,x) where $(u,y) \in \Sigma_e$ and x is a corresponding state trajectory. Denote by $\Sigma_i^+(0)(x_0)$, the set of all 3-

tuples (u,y,x) where u,y,x are functions defined on $[0,\infty)$ coinciding with the restriction of functions $\bar{u}$, $\bar{y}$, $\bar{x}$ on $(-\infty,\infty)$ such that $(\bar{u}, \bar{y}, \bar{x}) \in \Sigma_i$ and $\bar{x}(0) = x_o \cdot \Sigma_e^+(0)(x_o)$ is defined as the projection of $\Sigma_i^+(0)(x_o)$ onto the set of pairs (u,y). Note that $\Sigma_e^+(0)(x_o)$ is in fact the same object as the input-output map F defined by system Σ initialized at x_o.

Unfortunately neither the input-output behaviour Σ_e, or input-output map $\Sigma_e^+(0)(x_o)$, defines Σ uniquely. However for the systems Σ or Σ^L defined above there is a well understood theory relating initialized systems to their input-output maps, see Sussmann [22] and Jakubczyk [11]. The main object of interest is a so called minimal system, which is one having certain controllability and observability properties. The details of these definitions do not concern us in this paper, but minimal systems are important for the following reasons. Given an initialized system Σ with input-output map F, there exists an initialized minimal system Σ', generally evolving on a manifold M' of lower dimension than that of M, whose input-output map is again F. Furthermore, any two initialized minimal systems with the same input-output maps are isomorphic, which in particular implies that the state manifolds are diffomorphic. Thus minimal realizations of an input-output map are unique up to isomorphism. This isomorphism is described easily for two systems Σ and Σ' in terms of the differmorphism $\varnothing$: M → M' of the respective state manifolds.

$$\varnothing_* f(x,u) = f'(\varnothing(x),u), \; u \in \Omega, \; x \in M$$

$$h(x,u) = h'(\varnothing(x),u), \; u \in \Omega, \; x \in M$$

The existence and uniqueness of minimal realizations of input-output maps makes it possible to distinguish Hamiltonian input-output systems which we now describe. See Brockett [2] for some motivational ideas on mechanical/Hamiltonian systems.

We say Σ is a Hamiltonian input-output system if M can be given the structure of symplectic manifold (M,ω), with symplectic form ω, with respect to which for each fixed $u \in \Omega$, f(x,u) is a locally Hamiltonian vector field, and for each $x_o \in M$ there exists a neighborhood V of x_o such that on V, Σ can be written in the form

$$\Sigma_H \qquad \dot{x} = X_{H_u}(x), \;\; y = \frac{\partial H^T}{\partial u}(x,u)$$

$$x \in V \subset M, \;\; u \in \Omega \subset \mathfrak{R}^m, \;\; y \in \mathfrak{R}^m.$$

That is on V, for each fixed u, X_{H_u} is a Hamiltonian vector field with Hamiltonian function $x \to H_u(x) = H(x,u)$, $(\omega(X_{H_u},Z) = -dH_u(Z)$ for all vector fields Z on V). Note that we do not insist that f(x,u) is a global Hamiltonian vector field on M. In case of an input-output system Σ^L, this definition reduces to the fact that we may rewrite the system in the form

$$\Sigma_H^L \qquad \dot{x} = g_o(x) - \sum_{i=1}^{m} u_i X_{H_i}(x), \;\; x \in M, \;\; u \in \Omega \subset \mathfrak{R}^m$$

$$y_i = H_i(x), \; 1 \le i \le m, \;\; y \in \mathfrak{R}^m.$$

Here X_{H_u} are global Hamiltonian vector fields on M with Hamiltonian functions H_i, and g_o is a locally Hamiltonian vector field with locally defined Hamiltonian function H_o. See Van der Schaft [18] [19] for further discussion of these definitions.

This definition of a Hamiltonian input-output system is not totally accepted, but it does fit with many requirements. In fact Jakubczyk [13] and [14] deals primarily with a slight variant of the class defined by Σ_H, and considers output functions $y = H(x,u)$ instead of $y = \frac{\partial H^T}{\partial u}(x,u)$. We denote this class of systems by Σ_J. The class Σ^L_H, of which Σ_H is a generalization, may be motivated by the forced Lagrangian system, with configuration space $\mathfrak{R}^n$,

$$\text{(1)} \qquad \frac{d}{dt}\left(\frac{\partial L}{\partial \dot{q}_i}\right) - \frac{\partial L}{\partial q_i} = \begin{cases} u_i & i = 1, \dots m \\ 0 & i = m+1 \dots n \end{cases}$$

where $L(q,\dot{q})$ is a Lagrangian function and u_i represents external forces or controls. This system may be put into Hamiltonian form, assuming that the Legendre transformation is well defined, by setting $p_i = \frac{\partial L}{\partial \dot{q}_i}$, $1 \le i \le n$ and

$$H_o(p,q) = \sum_{i=1}^{n} p_i q_i - L(q,\dot{q}).$$

The result is the set of equations

$$\text{(2)} \qquad \begin{aligned} \dot{q}_i &= \frac{\partial H_o}{\partial p_i} & & i = 1 \dots n \\ \dot{p}_i &= -\frac{\partial H_o}{\partial q_i} + \begin{cases} u_i \\ 0 \end{cases} & & \begin{matrix} i = 1, \dots m \\ i = m+1 \dots n \end{matrix} \end{aligned}$$

Setting $H_i(q,p) = q_i$, $i = 1 \dots m$ and designating outputs by $y_i = q_i$, $i = 1 \dots m$ we obtain a generalized version of the introductory example

$$\text{(3)} \qquad \begin{aligned} \dot{q}_i &= \frac{\partial H_o}{\partial p_i} - \sum_{j=1}^{m} u_j \frac{\partial H_j}{\partial p_i} \\ \dot{p}_i &= -\frac{\partial H_o}{\partial q_i} + \sum_{j=1}^{m} u_j \frac{\partial H_j}{\partial q_i}, & & 1 \le i \le n \\ y_i &= H_i, & & 1 \le i \le m. \end{aligned}$$

This system of equations now represents a local coordinate version of the system Σ^L_H, in which the symplectic form ω is given in the standard form $\omega = \sum_{i=1}^{n} dp_i \wedge dq_i$. Motivation for the step in complexity from the equations (2) to the equations (3) and then Σ^L_H and Σ_H, is given in [18] and references therein.

The general problem addressed in this survey is then the characterization of Hamiltonian systems Σ_H, Σ_J, (Σ_H^L) within the class of systems Σ (Σ^L) via some suitable (external) input-output representation of the system. In section (2) we review some results in the linear theory. In section (3) we describe some explicit representations of the external behaviour of systems and a classical problem which motivated some of the results in section (4), concerning a variational characterization of Hamiltonian systems. Finally in section (5) we present a survey of the results by Jakubczyk and related results of the authors, giving a nonvariational characterization of Hamiltonian systems.

2. LINEAR THEORY.

The class of linear systems is usually described by the equations

$$\Sigma_L \qquad \begin{aligned} &\dot{x} = Ax + Bu, && u \in \Omega \subset \Re^m, \ x \in \Re^n. \\ &y = Cx, && y \in \Re^p. \end{aligned}$$

Within this class, the Hamiltonian systems are described by those systems of the form

$$\Sigma_{LH} \qquad \begin{aligned} &\dot{x} = Ax + Bu, && x \in \Re^{2n}, \quad u \in \Re^m \\ &y = B^TJx, && y \in \Re^m, \end{aligned}$$

where J is a full rank skew symmetric matrix, which in canonical symplectic coordinates, is given by the matrix

$$\begin{bmatrix} 0 & -I \\ I & 0 \end{bmatrix}$$

where I is the $n \times n$ identity matrix and A satisfies $A^TJ + JA = 0$. The corresponding Hamiltonian function $H_o(x) - \sum_{i=1}^{m} u_iH_i(x)$ of the Hamiltonian system Σ_H^L is defined by setting

$$H_o(x) = \frac{1}{2}x^TJax, \qquad -\sum_{i=1}^{m} u_iH_i(x) = x^TJBu.$$

The zero state response of the system Σ_{LH} may be expressed as

$$y(t) = \int_0^t W(t - \sigma)\, u(\sigma)\, d\sigma \tag{4}$$

where $W(t) = B^TJ\, e^{At}\, B$ satisfies

$$W(t) = -W(-t)^T. \tag{5}$$

THEOREM 1: **BROCKET AND RAHIMI [1].** *The zero state response (4) of a linear system Σ_L has a realization by a Hamiltonian system Σ_{LH} if and only if condition (5) is satisfied. We may assume that the Hamiltonian system Σ_{LH} is minimal.*

This characterization of Hamiltonian systems may be further explained in terms of the inputs and outputs as follows. Note first that for systems Σ_{LH} we may write

$$W(t - \sigma) = B^T J\, e^{At} J\, e^{A^T \sigma} J^T B$$

or, by setting $H(\sigma) = e^{A^T\sigma} J^T B$ we obtain

$$W(t - \sigma) = H(t)^T J\, H(\sigma). \tag{6}$$

Let the columns of $H(\sigma)$ be denoted by $h_j(\sigma)$, $j = 1 \ldots m$.

We shall give our explanation in terms of input-output maps of the form

$$y(t) = \int_{-\infty}^{t} W(t - \sigma)\, u(\sigma)\, d\sigma, \tag{7}$$

corresponding to a boundary condition for Σ_{LH} $\lim.\ t \to -\infty\ (x(t)) = 0$. We shall work only with controls u which satisfy

$$\int_{-\infty}^{\infty} H(\sigma)\, u(\sigma)\, d\sigma \;=\; \sum_{i=1}^{m} h_i(\sigma)\, u_i(\sigma)\, d\sigma \;=\; 0. \tag{8}$$

For minimal systems this corresponds to controls for which the state satisfies the additional constraint $\lim.\ t \to +\infty\ (x(t)) = 0$.

By (6) we may express the input-output map as

$$y(t) = \int_{\infty}^{t} H(t)^T J\, H(\sigma)\, u(\sigma)\, d\sigma\,.$$

From (5) and (8) this is simply

$$y(t) = -\int_{t}^{\infty} H(-\sigma)^T J\, H(-t)\, u(\sigma)\, d\sigma$$

or

$$y(t) = \int_{-\infty}^{t} H(\sigma)^T J^T H(-t)\, u(-\sigma)\, d\sigma. \tag{9}$$

For any scalar function $\bar{u}$ on $(-\infty, \infty)$ let

$$L_{ij}(\,\bar{u}\,)(\,t\,) = \int_{-\infty}^{t} h_i(t)^T J\, h_j(\sigma)\, \bar{u}(\sigma)\, d\sigma$$

and let P denote the operator of time reversal

$$(P\bar{u})(t) = \bar{u}(-t)$$

Equation (9) demonstrates the classical relationship

$$L_{ij}(\bar{u}) = P(L_{ji}(P\bar{u})), \quad 1 \leq i, j \leq n. \tag{10}$$

This property, of the linear Hamiltonian systems uses the time invariance property and would not be retained by the more general class of systems described by input-output maps

$$y(t) = \int_{-\infty}^{t} W(t,\sigma)\, u(\sigma)\, d\sigma \tag{11}$$

where the kernel satisfies a generalization of (6)

$$W(t,\sigma) = H(t)^T J\, H(\sigma), \tag{12}$$

and as a consequence satisfies a generalization of (5),

$$W(t,\sigma) + W(\sigma,t)^T = 0. \tag{13}$$

However input-output maps (11) defined by kernels (12), do satisfy another property. For inputs u_1 and u_2 which satisfy the condition (8) and corresponding outputs y_1 and y_2 we consider the expression

$$\int_{-\infty}^{\infty} \big(y_1(t)^T u_2(t) - y_2(t)^T u_1(t)\big)\, dt = \int_{-\infty}^{\infty} \int_{-\infty}^{t} \big(u_2(t)^T W(t,\sigma)\, u_1(\sigma) - u_1(t)^T W(t,\sigma)\, u_2(\sigma)\big)\, d\sigma\, dt$$

Now $$\int_{-\infty}^{\infty} \int_{-\infty}^{t} u_1(t)^T H(t)^T J\, H(\sigma) u_2(\sigma)\, d\sigma\, dt = -\int_{-\infty}^{\infty} \int_{t}^{\infty} u_1(t)^T H(t)^T J\, H(\sigma)\, u_2(\sigma)\, d\sigma\, dt$$

by (8), $$= -\int_{-\infty}^{\infty} \int_{-\infty}^{t} u_1(\sigma)^T H(\sigma)^T J\, H(t) u_2(t)\, d\sigma\, dt$$

$$= \int_{-\infty}^{\infty} \int_{-\infty}^{t} u_2(t)^T W(t,\sigma)\, u_1(\sigma)\, d\sigma\, dt\,.$$

Thus

$$\int_{-\infty}^{\infty} \big(y_1(t)^T u_2(t) - y_2(t)^T u_1(t)\big)\, dt = 0. \tag{14}$$

A major conjecture by Van der Schaft [18], stated that Hamiltonian systems in general could be characterized by variations of input-output behaviors in Σ^e which satisfy a relationship similar to that in equation (14). It was the principal concern of the authors in [5] to formulate and prove an exact statement of this conjecture.

3. VARIATIONS AND THE EXTERNAL REPRESENTATION OF SYSTEMS.

We introduce three representations of the external behavior of input-output systems, two of which describe the input-output map, while the remaining one describes the input-output behavior. The first description of the input-output map that we give, was introduced by Jakubczyk [11]. We use the notation

$$(t_1,u_1)(t_2,u_2)(\ldots\ldots\ldots)(t_k,u_k)u_{k+1}$$

to denote the piecewise constant control consisting of the concatenation of constant controls $u_i \in \Omega$, on time intervals of length t_i, for $1 \le i \le k$, and finishing with a final value of $u_{k+1} \in \Omega$. We represent the evaluation of the input-output map F, of an initialized system Σ, $x(0) = x_o$, on such a control by the expression

$$(15)\qquad \begin{aligned} &F(\,(t_k,u_k)\,(t_{k-1},u_{k-1})\,\ldots\,\ldots\,(t_1,u_1)\,u_o\,) \\ &= h^{u_o} \circ \gamma^{u_1}_{t_1} \circ \ldots \circ \gamma^{u_k}_{t_k}(x_o) \end{aligned}$$

where $h^u(x) = h(x,u)$ and $(t,x) \to \gamma^u_t\,(x)$ is the flow of the complete vector field f^u, $f^u(x) = f(x,u)$, $u \in \Omega$.

We write

$$Q(\,u_o \ldots u_k\,) = \frac{\partial}{\partial t_1} \cdots\cdots \frac{\partial}{\partial t_k}\Big|_{t_i=0} F(\,(t_k,u_k)\,\ldots\,(t_1,u_1)\,u_o)$$

from which it is clear that the components of Q are given by

$$(16)\qquad Q_i\,(\,u_o \cdots u_k\,) = f^{u_k}\,(\,f^{u_{k-1}}\,(\,\ldots\,f^{u_1}\,(h^{u_o}_i)\,\ldots\,)\,(x_o),\ 1 \le i \ge p,$$

where f^u is viewed as a differential operator.

We may now formally represent the input-output map of Σ, initialized at x_o by the expression

$$(17)\qquad F = \sum_{\omega \in \Omega^*} Q(\omega)\,\omega$$

where Ω^* is the free monoid generated by Ω.

Jakubczyk [12] is able to establish necessary and sufficient conditions on the coefficients $Q(\omega)$, $\omega \in \Omega^*$, so that the expressions (17) are in one to one correspondence with the initialized systems Σ, at least in the case Ω is compact and convex. Clearly a necessary condition is that for every $u_o, \ldots\ldots u_k \in \Omega$ the series

$$\sum_{i_1 \ldots i_k} Q\big(\,u_o\,u_1^{i_1} \ldots\ldots u_k^{i_k}\big)\,\frac{t_1^{i_1}}{i_1!} \cdots\cdots \frac{t_k^{i_k}}{i_k!}$$

converges in a neighborhood of $0 \in \Re^k$ and has an analytic continuation to all of $\Re^k$, since this must represent the expression (15). (Here use the notation $u^k = u \ldots\ldots u$). However, the most restrictive condition is a rank condition which ensures that the input-output map represented by (17) can be realized by a finite dimensional system Σ.

While the input-output map of an initialized system Σ^L can be represented as above, a more dedicated representation is given by the Volterra series which we denote by

$$y(t) = W_0(t,x_0) + \int_0^t W_1(t,\sigma_1,x_0)\,(u(\sigma_1))\,d\sigma_1 + \dots\dots \tag{18}$$

$$+ \int_0^t \int_0^{\sigma_1} \cdots \int_0^{\sigma_{k-1}} W_k\,(t, \sigma_1 \dots \sigma_k, x_0)\,(u(\sigma_1) \dots u(\sigma_1))\,d\sigma_1 \dots d\sigma_k + \dots .$$

$W_k\,(t,\sigma, \dots \sigma_k, x_o)$ is a k-linear $\Re^P$ valued mapping on $\Re^m$. To describe the relation between these kernels and the system data we let $(t,x) \to \Upsilon(t)(x)$ denote the flow of the vector field g_o, and set

$$G_i(\sigma)(x) = \Upsilon(-\sigma)_*\, g_i(\Upsilon(\sigma)x)$$

The components of W_k are now given by

$$W_k^{i_o\, i_1 \dots i_k}\,(t, \sigma_1 \dots \sigma_k, x_o) \tag{19}$$

$$G_{i_k}(\sigma_k)(x_o)\,(G_{i_{k-1}}(\sigma_{k-1})\,(\dots\dots G_{i_1}(\sigma_1)\,(H_{i_o} \circ \Upsilon(t))\,\dots\dots)$$

See Brockett [3], and Lesiak and Krener [15] and Jakubczyk [12] for details describing the relationship between the system Σ^L and the Volterra series (18). Note however, that as a result of the analyticity assumption for the systems Σ^L, the kernels W_k are analytic in the region $t \geq \sigma_1 \geq \ \geq \sigma_k \geq 0$ and have analytic continuations to $\Re^{k+1}$.

The final representation we consider does not deal with the input-output map but rather with the input-output behavior. It takes the form of a system of implicit ordinary differential equations in the input and output variables, the solutions of which gives the input-output behavior set Σ_e. In general we express them in the form $\left(y^{(k)} = \frac{d^k}{dt^k} y \right)$.

$$T_k(\, y, y^{(1)}, \dots y^{(\ell_k)}, u, u^{(1)}, \dots u^{(r_k)}) = 0,\ \ell_k > r_k,\ k = 1, 2 \dots \tag{20}$$

For example, for a linear system Σ_L the input -output behavior Σ_e may be described as the solutions to a set of linear equations

$$\sum_{k=0}^{n} t_{1k} y^{(k)} = \sum_{k=0}^{r} t_{2k} u^{(k)}, \quad n > r$$

where t_{ik}, $i = 1, 2$ are matrices. For nonlinear systems the form of the mapping T_k is still under active investigation, but see Van der Schaft [20] [21], Fliess [8], [9], Glad [10], Crouch and Lamnabhi [4], for a preliminary discussion.

One example of such a system of equations occurs when dealing with forced Newtonian equations, where the configuration space is $\Re^m$, and written as

$$N_k(q, \dot{q}, \ddot{q}) = F_k, \quad k = 1, \dots m.$$

Regarding the configuration variables as outputs $q_i = y_i$, $1 \le i \le m$, and the external forces as inputs $F_k = u_k$, $1 \le k \le m$ we obtain equations in the form of the equations (20).

$$N_k(y, \dot{y}, \ddot{y}) - u_k = 0, \quad k = 1 \dots m, \ell_k = 2, r_k = 0. \tag{21}$$

The classical inverse problem in Newtonian mechanics asks when there exists a Lagrangian function $L(q,\ddot{q})$ such that

$$\frac{d}{dt}\left(\frac{\partial}{\partial \dot{q}_i} L\right) - \frac{\partial L}{\partial q_i} = N_i(q,\dot{q},\ddot{q}), \quad 1 \le i \le m$$

We note, as in Takens [23], that using (1) and (3), a resolution of this problem solves the problem posed in this paper for a restricted class of system; namely, when do the set of input-output behaviors generated by the equations (21) arise from a Hamiltonian system of equations?

The solution to the problem given in Santilli [16], along with an extensive discussion of more involved problems. We briefly sketch the main idea, as it motivates the analysis in the following section. In general, if $t \to r(t)$ is a piecewise continuous mapping $r: \Re \to \Re^k$, we define a variation of r as a mapping $(t,\varepsilon) \to r(t,\varepsilon)$ from $\Re^2$ into $\Re^k$ such that

(i) $r(t,0) = r(t), \; t \in \Re$

(ii) $t \to \frac{\partial r}{\partial \varepsilon}(t,0) = \delta\, r(t)$ is piecewise continuous.

We call δr the variational field along r, but sometimes simply refer to δr as a variation of r. If r takes its values in a manifold M and is absolutely continuous then we assume δr is an absolutely continuous and $\delta r(t) \in T_{r(t)}M$, the tangent space to M at r(t).

Now assume that the Newtonian equations, $q \in \Re^m$, $N_i(q,\dot{q},\ddot{q}) = 0$, $1 \le i \le m$, possess a solution q for which there exists a variation $(t,\varepsilon) \to q(t,\varepsilon)$, such that for each ε, $t \to q(t,\varepsilon)$ is also a solution of the equations. By differentiation we see that

$$0 = \hat{M}(r(t))\, \delta r(t) \equiv \frac{\partial N}{\partial q}(r(t))\, \delta\, q(t) + \frac{\partial N}{\partial \dot{q}}(r(t))\, \delta\, \dot{q}(t) + \frac{\partial N}{\partial \ddot{q}}(r(t))\, \delta\, \ddot{q}(t)$$

where $$r(t)^T = (q(t),\dot{q}(t),\ddot{q}(t)), \; \delta\, r(t)^T = (\delta q(t),\delta\dot{q}(t),\delta\ddot{q}(t))$$

and $$\delta\dot{q}(t) = \frac{\partial}{\partial \varepsilon}\dot{q}(t,0), \quad \delta\ddot{q}(t) = \frac{\partial}{\partial \varepsilon}\ddot{q}(t,0).$$

These equations are known as the variational equations. Given two variations q_1 and q_2 of the same solution q we obtain $\delta\, r_i(t)^T = (\delta q_i(t),\delta\dot{q}_i(t), \delta\ddot{q}_i(t))$ and $M(r(t))\, \delta r_i(t) \equiv 0$, $i = 1, 2$.

Moreover, there exists a unique $m \times 3m$ matrix function of r, M^* such that

(22) $$\delta q_2(t)^T M(r(t))\, \delta r_1(t) - \delta q_1(t)^T M^*(r(t))\, \delta r_2(t) = \frac{d}{dt} Q\big(r(t), \delta r_1(t), \delta r_2(t) \big),$$

for some unique function Q. The equations

$$M^*(r(t))\, \delta r(t) \equiv 0$$

are known as the adjoint variational equations. There exists a function $L(q,\dot{q})$ which solves the inverse problem in Classical Mechanics if and only if the variational equations are self adjoint, that is,

$$M(r) = M^*(r).$$

Clearly we can form the variational equations corresponding to the set of equations (20), in which case we obtain equations of the form

$$\sum_{j=0}^{\ell_k} \frac{\partial T_k}{\partial y^{(j)}} \delta y^{(j)} + \sum_{j=0}^{r_k} \frac{\partial T_k}{\partial u^{(j)}} \delta u^{(j)} = 0.$$

One might conjecture that a solution to the full characterization of Hamiltonian systems now involves the correct notion of self adjointness for the variational equations above. To do this, however, we resort to the state space description of the system. However, preliminary work along this direct line of attack was begun in [4].

4. VARIATIONAL CRITERIA FOR HAMILTONIAN INPUT-OUTPUT SYSTEMS.

In this section we review the variational criteria for Hamiltonian input-output systems obtained by the authors in [5]. To simplify the exposition, only the class of systems Σ^L is considered, although it is applicable to the wider class Σ. Take an arbitrary piecewise constant input $u(t)$, $t \in [0,T]$ such that the solution $x(t)$ of Σ^L remains within one coordinate neighborhood of M. This also yields an output $y(t)$, $t \in [0,T]$. Along this input-state-output trajectory (u,x,y) the variational system is given by

(23) $$\begin{aligned} \dot{v}(t) &= Dg_o(x(t))\, v(t) + \sum_{j=1}^{m} u_j(t)\, Dg_j(x(t))\, v(t) + \sum_{j=1}^{m} u_j^v(t)\, g_j(x(t)) \\ y_j^v(t) &= DH_j(x(t))\, v(t), \quad j = 1, \dots, p, \; v(0) = 0 \in \Re^n. \end{aligned}$$

where D denotes taking the Jacobian matrix. Furthermore, $u^v = (u_1^v, \dots, u_m^v)$ and $y^v = (y_1^v, \dots, y_p^v)$ denote the inputs and outputs of the variational system, and are called the variational inputs and outputs.

The rationale or this definition arises from the fact that if $(u(t,\varepsilon), x(t,\varepsilon), y(t,\varepsilon))$, $t \in [0,T]$ is a variation of the trajectory (u,x,y) such that for each ε, $t \to (u(t,\varepsilon), x(t,\varepsilon), y(t,\varepsilon))$ is also a trajectory of Σ^L then the variational field along (u,x,y) may be identified with a trajectory (u^v, v, y^v) of the variational system along (u,x,y).

Along this same trajectory $(u(t), x(t), y(t))$, $t \in [0,T]$, the adjoint system is defined as the dual linear time-varying system

$$
(24)\quad
\begin{aligned}
-\dot{p}(t) &= Dg_o^T(x(t))\, p(t) + \sum_{j=1}^{m} u_j\, Dg_j^T(x(t))\, p(t) + \sum_{j=1}^{p} u_j^a(t)\, DH_j^T(x(t)) \\
y_j^a(t) &= g_j^T(x(t))\, p(t), \qquad j = 1, \ldots, m,\ p(0) = 0 \in \mathfrak{R}^n.
\end{aligned}
$$

with inputs $u^a = (u_1^a, \ldots u_p^a)$ and outputs $y^a = (y_1^a, \ldots y_m^a)$. For any input functions $u^v(t)$ and $u^a(t)$ it follows from (23) and (25) that

$$
(25)\qquad \frac{d}{dt}\, p^T(t)\, v(t) = (y^a(t))^T\, u^v(t) - (y^v(t))^T\, u^a(t)
$$

Moreover, if a system with inputs u^a and outputs y^a satisfies (25) for any u^v and y^v then it is equal to the adjoint system [5]. Hence the adjoint system is uniquely determined by the variational system. The variational and adjoint systems are only defined locally along a trajectory (u(t), x(t), (y(t)), $t \in [a,b]$, such that x(t) remains within one coordinate neighborhood. However, global (and coordinate free) definitions can be given if we combine the original system together with all its variational or adjoint systems. Equations Σ^L together with (23) define the prolonged system, or prolongation, which has state space TM (local coordinates (x,v)), input space $T\,\mathfrak{R}^m$ (local coordinates (u,u^v)) and output space $T\,\mathfrak{R}^p$ (local coordinates (y,y^v)). Equations Σ^L together with (24) define the Hamiltonian extension, which has state space T^*M (local coordinates (x,p)), input space $\mathfrak{R}^m \times \mathfrak{R}^p$ (local coordinates (u,u^a)) and output space $\mathfrak{R}^p \times \mathfrak{R}^m$ (local coordinates (y,y^a)).

The input-output map of the variational system along a trajectory (u(t), x(t), y(t)) of Σ^L is given by

$$
(26)\qquad y^v(t) = \int_0^t W_v(t,\sigma,u)\, u^v(\sigma)\, d\sigma, \qquad t \geq 0,
$$

where $W_v(t,\sigma,u)$ is the $p \times m$ matrix with (i,j)-th element

$$
(27)\qquad DH_i(x(t))\, \Phi^u(t,\sigma)\, g_j(x(\sigma))
$$

and the transition matrix $\Phi^u(t,\sigma)$ is the unique solution of the equation

$$
(28)\qquad \frac{\partial}{\partial t}\Phi^u(t,\sigma) = \left[Dg_o(x(t)) + \sum_{j=1}^{m} u_j(t)\, Dg_j(x(t)) \right] \Phi^u(t,\sigma)
$$

and $\Phi^u(\sigma,\sigma)$ is $k \times k$ identity matrix. It is easily seen that $W_v(t,\sigma,u)$ exists for all t, $\sigma \geq 0$ and also can be defined in a coordinate free way [5]. Similarly, the input-output map of the adjoint system is given by

$$
(29)\qquad y^a(t) = \int_0^t W_a(t,\sigma,u)\, u^a(\sigma)\, d\sigma, \quad t \geq 0
$$

where $W_a(t,\sigma,u)$ is determined from $W_v(t,\sigma,u)$ by the relation

$$W_a(t,\sigma,u) = - W_v^T(\sigma,t,u) \text{ for all } u.$$

DEFINITION: *A variational system along an input u is called self-adjoint if*

(30) $$W_v(t,\sigma,u) = W_a(t,\sigma,u) \; (= - W_v^T(t,\sigma,u)) \textit{ for all } t, \sigma \geq 0.$$

(in particular $p = m$).

We note that this condition is similar to the generalized condition (13) for linear systems. The following result was obtained by the authors and represents generalization of the self adjointness criteria in the inverse problem of classical mechanics described in the previous section. Note that the definition of minimal used here is slightly stronger than is usually necessary in applications of the existence and uniqueness theorems for minimal realizations of input-output maps, described in the introduction.

THEOREM 2: [5] [17]. *A minimal system Σ^L is Hamiltonian if and only if all the variational systems along any piecewise constant input are self-adjoint.*

Although this result characterizes Hamiltonian systems Σ_H^L within the class Σ^L, the characterization relies on the state space description. The original problem posed was to obtain the characterization in terms of an external representation of the system, so we now turn our attention to this specific problem. We restrict ourselves to piecewise constant inputs and piecewise constant variations, so for example any variational field along an input $\overline{u}$ may be generated by a variation

$$u(t,\epsilon) = \overline{u}(t) + \varepsilon \, \delta u(t)$$

where $\delta u(t)$ is piecewise constant. The main technical concept is now introduced.

DEFINITION. $(\delta u, \delta y)$ *is called an admissible variation of compact support of* $(\overline{u},\overline{y}) \in \Sigma_e^+(0)\,(x_o)$ *if*

(i) $\delta u(0) = 0$, *and* $\text{supp}\, \delta u$ *is compact*

(ii) $\text{supp}\, \delta y \subset \text{supp}\, \delta u$

(iii) *Let* $\text{supp}\, \delta u \subset (0,T)$ *and let* $(\overline{u}', \overline{y}') \in \Sigma_e^+(0)\,(x_o)$ *be such that* $\overline{u}'(t) = \overline{u}(t)$ *and hence* $\overline{y}'(t) = \overline{y}(t)$ *for* $t \in [0,T]$. *Define a variation of* $u'(u,\varepsilon)$ *of* $\overline{u}'$ *by setting* $u'(t,\varepsilon) = \overline{u}'(t) + \varepsilon\delta\, u(t)$. *This yields a variation* $(u'(t,\varepsilon), y'(t,\varepsilon))$ *of* $(\overline{u}',\overline{y}')$. *We require that the resulting (infinitesimal) variation* $(\delta u', \delta y')$ *of* $(\overline{u}',\overline{y}')$ *also satisfies (ii), i.e.,* $\text{supp}\, \delta y' \subset \text{supp}\, \delta u' = \text{supp}\, \delta u$.

Admissible variations $(\delta u,\delta y)$ of $(\bar{u},\bar{y})$ with supp $\delta u \subset (0,T)$ can be fully characterized in terms of the $W_v(t,\sigma,\bar{u})$ of the variational system (23) along $(\bar{u}, \bar{x}, \bar{y})$ as defined in (27) and (28). Since the transition matrix $\Phi^{\bar{u}}$ satisfies $\Phi^{\bar{u}}(t,\sigma) = \Phi^{\bar{u}}(t,\sigma)\,\Phi^{\bar{u}}(0,\sigma)$, we may write

$$W_v(t,\sigma,\bar{u}) = G(t,\bar{u})\,H(\sigma,\bar{u})$$

with $G(t,\bar{u})$ an $m \times k$ matrix and $H(\sigma,\bar{u})$ a $k \times m$ matrix. If Σ^L is a minimal system then as shown in [5] a variation $(\delta u,\delta y)$ of $(\bar{u},\bar{y}) \in \Sigma_e^+(0)(x_0)$ with supp $\delta u \subset (0,T)$ is admissible if and only if

$$\int_0^t H(\sigma,\bar{u})\,\delta u(\sigma)\,d\sigma = 0.$$

It follows that given a minimal Hamiltonian system Σ_H^L, with initial state x_0, any admissible variation $(\delta u,\delta y)$ of compact support of some trajectory $(u,y) \in \Sigma_e^+(0)(x_0)$ gives rise to state trajectories of the variational and adjoint variational systems also of compact support. This observation, together with equation (25), (which in this context generalizes equation (22) in the treatment of the inverse problem in classical mechanics) demonstrates the necessity in the following result, which constitutes one of the main results in [5].

***THEOREM 3:* [5],[17].** *Consider a minimal system Σ^L. The system is Hamiltonian if and if for any $(u,y) \in \Sigma_e^+(0)(x_0)$ and admissible variations $(\delta_i u,\delta_i y)$ of (u,y) with compact support, $i = 1, 2$, we have*

$$\int_0^\infty \left[\delta_2^T y(t)\,\delta_1 u(t) - \delta_1^T y(t)\,\delta_2 u(t)\right] dt = 0$$

Note that for minimal Hamiltonian realizations the internal energy H_0 need not be globally defined. On the other hand there always exists a Hamiltonian realization Σ_H^L for which H_0 is globally defined.

As the complexity of the statement of theorem (3.2) suggests this characterization of Hamiltonian systems is not particularly well suited to the input-output map representation. As we now show it is far better suited to the input-output behavior representation introduced in section 1. We restate our principal definition in this context.

DEFINITION. $(\delta u, \delta y)$ *is called an admissible variation of compact support of* $(\bar{u},\bar{y}) \in \Sigma_e$ *if*

(i) supp δu *is compact*

(ii) supp $\delta y \subset$ supp δu

(iii) *Suppose* supp $\delta u \subset [T_1,T_2]$, $x(T_1) = x_{T_1}$, $\bar{x}(T_2) = x_{T_2}$. *Let* $(\bar{u}',\bar{x}',\bar{y}') \in \Sigma_i$ *be such that it coincides with* $(\bar{u},\bar{x},\bar{y})$ *for* $t \in [T_1,T_2)$. *Define a variation* $u'(t,\varepsilon)$ *of* $\bar{u}(t)$ *by setting*

$u'(t,\varepsilon) = \bar{u}(t) + \varepsilon\, \delta\, u(t)$. *This yields a variation* $(u'(t,\varepsilon), y'(t,\varepsilon))$ *of* $(\bar{u}', \bar{y}')$. *We require that the resulting (infinitesimal) variation* $(\delta u', \delta y')$ *of* $(\bar{u}', \bar{y}')$ *also satisfies (ii), i.e.* $\text{supp}\ \delta y' \subset \text{supp}\ \delta u' = \text{supp}\ \delta u$.

The direct analogue of theorem (3), proved in [5], is now given in the following form.

THEOREM 4 : **[5],[17].** *Consider a minimal non-initialized system* Σ^L. *Every variational system is self-adjoint (or equivalently (Theorem 2), the system is Hamiltonian), if and only if for any* $(u,y) \in \Sigma_e$ *all admissible variations* $(\delta_i u, \delta_i y)$ *of* (u,y) *with compact support,* $i = 1,2$, satisfy

$$\int_{-\infty}^{+\infty} \left[\delta_2^T y(t)\, \delta_1 u(t) - \delta_1^T y(t)\, \delta_2 u(t) \right] dt = 0 \tag{31}$$

In a formal setting, theorem (4) can be improved as follows. Consider the "manifold" of maps $N_{M,m}$, defined as the union of all behavior sets Σ_e as Σ^L ranges over all non-initialized minimal systems, with state space M and input space R^m. On this manifold we suppose the "tangent space" to it at (u,y), denoted $T_{(u,y)} N_{M,m}$ as suggested by (31), i.e.,

$$\mu_{(u,y)}\big((\delta_1 u, \delta_1 y), (\delta_2 u, \delta_2 y)\big) = \int_{-\infty}^{+\infty} \left[\delta_2^T y(t)\, \delta_1 u(t) - \delta_1^T y(t)\, \delta_2 u(t) \right] dt \tag{32}$$

Consider now a Hamiltonian system Σ on M with m inputs. Then Σ_e is a "submanifold" of $N_{M,m}$. Now theorem (4) implies that the symplectic form μ is zero restricted to Σ_e, or equivalently, Σ_e is an isotropic submanifold of $N_{M,m}$. On the other hand the following formal result was also proved in [5].

THEOREM 5: **[5], [17].** *A minimal non-initialized system* Σ^L *on* M *is Hamiltonian if and only if* Σ_e *is a Lagrangian submanifold of* $N_{M,m}$.

Note that the statement of theorem (4), and equation (31) constitute a generalization of the result (14) obtained for linear systems, as conjectured by Van der Schaft [18].

5. NON-VARIATIONAL CHARACTERIZATION OF HAMILTONIAN SYSTEMS.

In this section we look at characterizations of Hamiltonian systems which do not involve variations. We consider first systems Σ^L and their input-output maps given in terms of the Volterra series (18) and (19). If we look at the Volterra kernels specifically for a Hamiltonian system Σ_H^L then, as already established in Crouch [7], they take a special form, described as follows. We let

$$\{h,f\} = \omega(X_h, X_g)$$

denote the Poisson bracket of two functions h and f on M, where X_h and X_g are the respective Hamiltonian vector fields. By making the following definition in terms of the Hamiltonian system Σ^L_H,

$$H_i(\sigma)(x) = H_i \circ \gamma(\sigma)(x) \qquad \leq i \leq m$$

where γ is the flow of vector field g_o, it follows that the expressions (19), for the Volterra kernels, have the form,

$$W_k^{i_o, i_1, \ldots i_k}(t, \sigma, \ldots \sigma_k, x_o) = \{ H_{i_k}(\sigma_k), \{H_{i_{k-1}}(\sigma_{k-1}), \ldots\ldots \{H_{i_1}(\sigma_1), H_{i_o}(t)\} \ldots\}\} (x_o). \tag{33}$$

We conclude that for such systems, the Volterra kernels obey certain symmetry properties, corresponding to iterated Poisson brackets. Already in Crouch and Irving [6], it was shown that conversely, the presence of these symmetry properties did indeed guarantee a Hamiltonian realization of the input-output map, in case that the input-output map was described by a finite Volterra series. The following result generalizes this result, using the variational properties of Hamiltonian systems as a method of proof. The statement involves some notation as follows. Set

$$W_n(\sigma_o, \sigma_1 \ldots [\sigma_r, \sigma_{r+1}]\, \sigma_{r+2} \ldots \sigma_n, x_o) =$$

$$W_n(\sigma_o, \sigma_1 \ldots \sigma_r, \sigma_{r+1}, \sigma_{r+2} \ldots \sigma_n, x_o) - W_n(\sigma_o, \sigma_1 \ldots \sigma_{r+1}, \sigma_r, \sigma_{r+2} \ldots \sigma_n, x_o)$$

and

$$W_n(\sigma_o, \sigma_1 \ldots [\sigma_k \ldots \sigma_r], \sigma_{r+1} \ldots \sigma_n, x_o) =$$

$$W_n(\sigma_o, \sigma_1 \ldots [\sigma_k, \sigma_{k+1} \ldots \sigma_{r-1}], \sigma_r, \sigma_{r+1} \ldots \sigma_n, x_o) \tag{34}$$

$$- W_n(\sigma_o, \sigma_1 \ldots \sigma_r[\sigma_k \ldots \sigma_{r-1}], \sigma_{r+1} \ldots \sigma_n, x_o).$$

We note that in the context of the analytic data, assumed for the systems Σ and Σ^L, these definitions are unambiguous even when restricted to the domains $\sigma_o \geq \sigma_1 \ldots \geq \sigma_n$, since the analytic continuations exist and uniquely define the kernels outside the domain.

THEOREM 6: **[5].** *Consider an input-output map which has a minimal realization by a system Σ^L and which is represented by a Volterra series (18). Then all variational systems of Σ^L are self adjoint if and only if the Volterra kernels satisfy*

$$W_n([\sigma_o, \sigma_1 \ldots \sigma_k]\, \sigma_{k+1} \ldots \sigma_n) = (k+1)\, W_n(\sigma_o \ldots \sigma_k \ldots \sigma_n) \qquad \text{for } k \geq 1, n \geq 2. \tag{35}$$

That the conditions (35) are necessary and sufficient to characterize Hamiltonian systems, by theorem 2, is basically a result of the expressions (33) and the Dynkin, Specht, Werer criteria for Lie elements in a Lie algebra.

We now turn to systems Σ and the representation of their input-output maps given as the formal series (17). Jakubczyk was able to characterize which series correspond to the Hamiltonian systems Σ_J. Notation as in (34) is used to define the elements

$$Q(u_o \cdots [u_k \cdots u_r] u_{r+1} \cdots u_n).$$

***THEOREM 7:* JAKUBCZYK, [13],[14].** *Suppose that an input-output map* $F = \sum_{\omega \in \Omega^*} Q(\omega)\, \omega$, *has a realization by a minimal system* Σ, *then it also has a realization by a Hamiltonian system* Σ_J, *if and only if either of the following hold:*

(*i*) $Q ([u_1 \ldots\ldots u_k]) = k\, Q (u_1 \ldots\ldots u_k), k \geq 2, u_i \in \Omega.$

(*ii*) $Q(v_o \cdots v_r [u_o \cdots u_k]) + (-1)^{k+r} Q(u_o \cdots u_k [v_o \cdots v_k]) = 0.$

for $k, r \geq 1, u_i, v_i \in \Omega.$

Condition (36i) plays exactly the same role as does the condition (35). To explain the alternate condition (36ii) it is interesting to review the formal momentum map with which Jakubczyk constructs realizations of Hamiltonian systems. Let A and G denote the free algebra and Lie algebra over $\Re$, generated by Ω. Let A^* and G^* denote the respective dual spaces, identified with the algebra and of formal power series, and formal Lie series respectively, in elements of Ω. The duality between A^* and A may be expressed in the form

$$\left\langle \sum_{\omega \in \Omega^*} a_\omega \omega, \sum_{\omega \in \Omega^*} b_\omega \omega \right\rangle = \sum_{\omega \in \Omega^*} a_\omega b_\omega, \quad a_\omega, b_\omega \in \Re.$$

which is well defined because the sum on the right is finite.

Assume now that we are given a space of functions $\{h^\alpha, \alpha \in \Omega\}$ on a symplectic manifold (M,ω). Let H denote the corresponding Lie algebra of functions on M, generated by this set under Poisson bracket. Construct a map λ: $G \to F$ defined by $\lambda(\alpha) = h^\alpha$ for $\alpha \in \Omega$ and extended to G by insisting that λ is a Lie algebra Homomorphism, e.g., $\lambda(\alpha_1\alpha_2 - \alpha_2\alpha_1) = \{h^{\alpha_1}, h^{\alpha_2}\}$.

The formal momentum mapping μ: $M \to G^*$ is now defined by setting

$$\langle \mu(x),\omega \rangle = \lambda(\omega)(x), \quad x \in M, \omega \in G$$

and writing

$$\mu(x) = \sum_{\omega \in \Omega^*} \langle \mu(x),\omega \rangle\, \omega. \tag{37}$$

A skew symmetric bilinear form on G^* may be defined by setting

$$\Omega_p(\omega,\omega') = \langle p, [\omega,\omega'] \rangle, \quad \omega, \omega' \in G, p \in G^*.$$

In particular, if $p = \mu(x_o)$

(38) $$\Omega_{\mu(x_o)}(\omega,\omega') = \lambda([\omega,\omega'])(x_o) = \{\lambda(\omega),\lambda(\omega')\}\ (x_o).$$

Now on initialized Hamiltonian system Σ_J, $x(0) = x_o$, gives rise to exactly the space of functions $\{h^{\alpha}; \alpha \in \Omega\}$ as above, and as is clear from the expressions (16) in this case, the formal momentum $\mu(x_o)$ as defined in equation (37), is nothing other than the representation (17) of the input-output map of the initialized system with

$$\begin{aligned} Q(u_o,u_1 \dots u_k) &= \{h^{u_k}, \{h^{u_{k-1}}, \dots\{h^{u_1},h^{u_o}\} \dots \}\ (x_o) \\ &= \lambda(\ [u_k\ [\dots\ [u_1,u_o]\ \dots\ ]\ )\ (x_o) \end{aligned}$$

Condition (36ii) is simply a statement of the skew symmetry of the bilinear form $\Omega_{\mu(x_o)}$ as expressed in (38).

Jakubczyk [13] further shows that any formal momentum $p \in G^*$ defines the input-output map of a system Σ_J if and only if the rank of Ω_p is finite and the coefficients $\langle p,\omega\rangle$ $\omega \in \Omega^*$ define analytic maps, as described in section (3). A realization of the system is given in analogy with the usual construction: the co-adjoint orbit of p under G in G^* is the state space, which is finite dimensional in case Ω_p has finite rank and the system of Hamiltonian vector fields are generated by the Hamiltonian functions on G^*

$$p \to \langle p,u\rangle,\ \ u \in \Omega.$$

6. REFERENCES.

1. R.W. Brockett and A. Rahimi, "Lie Algebras and Linear Differential Equations," in ordinary Differential Equations, ed. L. Weiss, Academic Press, New York, 1972.

2. R.W. Brocket, "Control Theory and Analytical Mechanics," in Geometric Control Theory, eds. C. Martin and R. Hermann in vol. VI of Lie Groups; History, Frontiers and Applications, Math Science Press, Brokline (1977).

3. R.W. Brockett, "Volterra Series and Geometric Control Theory," Automatica, Vol. 12, pp. 167-176, (1976).

4. P.E. Crouch and F. Lamnabhi-Lagarrigue, "State Space Realizations of Nonlinear Systems Defined by Input-Output Differential Equations," in Analysis and Optimization Systems, Lecture Notes in Control and Information Sciences, Vol. 111, pp. 138-149, (1988).

5. P.E.Crouch and A.J. van der Schaft, "Variational and Hamiltonian Control Systems," Lecture Notes in Control and Information Sciences, Vol. 101, (1987).

6. P.E. Crouch and M. Irving, "On Finite Volterra Series Which Admit Hamiltonian Realizations," Math Systems Theory, Vol. 17, pp. 293-318, (1984).

7. P.E. Crouch, "Geometric Structures in Systems Theory," I.E.E. Proc., Vol. 128, Pt.D., pp. 242-252, (1981).

8. M. Fliess, "A Note on the Invertibility of Nonlinear Input-Output Differential Systems," Systems and Control Setters, Vol. 8, pp. 147-151, (1986).

9. M. Fliess, "Nonlinear Control Theory and Differential Algebra," Proc. Modelling Adaoptive Central Conf., Sopron, Hungary, (1986).

10. T. Glad, "Nonlinear State Space and Input-Output Descriptions Using Differential Polynomials," Report LiTH.-ISY-I-0920, Linköpine University, (1988).

11. B. Jakubczyk, "Existence and Uniqueness of Realizations of Nonlinear Systems," S.I.A.M J. Control and Optimization, Vol. 18, pp. 455-471, (1980).

12. B. Jakubczyk, "Realizations of Nonlinear Systems; Three Approaches," in Proceedings of Conference on the Algebraic and Geometric Methods in Nonlinear Control Theory, Paris (1985), M. Flies, M. Hazewiukel, eds., Rudel, Nordrect, pp. B-31, (1987).

13. B. Jakubczyk, "Poisson Structures and Relations on Vector Fields and their Hamiltonians," Bull. Pol. Acad. Sci., Ser. Math., (to appear) (1986).

14. B. Jakubczyk, "Existence of Hamiltonian Realizations of Nonlinear Causal Operators," Bull. Pol. Acad. Sci., Ser. math. (to appear) (1986).

15. C.M. Lesiak and A.J. Krener, "The Existence and Uniqueness of Volterrra Series for Nonlinear Systems," I.E.E.E. Trans. Automatic control, Vol. AC-23, pp. 1090-1095, (1978).

16. R.M. Santilli, "Foundations of Theoretical Mechanics I," Springer, New York (1983).

17. A.J. Van der Schaft, and P.E. Crouch, "Hamiltonian and Self-Adjoint Control Systems," Systems and Control Letters, Vol. 8 (1987) pp. 289-295.

18. A.J. Van der Schaft, "System Theoretic Descriptions of Physical Systems," Doct. Diss. Univ. of Groningen, 1983, appeared as CWI Tract No. 3, CWI, Amsterdam (1984).

19. A.J. Van der Schaft, "Hamiltonian Dynamics with External Forces and Observations," Mathematical Systems Theory, 15 (1982), pp. 145-168.

20. A.J. Van der Schaft, "Representing a Nonlinear State Space System as a Set of Higher-Order Differential Equations in the Inputs and Outputs," Memorandum No. 698, University of Twente, Faculty of Applied Mathematics, (1988).

21. A.J. Van der Schaft, "Transformations of Nonlinear Systems Under External Equivalence," Memorandum No. 700, University of Twente, Faculty of Applied Mathematics, (1988).

22. H.J. Sussman, "Existence and Uniqueness of Minimal Realizations of Nonlinear Systems," mathematical Systems Theory, 10 (1977), pp. 263-284.

23. F. Takens, "Variational and Conservative Systems," Report ZA 7603, Univ. of Groningen (1976).

DEPARTMENT OF ELECTRICAL AND COMPUTER ENGINEERING
ARIZONA STATE UNIVERSITY
TEMPE, AZ 85287 U.S.A.

DEPARTMENT OF APPLIED MATHEMATICS
TWENTE UNIVERSITY OF TECHNOLOGY
P.O. BOX 217
7500 A.E. ENSCHEDE
THE NETHERLANDS

Contemporary Mathematics
Volume **97**, 1989

ROBUSTNESS OF DISTRIBUTED PARAMETER SYSTEMS

Ruth F. Curtain[1]

ABSTRACT. Two robust stabilization problems are discussed for a large class of infinite-dimensional linear systems. If a controller K stabilizes the nominal sytem G, it is important to know whether K stabilizes perturbations of G; this is the robustness issue. The first problem considers the case of additive perturbations, $G+\Delta G$, where ΔG is in L_∞ and the second problem allows for unstable perturbations of the class $(M+\Delta_M)^{-1}(N+\Delta_N)$, where $G=M^{-1}N$ is a normalized co-prime factorization for G and Δ_M and Δ_N are stable transfer functions. Both problems have elegant mathematical solutions with close connections to H_∞-theory. Applications to partial differential systems are discussed.

1. ROBUST STABILIZATION UNDER ADDITIVE PERTURBATIONS. Since in practice one nearly always has only an approximate model of a system it is important to know that the controller one designs for this *nominal* model will also stabilize other systems which are "close" to the nominal model; this is the robustness issue. Recently there has been a spate of papers proving that certain clever velocity feedback stabilizing schemes for various p.d.e. models of undamped flexible beams are embarassingly non-robust; they can be de-stabilized by arbitrary small delays. ([18]). At present this phenomenon does not appear to be well-understood in p.d.e. circles, but it is striking that in all the examples a frequency domain analysis indicates that the open-loop transfer function is improper. (cf. [7]). In engineering circles it has been known for a long time that even finite-dimensional improper systems can be de-stabilized by arbitrarily small delays ([34], [3] and [7]). This suggests that it may be wise to make sure that the nominal plant is proper before designing a stabilizing controller.

An appropriate class of infinite-dimensional systems was introduced by Callier and Desoer in [8]-[10].

1980 Mathematics Subject Classification (1985 Revision). 93-06, 93c25.

Partly supported by the Netherlands Foundation for the Technical Sciences (STW) project no. GWI33.0533.

DEFINITION 1.1 $\mathcal{A}(\mu)$, $\hat{\mathcal{A}}(\mu)$

For $\mu\in\mathbb{R}$, we say that $f\in\mathcal{A}(\mu)$ if

$$(1.1)\qquad f(t)=\begin{cases} 0 & ;t<0 \\ f_a(t)+\sum_{i=0}^{\infty} f_i\delta(t-t_i) \end{cases}$$

where $f_a(t)e^{-\mu t}\in L_1(0,\infty)$, f_i and t_i are real numbers, $0=t_0<t_1<t_2,\dots,$ δ represents the delta distribution and $\sum_{i=0}^{\infty}|f_i|e^{-\mu t_i}<\infty$.

$\mathcal{A}(\mu)$ is a commutative convolution Banach algebra with identity.

We say that $f\in\mathcal{A}_-(\mu)$ if there exists a $\mu_1<\mu$ for which $f\in\mathcal{A}(\mu_1)$, and so $\mathcal{A}_-(\mu)$ is a subalgebra of $\mathcal{A}(\mu)$.

$\hat{\mathcal{A}}_-(\mu)$ denotes the class of Laplace transforms of $\mathcal{A}_-(\mu)$.

Intuitively, $\hat{\mathcal{A}}_-(0)$ would correspond to the class of *stable systems*: it is a proper subset of systems with a bounded and analytic transfer function in $Re\, s\geq 0$ (the H^∞ class). However, although $\hat{\mathcal{A}}_-(\mu)$ is a commutative domain with identity, its field of fractions is not a Bezout domain and so not all elements will have co-prime factorizations which is so important for controller synthesis (see [33]). This motivated Callier and Desoer to introduce the subset $\hat{\mathcal{B}}(\mu)$, of the field of fractions of $\hat{\mathcal{A}}(\mu)$, which does admit left and right co-prime factorizations.

DEFINITION 1.2 The Callier–Desoer class $\hat{\mathcal{B}}(\mu)$

$$(1.2)\qquad \hat{\mathcal{B}}(\mu)=[\hat{\mathcal{A}}_-(\mu)].[\hat{\mathcal{A}}_-^\infty(\mu)]^{-1}$$

where $\hat{\mathcal{A}}_-^\infty(\mu)=\{\hat{f}\in\hat{\mathcal{A}}_-(\mu)$ and $\hat{f}$ is bounded away from zero at ∞ in $\mathbb{C}_\mu^+=\{s:Re\, s\geq\mu\}\}$. $\hat{f}$ is bounded away from zero at ∞ in $\mathbb{C}_\mu^+$ if there exist $\eta>0$ and $\rho>0$ such that $|\hat{f}(s)|\geq\eta$ for all $s\in\mathbb{C}_\mu^+$ such that $|s-\mu|\geq\rho$.

Alternatively, $\mathcal{B}(\mu)$ is the field of fractions of $\hat{\mathcal{A}}_-(\mu)$ with respect to $\hat{\mathcal{A}}_-^\infty(\mu)$.

Following the convenient notation from [33], we denote by $\mathcal{M}(\hat{\mathcal{B}}(\mu))$ and $\mathcal{M}(\hat{\mathcal{A}}_-(\mu))$ the class of matrix-valued transfer functions whose elements are in $\hat{\mathcal{B}}(\mu)$ and $\hat{\mathcal{A}}_-(\mu)$ respectively; intuitively these correspond to our class of transfer functions and the transfer functions which are stable with respect to the stability margin μ respectively.

Properties of the Callier–Desoer class have been studied at length in [8–10]. The following theorem reveals that the transfer functions belonging to $\mathcal{M}(\hat{\mathcal{B}}(\mu))$ have a very simple decomposition into the sum of a stable (wrt. $\mathbb{C}_\mu^+$) infinite dimensional part and a totally unstable (wrt. $\mathbb{C}_\mu^-$) finite dimensional part.

THEOREM 1.3 ([8], Theorem 3.3., [10], Lemma 2.1)

(a) $\hat{\mathcal{B}}(\mu) = [\hat{\mathcal{A}}_{-}(\mu)].[\mathcal{R}^{\infty}(\mu)]^{-1}$

where $\mathcal{R}^{\infty}(\mu)$ denotes the class of proper rational functions with complex coefficients which are holomorphic in $\mathbb{C}_{\mu}^{+}$.

(b) If $G \in \mathcal{M}(\hat{\mathcal{B}}(\mu))$, then $G = G_{+} + G_{-}$ where $G_{-} \in \mathcal{M}(\hat{\mathcal{A}}_{-}(\mu))$ and G_{+} is a strictly proper $p \times m$ valued rational function; G_{+} is zero if and only if $G \in \mathcal{M}(\hat{\mathcal{A}}_{-}(\mu))$. If $G_{+} \neq 0$, then it is the sum of the principal parts of the Laurent expansions of G at its poles in $\mathbb{C}_{\mu}^{+}$.

(c) $G \in \mathcal{M}(\hat{\mathcal{B}}(\mu))$ if and only if G has the decomposition $G = G_{+} + G_{-}$, where $G_{-} \in \mathcal{M}(\hat{\mathcal{A}}_{-}(\mu))$ and G_{+} is a strictly proper $p \times m$ valued rational function with all its poles in $Re\,\lambda \geq \mu$.

The above theorem shows that the Callier–Desoer class is a subset of all linear infinite-dimensional systems. In particular, it excludes systems with infinitely many unstable poles and it excludes systems with a distribution impulse response; examples of such systems are some neutral systems and some mathematical models of undamped flexible systems, [11], [18]. However, the Callier–Desoer class does include many infinite-dimensional systems of interest including delay systems of neutral and retarted type, parabolic and damped hyperbolic pde systems (see [29], [18], [6], [16], [17], [13]). We include an example of a flexible beam, which belongs to the Callier–Desoer class.

EXAMPLE 1.4 Euler–Bernoulli beam with Kelvin–Voigt Damping

Consider the following p.d.e. describing the deflection $w(x,t)$ at position x at time t.

$$w_{tt} + \alpha A w_t + A w = \delta(0)u_1(t) - \delta'(0)u_2(t)$$

$$w_{xx}(-1) = 0 = w_{xx}(1); w_{xxx}(-1) = 0 = w_{xxx}(1)$$

$$-1 \leq x \leq 1; t \geq 0$$

$$y_1(t) = w(0,t); y_2(t) = w_x(0,t)$$

where A is the positive, self adjoint operator

$$\begin{cases} A = \dfrac{d^4}{dx^4} \\[2ex] D(A) = \left\{ \begin{array}{l} h \in L_2(-1,1) : w_x, w_{xx}, w_{xxx}, w_{xxxx} \in L_2(-1,1) \\ w_{xx}(-1) = w_{xx}(1) = 0, w_{xxx}(-1) = 0 = w_{xxx}(1) \end{array} \right\} \end{cases}$$

Its transfer function is

$$G(s) = \frac{\mu}{2s^2}(1+\cos\mu\,\cosh\mu)\begin{bmatrix} \frac{1}{\cosh\mu\,\sin\mu+\sinh\mu\,\cos\mu} & 0 \\ 0 & \frac{\mu^2}{\cosh\mu\,\sin\mu-\sinh\mu\,\cos\mu} \end{bmatrix}$$

where $\mu^4 = \frac{-s^2}{\alpha+s}$.

We shall be considering stabilizing plants $G\in M(\hat{\mathcal{B}}(0))$ by means of a feedback configuration (figure 1) where the controller $K\in M(\hat{\mathcal{B}}(0))$.

DEFINITION 1.5. The feedback system (G,K) of figure 1 with $G,K\in M(\hat{\mathcal{B}}(0))$ is said to be *internally stable* if and only if

(a) $S=(I-GK)^{-1}$, KS, SG, $I-KSG\in M(\hat{\mathcal{A}}_{-}(0))$.

(b) $det(I-GK)(\infty)\neq 0$.

Here we are interested in designing controllers which have a guaranteed robustness margin. In [21] Glover developed a theory for such controllers for finite-dimensional systems and this was extended in [16, 17] to the Callier–Desoer class. Here we present a somewhat simplified account of the theory from [17], which considers robustness of the controller system of figure 1 under unstructured, additive, infinite-dimensional perturbations, Δ.

The precise conditions on G and Δ are assumed to be the following:

(1.3) The nominal system $G\in M(\hat{\mathcal{B}}(0))$, is strictly proper and has no poles on the imaginary axis.

(1.4) The permissible perturbations $\Delta\in M(\hat{\mathcal{B}}(0))$, are proper and are such that Δ and $\Delta+G$ have equal numbers of poles in $Re\, s\geq 0$.

and we are interested in how large the perturbations Δ can be without destroying the stability of the closed loop system of figure 2.

DEFINITION 1.6. The feedback system (G,K) of figure 2 with $G,K\in M(\hat{\mathcal{B}}(0))$ is said to be *additively robustly stable* with *robustness margin* ε if the feedback system of figure 2.1 is internally stable for all (G,Δ) satisfying (1.3) and (1.4) and $\|\Delta\|_\infty<\varepsilon$.

Robust stabilizability was studied in [12] where using Nyquist arguments they proved the following theorem.

THEOREM 1.7. If $G,K\in M(\hat{\mathcal{B}}(0))$ and (1.3), (1.4) are satisfied, (G,K) is additively robustly stable if and only if (G,K) is internally stable and

(1.5) $$\|K(I-GK)^{-1}\|_\infty \leq \frac{1}{\varepsilon}$$

This theorem can be used to compare the robustness of different controllers as in [5], but more interesting is that it can be used to obtain the following powerful result.

THEOREM 1.8 [17]. If $G,K\in M(\hat{\mathcal{B}}(0))$ and (1.3) is satisfied, given an $\varepsilon>0$ there exists a compensator K which stabilizes $G+\Delta$ for all Δ satisfying (1.4) and $\|\Delta\|_\infty<\varepsilon$ provided that

(1.6) $\sigma_{min}(G_u) \geq \varepsilon$

where $\sigma_{min}(G_u)$ is the smallest Hankel singular value of the finite–dimensional unstable part of G.

We recall from Theorem 1.3(b) that Callier–Desoer systems can be decomposed into a stable infinite–dimensional part and an unstable finite–dimensional part. Theorem 1.8 says that additive robustness is purely a property of this unstable *finite-dimensional* part and this is usually more accurately modelled than the infinite–dimensional part. Moreover, $\sigma_{min}(G_u)$ can be easily calculated a–priori yielding valuable information about the maximum acheivable robustness for a nominal model G. The same theorem can be applied to $G(s-\beta)$ giving information about robustness with respect to an arbitrary stability margin β.

In addition to the existence result in [CG4] there are also explicit formulas for a controller K which acheives the maximum robustness margin $\sigma_{min}(G_u)$. These depend on the stable part of G, G_s, and so this K is *infinite-dimensional*, which is undesirable for applications. However if one replaces G_s by a finite–dimensional L_∞–approximation, G_s^k, say, then one can design a finite- dimensional controller (of the order of k plus the McMillan degree of G_u) which has a robustness margin of at least $\sigma_{min}(G_u) - \|G_s - G_s^k\|_\infty$. So in combination with a theory for good L_∞–approximations (for example, that described in section 4) we have a practical design technique for additively robust finite–dimensional controllers for infinite–dimensional systems of the Callier–Desoer class. ([14]). It has been applied to delay systems in [16] and in comparison studies on beam models in [6]. Finite–dimensional controllers for infinite- dimensional systems based on reduced order models have appeared elsewhere in the literature [1], but [16,17] was the first result to give a–priori estimates for the acheivable robustness and to clarify the role of approximation in controller design.

In particular, Theorem 1.8 explains the mysteries surrounding *"spillover"* which sometimes occurs in controller design based on reduced order models ([1]). In reduced order model design one usually designs a finite–dimensional controller K_f based on some low–order approximation G_f of the infinite–dimensional system G and then hopes that K_f will also stabilize G. Theorems 1.7 and 1.8 now applied to G_f tell us that K_f will stabilize G if it stabilizes G_f and

(1.7) $\|G - G_f\|_\infty \;\leq\; \|K_f(I - G_f K_f)^{-1}\|_\infty^{-1}$

An intelligent choice for K_f would be the maximally robust controller for G_f, for then Theorem 1.8 tells us that K_f stabilizes G if

(1.8) $\|G-G_f\|_\infty \leq \sigma_{min}(G_f^u)$

where G_f^u is the unstable part of G_f. (1.8) gives an a-priori criterion for choosing the reduced order model G_f for G; it clarifies the trade-off between the robustness margin, the order of the controller and the order of the reduced-order model. (see [4]).

2. ROBUST STABILIZATION UNDER NORMALIZED CO-PRIME FACTOR PERTURBATIONS. In section 1 we outlined a theory for robust stability with respect to unstructured additive perturbations. Here we consider robust stabilization with respect to additive, stable perturbations in a co-prime factorization of the system, which allows for unstable perturbations via the denominator factor. This type of robustness was advocated by Vidyasagar in [33], where he showed that this family of perturbations is particularly appropriate for feedback systems analysis. Very recently in [22,23] a very pleasing, explicit solution to this problem was obtained for normalized co-prime factorizations of finite-dimensional systems. The problem can just as easily be formulated for systems $G\in M(\hat{\mathcal{B}}(0))$ after we have defined normalized left co-prime factorizations.

DEFINITION 2.1. If $G\in M(\hat{\mathcal{B}}(\mu))$, then $\tilde{M}^{-1}\tilde{N}$ is called a *left co-prime facotrization* of G if and only if

(i) $\tilde{M}$ and $\tilde{N}\in M(\hat{\mathcal{A}}_-(\mu))$ and $det\, \tilde{M}\in\hat{\mathcal{A}}_-(\mu)$.

(ii) $G=\tilde{M}^{-1}\tilde{N}$

(iii) $\tilde{M}$ and $\tilde{N}$ are left co-prime in the sense that there exist X and Y in $M(\hat{\mathcal{A}}_-(\mu))$ such that

(2.1) $\tilde{N}Y-\tilde{M}X=I$

$\tilde{M}^{-1}\tilde{N}$ is called a *normalized* left co-prime factorization of G if it is a left coprime factorization and

(2.2) $\tilde{N}(s)\tilde{N}^t(-s)+\tilde{M}(s)\tilde{M}^t(-s)=I$ in $Re\, s>$).

All transfer functions in $M(\hat{\mathcal{B}}(0))$ have left co-prime factorizations.

We are interested in the robustness of internal stability under certain perturbations in the system G with respect to a given normalized left co-prime factorization $G=\tilde{M}^{-1}\tilde{N}$ (1). The perturbations we consider are of the form.

(2.3) $G_\Delta=(\tilde{M}+\Delta_M)^{-1}(\tilde{N}+\Delta_N)$

where Δ_N, Δ_M are unknown (stable) transfer functions in $M(\hat{\mathcal{A}}_-(0))$ which represent the uncertainty in our nominal system model. Notice that there are no restrictions on the number of poles in G_Δ, in contrast to the additive perturbation theory discussed in section 1. ((1.4)).

The robust design objective is to find a feedback controller $K\in M(\hat{\mathcal{B}}(0))$

which stabilizes not only the nominal system G, but also the family of perturbed systems defined by

(2.4) $\mathcal{G}_\epsilon=\{G_\Delta=(\tilde{M}+\Delta_M)^{-1}(\tilde{N}+\Delta_N)$ such that Δ_M, $\Delta_N\in\mathcal{M}(\hat{\mathcal{A}}_-(0))$ and $\|[\Delta_M,\Delta_N]\|_\infty<\varepsilon\}$ (see figure 3)

This leads to the following definition of factor robust stability.

DEFINITION 2.2. Suppose that the system $G\in\mathcal{M}(\hat{\mathcal{B}}(0))$ has the left normalized co-prime factorization of Definition 2.1, then the feedback system $(\tilde{M},\tilde{N},K,\varepsilon)$ of figure 3 is *factor robustly stable* if and only if (G_Δ,K) is internally stable for all $G_\Delta\in\mathcal{G}_\epsilon$. If there exists a K such that $(\tilde{M},\tilde{N},K,\varepsilon)$ is factor robustly stable, then (G,ε) is said to be factor robustly stabilizable with *factor robustness margin* ε.

Appealing again to the results of [12] applied to figure 3 (cf. Theorem 1.6), we obtain the following characterization of factor robust stabilizability.

LEMMA 2.3. Suppose that $G\in\mathcal{M}(\hat{\mathcal{B}}(0))$ has the normalized left co-prime factorization $G=\tilde{M}^{-1}\tilde{N}$, then (G,ε) is factor robustly stabilizable if and only if

(2.5) $$\inf_K \left\| \begin{bmatrix} K(I-GK)^{-1}\ \tilde{M}^{-1} \\ (I-GK)^{-1}\ \tilde{M}^{-1} \end{bmatrix} \right\|_\infty \leq \frac{1}{\varepsilon}$$

where the infinum is taken over all stabilizing controllers $K\in\mathcal{M}(\hat{\mathcal{B}}(0))$.

(2.5) has the form of an H_∞-optimization problem and it can be converted to the general formulation of [19].

(2.6) $$\inf_K \|\mathcal{F}_L(P,K)\|_\infty \leq \frac{1}{\varepsilon}$$

where

(2.7) $$P=\begin{bmatrix} P_{11}\ P_{12} \\ P_{21}\ P_{22} \end{bmatrix} = \begin{bmatrix} 0 & I \\ \tilde{M}^{-1} & G \\ \tilde{M}^{-1} & G \end{bmatrix}$$

and $\mathcal{F}_L(P,K)$ denotes a *linear fractional transformation*

(2.8) $$\mathcal{F}_L(P,K)=P_{11}+P_{12}K(I-P_{22}K)^{-1}P_{21}$$

Now (2.6) is a general H^∞-problem and in [22] it was reduced to the general distance problem.

(2.9) $$\inf_{J\in\mathcal{M}(\hat{\mathcal{A}}_-(0))} \left\| \begin{bmatrix} R_1+J \\ I \end{bmatrix} \right\|_\infty$$

where R_1 is the antistable system given by

(2.10) $$R_1^t(-s)=C(I+PQ)(s-A+BB^*Q)^{-1}B$$

and P and Q are the unique solutions of the filter and control Riccati equations.

(2.11) $$AP+PA^*-PC^*CP+BB^*=0$$

(2.12) $$A^*Q+QA-QBB^*Q+C^*C=0$$

The solution to (2.4) was shown to be $(1+\lambda_{max}(PQ))^{\frac{1}{2}}$ and the acheivable factor robustness margin was

(2.13) $$\varepsilon_{max}=[1+\lambda_{max}(PQ)]^{-\frac{1}{2}}=(1-\|[\tilde{N},\tilde{M}]\|_H^2)^{\frac{1}{2}}$$

Explicit formulas for factor robust controllers were given in terms of the original system parameters and P and Q.

In [15] it was shown that this result could be generalized to the following subclass of the Callier–Desoer class.

DEFINITION 2.4 The Pritchard–Salamon class

Suppose there exist two separable Hilbert spaces V and W with continuous, dense injection $W \hookrightarrow V$ and that A generates a C_0–semigroup on both spaces which we denote by the same symbol $S(t)$. In addition we assume that

(2.14) $$Z=D_V(A)\hookrightarrow W; \qquad Y=D_{W^*}(A^*)\hookrightarrow V^*$$

We suppose that $B\in\mathcal{L}(\mathbb{R}^m,V)$ induces a smooth controllability map with respect to W on $[0,t_1]$ for all finite t_1

(2.15) $$\left\|\int_0^{t_1} S(t_1-s)Bu(s)ds\right\|_W \leq \beta\|u\|_{L_2(0,t_1;\mathbb{R}^m)}$$

and $C\in\mathcal{L}(W,\mathbb{R}^p)$ induces a smooth observability map with respect to V on $[0,t_1]$ for all finite t_1,

(2.16) $$\|Cs(\cdot)x\|_{L_2(0,t_1;\mathbb{R}^p)}\leq\gamma\|x\|_V$$

Furthermore we suppose that (A,B) is exponentially stabilizable and (C,A) is exponentially detectable.

In [13] it was shown that example 1.4 satisfies the conditions of Definition 2.4. The Pritchard–Salamon class was introduced in [29] for the study of the linear quadratic control problem and they showed that for these systems, both control and filter Riccati equations (2.11) and (2.12) have unique solutions when correctly interpreted. This is of course the key in extending the solution to the infinite–dimensional case, although there are other technicalitites involved (see [15]). In particular ε_{max} cannot be exactly acheived, but there exist controllers which acheive a robustness margin arbitrarily close to it.

Although this is a very pleasing result, the solution is at present primarily of theoretical value. The numerical aspects are not so straightforward as everything depends on P and Q and this means that approximating schemes will have to take account of the convergence of the Riccati equations. While there exists an extensive literature on numerical solutions of Riccati equations in infinite–dimensions, it has two major disadvantages. All existing schemes are realisation dependent and most are for systems with bounded inputs and outputs [2], [20], [25], [26]. More work needs to be done on these numerical aspects.

3. CONCLUSIONS. We have presented two different types of theories for robust stabilization. The first was robustness with respect to additive perturbations and has a complete solution for the class of transfer functions in the Callier–Desoer class. It also generates a practical approach to the design of finite–dimensional controllers for infinite–dimensional systems based on reduced order models with 'a–priori' trade–offs between controller order and robustness.

The second robustness theory allowed for very general unstable perturbations and while a complete theoretical solution was obtained, several nontrivial numerical problems remain to be solved.

Both theories assume a proper transfer function with an integrable impulse response. From finite–dimensional considerations the assumption of a proper transfer function seems unavoidable ([34]), but there do exist infinite–dimensional systems with a distribution impulse response [31], [32]. If, however, we model the inputs or outputs with dynamics (as is often the case with physical actuators), the resulting impulse response is smoothed to an integrable function. So the class of systems covered by our theory is wider than one may think for real systems. So the most serious limitation is that the systems may have finitely many unstable poles, whereas some p.d.e. models of flexible systems have infinitely many unstable poles ([11], [18]). If one again assumes that one is interested in physically realizable controllers, then these are necessarily band–limited and in [24] it is argued that systems with infinitely many poles cannot be stabilized by band–limited controllers. Consequently we claim that from a practical point of view the Callier–Desoer class is not that restrictive. The exponentially stabilizable and detectable Pritchard–Salamon class is a subclass of the Callier–Desoer class, but it does include many classes of p.d.e. and delay ssytems ([29], [13]).

The most serious restriction of this theory is that they both assume unstructured perturbations, that is, the perturbed systems lie in some ball around the nominal system. For many applications, for example flexible systems, one knows something about the structure of the uncertainties in the model, which one would like to exploit. A theory for robustness under structured perturbations is very important and is a current area of research. (see [30]).

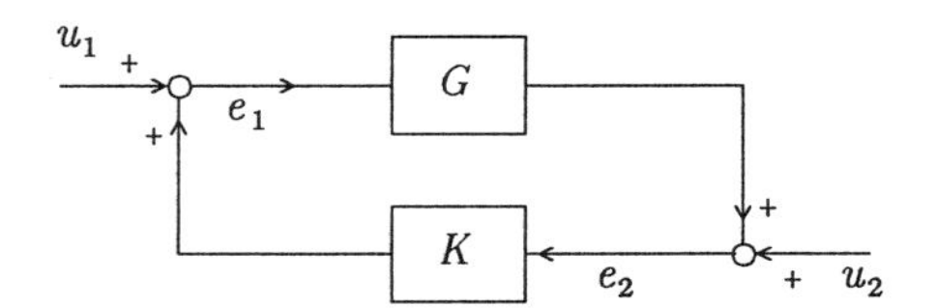

Figure 1. Internal stability

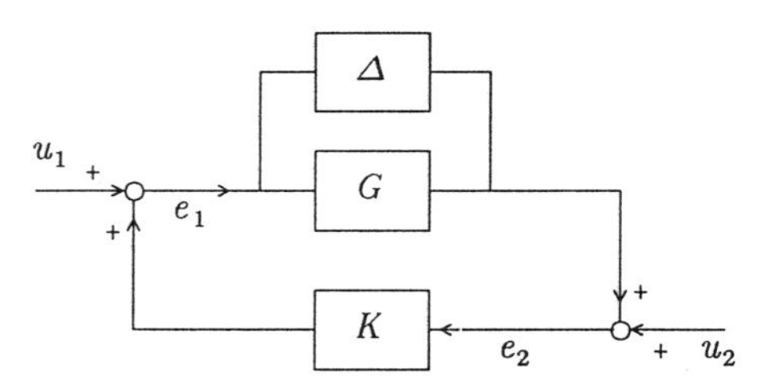

Figure 2. Additive robust stability

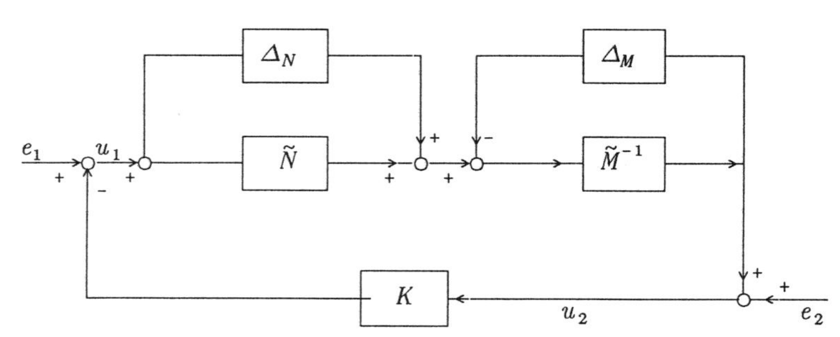

Figure 3. Left coprime factor perturbations

BIBLIOGRAPHY

1. M. Balas, Towards a (more) practical control theory for distributed parameter systems in "Control and Dynamic Systems: Advances in Theory and Applications, Vol.18, C.T. Leondes (ed.), Academic Press, New York, 1980.

2. H.T. Banks and K. Kunisch, "The Linear Regulator Problem for Parabolic Systems", SIAM J. Control and Optimization, **22**, p. 684–698, 1984.

3. J.F. Barman, F.M. Callier and C.A. Desoer, L^2–Stability and L^2–Instability of Linear Time–Invariant Distributed Feedback Systems Perturbed by a Small Delay in the Loop, IEEE Trans. AC, vol 18, p.479–484, 1973.

4. J. Bontsema and R.F. Curtain, "A Note on Spillover and Robustness of Flexible Systems", IEEE Trans. Autom. Contr. **33**, p.567–569, 1988.

5. J. Bontsema and R.F. Curtain, "Comparison of Some Controllers on a Flexible Beam", pp.165–166, Proc. 25th IEEE Conf. on Decision and Control, 1986, Athens, IEEE Control Systems Soc, NY, 1986.

6. J. Bontsema, R.F. Curtain and J.M. Schumacher, "Robust Control of Flexible Structures: A Case Study". Automatica, Vol.24, No.2 pp.177–186, 1988.

7. J. Bontsema and S.A. De Vries, "Robustness of Flexible Systems Against Small Time Delays", 27th CDC, 1988.

8. F.M. Callier and C.A. Desoer, "An algebra of transfer functions for distributed linear time–invariant systems", IEEE Trans. Circuits and Systems **25** (1978), p. 651–663, (Corrections: **26** (1979), p.320).

9. F.M. Callier and C.A. Desoer, "Simplifications and New Connections on an algebra of transfer functions of distributed linear time–invariant systems", IEEE Trans. Circuits of Systems, CAS–27 , p. 320–323, 1980.

10. F.M. Callier and C.A. Desoer, "Stabilization, Tracking and Distributed Rejection in Multivariable Convolution Systems", Ann. Soc. Sci. Bruxelles, **94**, 1980, p. 7–51.

11. G. Chen, M.C. Delfour, A.M. Krall and G. Payre, "Modelling, stabilization and control of serially connected beams", SIAM J. Control and Optim., (1987), **25**, pp. 526–546.

12. M. Chen and C.A. Desoer, "Necessary and Sufficient Conditions for Robust Stability of Distributed Feedback Systems", Int. J. Control, Vol. 35, p. 255–267, 1982.

13. R.F. Curtain, "Equivalence of Input–Ouput Stability and Exponential Stability for Infinite–dimensional systems", J. Math. Systems Theory, (to appear), 1988.

14. R.F. Curtain, "L_∞–Approximations of Complex Functions and Robust Controllers for Large Flexible Space Structures", Nieuw Archief voor Wiskunde, (to appear).

15. R.F. Curtain, "Robust Stabilizability of Normalized Co–prime Factors; the Infinite–Dimensional Case". Report TW 291, University of Groningen, NL, 1988.

16. R.F. Curtain and K. Glover, "Robust Stabilization of Infinite–

dimensional Systems by Finite Dimensional Controllers: Derivations and Examples", pp. 113–128, Modelling, Identification and Robust Control, (Proc. MTNS 1985, Sweden) Ed. C.I. Byrnes & A. Lindquist, Elsevier Science Publishers, North Holland, 1986.

17. R.F. Curtain and K. Glover, "Robust Stabilization of infinite dimensional systems by finite dimensional controllers", Systems and Control Letters 7 (1986) p 41–47.

18. R. Datko, J. Lagnese and M.P. Polis, "An Example on the Effect of Time Delays in Boundary Feedback Stabilization of Wave Equations". SIAM J. Control and Optimization, vol.24, p.152–156, 1986.

19. J. Doyle, "Lecture Notes in Advances in Multivariable Control", Office of Naval Research/Honeywell Workshop, Minneapolis, MN, 1984.

20. J.S. Gibson, "Linear Quadratic Control of Hereditary Differential Systems: Infinite Dimensional Riccati Equations and Numerical Approximations", SIAM J. Control, Opt. **21**, p. 95–139, 1983.

21. K. Glover, "Robust stabilization of multivariable linear systems: Relations to approximation", Int. J. Control, **43**, pp. 741–766, 1986.

22. K. Glover and D. McFarlane, "Robust Stabilization of Normalized Co-prime Factors: An Explicit H_∞-solution", to appear in the Proc. of the ACC, 1988.

23. K. Glover and D. McFarlane, "Robust Stabilization of Normalized Co-prime Factor Plant Descriptions with H_∞-bounded Uncertainty", submitted to IEEE Trans. on Automatic Control, 1988.

24. A.J. Helmicki, C.A. Jacobson and C.N. Nett, "Fundamentals of Practical Controller Design for LTI Plants with Distributed Parameters": Part 1, Modelling and Well-Posedness. Proc. 1987 American Control Conference, p.1203–1208, June 1987.

25. K. Ito, "Strong Convergence and Convergence Rates of Approximating Solutions for Algebraic Riccati Equations in Hilbert Spaces", p. 153–166, Distributed Parameter Systems (Proc. 3rd Int. Conf. Vorau, Styria, July 6–12, 1986), LNCIS 102, Springer, Verlag, 1987.

26. F. Kappel and D. Salamon, "Spline Approximation for Retarded Systems and the Riccati Equation", SIAM J. Control and Opt. **25**, p. 1082–117, 1987.

27. C.N. Nett, C.A. Jacobson, M.J. Balas, "Fractional Representation Theory: Robustness with Applications to Finite-Dimensional Control of a Class of Linear Distributed Systems", Proc. IEEE Conf. on Decision and Control, p. 269–280, 1983.

28. C.N. Nett, C.A. Jacobson, M.J. Balas, "A connection between state space and doubly coprime fractional representations", Trans. IEEE, 1984, **9**, p 831–832.

29. A.J. Pritchard and D. Salamon, "The linear quadratic optimal control problem for infinite dimensional systems with unbounded input and output operators", SIAM J. Control and Optimiz., 1987, **25**, p. 121–144.

30. A.J. Pritchard and S. Townley, "A Stability Radius for Infinite-Dimensional Systems", p.272–291, Distributed Parameter Systems (Proc.

3rd Int. Conf. Vorau, Styria, July 6–12, 1986), LNCIS 102, Springer Verlag, 1987.

31. D. Salamon, "Control and Observation of Neutral Systems", Research Notes in Mathematics no.91, Pitman Advanced Publishing Program, Boston, 1984.

32. D. Salamon, "Infinite–dimensional linear systems with unbounded control and observation: a functional analytic approach", Trans. Amer. Math. Soc. 1987, **300**, pp.383–431.

33. M. Vidyasagar, "Control System Synthesis: A Factorization Approach", 1985, MIT Press, Cambridge, Mass., USA.

34. J.C. Willems, "The Analysis of Feedback Systems", MIT Press, Cambridge, MA, 1971.

DEPARTMENT OF MATHEMATICS
UNIVERSITY OF GRONINGEN
P.O.BOX 800
9700 AV GRONINGEN
THE NETHERLANDS

Contemporary Mathematics
Volume 97, 1989

On the Relationship Between Discrete-Time Optimal Control and Recursive Dynamics For Elastic Multibody Chains

C. J. DAMAREN AND G. M. T. D'ELEUTERIO
Institute for Aerospace Studies
University of Toronto
Downsview, Ontario, Canada M3H 5T6

Abstract

A recursive algorithm for determining the motion of a topological chain of interconnected elastic bodies with arbitrary joint constraints is presented. The system of equations arising from a Newton-Euler formulation of the dynamics is shown to be identical in form to the two-point boundary value problem involved in a discrete-time optimal control problem. The recursive techniques that are used to determine the joint and elastic coordinate accelerations are analogous to the procedures that yield the optimal control policy in the latter problem. For each relevant quantity in the multibody dynamics problem, there is a corresponding element in the optimal control problem. The relationship between the two is further uncovered by identifying the performance index as Gibbs' function for the elastic chain.

1 Introduction

The equations of multibody dynamics are a source of significant challenge for the dynamicist. Their structure is such that a single, definitive solution procedure does not present itself. With the increasing use of robotic systems in terrestial and space applications, the need for these techniques is great. The use of lightweight materials and the requirement of faster operating speeds highlights the added complexity of structural flexibility. Given the external influences and control forces acting on a flexible multibody system, the ability to determine its subsequent motion is of paramount importance.

Here, we shall consider the dynamics of a topological chain of elastic bodies. The equations of motion are written using the Newton-Euler formulation described by SINCARSIN & HUGHES [1989]. Structural flexibility is modeled using linear elasticity which is assumed to be adequate for the anticipated applications where large elastic deformations must be avoided. The elastic influences enter the equations in discretized form as would be provided by a Ritz-based scheme such as the Finite-Element Method. The joints connecting neighbouring bodies can possess as many as six (free) degrees of freedom. However, for the present discussion, relative translations between bodies are restricted to be small.

The fundamental drawback with a Newton-Euler formulation is that the interbody forces (and torques) must be known. The components associated with the free joint degrees

0271-4132/89 $1.00 + $.25 per page

of freedom, the control forces, are prescribed whereas the constraint forces provided by the joint are not known *a priori*. Global solutions of the problem consolidate the dynamics of the chain of bodies into a single system of equations from which the constraint forces can be eliminated. This line of attack is taken by WALKER & ORIN [1982] and HUGHES & SINCARSIN [1989]. Recursive algorithms, which circumvent the consolidation of the equations by stepping from body to body in a systematic manner, have the benefit of potentially reducing the number of calculations required. This number grow linearly with the number of bodies in the chain whereas global procedures typically involve $O(M^3)$ calculations where M is the total (rigid and elastic) number of coordinates.

The recursive path to a solution of multibody dynamics problems was pioneered by ARMSTRONG [1979] and FEATHERSTONE [1983] among others. The difference between the two is the way in which the constraint forces are eliminated. A substantial generalization of Armstrong's method is presented in some detail by D'ELEUTERIO [1989]. Readers interested in a recursive solution employing a variational approach can consult BAE & HAUG [1987] who also consider closed-loop topologies. RODRIGUEZ [1987] has developed a recursive technique for hinged rigid bodies that exploits the relationship between multibody dynamics and the equations that occur in discrete-time optimal estimation and smoothing problems.

The algorithm presented here extends Featherstone's idea to encompass elastic bodies with arbitrary joint constraints. The key lies in the assumption that the interbody forces can be expressed as an affine function of certain acclerations. We will show that the kinematical constraints and body motion equations constitute a two-point boundary value problem that is identical in form to that arising in a discrete-time optimal *control* problem. In fact, a one-to-one correspondence can be made between relevant quantities in the multibody dynamics problem and the entities of the control problem. Elimination of the constraint forces in the former problem is analogous to eliminating the adjoint variables in the latter one. The solution for the optimal control policy is precisely that which yields the joint and elastic accelerations. Furthermore, the performance index will be identified as GIBBS' [1879] function for elastic multibody chains.

These results will extend those of D'ELEUTERIO & DAMAREN [1989] for chains of rigid bodies to encompass the case of elastic bodies. Following the lead of RODRIGUEZ [1987], the recursive algorithm will be used to write a literal inverse for the global mass matrix (with respect to rigid *and* elastic degrees of freedom), which provides insight into the relationship existing between the recursive and global approaches.

2 Equations of Motion

We begin by considering a chain of contiguous elastic bodies $\mathcal{E}_0$, $\mathcal{E}_1, \ldots, \mathcal{E}_N$ as shown in Figure 1. Interbody joints may permit arbitrary relative (rotational and/or translational) motion. Each joint therefore possesses at least one degree of freedom and at most six. For the present development, interbody translations are assumed small; however, the extension to large translations can be taken into account by a straightforward transformation which renders the form of the motion equations unchanged [D'ELEUTERIO 1989]. Only a brief overview of the formulation will be given here. Full details are provided by SINCARSIN & HUGHES [1989].

Let us attach to $\mathcal{E}_n$ a reference frame $\mathcal{F}_n$ at O_n. The absolute velocity (with respect to inertial space, $\mathcal{F}_I$) of O_n is denoted by $\mathbf{v}_n$ and the absolute angular velocity of $\mathcal{F}_n$ by $\boldsymbol{\omega}_n$. (See Figure 2.) Both $\mathbf{v}_n$ and $\boldsymbol{\omega}_n$ are expressed in $\mathcal{F}_n$.

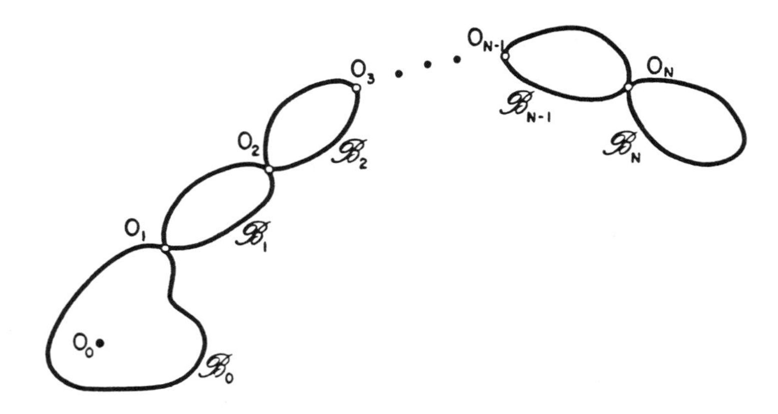

Figure 1: A Chain of Elastic Bodies

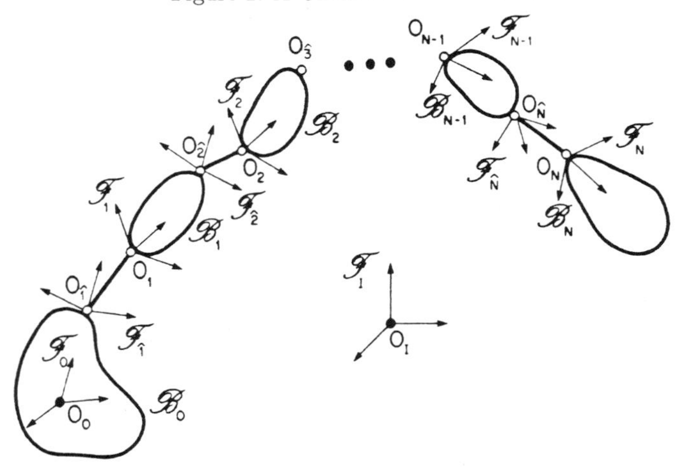

Figure 2: Reference Frames (with Interbody Translations)

In the spirit of the Finite-Element Method, the velocity distribution of $\mathcal{E}_n$ may be expressed as

$$\mathbf{v}(\mathbf{r}_n, t) = \mathbf{v}_n(t) - \mathbf{r}_n^{\times}\boldsymbol{\omega}_n(t) + \boldsymbol{\Delta}_n(\mathbf{r}_n)\dot{\boldsymbol{q}}_{n,e}(t) \tag{1}$$

where $\boldsymbol{\Delta}_n(\mathbf{r}_n)$ is a matrix of basis functions which assumes $\mathcal{E}_n$ to be constrained (cantilevered) at O_n, *i.e.*,

$$\boldsymbol{\Delta}_n(\mathbf{0}) = \mathbf{O}, \qquad \boldsymbol{\nabla}^{\times}\boldsymbol{\Delta}_n(\mathbf{0}) = \mathbf{O}$$

and

$$\boldsymbol{q}_{n,e} \triangleq [\ q_{1n} \quad q_{2n} \quad \cdots\ ]^T$$

represents the elastic coordinates. Also, $(\cdot)^{\times}$ is the matrix operation corresponding to the vector cross product.

A succinct form for the motion equations can be had if we introduce

$$\boldsymbol{v}_n \triangleq \begin{bmatrix} \mathbf{v}_n \\ \boldsymbol{\omega}_n \end{bmatrix} \tag{2}$$

as the *generalized velocity* of $\mathcal{E}_n$ at O_n. We shall accordingly need an accompanying definition for a *generalized force* acting at O_n, namely,

$$\boldsymbol{f}_n^{n-1} \triangleq \begin{bmatrix} \mathbf{f}_n^{n-1} \\ \mathbf{g}_n^{n-1} \end{bmatrix} \tag{3}$$

where $\mathbf{f}_n^{n-1}$ and $\mathbf{g}_n^{n-1}$ are the reaction forces and torques on $\mathcal{E}_n$ due to $\mathcal{E}_{n-1}$ as expressed in $\mathcal{F}_n$. The reader may already be familiar with these or similar quantities from screw calculus; indeed, (2) and (3) may be directly compared to *twist velocities* and *wrenches.*

The equation of motion for $\mathcal{E}_n$ is obtained using a Newton-Euler formulation and is best presented in two parts. The *rigid* part of the motion equation may be written as

$$\mathcal{M}_{n,rr}\dot{\boldsymbol{v}}_n + \mathcal{M}_{n,re}\ddot{\boldsymbol{q}}_{n,e} = \boldsymbol{f}_{nT,r} + \boldsymbol{f}_{nI,r} \tag{4}$$

Herein

$$\mathcal{M}_{n,rr} \triangleq \begin{bmatrix} m_n\mathbf{1} & -\mathbf{c}_n^{\times} \\ \mathbf{c}_n^{\times} & \mathbf{J}_n \end{bmatrix}$$

is the (constant) rigid mass matrix associated with $\mathcal{E}_n$, that is, m_n, $\mathbf{c}_n$ and $\mathbf{J}_n$ are the zeroeth (mass), first and second moments of inertia (about O_n) of $\mathcal{E}_n$, and

$$\mathcal{M}_{n,re} \triangleq \begin{bmatrix} \mathbf{P}_n \\ \mathbf{H}_n \end{bmatrix}$$

where

$$\mathbf{P}_n \triangleq \int_{\mathcal{E}_n} \boldsymbol{\Delta}_n(\mathbf{r}_n)dm, \qquad \mathbf{H}_n \triangleq \int_{\mathcal{E}_n} \mathbf{r}_n^{\times}\boldsymbol{\Delta}_n(\mathbf{r}_n)dm$$

In addition, $\boldsymbol{f}_{nT,r}$ is the total external (generalized) force acting on $\mathcal{E}_n$, including interbody forces, and

$$\boldsymbol{f}_{nI,r} = (\boldsymbol{v}_n^{\times})^T\mathcal{M}_{n,rr}\boldsymbol{v}_n \tag{5}$$

where

$$\boldsymbol{v}_n^{\times} \triangleq \begin{bmatrix} \boldsymbol{\omega}_n^{\times} & \mathbf{v}_n^{\times} \\ \cdot & \boldsymbol{\omega}_n^{\times} \end{bmatrix}$$

is a compact representation for the nonlinear inertial terms.

The corresponding *elastic* equation of motion for $\mathcal{E}_n$ is

$$\mathcal{M}_{n,re}^T\dot{\boldsymbol{v}}_n + \mathcal{M}_{n,ee}\ddot{\boldsymbol{q}}_{n,e} + \mathcal{K}_{n,ee}\boldsymbol{q}_{n,e} = \boldsymbol{f}_{nT,e} + \boldsymbol{f}_{nI,e} \tag{6}$$

where

$$\mathcal{M}_{n,ee} \triangleq \int_{\mathcal{E}_n} \mathbf{\Delta}_n^T(\mathbf{r}_n)\mathbf{\Delta}_n(\mathbf{r}_n)dm, \qquad \mathcal{K}_{n,ee} \triangleq \int_{\mathcal{E}_n} \mathbf{\Delta}_n^T(\mathbf{r}_n)\mathbf{K}_n\mathbf{\Delta}_n(\mathbf{r}_n)dV$$

are the mass and stiffness matrices for $\mathcal{E}_n$. The stiffness operator $\mathbf{K}_n$ for $\mathcal{E}_n$ implies the constrained boundary conditions at O_n. It should be observed that $\mathcal{M}_{n,ee}$ and $\mathcal{K}_{n,ee}$ are independent of chain configuration and are symmetric and positive-definite. Also,

$$\boldsymbol{f}_{nT,e} \triangleq \int_{\mathcal{E}_n} \mathbf{\Delta}_n^T(\mathbf{r}_n)\mathbf{f}_{nT}(\mathbf{r}_n, t)dV$$

where $\mathbf{f}_{nT}$ is the total external force distribution acting on $\mathcal{E}_n$. The nonlinear inertial term $\boldsymbol{f}_{nI,e}$ can be written as

$$\boldsymbol{f}_{nI,e} \triangleq -\mathbf{P}_n^T\boldsymbol{\omega}_n^\times\mathbf{v}_n - \boldsymbol{f}_{nR,e}$$

where

$$[\boldsymbol{f}_{nR,e}]_i \triangleq [\mathbf{I}_n]_{ijk}[\boldsymbol{\omega}_n]_j[\boldsymbol{\omega}_n]_k$$

$$[\mathbf{I}_n]_{ijk} \triangleq (\delta_{jp}\delta_{pk} - \delta_{jk}\delta_{pq}) \int_{\mathcal{E}_n} [\mathbf{\Delta}_n]_{ji}[\mathbf{r}_n]_k dm$$

and δ_{jk} is the Kronecker delta. We hasten to mention that $\mathbf{I}_n$ is an inertial body quantity which is also independent of chain configuration.

To summarize, then, the equations of motion for $\mathcal{E}_n$ are

$$\begin{aligned} \mathcal{M}_{n,rr}\dot{\boldsymbol{v}}_n + \mathcal{M}_{n,re}\ddot{\boldsymbol{q}}_{n,e} &= \boldsymbol{f}_{nT,r} + \boldsymbol{f}_{nI,r} \\ \mathcal{M}_{n,re}^T\dot{\boldsymbol{v}}_n + \mathcal{M}_{n,ee}\ddot{\boldsymbol{q}}_{n,e} + \mathcal{K}_{n,ee}\boldsymbol{q}_{n,e} &= \boldsymbol{f}_{nT,e} + \boldsymbol{f}_{nI,e} \end{aligned} \tag{7}$$

It should be added that these equations neglect terms which involve the product of rigid quantities $\{\boldsymbol{v}_n, \dot{\boldsymbol{v}}_n\}$ and elastic quantities $\{\boldsymbol{q}_{n,e}, \dot{\boldsymbol{q}}_{n,e}\}$ as well as terms containing products of elastic quantities. The only nonlinear inertial terms which have been included are those arising from rigid-body motion. The other terms can be incorporated without much difficulty, only a little tedious effort. It is important to note, however, that the development of the recursive simulation procedure which follows would be unaffected by the addition of such terms. For a rate-linear model, $\boldsymbol{f}_{nI} \equiv \mathbf{0}$.

Yet one more notational simplification shall be made before continuing. Defining

$$\dot{\boldsymbol{q}}_n \triangleq \begin{bmatrix} \boldsymbol{v}_n \\ \dot{\boldsymbol{q}}_{n,e} \end{bmatrix} \tag{8}$$

the motion equations can be rendered as

$$\mathcal{M}_n\ddot{\boldsymbol{q}}_n + \mathcal{K}_n\boldsymbol{q}_n = \boldsymbol{f}_{nT} + \boldsymbol{f}_{nI} \tag{9}$$

wherein the coefficient matrices and the force terms can be inferred from (7). This form of the equation of motion will prove extremely useful.

Interbody Geometrical Constraints

We now proceed to take into account the constraints imposed by the joints between contiguous bodies. The generalized velocity $\boldsymbol{v}_n$ of $\mathcal{E}_n$ may be expressed recursively as

$$\boldsymbol{v}_n = \boldsymbol{\mathcal{T}}_{n,n-1}\boldsymbol{v}_{n-1} + \boldsymbol{\mathcal{S}}_{n,n-1}\dot{\boldsymbol{q}}_{n-1,e} + \boldsymbol{v}_{n,\text{int}} \tag{10}$$

where

$$\boldsymbol{v}_{n,\text{int}} \triangleq \begin{bmatrix} \mathbf{v}_{n,\text{int}} \\ \boldsymbol{\omega}_{n,\text{int}} \end{bmatrix}$$

is the *relative* interbody generalized velocity of $\mathcal{E}_n$ to $\mathcal{E}_{n-1}$. More precisely, $\mathbf{v}_{n,\text{int}}$ is the velocity of O_n relative to $O_{\hat{n}}$, at the outboard end of $\mathcal{E}_{n-1}$, and $\boldsymbol{\omega}_{n,\text{int}}$ is the angular velocity of $\mathcal{F}_n$ with respect to $\mathcal{F}_{\hat{n}}$, attached to $\mathcal{E}_{n-1}$ at $O_{\hat{n}}$. (See Figure 2.) We also have that

$$\boldsymbol{\mathcal{T}}_{n,n-1} \triangleq \begin{bmatrix} \mathbf{C}_{n,n-1} & -\mathbf{C}_{n,n-1}\mathbf{r}_{n-1}^{n\,\times} \\ \cdot & \mathbf{C}_{n,n-1} \end{bmatrix}$$

is a *generalized transformation matrix* and

$$\boldsymbol{\mathcal{S}}_{n,n-1} \triangleq \begin{bmatrix} \mathbf{C}_{n,n-1}\boldsymbol{\Delta}_{n-1}(\mathbf{r}_{n-1}^{n}) \\ \frac{1}{2}\mathbf{C}_{n,n-1}\boldsymbol{\nabla}^{\times}\boldsymbol{\Delta}_{n-1}(\mathbf{r}_{n-1}^{n}) \end{bmatrix}$$

is a *generalized influence matrix*; $\mathbf{C}_{n,n-1}$ is the rotation matrix from $\mathcal{F}_{n-1}$ to $\mathcal{F}_n$ and $\mathbf{r}_{n-1}^{n}$ is the position of $O_{\hat{n}}$ with respect to O_{n-1}. (For small interbody translations, O_n and $O_{\hat{n}}$ are essentially coincident.) The geometric constraints imposed by the joints can thus be expressed formally as

$$\boldsymbol{v}_{n,\text{int}} = \boldsymbol{\mathcal{P}}_{n,r}\boldsymbol{v}_{n\gamma} \tag{11}$$

where $\boldsymbol{\mathcal{P}}_{n,r}$ is a projection matrix and $\boldsymbol{v}_{n\gamma}$ is the column of free joint (rate) variables.

Given the form (9) for the motion equation, we would prefer to deal with $\dot{\boldsymbol{q}}_n$ generally instead of $\boldsymbol{v}_n$ explicitly. To this end, we recognize that (10) is contained in

$$\dot{\boldsymbol{q}}_n = \boldsymbol{\mathcal{U}}_{n,n-1}\dot{\boldsymbol{q}}_{n-1} + \dot{\boldsymbol{q}}_{n,\text{int}} \tag{12}$$

where

$$\dot{\boldsymbol{q}}_{n,\text{int}} \triangleq \begin{bmatrix} \boldsymbol{v}_{n,\text{int}} \\ \dot{\boldsymbol{q}}_{n,e} \end{bmatrix}, \quad \boldsymbol{\mathcal{U}}_{n,n-1} \triangleq \begin{bmatrix} \boldsymbol{\mathcal{T}}_{n,n-1} & \boldsymbol{\mathcal{S}}_{n,n-1} \\ \cdot & \cdot \end{bmatrix}$$

Furthermore, setting

$$\dot{\boldsymbol{q}}_{n\gamma} \triangleq \begin{bmatrix} \boldsymbol{v}_{n\gamma} \\ \dot{\boldsymbol{q}}_{n,e} \end{bmatrix}, \boldsymbol{\mathcal{P}}_n \triangleq \begin{bmatrix} \boldsymbol{\mathcal{P}}_{n,r} & \cdot \\ \cdot & \mathbf{1} \end{bmatrix}$$

leads to

$$\dot{\boldsymbol{q}}_{n,\text{int}} = \boldsymbol{\mathcal{P}}_n\dot{\boldsymbol{q}}_{n\gamma} \tag{13}$$

which serves to replace (11).

Interbody Forces

Turning our attention now to the forces acting on $\mathcal{E}_n$, we note that the rigid component is given by

$$f_{nT,r} = \mathcal{T}_{n+1,n}^T f_{n+1}^n - f_n^{n-1} + f_{n,\text{ext},r} \tag{14}$$

where $f_{n,\text{ext},r}$ is due to solely external influences. Also,

$$f_{nT,e} = \mathcal{S}_{n+1,n}^T f_{n+1}^n + f_{n,\text{ext},e} \tag{15}$$

where again $f_{n,\text{ext},e}$ arises strictly from external sources. The generalized interbody forces f_n^{n-1} can moreover be expressed as a sum of control forces $f_{n,c,r}$ and constraint forces $f_{n,\square}$, *i.e.*,

$$f_n^{n-1} = -\mathcal{P}_{n,r} f_{n,c,r} - \mathcal{Q}_{n,r} f_{n,\square} \tag{16}$$

The projection matrix $\mathcal{Q}_{n,r}$ is the orthogonal complement of $\mathcal{P}_{n,r}$. (A subscript 'r' is not needed on $f_{n,\square}$ since the constraint forces can only be associated with the joints.)

We are able to combine (14) and (15) as follows

$$f_{nT} = \mathcal{U}_{n+1,n}^T f_{n+1,\text{int}} - f_{n,\text{int}} + f_{n,\text{ext}} \tag{17}$$

which introduces

$$f_{n,\text{int}} \triangleq \begin{bmatrix} f_n^{n-1} \\ \mathbf{0} \end{bmatrix}, \qquad f_{n,\text{ext}} \triangleq \begin{bmatrix} f_{n,\text{ext},r} \\ f_{n,\text{ext},e} \end{bmatrix}$$

Thus, $f_{n,\text{int}}$ extends the role of the interbody force f_n^{n-1}. In fact, we can write

$$f_{n,\text{int}} = -\mathcal{P}_n f_{n,c} - \mathcal{Q}_n f_{n,\square} \tag{18}$$

where

$$f_{n,c} \triangleq \begin{bmatrix} f_{n,c,r} \\ \mathbf{0} \end{bmatrix}, \qquad \mathcal{Q}_n \triangleq \begin{bmatrix} \mathcal{Q}_{n,r} \\ \mathbf{0} \end{bmatrix}$$

the latter being the orthogonal complement of $\mathcal{P}_n$. Although the location of the control actuators is restricted to the joints, control inputs on the bodies themselves can be easily facilitated [SINCARSIN & HUGHES 1989].

Projection Matrices

A few words are perhaps in order regarding the projection matrices. First, as a simple yet very important example, consider a joint with a single rotational degree of freedom about, say, the third axis of an appropriately chosen reference frame. The corresponding projection matrix $\mathcal{P}_{n,r}$ is

$$\mathcal{P}_{n,r} = [\ 0 \quad 0 \quad 0 \quad 0 \quad 0 \quad 1\ ]^T$$

We may also add that $\boldsymbol{v}_{n\gamma} = \dot{\gamma}_3$, where γ_3 is the angle of rotation.

In general, $\mathcal{P}_{n,r}$ is not constant, as above, but rather is dependent of configuration. Contemplation of a universal joint will quickly reveal this fact. The columns of $\mathcal{P}_{n,r}$ are in general not orthonormal but

$$\mathcal{P}_{n,r}^T \mathcal{P}_{n,r} = \mathcal{I}_{n,r} \tag{19}$$

where $\mathcal{I}_{n,r}$ is nonsingular.

The complementary projection matrix $\mathcal{Q}_{n,r}$ satisfies

$$\mathcal{P}_{n,r}^T \mathcal{Q}_{n,r} = \mathbf{O} \tag{20}$$

Without loss in generality, the columns of $\mathcal{Q}_{n,r}$ can be taken as orthonormal. Relations (19) and (20), of course, also apply to $\mathcal{P}_n$ and $\mathcal{Q}_n$ with $\mathcal{I}_n = \text{diag}\{\mathcal{I}_{n,r},\ \mathbf{1}\}$ in (19).

A More Convenient Form for the Motion Equations

It will be prove substantially expedient if we observe that $\ddot{\boldsymbol{q}}_n$ can be parsed as[1]

$$\ddot{\boldsymbol{q}}_n = \boldsymbol{y}_n + \boldsymbol{y}_{n,\text{non}} \tag{21}$$

where $\boldsymbol{y}_n$ obeys the recursive relation

$$\boldsymbol{y}_n = \mathcal{U}_{n,n-1}\boldsymbol{y}_{n-1} + \mathcal{P}_n \ddot{\boldsymbol{q}}_{n\gamma} \tag{22}$$

Substituting (21) into (10) and differentiating, while observing (22), uncovers a recurrence for $\boldsymbol{y}_{n,\text{non}}$ as well:

$$\boldsymbol{y}_{n,\text{non}} = \mathcal{U}_{n,n-1}\boldsymbol{y}_{n-1,\text{non}} + \dot{\mathcal{U}}_{n,n-1}\dot{\boldsymbol{q}}_{n-1} + \dot{\mathcal{P}}_n \dot{\boldsymbol{q}}_{n\gamma} \tag{23}$$

The quantities $\boldsymbol{y}_{n,\text{non}}$ are effectively responsible for nonlinear interbody kinematic effects. In a rate-linear model, only $\boldsymbol{y}_n$ would remain.

Employing (21) in the motion equation (9) yields

$$\mathcal{M}_n \boldsymbol{y}_n + \mathcal{K}_n \boldsymbol{q}_n = \boldsymbol{f}_{nT} + \boldsymbol{f}_{nI} + \boldsymbol{f}_{n,\text{non}} \tag{24}$$

where

$$\boldsymbol{f}_{n,\text{non}} \triangleq -\mathcal{M}_n \boldsymbol{y}_{n,\text{non}}$$

Upon noting (17), we can in fact write

$$\mathcal{M}_n \boldsymbol{y}_n = \mathcal{U}_{n+1,n}^T \boldsymbol{f}_{n+1,\text{int}} - \boldsymbol{f}_{n,\text{int}} + \boldsymbol{f}_{n,\text{net}} \tag{25}$$

where

$$\boldsymbol{f}_{n,\text{net}} \triangleq \boldsymbol{f}_{n,\text{ext}} + \boldsymbol{f}_{nI} + \boldsymbol{f}_{n,\text{non}} - \mathcal{K}_n \boldsymbol{q}_n$$

The result (25) is a much more convenient form of the motion equation as shall soon be evident.

Kinematical Equations

Lest it appear forgotten, we should point out that the rotational kinematical equations accompanying the dynamical equations (25) are

$$\dot{\mathbf{C}}_{n,n-1} = -\boldsymbol{\omega}_{n,\text{int}}^{\times}\mathbf{C}_{n,n-1} - \mathbf{C}_{n,n-1}(\boldsymbol{\Theta}_{n-1}\dot{\boldsymbol{q}}_{n-1,e})^{\times} \tag{26}$$

where

$$\boldsymbol{\Theta}_{n-1} \triangleq \frac{1}{2}\boldsymbol{\nabla}^{\times}\boldsymbol{\Delta}_{n-1}(\mathbf{r}_{n-1}^{n}) \tag{27}$$

The second term in (26) takes into account the effect due to deformation in the bodies. Interbody translation is given by the integration of $\mathbf{v}_{n,\text{int}}$.

[1]The authors acknowledge the insight of Dr. D.F. Golla here.

3 Recursive Simulation Dynamics

The recursive simulation procedure introduced here can be described as a generalization of Featherstone's method to elastic multibody chains with arbitrary interbody constraints. The procedure may, in fact, be directly compared to a similar generalization of Armstrong's recursive method [D'ELEUTERIO 1989]. The relationship between the two shall be shown in due course. The present approach has an attractive and revealing analogy in discrete-time optimal control.

Recursion for $\boldsymbol{f}_{n,\text{int}}$

Our recursive method is founded on the result that $\boldsymbol{f}_{n,\text{int}}$ can be written as as

$$-\boldsymbol{f}_{n,\text{int}} = \boldsymbol{\Psi}_n \boldsymbol{y}_n + \boldsymbol{\psi}_n \tag{28}$$

The proof of (28) is by induction and can be immediately reasoned by comparing the present scenario to that for rigid multibody chains [D'ELEUTERIO & DAMAREN 1989]; however, in the interest of completeness, we shall detail the proof here:

Step I. For $\mathcal{E}_N$, (25) becomes

$$\boldsymbol{\mathcal{M}}_N \dot{\boldsymbol{y}}_N = -\boldsymbol{f}_{N,\text{int}} + \boldsymbol{f}_{N,\text{net}} \tag{29}$$

it which it has been observed that

$$\boldsymbol{f}_{N+1,\text{int}} \equiv \boldsymbol{0}$$

since $\mathcal{E}_N$ is the (free) terminal body. It is clear that if we set

$$\boldsymbol{\Psi}_N = \boldsymbol{\mathcal{M}}_N, \qquad \boldsymbol{\psi}_N = -\boldsymbol{f}_{N,\text{net}} \tag{30}$$

(28) is satisfied for $n = N$.

Step II. We assume that

$$-\boldsymbol{f}_{n+1,\text{int}} = \boldsymbol{\Psi}_{n+1} \boldsymbol{y}_{n+1} + \boldsymbol{\psi}_{n+1} \tag{31}$$

Step III. It shall be shown that (28) follows from (31). From (22),

$$\boldsymbol{y}_{n+1} = \boldsymbol{\mathcal{U}}_{n+1,n} \boldsymbol{y}_n + \boldsymbol{\mathcal{P}}_{n+1} \ddot{\boldsymbol{q}}_{n+1,\gamma} \tag{32}$$

Substituting (32) and (18) in (31) and premultiplying by $\boldsymbol{\mathcal{P}}_{n+1}^T$ gives

$$\boldsymbol{\mathcal{I}}_{n+1} \boldsymbol{f}_{n+1,c} = \boldsymbol{\Psi}_{n+1,PP} \ddot{\boldsymbol{q}}_{n+1,\gamma} + \boldsymbol{\mathcal{P}}_{n+1}^T \boldsymbol{\Psi}_{n+1} \boldsymbol{\mathcal{U}}_{n+1,n} \boldsymbol{y}_n + \boldsymbol{\psi}_{n+1,P} \tag{33}$$

where, in general,

$$\boldsymbol{\Psi}_{nPP} \triangleq \boldsymbol{\mathcal{P}}_n^T \boldsymbol{\Psi}_n \boldsymbol{\mathcal{P}}_n, \qquad \boldsymbol{\psi}_{nP} \triangleq \boldsymbol{\mathcal{P}}_n^T \boldsymbol{\psi}_n$$

Solving for $\ddot{q}_{n+1,\gamma}$ from (33), inserting back into (32) and using the result with (31) in (25) eventually leads to

$$\begin{aligned} -\boldsymbol{f}_{n,\text{int}} &= \{\boldsymbol{\mathcal{M}}_n + \boldsymbol{\mathcal{U}}_{n+1,n}^T(\boldsymbol{\Psi}_{n+1} - \boldsymbol{\Psi}_{n+1}\boldsymbol{\mathcal{P}}_{n+1}\boldsymbol{\Psi}_{n+1,PP}^{-1}\boldsymbol{\mathcal{P}}_{n+1}^T\boldsymbol{\Psi}_{n+1})\boldsymbol{\mathcal{U}}_{n+1,n}\}\boldsymbol{y}_n \\ &\quad + \{\boldsymbol{\mathcal{U}}_{n+1,n}^T\boldsymbol{\Psi}_{n+1}\boldsymbol{\mathcal{P}}_{n+1}\boldsymbol{\Psi}_{n+1,PP}^{-1}(\boldsymbol{\mathcal{I}}_{n+1}\boldsymbol{f}_{n+1,c} - \boldsymbol{\psi}_{n+1,P}) \\ &\qquad + \boldsymbol{\mathcal{U}}_{n+1,n}^T\boldsymbol{\psi}_{n+1} - \boldsymbol{f}_{n,\text{net}}\} \end{aligned} \tag{34}$$

Whence, we can identify

$$\begin{aligned} \boldsymbol{\Psi}_n &= \boldsymbol{\mathcal{M}}_n + \boldsymbol{\mathcal{U}}_{n+1,n}^T(\boldsymbol{\Psi}_{n+1} - \boldsymbol{\Psi}_{n+1}\boldsymbol{\mathcal{P}}_{n+1}\boldsymbol{\Psi}_{n+1,PP}^{-1}\boldsymbol{\mathcal{P}}_{n+1}^T\boldsymbol{\Psi}_{n+1})\boldsymbol{\mathcal{U}}_{n+1,n} \\ \boldsymbol{\psi}_n &= \boldsymbol{\mathcal{U}}_{n+1,n}^T\boldsymbol{\Psi}_{n+1}\boldsymbol{\mathcal{P}}_{n+1}\boldsymbol{\Psi}_{n+1,PP}^{-1}(\boldsymbol{\mathcal{I}}_{n+1}\boldsymbol{f}_{n+1,c} - \boldsymbol{\psi}_{n+1,P}) + \boldsymbol{\mathcal{U}}_{n+1,n}^T\boldsymbol{\psi}_{n+1} - \boldsymbol{f}_{n,\text{net}} \end{aligned} \tag{35}$$

Step IV. By induction, then, (28) is proven. □

The matrix $\boldsymbol{\Psi}_n$ is an *augmented mass matrix*. In fact, it is the mass matrix (about O_n), associated with the constrained joint coordinates, of the part of the chain from $\mathcal{E}_n$ to $\mathcal{E}_N$. Like $\boldsymbol{\mathcal{M}}_n$, $\boldsymbol{\Psi}_n$ is symmetric and positive-definite; however, it is, along with $\boldsymbol{\psi}_n$, configuration-dependent.

Recursion for $\ddot{q}_{n\gamma}$

It is evident from the foregoing that $\boldsymbol{\Psi}_n$ and $\boldsymbol{\psi}_n$ can be evaluated recursively inward, *i.e.*, from $\mathcal{E}_N$ to $\mathcal{E}_0$. With these quantities in hand, one can then solve for $\ddot{q}_{n\gamma}$ by outward recursion, from $\mathcal{E}_0$ to $\mathcal{E}_N$. This is can be seen by rewriting (33) for $\mathcal{E}_n$ instead of $\mathcal{E}_{n+1}$ as

$$\ddot{\boldsymbol{q}}_{n\gamma} = \boldsymbol{\Psi}_{nPP}^{-1}(\boldsymbol{\mathcal{I}}_n\boldsymbol{f}_{n,c} - \boldsymbol{\mathcal{P}}_n^T\boldsymbol{\Psi}_n\boldsymbol{\mathcal{U}}_{n,n-1}\boldsymbol{y}_{n-1} - \boldsymbol{\psi}_{nP}) \tag{36}$$

Note that all the quantities on the right-hand side are known since $\boldsymbol{y}_{n-1}$ (and $\boldsymbol{y}_{n-1,\text{non}}$, which does not explicitly appear) can be computed recursively from its inboard neighbor according to (22).

Comment

Before proceeding onward, it is worth pointing out that

$$\boldsymbol{\Psi}_n - \boldsymbol{\Psi}_n\boldsymbol{\mathcal{P}}_n\boldsymbol{\Psi}_{nPP}^{-1}\boldsymbol{\mathcal{P}}_n^T\boldsymbol{\Psi}_n = \boldsymbol{\mathcal{Q}}_n\boldsymbol{\Phi}_n\boldsymbol{\mathcal{Q}}^T \tag{37}$$

where

$$\boldsymbol{\Phi}_n = \boldsymbol{\Psi}_{nQQ} - \boldsymbol{\Psi}_{nPQ}^T\boldsymbol{\Psi}_{nPP}^{-1}\boldsymbol{\Psi}_{nPQ}$$

and

$$\boldsymbol{\Psi}_{nPQ} \triangleq \boldsymbol{\mathcal{P}}_n^T\boldsymbol{\Psi}_n\boldsymbol{\mathcal{Q}}_n, \qquad \boldsymbol{\Psi}_{nQQ} \triangleq \boldsymbol{\mathcal{Q}}_n^T\boldsymbol{\Psi}_n\boldsymbol{\mathcal{Q}}_n$$

Showing (37) requires invoking the identity

$$\boldsymbol{\mathcal{P}}_n\boldsymbol{\mathcal{I}}_n^{-1}\boldsymbol{\mathcal{P}}_n^T + \boldsymbol{\mathcal{Q}}_n\boldsymbol{\mathcal{Q}}_n^T = \mathbf{1}$$

By virtue of (37), we can rewrite the first of (35) as

$$\boldsymbol{\Psi}_n = \boldsymbol{\mathcal{M}}_n + \boldsymbol{\mathcal{U}}_{n+1,n}^T \boldsymbol{\mathcal{Q}}_{n+1} \boldsymbol{\Phi}_{n+1} \boldsymbol{\mathcal{Q}}_{n+1}^T \boldsymbol{\mathcal{U}}_{n+1,n} \tag{38}$$

which is a more streamlined expression.

The significance of $\boldsymbol{\Phi}_n$, however, lies in the fact that

$$\boldsymbol{f}_{n,\square} = \boldsymbol{\Phi}_n \boldsymbol{\mathcal{Q}}_n^T \boldsymbol{y}_n + \boldsymbol{\phi}_n \tag{39}$$

where

$$\boldsymbol{\phi}_n = \boldsymbol{\mathcal{Q}}_n^T \boldsymbol{\psi}_n + \boldsymbol{\Psi}_{nPQ}^T \boldsymbol{\Psi}_{nPP}^{-1} (\boldsymbol{\mathcal{I}}_n \boldsymbol{f}_{n,c} - \boldsymbol{\psi}_{nP})$$

This result is the keystone to the generalization of Armstrong's method to elastic multibody chains with arbitrary joint constraints.

4 Discrete-Time Optimal Control

We now present an optimal control problem which generates a two-point boundary value problem (TPBVP) whose form is identical to the recursive multibody simulation dynamics. Consider the linear state equation

$$\mathbf{x}_{k+1} = \mathbf{A}_k \mathbf{x}_k + \mathbf{B}_k \mathbf{u}_k, \quad \mathbf{x}_{-1} = \mathbf{0} \tag{40}$$

where the states are denoted by $\mathbf{x}_k$ and $\mathbf{u}_k$ is a deterministic sequence of control inputs. The input matrices $\mathbf{B}_k$ are assumed to be monic (injective). Our goal is find the sequence $\mathbf{u}_{k-1}$, $k = 0, \ldots, N$ that minimizes the performance index

$$\mathcal{J} = \sum_{k=0}^{N} \frac{1}{2} \mathbf{x}_k^T \mathbf{M}_k \mathbf{x}_k + \mathbf{x}_k^T \mathbf{h}_k - \mathbf{u}_{k-1}^T \mathbf{t}_{k-1} \tag{41}$$

The matrices $\mathbf{M}_k$ are a sequence of positive-definite weighting matrices, and $\mathbf{h}_k$ and $\mathbf{t}_k$ are vector weighting sequences. Since $\mathbf{u}_N$ does not influence $\mathbf{x}_k$, $k \leq N$, we have assumed that $\mathbf{t}_N = \mathbf{0}$. This problem is slightly different than the standard "linear plant with quadratic costs" version that one typically encounters since the cost functional in the present case is linear in the control variable.

The minimization of $\mathcal{J}$ subject to the constraint (40) can be carried out by augmenting the performance index:

$$\mathcal{J}' \triangleq \sum_{k=0}^{N} \frac{1}{2} \mathbf{x}_k^T \mathbf{M}_k \mathbf{x}_k + \mathbf{x}_k^T \mathbf{h}_k - \mathbf{u}_{k-1}^T \mathbf{t}_{k-1} + \boldsymbol{\lambda}_k^T (\mathbf{x}_k - \mathbf{A}_{k-1} \mathbf{x}_{k-1} - \mathbf{B}_{k-1} \mathbf{u}_{k-1})$$

where $\boldsymbol{\lambda}_k$ are lagrange multipliers (adjoint variables). The necessary conditions for optimality,

$$\frac{\partial \mathcal{J}'}{\partial \boldsymbol{\lambda}_{k+1}} = \frac{\partial \mathcal{J}'}{\partial \mathbf{x}_k} = \frac{\partial \mathcal{J}'}{\partial \mathbf{u}_k} = \mathbf{0}$$

produce the two-point boundary value problem (TPBVP):

$$\mathbf{x}_{k+1} = \mathbf{A}_k\mathbf{x}_k + \mathbf{B}_k\mathbf{u}_k\ , \quad \mathbf{x}_{-1} = \mathbf{0} \tag{42}$$
$$\boldsymbol{\lambda}_k = \mathbf{A}_k^T\boldsymbol{\lambda}_{k+1} - \mathbf{M}_k\mathbf{x}_k - \mathbf{h}_k\ , \quad \boldsymbol{\lambda}_{N+1} = \mathbf{0} \tag{43}$$

$$\mathbf{t}_k = -\mathbf{B}_k^T\boldsymbol{\lambda}_{k+1} \tag{44}$$

Without loss in generality, $\boldsymbol{\lambda}_{N+1}$ has been set to $\mathbf{0}$ since $\mathbf{t}_N = \mathbf{0}$ in (44). Therefore, (43) shows that $\boldsymbol{\lambda}_N = -\mathbf{M}_N\mathbf{x}_N - \mathbf{h}_N$ which suggests an affine relationship between $\mathbf{x}_k$ and $\boldsymbol{\lambda}_k$:

$$\boldsymbol{\lambda}_k = -\mathbf{S}_k\mathbf{x}_k - \mathbf{r}_k \tag{45}$$

with $\mathbf{S}_N = \mathbf{M}_N$ and $\mathbf{r}_N = \mathbf{h}_N$.

Substituting (45) into (44) for $\boldsymbol{\lambda}_{k+1}$ and replacing $\mathbf{x}_{k+1}$ with the right side of (42) produces the feedback law

$$\mathbf{u}_k = -\mathbf{K}_k\mathbf{x}_k + \mathbf{R}_k^{-1}(\mathbf{t}_k - \mathbf{B}_k^T\mathbf{r}_{k+1}) \tag{46}$$

where

$$\mathbf{R}_k \triangleq \mathbf{B}_k^T\mathbf{S}_{k+1}\mathbf{B}_k\ , \quad \mathbf{K}_k \triangleq \mathbf{R}_k^{-1}\mathbf{B}_k^T\mathbf{S}_{k+1}\mathbf{A}_k \tag{47}$$

The matrix $\mathbf{R}_k$ will be invertible if $\mathbf{S}_{k+1}$ is positive-definite, which we show below.

The transformation (45) can be shown to hold for general $\boldsymbol{\lambda}_k$ using an inductive procedure not unlike the proof of (15). Therefore, let us substitute (45) for $\boldsymbol{\lambda}_k$ and $\boldsymbol{\lambda}_{k+1}$ in (43) and use (42) for $\mathbf{x}_{k+1}$ and (46) for $\mathbf{u}_k$ to yield

$$\begin{aligned} &[\mathbf{S}_k - \mathbf{A}_k^T(\mathbf{S}_{k+1} - \mathbf{S}_{k+1}\mathbf{B}_k\mathbf{R}_k^{-1}\mathbf{B}_k^T\mathbf{S}_{k+1})\mathbf{A}_k - \mathbf{M}_k]\mathbf{x}_k \\ = \;& -\mathbf{r}_k + (\mathbf{A}_k - \mathbf{B}_k\mathbf{K}_k)^T\mathbf{r}_{k+1} + \mathbf{K}_k^T\mathbf{t}_k + \mathbf{h}_k \end{aligned}$$

Since this must hold for general $\mathbf{x}_k$, the coefficient of $\mathbf{x}_k$ must vanish as well as the right hand side. Therefore,

$$\mathbf{S}_k = \mathbf{A}_k^T(\mathbf{S}_{k+1} - \mathbf{S}_{k+1}\mathbf{B}_k\mathbf{R}_k^{-1}\mathbf{B}_k^T\mathbf{S}_{k+1})\mathbf{A}_k + \mathbf{M}_k \quad (\mathbf{S}_N = \mathbf{M}_N) \tag{48}$$

which is a matrix difference equation of Riccati type and

$$\mathbf{r}_k = \boldsymbol{\Gamma}_{k+1,k}^T\mathbf{r}_{k+1} + \mathbf{K}_k^T\mathbf{t}_k + \mathbf{h}_k\ , \quad \boldsymbol{\Gamma}_{k+1,k} \triangleq \mathbf{A}_k - \mathbf{B}_k\mathbf{K}_k \quad (\mathbf{r}_N = \mathbf{h}_N) \tag{49}$$

which is a nonhomogeneous linear difference equation for $\mathbf{r}_k$.

The invertibility of $\mathbf{R}_k$ can be verified by noticing that $(\mathbf{A}_k - \mathbf{B}_k\mathbf{K}_k)^T\mathbf{S}_{k+1}\mathbf{B}_k = \mathbf{O}$ which follows from the definitions of $\mathbf{R}_k$ and $\mathbf{K}_k$ (47). Therefore, the matrix equation (48) can be rewritten compactly as

$$\mathbf{S}_k = \boldsymbol{\Gamma}_{k+1,k}^T\mathbf{S}_{k+1}\boldsymbol{\Gamma}_{k+1,k} + \mathbf{M}_k \tag{50}$$

Since $\mathbf{S}_N = \mathbf{M}_N$ is symmetic and positive-definite, so is $\mathbf{R}_{N-1}$, and therefore $\mathbf{S}_k$ and $\mathbf{R}_{k-1}$ are symmetric and positive-definite (using backwards induction).

Optimal Control Policy

The optimal control policy can now be summarized as follows: one solves the matrix equation (48) (or (50)) and the vector equation (49) backwards from $k = N$ to $k = 0$. This permits determination of the gains $\mathbf{R}_k$ and $\mathbf{K}_k$ at each step which can be stored for future use. Then, from (46), the initial control is given by

$$\mathbf{u}_{-1} = (\mathbf{R}_{-1})^{-1}[\mathbf{t}_{-1} - \mathbf{B}_{-1}^T\mathbf{r}_0]$$

since $\mathbf{x}_{-1} = \mathbf{0}$ and the next value of the state is $\mathbf{x}_0 = \mathbf{B}_{-1}\mathbf{u}_{-1}$. Subsequent controls can be calculated using (46) while propagating the state forward using the state equation (40). It is not clear from the above analysis whether the necessary conditions produce a global minimum. However, an alternative approach to the problem using BELLMAN's [1957] method of dynamic programming shows that not only does $\mathbf{u}_k$, as given by (46), yield a minimum, but it is unique as well.

The optimal control law, (46), is in feedback form. We would like to find a global description of the control that is written only in terms of the performance index variables. Let us begin by substituting (46) into the state equation (42) to give

$$\mathbf{x}_{k+1} = \mathbf{\Gamma}_{k+1,k}\mathbf{x}_k + \mathbf{B}_k\mathbf{R}_k^{-1}(\mathbf{t}_k - \mathbf{B}_k^T\mathbf{r}_{k+1}) \,, \quad \mathbf{x}_{-1} = \mathbf{0} \tag{51}$$

If we define the closed-loop transition matrix by

$$\mathbf{\Gamma}_{k,l} \triangleq \mathbf{\Gamma}_{k,k-1}\mathbf{\Gamma}_{k-1,k-2}\cdots\mathbf{\Gamma}_{l+1,l} \ (l < k) \,, \quad \mathbf{\Gamma}_{k,k} \triangleq \mathbf{1} \,, \quad \mathbf{\Gamma}_{k,l} \triangleq \mathbf{O} \ (k < l) \tag{52}$$

then the solution of (51) can be written as

$$\mathbf{x}_{k-1} = \sum_{m=0}^{k-1} \mathbf{\Gamma}_{k-1,m}\mathbf{B}_{m-1}\mathbf{R}_{m-1}^{-1}(\mathbf{t}_{m-1} - \mathbf{B}_{m-1}^T\mathbf{r}_m) \tag{53}$$

Furthermore, the solution of (49) is

$$\mathbf{r}_k = \sum_{m=k+1}^{N} \mathbf{\Gamma}_{m-1,k}^T\mathbf{K}_{m-1}\mathbf{t}_{m-1} + \sum_{m=k}^{N} \mathbf{\Gamma}_{m,k}^T\mathbf{h}_m \tag{54}$$

Now, substitute the second of these, (54), into the first to give

$$\begin{aligned}
\mathbf{x}_{k-1} \quad = \quad & \sum_{m=0}^{k-1} \mathbf{\Gamma}_{k-1,m}\mathbf{B}_{m-1}\mathbf{R}_{m-1}^{-1}\mathbf{t}_{m-1} \\
& - \sum_{m=0}^{k-1} \mathbf{\Gamma}_{k-1,m}\mathbf{B}_{m-1}\mathbf{R}_{m-1}^{-1}\mathbf{B}_{m-1}^T \sum_{l=m+1}^{N} \mathbf{\Gamma}_{l-1,m}^T\mathbf{K}_{l-1}\mathbf{t}_{l-1} \\
& - \sum_{m=0}^{k-1} \mathbf{\Gamma}_{k-1,m}\mathbf{B}_{m-1}\mathbf{R}_{m-1}^{-1}\mathbf{B}_{m-1}^T \sum_{l=m}^{N} \mathbf{\Gamma}_{l,m}^T\mathbf{h}_l
\end{aligned}$$

The feedback law (46) can be written for $\mathbf{u}_{k-1}$ in terms of $\mathbf{x}_{k-1}$ and $\mathbf{r}_k$ which are available above. If we do this and and define

$$\mathbf{G}_{k,m} \triangleq \mathbf{K}_k\mathbf{\Gamma}_{k,m+1}\mathbf{B}_m \quad (\mathbf{G}_{k,m} = \mathbf{O} \,, \quad k \le m) \tag{55}$$

then the result is

$$\begin{aligned}\mathbf{u}_{k-1} &= \mathbf{R}_{k-1}^{-1}\mathbf{t}_{k-1} - \sum_{m=0}^{k-1}\mathbf{G}_{k-1,m-1}\mathbf{R}_{m-1}^{-1}\mathbf{t}_{m-1} - \mathbf{R}_{k-1}^{-1}\sum_{m=k+1}^{N}\mathbf{G}_{m-1,k-1}^{T}\mathbf{t}_{m-1} \\ &+ \sum_{m=0}^{k-1}\mathbf{G}_{k-1,m-1}\mathbf{R}_{m-1}^{-1}\sum_{l=m+1}^{N}\mathbf{G}_{l-1,m-1}^{T}\mathbf{t}_{l-1} - \mathbf{R}_{k-1}^{-1}\mathbf{B}_{k-1}^{T}\sum_{m=k}^{N}\mathbf{\Gamma}_{m,l}^{T}\mathbf{h}_{m} \\ &+ \sum_{m=0}^{k-1}\mathbf{G}_{k-1,m-1}\mathbf{R}_{m-1}^{-1}\mathbf{B}_{m-1}^{T}\sum_{l=m}^{N}\mathbf{\Gamma}_{l,m}^{T}\mathbf{h}_{l}\end{aligned} \tag{56}$$

Let us form the global quantities

$$\mathbf{u} = \mathrm{col}\{\mathbf{u}_{k-1}\}\ ,\quad \mathbf{t} = \mathrm{col}\{\mathbf{t}_{k-1}\}\ ,\quad \mathbf{h} = \mathrm{col}\{\mathbf{h}_{k}\}$$

$$\mathbf{G} = \mathrm{matrix}\{\mathbf{G}_{k-1,m-1}\}\ ,\quad \mathbf{R} = \mathrm{diag}\{\mathbf{R}_{k-1}\}\ ,\quad \mathbf{L} = \mathrm{matrix}\{\mathbf{B}_{k-1}^{T}\mathbf{\Gamma}_{m,k}^{T}\} \tag{57}$$

where the indices k and m range from 0 to N. The matrix $\mathbf{G}$ is upper block triangular because of (55). With these definitions, (56) can be written compactly as

$$\mathbf{u} = (\mathbf{1}-\mathbf{G})\mathbf{R}^{-1}(\mathbf{1}-\mathbf{G}^{T})\mathbf{t} - (\mathbf{1}-\mathbf{G})\mathbf{R}^{-1}\mathbf{L}\mathbf{h} \tag{58}$$

which is the desired representation for the control variables, $\mathbf{u}$. Given the parameters in the performance index ($\mathbf{M}_k$, $\mathbf{t}_k$, and $\mathbf{h}_k$), the optimal control policy can be precomputed in its entirety without knowledge of the state $\mathbf{x}_k$. The coefficient matrix of $\mathbf{t}$ is symmetric and nonnegative-definitive. This open-loop form of the control law will have an interesting interpretation in the context of multibody dynamics.

5 Relationship Between Optimal Control and Recursive Dynamics

The TPBVP generated by the previous optimal control problem (42-44) is identical in form to that of the multibody dynamics problem (22), (25), and

$$\mathcal{I}_n \boldsymbol{f}_{n,\mathrm{c}} = -\mathcal{P}_n^T \boldsymbol{f}_{n,\mathrm{int}} \tag{59}$$

respectively. The latter equation (59) results from premultiplication of (18) by $\mathcal{P}_n^T$ while recognizing (19) and (20) *et seq.* Therefore, we shall make the one-to-one correspondences:

$$\begin{array}{cccccc} \mathbf{x}_k & \longleftrightarrow & \boldsymbol{y}_n & \boldsymbol{\lambda}_k & \longleftrightarrow & \boldsymbol{f}_{n,\mathrm{int}} \\ \mathbf{u}_k & \longleftrightarrow & \ddot{\boldsymbol{q}}_{n+1,\gamma} & \mathbf{h}_k & \longleftrightarrow & -\boldsymbol{f}_{n,\mathrm{net}} \\ \mathbf{A}_k & \longleftrightarrow & \mathcal{U}_{n+1,n} & \mathbf{t}_k & \longleftrightarrow & \mathcal{I}_{n+1}\boldsymbol{f}_{n+1,\mathrm{c}} \\ \mathbf{B}_k & \longleftrightarrow & \mathcal{P}_{n+1} & \mathbf{M}_k & \longleftrightarrow & \mathcal{M}_n \end{array}$$

where recursion in time (k) for the control problem is replaced with spatial recursion (n) at a given instant in time for the multibody dynamics problem. In the optimal control analogy, the states correspond to the accelerations $\boldsymbol{y}_n$, the adjoint variables are analogous to the interbody forces $\boldsymbol{f}_{n,\mathrm{int}}$, and the control inputs play the roles of the joint (and elastic) accelerations $\ddot{q}_{n,\gamma}$. Furthermore, the state and input matrices take the place of the interbody transformation and projection matrices, respectively.

The transformation for the adjoint variables (45) may be compared to the description (28) for the interbody forces which provides the identifications

$$\mathbf{S}_k \longleftrightarrow \Psi_n \;,\; \mathbf{r}_k \longleftrightarrow \psi_n \;,\; \mathbf{R}_k \longleftrightarrow \Psi_{n+1,PP} \tag{60}$$

With these in hand, the matrix equation (48) and the vector equation (49) are the same as the recursive relationships (35). The optimal feedback law which expresses $\mathbf{u}_{k-1}$ in terms of the state is precisely the analog of (36), the recursive expression for $\ddot{q}_{n\gamma}$ in terms of y_{n-1}.

Let us now write the performance index (41) in terms of multibody dynamics quantities:

$$\mathcal{J} = \sum_{n=0}^{N} \frac{1}{2} y_n^T \mathcal{M}_n y_n - f_{n,\text{net}}^T y_n - f_{n,\text{c}}^T \mathcal{I}_n \ddot{q}_{n\gamma}$$

We can say that the equations defining the dynamics of the elastic chain are the result of minimizing $\mathcal{J}$ (with respect to y_n and $\ddot{q}_{n\gamma}$) subject to the kinematical constraint equation (22). This is a generalization of GIBBS' [1879] formulation of the dynamics of a system of N particles with masses m_n, coordinates x_n, y_n, z_n, and subjected to forces X_n, Y_n, Z_n: minimize

$$\sum_{n=0}^{N} \frac{1}{2} m_n(\ddot{x}_n^2 + \ddot{y}_n^2 + \ddot{z}_n^2) - X_n\ddot{x}_n - Y_n\ddot{y}_n - Z_n\ddot{z}_n$$

subject to the kinematical constraints. In other words, we have found a variational principle from which the equations of motion can be derived.

In another work, RODRIGUEZ [1987] interpreted the equations of motion and constraint equations for a chain of hinged rigid bodies in terms of optimal filtering (the Kalman filter) and smoothing (the Bryson-Frazier smoother). In his formulation, the bodies in the chain are numbered inwardly (*i.e.*, the tip body is $\mathcal{B}_0$ and the root body is $\mathcal{B}_N$). With this convention, his analogy is the dual of that presented above. The role of the states is played by the interbody forces and the adjoint states ('costates') are the link accelerations, which is a juxtaposition of the results given above. The control torque associated with a hinge is interpreted as a measurement of the state whereas we have joint accelerations acting as control inputs in the state equation.

It was Rodriguez who pointed out that the inverse of the global mass matrix can be obtained using the procedures at the end of §4. Here, we extent this result to the elastic case. If we define the global quantities

$$\ddot{q}_\gamma \triangleq \text{col}\{\ddot{q}_{n\gamma}\} \;,\; f_\text{c} \triangleq \text{col}\{\mathcal{I}_n f_{n,\text{c}}\} \;,\; f_\text{net} \triangleq \text{col}\{f_{n,\text{net}}\}$$

then the analog of (58) becomes

$$\ddot{q}_\gamma = (\mathbf{1} - \mathbf{G})\Psi_{PP}^{-1}(\mathbf{1} - \mathbf{G}^T) f_\text{c} + (\mathbf{1} - \mathbf{G})\Psi_{PP}^{-1}\mathbf{L} f_\text{net} \tag{61}$$

where $\Psi_{PP} = \text{diag}\{\Psi_{n,PP}\}$ and $\mathbf{G}$ and $\mathbf{L}$ correspond to the original definitions, (57), expressed in terms of their multibody dynamics equivalents. The above equation can be compared to a global formulation of the dynamics (see HUGHES & SINCARSIN [1989], for example):

$$\mathcal{M}\ddot{q}_\gamma = f_\text{c} + d_\text{net} \tag{62}$$

Here, $\mathcal{M}$ is the global mass matrix and $\boldsymbol{d}_{\text{net}}$ incorporates miscellaneous terms stemming from the stiffness forces, external influences, and nonlinearties in the kinematics and dynamics (*i.e.*, the global equivalent of $\boldsymbol{f}_{n,\text{net}}$). A comparison of the two equations allows us to identify

$$(\mathbf{1}-\mathbf{G})\Psi_{PP}^{-1}(\mathbf{1}-\mathbf{G}^T) \equiv \mathcal{M}^{-1} \tag{63}$$

as a literal expression for the inverse of the global mass matrix.

We have now come full circle. The recursive algorithm has been connected with the global procedures that are commonly used. The above analysis seems to suggest that the recursive procedures developed in this paper are essentially an efficient means of solving the global system of equations (62) without actually assembling them and performing the inverse operation above or, for example, Gaussian elimination.

6 Summary of the Recursive Algorithm

We shall now consolidate the results of the previous sections and show how one can obtain the trajectory of the elastic chain. For each body, the control forces $\boldsymbol{f}_{n,c}(t)$ and the external force distribution $\boldsymbol{f}_{n,\text{ext}}(t)$ are given as well as the initial values of $\mathbf{C}_{n,n-1}$, $\boldsymbol{q}_{n,e}$, and $\dot{\boldsymbol{q}}_{n\gamma}$ which consists of the free joint velocities $\boldsymbol{v}_{n\gamma}$ and the elastic quantities $\dot{\boldsymbol{q}}_{n,e}$. Beginning with $t=0$, we proceed as follows:

Step 1. Outward Kinematics Recursion for $\dot{\boldsymbol{q}}_n$ and $\boldsymbol{f}_{n,\text{net}}$.
Do $n=0$ to N:
Generate $\mathcal{U}_{n,n-1}$ $(\mathcal{T}_{n,n-1}, \mathcal{S}_{n,n-1})$ using $\mathbf{C}_{n,n-1}$
$\dot{\boldsymbol{q}}_{n,\text{int}} = \mathcal{P}_n \dot{\boldsymbol{q}}_{n\gamma}$
$\dot{\boldsymbol{q}}_n = \mathcal{U}_{n,n-1}\dot{\boldsymbol{q}}_{n-1} + \dot{\boldsymbol{q}}_{n,\text{int}}$
Generate $\dot{\mathcal{U}}_{n,n-1}$ using $\dot{\mathbf{C}}_{n+1,n}$
$\boldsymbol{y}_{n,\text{non}} = \mathcal{U}_{n,n-1}\boldsymbol{y}_{n-1,\text{non}} + \dot{\mathcal{U}}_{n,n-1}\dot{\boldsymbol{q}}_{n-1} + \dot{\mathcal{P}}_n\dot{\boldsymbol{q}}_{n\gamma}$
$\boldsymbol{f}_{nI,r} = (\boldsymbol{v}_n^{\times})^T \mathcal{M}_{n,rr}\boldsymbol{v}_n$
$\boldsymbol{f}_{nI,e} = -\mathbf{P}_n^T\boldsymbol{\omega}_n^{\times}\mathbf{v}_n - \boldsymbol{f}_{nR,e}$, $[\boldsymbol{f}_{nR,e}]_i \triangleq [\mathbf{I}_n]_{ijk}[\boldsymbol{\omega}_n]_j[\boldsymbol{\omega}_n]_k$
$\boldsymbol{f}_{n,\text{non}} = -\mathcal{M}_n\boldsymbol{y}_{n,\text{non}}$
$\boldsymbol{f}_{n,\text{net}} = \boldsymbol{f}_{n,\text{ext}} + \boldsymbol{f}_{nI} + \boldsymbol{f}_{n,\text{non}} - \mathcal{K}_n\boldsymbol{q}_n$
Next n.

Step 2. Inward Recursion for Ψ_n and $\boldsymbol{\psi}_n$.
Set $\Psi_N = \mathcal{M}_N$ and $\boldsymbol{\psi}_N = -\boldsymbol{f}_{N,\text{non}}$
Do $n = N-1$ to 0:
$\Psi_{n+1,PP} = \mathcal{P}_{n+1}^T\Psi_{n+1}\mathcal{P}_{n+1}$, $\boldsymbol{\psi}_{n+1,P} = \mathcal{P}_{n+1}^T\boldsymbol{\psi}_{n+1}$
$\mathbf{K}_n = \Psi_{n+1,PP}^{-1}\mathcal{P}_{n+1}\Psi_{n+1}\mathcal{T}_{n+1,n}$
$\mathbf{\Gamma}_{n+1,n} = \mathcal{U}_{n+1,n} - \mathcal{P}_{n+1}\mathbf{K}_n$

$$\Psi_n = \mathbf{\Gamma}_{n+1,n}^T\Psi_{n+1}\mathbf{\Gamma}_{n+1,n} + \mathcal{M}_n \tag{64}$$

$$\boldsymbol{\psi}_n = \mathbf{\Gamma}_{n+1,n}^T\boldsymbol{\psi}_{n+1} + \mathbf{K}_n^T\boldsymbol{f}_{n+1,c} - \boldsymbol{f}_{n,\text{net}} \tag{65}$$

Next n.
$\Psi_{0PP} = \mathcal{P}_0^T\Psi_0\mathcal{P}_0$, $\boldsymbol{\psi}_{0P} = \mathcal{P}_0^T\boldsymbol{\psi}_0$

Step 3. Outward Dynamics Recursion for $\ddot{\boldsymbol{q}}_{n\gamma}$.

For $\mathcal{E}_0$, $\ddot{\boldsymbol{q}}_{0\gamma} = \boldsymbol{\Psi}_{0PP}^{-1}[\boldsymbol{f}_{0,c} - \boldsymbol{\psi}_{0P}]$, $\boldsymbol{y}_0 = \boldsymbol{\mathcal{P}}_0\ddot{\boldsymbol{q}}_{0\gamma}$

Do $n = 1$ to N:

$$\ddot{\boldsymbol{q}}_{n\gamma} = -\mathbf{K}_{n-1}\boldsymbol{y}_{n-1} + \boldsymbol{\Psi}_{n,PP}^{-1}(\boldsymbol{f}_{n,c} - \boldsymbol{\psi}_{n,P})$$
$$\boldsymbol{y}_n = \boldsymbol{\mathcal{U}}_{n,n-1}\boldsymbol{y}_{n-1} + \boldsymbol{\mathcal{P}}_n\ddot{\boldsymbol{q}}_{n\gamma}$$
$$\dot{\mathbf{C}}_{n,n-1} = -\boldsymbol{\omega}_{n,\text{int}}^{\times}\mathbf{C}_{n,n-1} - \mathbf{C}_{n,n-1}(\boldsymbol{\Theta}_{n-1}\dot{\boldsymbol{q}}_{n-1,e})^{\times}$$

Next n.

Step 4. Estimate $\dot{\boldsymbol{q}}_{n\gamma}(t + \Delta t)$, $\mathbf{C}_{n,n-1}(t + \Delta t)$, and $\boldsymbol{q}_{n,e}(t + \Delta t)$ using some quadrature scheme. If $t + \Delta t$ is the desired terminal time , stop. If not, replace t with $t + \Delta t$ and go back to Step 1.

This completes the summary of the recursive simulation procedure. If one desires a rate-linear simulation of the dynamics, one can ignore the contributions of $\boldsymbol{f}_{nI}$ and $\boldsymbol{f}_{n,\text{non}}$ to $\boldsymbol{f}_{n,\text{net}}$ in Step 1. In Step 2, we have used the definitions of $\mathbf{K}_n$ and $\boldsymbol{\Gamma}_{n+1,n}$ from section 4, as they lead to the most compact defintions of $\boldsymbol{\Psi}_n$ and $\boldsymbol{\psi}_n$. The expressions (64) and (65) are equivalent, however, to the expanded forms in (35). The third step yields the accelerations $\ddot{\boldsymbol{q}}_{n\gamma}$ which can be integrated in conjunction with the kinematical equation for the rotation matrix to provide the elastic coordinates and their rates as well as the (rigid) joint velocities and their orientation/position.

7 Concluding Remarks

A recursive procedure has been presented for the solution of the simulation dynamics problem for an elastic multibody chain. A very general formulation was used that clearly identifies the various nonlinearities involved. The process by which one determines the joint accelerations and elastic coordinate accelerations is identical in form to the calculation of the optimal control policy in a discrete-time problem. The underlying analogy that makes this possible yields great insight into the structure of the multibody dynamics problem. In particular, the key equations required in the solution procedure can be isolated. This should be of benefit when considering more complicated topologies such as trees and configurations which possess closed loops.

Apart from the sheer academic interest of the results we have presented, one must keep in mind the computational savings that can be gained from a recursive solution of the problem. This is especially true for systems composed of many interconnected bodies, each of which possesses several elastic degrees of freedom. We have also shown how a literal inverse for the *global* mass matrix can be constructed using recursive methods. An expression of this nature could be very useful in constructing a model for the development of control strategies. The determination of this inverse sheds some light on the relationship between global and recursive solutions of the equations of motion. In summary, one cannot underestimate the insight that can be gained when the relationship between two seemingly unrelated problems is uncovered.

References

1. ARMSTRONG, W.W., "Recursive Solution to the Equations of an n-Link Manipulator", Proc. 5th World Congress on Theory of Machines and Mechanisms, Montreal, 1979, pp. 1343-1346.

2. BAE, D. S. & HAUG, E. J., "A Recursive Formulation for Constrained Mechanical Systems, Part I—Open Loop", *Mechanics of Structures and Machines*, Vol. 15(3), 1987.

3. BAE, D. S. & HAUG, E. J., "A Recursive Formulation for Constrained Mechanical Systems, Part I—Closed Loop", *Mechanics of Structures and Machines*, Vol. 15(4), 1987.

4. BELLMAN, R., *Dynamic Programming*, Princeton University Press, New Jersey, 1957.

5. D'ELEUTERIO, G.M.T., "Dynamics of Elastic Multibody Equations: Part C—Recursive Dynamics", submitted to *Dynamics and Stability of Systems.*

6. D'ELEUTERIO, G.M.T., & DAMAREN, C.J., "Recursive Multibody Dynamics and Discrete-Time Optimal Control," NASA Conference on Space Telerobotics, Pasadena, CA, 31 January-2 February 1989.

7. FEATHERSTONE, R., "The Calculation of Robot Dynamics Using Articulated-Body Inertias", *Int. J. Robotics Research*, Vol. 2, 1, 1983, pp. 13-30.

8. GIBBS, J. W., "On the Fundamental Formulae of Dynamics", *American Journal of Mathematics, Pure and Applied*, **II**, 1879, 49-64.

9. HUGHES, P. C. & SINCARSIN, G. B., "Dynamics of an Elastic Multibody Chains: Part B – Global Dynamics", to appear in *Dynamics and Stability of Systems*, 4, 3, 1989.

10. RODRIGUEZ, G., "Kalman Filtering, Smoothing and Recursive Robot Arm Forward and Inverse Dynamics," *IEEE J. Robot. Autom.*, **RA-3**, 6, 1987, pp. 624-639.

11. SINCARSIN, G.B., & HUGHES, P.C., "Dynamics of Elastic Multibody Chains: Part A—Body Motion Equations", to appear in *Dynamics and Stability of Systems*, 4, 3, 1989.

12. WALKER, M. W. & ORIN, D. E., "Efficient Dynamic Computer Simulation of Robotic Mechanisms", *J. Dynamic Systems, Measurement and Control*, Vol. 104, 1982, pp. 205-211.

Contemporary Mathematics
Volume 97, 1989

SOME SOLVABLE STOCHASTIC CONTROL PROBLEMS IN COMPACT SYMMETRIC SPACES OF RANK ONE*

T. E. Duncan

ABSTRACT. A family of stochastic control problems for diffusion processes in compact symmetric spaces of rank one are described and explicitly solved. The basic problem is to control Brownian motion in a compact symmetric space by a drift so that it remains close to a fixed point in the space called the origin. The solution of the control problem is obtained by finding an explicit, smooth solution to the associated Hamilton-Jacobi or dynamic programming equation. All irreducible compact symmetric spaces of rank one and dimension > 2 are included and for each of these symmetric spaces a countably infinite family of different explicitly solvable control problems are given.

1. INTRODUCTION. While there are a number of quite general results on the existence or the existence and the uniqueness of optimal stochastic controls for the control of a wide range of stochastic systems, there are only a relatively few examples of controlled diffusions where the optimal control is expressed explicitly as a function of the state of the system. This is especially true for controlled diffusions that are described by nonlinear stochastic differential equations. Some examples of explicitly solvable control problems in Euclidean spaces besides the well known linear regulator problem are given in [1,2,14,18].

AMS(MOS) Classification - 93E20, 58G32, 60H07, 49C20, 49A22
* Research partially supported by NSF Grant ECS-8718026

For the control of nonlinear stochastic systems it is natural to consider systems that possess some inherent geometry. This consideration motivated the study of controlled diffusions in manifolds which was initiated in [4,5,6]. Since it is too optimistic to consider initially an arbitrary Riemannian manifold to find explicitly solvable stochastic control problems it seems natural to investigate control problems on manifolds that possess some special structural symmetries. Apparently the first example of a solvable stochastic control problem for a controlled diffusion in a noncompact manifold with nonzero curvature is given in [7]. This example in real hyperbolic space was extended to other cost functionals and to other noncompact symmetric spaces of rank one [9,10]. Since many physical phenomena evolve in compact manifolds it is natural also to investigate stochastic control problems in compact symmetric spaces. Probably the most important examples of compact symmetric spaces of rank one are the spheres. An example of a solvable stochastic control problem in spheres is given in [8]. While the spheres are compact analogues of the real hyperbolic spaces, the control problems in the spheres have added difficulties because they are not usually globally trivial and it is more difficult to find appropriate cost functionals that allow one to find explicit solutions to the associated control problems.

In this paper a countable family of nontrivially distinct stochastic control problems are formulated and explicitly solved for each compact symmetric space of rank one and dimension > 2. The results here include those in [8]. It is more difficult to obtain natural cost functions for the compact symmetric spaces than for the noncompact symmetric spaces. For example, the spherical functions that are used in [9,10] for cost functions lack a global monotonicity property for compact symmetric spaces. This phenomenon is easily seen by comparing the real exponential function and the complex exponential function. Thus it is necessary to exercise more care in the use of spherical functions for cost functions in compact symmetric spaces. Furthermore, the topology of these manifolds affects these optimization problems in contrast to the noncompact symmetric spaces. Since the compact symmetric spaces of rank one are typically not globally trivial manifolds it is necessary to verify that the (optimal) controls for the optimization problems in compact symmetric spaces can be pieced together smoothly. The controlled diffusion for a compact symmetric space of rank one in this paper is Brownian motion plus a control in the drift. This is the same model that is considered in [7-10].

The compact symmetric spaces of rank one are the spheres and the projective spaces. It is clear that these spaces can be employed as useful mathematical models for some physical problems.

2. PRELIMINARIES. Initially a few results from semisimple Lie algebras are reviewed. A good reference for this material is Helgason [15]. Let g be a semisimple Lie algebra over $\mathbf{R}$ and let g_c be its complexification. A direct sum decomposition $g = k + p$ into a subalgebra k and a vector space p is called a Cartan decomposition if there exists a compact real form u such that

$$\sigma(u) \subset u \quad , \quad u \cap g = k \quad , \quad \sqrt{-1}u \cap g = p$$

where σ is the conjugation of g_c with respect to g. Every semisimple Lie algebra g over $\mathbf{R}$ admits a Cartan decomposition. Let W_Σ be the group generated by the reflections of Σ where Σ is the set of roots of the pair (g,a_p) where a_p is a maximal abelian subspace in p, $g = k + p$ is a Cartan decomposition and the rank of the associated symmetric space is the dimension of a_p. Let Δ be the roots of (g_c,a_c) where $a = a_k + a_p$ is the extension of a_p to a Cartan subalgebra a of g and a_c is the complexification of a and is a Cartan subalgebra of g_c. Let Δ_p be the elements of Δ that do not vanish identically on a_p. The set of restrictions of Δ_p to a_p is Σ as defined above and thus the elements of Σ are often called the restricted roots.

An element $H \in a$ is called regular if $\lambda(H) \neq 0$ for all $\lambda \in \Sigma$, otherwise it is called singular. The subset $a' \subset a$ of regular elements consists of the complement of finitely many hyperplanes, and its components are called Weyl chambers. Fix a Weyl chamber a^+ and call a root λ positive if λ has positive values in a^+. Let $A^+ = \exp a^+$.

Let $g = k + p$ be the direct sum decomposition (that is Cartan decomposition) of the Lie algebra g of G into the Lie algebra k of K and its orthogonal complement p with respect to the Killing form of g. For rank one symmetric spaces there is a restricted root $\alpha \in \Sigma$ such that 2α is the only other possible element in Σ. Thus there is a one dimensional abelian subspace a of p such that

$$p = a + p_\alpha + p_{2\alpha} \tag{1}$$

where p_α and $p_{2\alpha}$ are the eigenspaces associated with α and 2α respectively. Define

$$m_\alpha = \dim p_\alpha$$

$$m_{2\alpha} = \dim p_{2\alpha}$$

A Riemannian manifold M is said to be two-point homogeneous if for any two-point pairs $p,q \in M$ and $p',q' \in M$ satisfying $d(p,q) = d(p',q')$, where $d(\cdot,\cdot)$ is the distance, there is an isometry g of M such that $g\cdot p = p'$ and $g\cdot q = q'$. By Wang's classification [19] the compact, two-point homogeneous spaces of dimension > 1 are the compact symmetric spaces of rank one.

Let M be a compact symmetric space of rank one. Let G be the identity component of the group I(M) of isometries of M. Fix an origin $o \in M$ and let K be the isotropy subgroup of G at o. Let k and g be the algebras of K and G, respectively. Since g is semisimple let p be the orthogonal complement of k in g with respect to the (negative of the) Killing form B of g. We can assume by scaling the distance function on M by a constant that the differential of the mapping $g \mapsto g\cdot o$ of G onto M gives an isometry of p (with the metric $-$ B) onto the tangent space T_oM. This is the metric that is called the one induced by (the negative of) the Killing form.

Let L be the diameter of M, that is, the maximal distance between any two points. If $x \in M$ then let A_x be the set of points of M of distance L from

x. A_x is a submanifold of M that is called the antipodal manifold associated with x. This submanifold is important in the subsequent analysis.

The following is a complete list of the irreducible compact symmetric spaces of rank one and their corresponding antipodal manifolds (p. 167 [16]):

M		A_o
Spheres	S^n (n = 1, 2, ...)	Point
Real projective spaces	$\mathbf{P}^n(\mathbf{R})$ (n = 2, 3, ...)	$\mathbf{P}^{n-1}(\mathbf{R})$
Complex projective spaces	$\mathbf{P}^n(\mathbf{C})$ (n = 4, 6, ...)	$\mathbf{P}^{n-2}(\mathbf{C})$
Quaternion projective spaces	$\mathbf{P}^n(\mathbf{H})$ (n = 8, 12, ...)	$\mathbf{P}^{n-4}(\mathbf{H})$
Cayley plane	$\mathbf{P}^{16}(\mathrm{Cay})$	S^8

The superscripts denote the real dimension. Some of the lowest possible dimensions are missing because

$$\mathbf{P}^1(\mathbf{R}) = S^1, \quad \mathbf{P}^1(\mathbf{C}) = S^2, \quad \mathbf{P}^1(\mathbf{H}) = S^4$$

The multiplicities m_α and $m_{2\alpha}$ for the rank one symmetric spaces were given by Cartan [3] and are well known (e.g. p. 168 [16]).

$M = S^n$ or $M = \mathbf{P}^n(\mathbf{R})$

$$m_\alpha = n - 1 \qquad m_{2\alpha} = 0 \qquad \gamma = \frac{\pi}{2L}$$

$M = \mathbf{P}^n(\mathbf{C})$

$$m_\alpha = n - 2 \qquad m_{2\alpha} = 1 \qquad \gamma = \frac{\pi}{2L}$$

$M = \mathbf{P}^n(\mathrm{H})$

$$m_\alpha = n - 4 \qquad m_{2\alpha} = 3 \qquad \gamma = \frac{\pi}{2L}$$

$M = \mathbf{P}^{16}(\mathrm{Cay})$

$$m_\alpha = 8 \qquad m_{2\alpha} = 7 \qquad \gamma = \frac{\pi}{2L}$$

In the Killing form metric the diameter of M can be expressed in terms of m_α and $m_{2\alpha}$ as

$$L^2 = m_\alpha\left(\frac{\pi^2}{2}\right) + 2m_{2\alpha}\pi^2 \tag{2}$$

The spheres and the projective spaces on **R**, **C** and **H** can be expressed by the classical groups as

$$P^n(\mathbf{R}) = SO(n+1)/O(n) \tag{3}$$

$$S^n = SO(n+1)/SO(n) \tag{4}$$

$$P^n(\mathbf{C}) = SU(n+1)/S(U(1) \times U(n)) \tag{5}$$

$$P^n(\mathbf{H}) = Sp(n+1)/Sp(n) \times Sp(1) \tag{6}$$

These spaces with the metric induced from the negative of the Killing form have constant, positive sectional curvature.

Let $(\theta_1, \dots, \theta_{n-1})$ be Cartesian coordinates on an open subset of the unit sphere $S_1(0)$ in T_xM where M is a compact symmetric space of rank one. The mapping $\exp_x\colon T_xM \to M$ is a diffeomorphism of the ball $B_L(0) = \{ y \in T_xM\colon |y| < L\}$ onto the open set $M \backslash A_x$. The inverse of the mapping

$$(\theta_1, \dots, \theta_{n-1}, r) \mapsto \exp_x(r\theta_1, \dots, r\theta_{n-1}) \tag{7}$$

is called a system of geodesic polar coordinates at $x \in M$ where $r \in (0, L)$ and $r = |Y|$ for $Y \in B_L(0)$. Notice that only $x \in M$ and the antipodal manifold A_x are excluded in the system of local coordinates.

Let $M \cong G/K$ be an irreducible compact symmetric space of rank one with the metric induced by the negative of the Killing form. The Laplace-Beltrami operator Δ_M on M can be expressed in geodesic polar coordinates for $r \in (0, L)$ (p. 165 [16]) as

$$\Delta_M = \frac{\partial^2}{\partial r^2} + \gamma[m_\alpha \cot(\gamma r) + 2m_{2\alpha} \cot(2\gamma r)] \frac{\partial}{\partial r} + \Delta_{S_r} \tag{8}$$

where Δ_{S_r} is the Laplace-Beltrami operator on $S_r(o)$, the sphere in M with center o and radius r. The radial part $R(\Delta_M)$ of Δ_M is

$$R(\Delta_M) = \frac{\partial^2}{\partial r^2} + \gamma[m_\alpha \cot(\gamma r) + 2m_{2\alpha}\cot(2\gamma r)]\frac{\partial}{\partial r} \tag{9}$$

Let G be a connected Lie group and K be a compact subgroup. Let π: $G \to G/K$ be the canonical map. Let $\mathbf{D}(G)$ be the set of all left invariant differential operators on G, $\mathbf{D}_K(G)$ be the subspace of $\mathbf{D}(G)$ that are also right invariant under K and $\mathbf{D}(G/K)$ be the algebra of differential operators on G/K that are invariant under all the translations $\tau(g)$: $xK \to gxK$ of G/K. Let ϕ be a complex-valued function on G/K of class C^∞ that satisfies $\phi(\pi(e)) = 1$. ϕ is called a spherical function if

i) $\phi^{\tau(k)} = \phi$ for all $k \in K$

ii) $D\phi = \lambda_D\phi$ for each $D \in \mathbf{D}(G/K)$ where $\lambda_D \in \mathbf{C}$.

Sometimes it is convenient to consider $\tilde{\phi} = \phi\circ\pi$ on G. $\tilde{\phi}$ is spherical on G if and only if ϕ is spherical on G/K.

For rank one symmetric spaces the spherical functions are determined from an eigenvalue problem for the Laplace-Beltrami operator where the Riemannian metric is induced from the Killing form. This operator is often called the Casimir operator. Computing the eigenvalue problem following Harish-Chandra (p. 302 [13]) for a noncompact symmetric space of rank one we have

$$R(\Delta_{G/K})\phi_\lambda = \lambda_\Delta\phi_\lambda \tag{10}$$

where $\lambda_\Delta = -(\langle\lambda,\lambda\rangle + \langle\rho,\rho\rangle)$ and $R(\Delta_{G/K})$ is the radial part of $\Delta_{G/K}$. Since ρ is one-half of the sum of the positive roots with their multiplicities we have

$$\rho = \frac{1}{2}(m_\alpha + 2m_{2\alpha})\,\alpha \tag{11}$$

and

$$-(\langle\lambda,\lambda\rangle + \langle\rho,\rho\rangle) = -\frac{1}{2}(m_\alpha + m_{2\alpha})^{-1}[\lambda(H_o)^2 + \rho(H_o)^2] \tag{12}$$

where $\alpha(H_o) = 1$ so that $\langle H_o, H_o\rangle = 2(m_\alpha + 4m_{2\alpha})$. Since $\coth 2\gamma r = \frac{1}{2}(\coth\gamma r + \tanh\gamma r)$ we have

$$\Delta_{G/K} = \frac{\partial^2}{\partial r^2} + \gamma[(m_\alpha + m_{2\alpha}) \coth \gamma r + m_{2\alpha} \tanh \gamma r] \frac{\partial}{\partial r} + \Delta_{S_r} \tag{13}$$

and

$$R(\Delta_{G/K}) = \frac{d^2}{dr^2} + \gamma[(m_\alpha + m_{2\alpha}) \coth \gamma r + m_{2\alpha} \tanh \gamma r] \frac{d}{dr} \tag{14}$$

Now make the substitution $z = -(\sinh \gamma r)^2$ to obtain

$$R(\Delta_{G/K}) = 4\gamma^2[z(z - 1) \frac{d^2}{dz^2} + \frac{1}{2} ((m_\alpha + m_{2\alpha} + 2)z - (m_\alpha + m_{2\alpha} + 1)) \frac{d}{dz}] \tag{15}$$

Hence, after cancelling a common factor, the eigenvalue problem (10) is

$$\begin{aligned} z(z-1) \frac{d^2\phi_\lambda}{dz^2} + \frac{1}{2} [(m_\alpha + m_{2\alpha} + 2)z - (m_\alpha + m_{2\alpha} + 1)] \frac{d\phi_\lambda}{dz} \\ + \frac{1}{4} [\lambda(H_o)^2 + (\frac{1}{2} m_\alpha + m_{2\alpha})^2]\phi_\lambda = 0 \end{aligned} \tag{16}$$

To identify (16) with a well known differential equation let

$$a = \frac{1}{4} [m_\alpha + 2m_{2\alpha} + 2i\lambda(H_o)] \tag{17}$$

$$b = \frac{1}{4} [m_\alpha + 2m_{2\alpha} - 2i\lambda(H_o)] \tag{18}$$

$$c = \frac{1}{2} (m_\alpha + m_{2\alpha} + 1) \tag{19}$$

Then the above differential equation can be written as

$$z(z - 1) \frac{d^2\phi_\lambda}{dz^2} + [(a + b + 1)z - c] \frac{d\phi_\lambda}{dz} + ab\phi_\lambda = 0 \tag{20}$$

It is clear that z can be considered as a coordinate function on A^+. Thus $\phi_\lambda = g(z)$ is an analytic function on $z < 0$. Since the Weyl group has two elements $\{1, w\}$ and $wH = -H$ it follows that $\phi_\lambda(h) = \phi_\lambda(h^{-1})$ for $h \in A$ so ϕ_λ can be expanded in even powers in a neighborhood of $h = 1$ and thus g is analytic also at $z = 0$ with $g(0) = \phi_\lambda(1)$. Since $c > 0$ it follows by substitution that $g(z)$ is the hypergeometric function $F(a,b,c,z)$. The function $F(a,b,c,z)$ is a single-valued analytic function of z in the whole z-plane with a branch-cut along the positive real axis from one to infinity. Since the variable z can be considered as a coordinate function on A^+ or equivalently $\mathfrak{a}^+$ and φ_λ and F are holomorphic on all of $\mathfrak{a}_c$ we can assume that z is a coordinate function of $\exp(i\mathfrak{a})$ so that we have an expression for the spherical function on a simply connected, compact symmetric space of rank one.

For the subsequent applications it is useful to review some properties of the hypergeometric functions. Let a be a complex number and define (p. 56 [11])

$$(a)_n = \frac{\Gamma(a + n)}{\Gamma(a)} \tag{21}$$

so that $(a)_0 = 1$ $(a)_n = a(a + 1) \cdots (a + n - 1)$ $n = 1, 2, \ldots$.

If $c \neq 0, -1, -2, \ldots$ then

$$u_1(z) = \sum_{n=0}^{\infty} \frac{(a)_n(b)_n z^n}{(c)_n n!} \triangleq F(a,b,c,z) \tag{22}$$

is a solution of (20) which is regular at $z = 0$.

If $i\lambda \in \mathfrak{a}^*$ and a is a negative integer $-m$ so that $b > 0$ and $c > 0$ then the hypergeometric function is clearly a polynomial over $\mathbf{R}$ expressed as

$$F(-m,b,c,z) = \sum_{n=0}^{m} \frac{(-m)_n(b)_n}{(c)_n} \frac{z^n}{n!} \tag{23}$$

Each term of the polynomial is positive for $z < 0$ from the definition of $(a)_n$. The roles of a and b can be interchanged because $F(a,b,c,z) = F(b,a,c,z)$.

The Jacobi polynomial $P_n^{(\alpha,\beta)}(\cdot)$ (p. 168 [11]) where $\alpha > -1$ and $\beta > -1$ satisfies the differential equation

$$(1 - x^2)y'' + [\beta - \alpha - (\alpha + \beta + 2)x]y' + n(n + \alpha + \beta + 1)y = 0 \tag{24}$$

so that it is easily related to the hypergeometric function by the equation (p. 170 [11])

$$P_n^{(\alpha,\beta)}(x) = \binom{n + \alpha}{n} F(-n, n + \alpha + \beta + 1, \alpha + 1, \tfrac{1}{2} - \tfrac{1}{2} x) \tag{25}$$

Thus the subsequent results could be phrased in terms of the Jacobi polynomials instead of the hypergeometric functions.

Let M be a pseudo-Riemannian manifold with g the pseudo-Riemannian structure on M. Let $\varphi: q \mapsto (x_1(q), \ldots, x_m(q))$ be a coordinate system defined on an open set $U \subset M$. As usual we define the functions g_{ij}, g^{ij} and $\bar{g}$ on U by

$$g_{ij} = g\left(\frac{\partial}{\partial x_i}, \frac{\partial}{\partial x_j}\right) \tag{26}$$

$$\sum_j g_{ij} g^{jk} = \delta_{ij} \tag{27}$$

$$\bar{g} = |\det(g_{ij})| \tag{28}$$

If F is a C^∞-function on M then it gives rise to a vector field grad f which in terms of the coordinates on U is

$$\text{grad } f = \sum g^{ij} \partial_i f \frac{\partial}{\partial x_j} \tag{29}$$

and if X is a vector field on M then the divergence of X is the function on M that is given on U by

$$\text{div } X = \frac{1}{\sqrt{\bar{g}}} \sum_i \partial_i(\sqrt{\bar{g}}\, X_i) \tag{30}$$

where $X = \sum_i X_i \frac{\partial}{\partial x_i}$ on U. The Laplace-Beltrami operator L on M is defined on a C^∞-function f on M as

$$Lf = \text{div grad } f \tag{31}$$

which in terms of local coordinates is

$$Lf = \frac{1}{\sqrt{\bar{g}}} \sum_k \partial_k\Big(\sum_i g^{ik} \sqrt{\bar{g}}\, \partial_i f\Big) \tag{32}$$

Let g_1 be a Riemannian metric on M and let $g_2 = cg_1$ where c is a positive constant. Let L_1 and L_2 be the Laplace-Beltrami operators associated with g_1 and g_2 respectively. It is clear from (32) that L_1 and L_2 are related by the equation

$$L_2 = cL_1 \tag{33}$$

If $R(L_1)$ and $R(L_2)$ are the radial parts of L_1 and L_2 respectively then the analogue of (33) holds for the radial parts, that is,

$$R(L_2) = cR(L_1) \tag{34}$$

Furthermore, if φ is an eigenfunction of $R(L_1)$ with eigenvalue λ_1 then φ is an eigenfunction of $R(L_2)$ with eigenvalue $c\lambda_1$.

3. THE CONTROL PROBLEM. Let G/K be an irreducible, simply connected compact symmetric space of rank one and dimension > 2. Recall that the diameter of G/K denoted by L is the maximal distance between any two points and A_o is the antipodal manifold associated with the origin.

To exhibit a family of cost functionals, the hypergeometric functions that are polynomials are used. Let $m \in \mathbb{Z}^+$ be fixed. The derivative of the hypergeometric function

$$G(m,r) = F(\tfrac{1}{2} m_\alpha + m_{2\alpha} + m, -m, \tfrac{1}{2} (m_\alpha + m_{2\alpha} + 1), \sin^2 \gamma r) \tag{35}$$

with respect to r is negative for r positive and sufficiently small and is zero for $r = 0$. Thus there is a maximal interval $[0, \delta]$ where $G(m,\cdot)$ is strictly decreasing for $r \in [0, \delta]$. Choose $c > 0$ such that

$$\sqrt{c}L \leq \delta \tag{36}$$

where L is the diameter of G/K. Let $k_o(m, c)$ be chosen such that

$$G(m, \sqrt{c}L) + k_o(m, c) = 0 \tag{37}$$

The function $\tilde{G}(m,r)$ given by

$$\tilde{G}(m,r) = G(m,r) + k_o(m,c) \tag{38}$$

is used in the cost functional for the stochastic control problem.

Let $c > 0$ be fixed and satisfy (36). Let G/K be the Riemannian symmetric space with the metric induced from $-cB(\cdot,\cdot)$ where $B(\cdot,\cdot)$ is the Killing form. Let $(Z(t), t \geq 0)$ be the controlled diffusion with the infinitesimal generator

$$\frac{1}{2} c\Delta_{G/K} + u \frac{\partial}{\partial r} \tag{39}$$

where r is the local coordinate from the geodesic polar coordinates (7) and $\Delta_{G/K}$ is the Laplace-Beltrami operator on G/K with the metric induced from the negative of the Killing form. From (33) it follows that $c\Delta_{G/K}$ is the Laplace-Beltrami operator for G/K with metric $-cB$. While controls could also be assumed in the $(\theta_1, \ldots, \theta_{n-1})$ coordinates it will become apparent from the form of the cost function that these additional controls are superfluous. For the analysis of the controlled diffusion it suffices to describe the local stochastic differential equation for the radial part of the infinitesimal generator (39). This equation is

$$dX(t) = [\frac{c}{2}\gamma(m_\alpha \cot \gamma X(t) + 2m_{2\alpha} \cot 2\gamma X(t)) + U(t)]dt + \sqrt{c}\, dB(t) \tag{40}$$

where $X(0) = \alpha \in (0, \sqrt{c}\, L)$, $X(t) \in (0, \sqrt{c}\, L)$ and $(B(t), t \geq 0)$ is a real-valued standard Brownian motion.

Recall that the eigenvalue in (10) is

$$\begin{aligned} \lambda_\Delta = -(\langle\lambda,\lambda\rangle + \langle\rho,\rho\rangle) &= -\frac{1}{2} (m_\alpha + m_{2\alpha})^{-1} [\lambda(H_o)^2 + \rho(H_o)^2] \\ &= -\gamma^2[\lambda(H_o)^2 + \rho(H_o)^2] \end{aligned} \tag{41}$$

For the choice of $m \in \mathbf{Z}^+$ in (18) we have

$$\lambda(H_o)^2 = - (2m + \frac{1}{2} m_\alpha + m_{2\alpha})^2 \tag{42}$$

so that

$$\lambda_\Delta = \gamma^2(4m^2 + 2mm_\alpha + 4mm_{2\alpha}) \tag{43}$$

Let $T > 0$ be chosen such that the Riccati differential equation

$$g' + \frac{c}{2}\gamma^2(4m^2 + 2mm_\alpha + 4mm_{2\alpha})g - \frac{1}{4}g^2 - 1 = 0 \tag{44}$$
$$g(T) = 0$$

has one and only one solution in $[0, T]$ and

$$\sup_{0 \le t \le T} |g(t)| \le \frac{2mc(m_\alpha + m_{2\alpha} - 1)(\frac{1}{2}m_\alpha + m_{2\alpha} + m)}{(m_\alpha + 4m_{2\alpha})(m_\alpha + m_{2\alpha} + 1)} \tag{45}$$

Notice that this Riccati equation may not have a global solution. However the existence and the uniqueness of a local solution follow easily. The inequality (45) is not vacuous because $\dim(G/K) > 2$.

The cost functional given in the geodesic polar coordinates at the origin is

$$J_m(U) = E_\alpha \int_0^T -\tilde{G}(m, X(t)) + h(m, X(t))\, U^2(t) + \tilde{k}(t, m, c)\, dt \tag{46}$$

where $X(0) = \alpha \in (0, \sqrt{c}\,L)$, $X(t) \in [0, \sqrt{c}\,L]$, $\tilde{G}$ is given by (38),

$$h(m, r) = \frac{[\tilde{G}_r(m, r)]^2}{\tilde{G}(m, r)} \tag{47}$$

$$\tilde{k}(t, m, c) = k_o(m, c)\, \frac{c}{2}\gamma^2(4m^2 + 2mm_\alpha + 4mm_\alpha)g(t) \tag{48}$$

where g is the solution of (44). The term $\tilde{k}$ in (46) does not affect the optimization problem, but its introduction here is convenient for subsequent calculations.

An admissible control at time t is a measurable function of $Z(t)$ that is smooth on $G/K \setminus (\{o\} \cup A_o)$ such that the solution of (40) exists and is unique in a sample path sense. By the K-invariance of the cost functional it suffices to consider controls at time t that are measurable functions of $X(t)$ which is determined by (40).

The main result of this paper is the following theorem that explicitly solves the stochastic optimal control problem (39-40) and (46) by exhibiting an optimal control.

THEOREM. The stochastic control problem described by (39-40) and (46) has an optimal control U^* that in geodesic polar coordinates at the origin is

$$U^*(s, r) = -\frac{1}{2}\frac{\tilde{G}(m,r)}{\tilde{G}_r(m,r)}\, g(s) \tag{49}$$

where $s \in [0, T]$, $r \in (0, \sqrt{c}\, L)$, g is the unique solution of (44) and U^* is extended by continuity to be zero on the antipodal manifold.

PROOF. It is well known (e.g. [12]) that the Hamilton-Jacobi or dynamic programming equation for a stochastic optimal control problem of diffusion type is

$$0 = \frac{\partial W}{\partial s} + \min_{v \in U} [A^{v}(s)W + L(s, x, v)] \tag{50}$$

where A^{v} is the infinitesimal generator of the controlled diffusion using the control v and L is the cost function. To apply a verification theorem (p. 159 [12]) to the control problem here it is assumed that the solution W of (50) with the boundary condition $W(s, y) = 0$ for $(s, y) \in \{T\} \times G/K$ is $C^{1,2}[(0, T) \times G/K]$ and continuous on $[0, T] \times G/K$.

Since the cost (46) as a function of the state with the control fixed is K-invariant and the control appears in (46) in a K-invariant way, the solution W of (50) is K-invariant and (50) can be written using only the radial part of (39) as

$$\begin{aligned} 0 = \frac{\partial W}{\partial s} + \min_{v \in U} [&\frac{c}{2}\frac{\partial^2 W}{\partial r^2} + \frac{c}{2}(\gamma m_{\alpha} \cot \gamma r + 2\gamma m_{2\alpha} \cot 2\gamma r)\frac{\partial W}{\partial r} + v\frac{\partial W}{\partial r} \\ &- \tilde{G}(m, r) + h(m, r)v^2 + \tilde{k}(t, m, c)] \end{aligned} \tag{51}$$

Performing the minimization in (51), it is clear that the control obtained by this minimization is

$$U^*(s, r) = \frac{-1}{2h(m, r)} \frac{\partial W}{\partial r}(s, r) \tag{52}$$

Assume a solution W of (51) as

$$W(s, r) = \tilde{G}(m, r)\; g(s) \tag{53}$$

Substitute (53) into (51) to obtain

$$\begin{aligned} 0 &= g' \tilde{G} + \frac{c}{2}\gamma^2(4m^2 + 2mm_\alpha + 4mm_{2\alpha})g\, \tilde{G} - \frac{1}{4} g^2 \tilde{G} - \tilde{G} \\ &= \tilde{G}(g' + \frac{c}{2}\gamma^2(4m^2 + 2mm_\alpha + 4mm_{2\alpha})g - \frac{1}{4} g^2 - 1) \end{aligned} \tag{54}$$

If g in (53) satisfies (44) then (53) is a smooth solution to (51) that satisfies the boundary condition $W(s, y) = 0$ for $(s, y) \in \{T\} \times (0, L)$. Since $\tilde{G}$ is zero for $r = \sqrt{c}L$, W given by (53) also satisfies the equation at $r = \sqrt{c}\,L$. However it is necessary to investigate the solution of (51) with the control (52) to verify that there is a unique strong solution, that is, there is one and only one solution to this equation in a sample path sense.

The optimal control (52) can be extended by continuity to be zero on the antipodal manifold. Thus the vector field that represents the control (52) can be extended to be a continuous function on $G/K \setminus \{o\}$. Let N be a neighborhood of the origin. Then it is easy to show that the vector field induced by the control (52) satisfies a global Lipschitz condition on $G/K \setminus N$. Thus the only point that remains for investigation is the origin. It is shown that

$$P(X(t) = 0 \text{ for some } t \in [0, T]) = 0 \tag{56}$$

where $(X(t),\ t \in [0, T])$ is the solution of (40) using the control (52). This verification (56) follows by comparison of $(X(t),\ t \in [0, T])$ with the two-dimensional Bessel process as in [7, 10].

It is well known (e.g. p. 61 [17]) that the two-dimensional Bessel process with the infinitesimal generator

$$\frac{1}{2}\left(\frac{d^2}{dr^2} + \frac{1}{r}\frac{d}{dr}\right) \tag{57}$$

does not hit $r = 0$ at a positive time. This two-dimensional Bessel process can be defined as the solution of the scalar stochastic differential equation

$$dY(t) = \frac{1}{2Y(t)}\,dt + dB(t) \tag{58}$$

where $(B(t), t \geq 0)$ is a real-valued standard Brownian motion. Multiplying the infinitesimal generator (57) by c the associated stochastic differential equation is

$$dY(t) = \frac{c}{2Y(t)}\,dt + \sqrt{c}\,dB(t) \tag{59}$$

It is easier to compare the solution of (40) with the solution of (59).

Define the stopping time $\mathcal{T}$ with respect to the natural family of σ-algebras of the solution of (40) with the control (52) as

$$\mathcal{T} = \begin{cases} \inf\{t : X(t) = 0 \ \text{ for } t \in [0, T]\} \\ +\infty \quad \text{if the above set is empty} \end{cases}$$

Assume that

$$P(\mathcal{T} < \infty) > 0 \tag{60}$$

A contradiction will be obtained. It is elementary to verify from (60) that there is a $t_0 \in \mathbf{R}$ such that

$$P(X(t) \in (0, \delta] \ \text{ for all } t \in [t_0, \mathcal{T})) > 0 \tag{61}$$

where $\delta > 0$ is determined from the inequality

$$\frac{c}{x} < (c\gamma m_\alpha \cot \gamma x + 2c\gamma m_{2\alpha} \cot 2\gamma x + 2U^*(s, x)) \tag{62}$$

for $x \in (0, \delta]$ where the bound on $|g|$ given in (45) is used. Comparing the solutions that are obtained by Picard iteration of (40) with the control (52) and (59) both with the initial condition $X(t_0)$ it easily follows that there is a contradiction to (60) using (62). Thus

$$P(\mathcal{T} = +\infty) = 1 \tag{63}$$

and the solution of (40) using the control (52) exists and is unique in a sample path sense. This completes the proof.

An important difference between the solvable stochastic control problems in compact symmetric spaces given here and the solvable stochastic control problems in noncompact symmetric spaces [7, 9, 10] is that the Riccati equation (44) may have explosions while the corresponding Riccati equation in [10] admits a global solution. In this way the topology of the manifold enters into the solution of the stochastic control problems.

REFERENCES

[1] V. E. Benes, L. A. Shepp and H. S. Witsenhausen, Some solvable stochastic control problems, Stochastics 4(1980), 39-83.

[2] A. Bensoussan and J. H. Van Schuppen, Optimal control of partially observable stochastic systems with an exponential-of-integral performance index, SIAM J. Control Optim. 23(1985), 599-613.

[3] É. Cartan, Sur certaines formes riemanniennes remarquables des geometries a groupe fondamental simple, Ann. Sci. École. Norm. Sup. 44(1927), 345-467.

[4] T. E. Duncan, Some stochastic systems on manifolds, Lecture Notes in Econ. and Math. Systems 107(1975), 262-270, Springer-Verlag, New York.

[5] T. E. Duncan, Dynamic programming optimality criteria for stochastic systems in Riemannian manifolds, Appl. Math. Optim. 3(1977), 191-208.

[6] T. E. Duncan, Stochastic systems in Riemannian manifolds, J. Optimization Theory Appl. 27(1979), 399-426.

[7] T. E. Duncan, A solvable stochastic control problem in hyperbolic three space, Systems and Control Letters, 8(1987), 435-439.

[8] T. E. Duncan, A solvable stochastic control problem in spheres in Geometry of Random Motion (R. Durrett and M. Pinsky, eds.) Contemporary Mathematics, 73(1988), 49-54, Amer. Math. Soc., Providence.

[9] T. E. Duncan, Some solvable stochastic control problems in real hyperbolic spaces in Analysis and Control of Nonlinear Systems (C. I. Byrnes, C. F. Martin and R. E. Saeks, eds.) Elsevier Science Publishers (North-Holland) 1988, 563-570.

[10] T. E. Duncan, Some solvable stochastic control problems in noncompact symmetric spaces of rank one, to appear in Stochastics.

[11] A. Erdelyi, W. Magnus, F. Oberhettinger and F. G. Tricomi, Higher Transcendental Functions (Bateman Manuscript Project) Vols. I and II, McGraw-Hill, 1953.

[12] W. H. Fleming and R. W. Rishel, Deterministic and Stochastic Optimal Control, Springer-Verlag, 1975.

[13] Harish-Chandra, Spherical functions on a semi-simple Lie group I., Amer. J. Math. 80(1958), 241-310.

[14] U. G. Haussmann, Some examples of optimal stochastic controls or: the stochastic maximum principle at work, SIAM Review 23(1981), 292-307.

[15] S. Helgason, Differential Geometry, Lie Groups and Symmetric Spaces, Academic Press, 1978.

[16] S. Helgason, Groups and Geometric Analysis, Academic Press, 1984.

[17] K. Itô and H. P. McKean, Jr., Diffusion Processes and their Sample Paths, Springer-Verlag, 1965.

[18] R. C. Merton, Optimum consumption and portfolio rules in a continuous-time model, J. Economic Theory 3(1971), 373-413.

[19] H. C. Wang, Two-point homogeneous spaces, Ann. of Math. 55(1952), 177-191.

DEPARTMENT OF MATHEMTAICS
UNIVERSITY OF KANSAS
LAWRENCE, KANSAS 66045

Contemporary Mathematics
Volume 97, 1989

SLEW-INDUCED DEFORMATION SHAPING
ON SLOW INTEGRAL MANIFOLDS*

T.A.W. Dwyer, III

Abstract

Computed torques for pointing and tracking require compensation for slew-induced structural, forebody/aftbody, or optical train alignment deformations, hereafter called deformations. Thus even if only line-of-sight variables are to be commanded, yet full state feedback is needed. The solution proposed here is to decouple by feedforward of the line-of-sight slew dynamics into the deformation control loop. It is also shown in this article how arbitrarily few actuators are needed for such deformation shaping, at the cost of higher differentiability of the reference line-of-sight dynamics. The low rates, single axis case is developed here in detail, and its extendability to high rates and multiple axes by global feedback linearization is outlined.

Introduction

In recent work on the control of robots with elastic joints, the state of the coupled rigid and elastic dynamics is forced to evolve in a "slow" integral manifold, so that the elastic distortion to the computed torque becomes a transient, i.e., the "transversal" (off-manifold) dynamics. That approach is followed here in treating the problem of slewing with deformable bodies, to show how a family of control-dependent slow manifolds can be constructed, on each of which the slew-induced deformations are expressible in terms of (higher derivatives of) the slewing angle. This relieves the bandwidth requirements of the primary slew actuator, but generally requires that slew-generated "deformation shaping" forces or moments also be applied, such as through available structural actuators. In the absence of sufficient natural damping, it is shown how a "fast" correction can also be applied,

---

*Research supported in part by SDIO and managed by AFOSR (AFSC) under contract F49620-87-C-0103, and by NASA Grant NAG-1-613.

involving an on-line synthesis of the "transversal" part of the deformation. This requires higher control bandwidth than if only the slow control were used, but only from the deformation shaping actuators, not from the slew actuator. The role of "prior" or "posterior" global feedback linearization for the extension of the method to the multiaxial or high rates situation is then outlined, as well as the interpretation of the slow manifolds as sliding surfaces for robust variable structure control implementation.

Deformable Rotational Dynamics

The model for a rotating body to be used here is given by Eqns. (1), (2).

$$J\ddot{\theta} + c\dot{\theta} + k\theta + \underline{n}^T\ddot{\underline{y}} = \tau + \underline{e}^T\underline{f} \tag{1}$$

$$M\ddot{\underline{y}} + C\dot{\underline{y}} + K\underline{y} + \underline{d}\ddot{\theta} = E\underline{f} \tag{2}$$

where rotation about an axis through the undeformed center of mass, measured by the angle θ in radians is presumed, entailing deformations represented by an nx1 column matrix $\underline{y}(t)$ of generalized coordinates. A torque τ about the same axis with respect to which the attitude angle θ is measured entails matching rotation-deformation coupling coefficients, $\underline{n} = \underline{d}$. Deformation shaping forces or moments are encoded by the mx1 matrix $\underline{f}$ of generalized forces. To fix ideas, the central hub with symmetrically placed and antisymmetrically deforming cantilevered appendages developed in [1] can be considered, wherein the coefficient matrices as well as deformations are expressed in terms of assumed mode shapes.

Elimation of $\ddot{\underline{y}}$ from Eqn. (1) and of $\ddot{\theta}$ from Eqn. (2) yields

$$J'\ddot{\theta} + c\dot{\theta} + k\theta = \underline{c}^T\dot{\underline{y}} + \underline{k}^T\underline{y} + \tau + \underline{e}'^T\underline{f} \tag{3}$$

$$M'\ddot{\underline{y}} + C\dot{\underline{y}} + K\underline{y} = \underline{c}'\dot{\theta} + \underline{k}'\theta - \underline{d}'\tau + E'\underline{f} \tag{4}$$

where $J' = J - \underline{n}^TM^{-1}\underline{d}$, $\underline{c}^T = \underline{n}^TM^{-1}C$, $\underline{k}^T = \underline{n}^TM^{-1}K$, $\underline{e}'^T = \underline{e}^T - \underline{n}^TM^{-1}E$, $M' = M - \underline{d}J^{-1}\underline{n}^T$, $\underline{c}' = \underline{d}J^{-1}c$, $k' = \underline{d}J^{-1}k$, $\underline{d}' = \underline{d}J^{-1}$ and $E' = E - \underline{d}J^{-1}\underline{e}^T$.

More general models, involving either coupling nonlinearities arising from gyroscopic effects or full spatially distributed deformation dynamics or both, are found in [2], [3], [4], [5], but then time-varying dynamics or unbounded operator coefficients arise, for which the present method is still under development.

High Bandwidth Computed Torque

The standard "proportional plus derivative" control law given by Eqn. (3) requires feedforward of measured structural acceleration to the pointing and tracking torque control,

$$\tau = \hat{J}\,\ddot{\theta}^* + \hat{c}\,\dot{\theta}^* + \hat{k}\,\theta^* + \hat{\underline{n}}^T\,\ddot{\hat{\underline{y}}}$$

$$- a_1(\dot{\theta} - \dot{\theta}^*) - a_o(\theta - \theta^*) \tag{5}$$

where hats denote estimates and θ^* is the commanded angular trajectory. With a good model and good sensors the closed loop dynamics is then approximated by Eqn. (4),

$$J\,\Delta\ddot{\theta} + (c + a_1)\Delta\dot{\theta} + (\kappa + a_o)\Delta\theta = \underline{0} \tag{6}$$

where $\Delta\theta : = \theta - \theta^*$ and no structural controls 'per se' are applied: $\underline{f} = \underline{0}$.

Besides the (everywhere present) question of model accuracy, the need for high bandwidth slew torque actuators and for full order structural sensing arises.

Exact Slow Manifolds

Following [6] - [8] in part, it is shown here how deformation-induced line of sight disturbances can be reduced to transients: Eqns. (1), (2) are first put into singular perturbation form: Let $\overline{K}$ be a normalized stiffness matrix, so that one has K expressed in terms of $\overline{K}$ and a perturbation parameter $\varepsilon > 0$:

$$\overline{K} = \varepsilon K \tag{7}$$

Without loss of generality one may set $M = I =$ identity and $K = \text{diag}\,\{\omega_1^{\,2}, \omega_2^2, \ldots\}$ with $\omega_1 \leqq \omega_2 \leqq \ldots$, then set $\varepsilon := \omega_1^{\,-2}$, so that $\overline{K} = \text{diag}\,\{1, (\omega_2/\omega_1)^2, \ldots\}$. Whatever the physical interpretation, one may seek polynomial controls in the parameter ε,

$$\tau = \tau_o + \varepsilon\tau_1 + \ldots + \varepsilon^p\tau_p \tag{8}$$

$$\underline{f} = \underline{f}_o + \varepsilon \underline{f}_1 + \ldots + \varepsilon^p \underline{f}_p \tag{9}$$

such that the deformations can be expressed as perturbations from the rigid body limit as $\varepsilon \to 0$, i.e.,

$$\underline{y} = \varepsilon \underline{z} + \underline{y}' \tag{10}$$

with the normalized deformation $\underline{z}$ (which has units of force) also a polynomial in ε,

$$\underline{z} = \underline{z}_o + \varepsilon \underline{z}_1 + \ldots + \varepsilon^{p-1} \underline{z}_{p-1} \tag{11}$$

and $\underline{y}'(t) \to \underline{0}$ as $t \to \infty$. Indeed, insertion of Eqns. (7) and (10) into Eqns. (1) and (2) brings the latter to singular perturbation form,

$$J \ddot{\theta} + c\dot{\theta} + k\theta + \varepsilon \underline{n}^T \ddot{\underline{z}} = \tau + \underline{e}^T \underline{f} + \nu \tag{12}$$

$$\varepsilon\{M\ddot{\underline{z}} + C\dot{\underline{z}}\} + \overline{K}\underline{z} + \underline{d}\,\ddot{\theta} = E\underline{f} + \underline{\nu}' \tag{13}$$

where the disturbances $\nu, \underline{\nu}'$ are given by Eqns. (14), (15),

$$\nu = -\underline{n}^T \ddot{\underline{y}}' \tag{14}$$

$$\underline{\nu}' = \underline{d} J^{-1} \nu \tag{15}$$

and $\underline{y}'$ is governed by

$$M' \ddot{\underline{y}}' + C\, \dot{\underline{y}}' + K\, \underline{y}' = \underline{0} \tag{16}$$

with M' as in Eqn. (4).

If $\varepsilon = 0$ then $\underline{z} = \underline{z}_o$ lies in the "rigid body manifold":

$$\underline{z}_o = \overline{K}^{-1} \left(E'\underline{f} - \underline{d}'\tau + \underline{c}'\,\dot{\theta} + \underline{k}'\theta\right) = h_o(\theta, \dot{\theta}, \tau, \underline{f}) \tag{17}$$

Then by Tychonov's theorem of singular perturbations [6], for (in general small enough) $\varepsilon > 0$ there is also a neighboring "slow manifold",

$$\underline{z} = z_o + \varepsilon \underline{z}_1 + \ldots = h_\varepsilon(\theta, \dot{\theta}, \underline{f}, \tau, \dot{\underline{f}}, \dot{\tau}, ..) \tag{18}$$

where if $\underline{z}(t_o) = h_\varepsilon$ (at t_o) then $\underline{z}(t) = h_\varepsilon$ (at t) for all $t \geq t_o$.

In particular, integral manifolds corresponding to polynomial controls in ε can be found that yield polynomial representations in ε (in which case ε need not be "small"). No appeal to time scale separation will be needed "a priori", provided the off-manifold correction $\underline{y}'$ is explicitly carried along in all computations: indeed, the insertion of the expansions (8), (9) and (11) into Eqns. (12) and (13), followed by collection of equal powers of ε, yield the following recursive formulas,

$$J\ddot{\theta} + c\dot{\theta} + k\theta = \tau_o + \underline{e}^T\underline{f}_o + \nu \tag{19}$$

$$\overline{K}\,\underline{z}_o + \underline{d}\,\ddot{\theta} = E\,\underline{f}_o + \underline{\nu}' \tag{20}$$

$$\underline{n}^T\ddot{\underline{z}}_j = \tau_{j+1} + \underline{e}^T\,\underline{f}_{j+1} \tag{21}$$

$$M\ddot{\underline{z}}_j + C\dot{\underline{z}}_j + \overline{K}\,\underline{z}_{j+1} = E\,\underline{f}_{j+1} \tag{22}$$

for j = 0, ..., p-2, and

$$\underline{n}^T\ddot{\underline{z}}_{p-1} = \tau_p + \underline{e}^T\,\underline{f}_p \tag{23}$$

$$M\ddot{\underline{z}}_{p-1} + C\dot{\underline{z}}_{p-1} = E\,\underline{f}_p \tag{24}$$

where $\underline{z}_p = \underline{0}$ is presumed, so that if one sets $\tau_j = 0$ and $\underline{f}_j = \underline{0}$ for $j \geq p + 1$ then also $\underline{z}_j = \underline{0}$ for $j \geq p$. Thus, the postulated polynomial solution given by Eqn. (11) is verified. In particular, <u>no convergence considerations arise and no truncation is needed</u> (hence, the term "exact slow manifolds"). This is possible if and only if Eqns. (19) - (24) can be solved for $\tau_j, \underline{f}_j$, j = 0, ...,p, as discussed next.

Deformation Suppression

The simplest choice of slow manifold is $\underline{z} = \underline{0}$, which exists if and only if it is possible to solve Eqn. (22) below for τ_o and $\underline{f}_o$,

$$\phi_o(s)\theta = \varepsilon_o\begin{pmatrix}\tau_o\\ \underline{f}_o\end{pmatrix} + \begin{pmatrix}\nu\\ \underline{\nu}'\end{pmatrix} \tag{25}$$

where

$$\phi_o(s) = \begin{pmatrix}Js^2 + cs + k\\ \underline{d}s^2\end{pmatrix} \tag{26}$$

and

$$\varepsilon_o = \begin{bmatrix} 1 & \underline{e}^T \\ \underline{0} & E \end{bmatrix} \tag{27}$$

obtained by setting $\underline{z}_o = \underline{0}$ in Eqns. (19) and (20), and then Laplace-transforming (the same symbols $\theta, \tau, \underline{f}, \underline{z}, \underline{y}$ will be used both in the time and frequency domains).

Noting that $\dim \varepsilon_o = (1+n) \times (1+m)$ (where $n = \dim \underline{y}$ and $m = \dim \underline{f}$), it is seen that ε_o is square if and only if $m=n$ ("independent modal space control"). Then ε_o is invertible if and only if E is invertible; indeed then Eqns. (19), (20) yield

$$(J - \underline{e}^T E^{-1} \underline{d})\ddot{\theta} + c\dot{\theta} + k\theta = \tau_o + \nu_o \tag{28}$$

$$E^{-1} \underline{d}\, \ddot{\theta} = \underline{f}_o + \underline{\nu}'_o \tag{29}$$

where

$$\nu_o = \nu + \underline{e}^T \underline{\nu}'_o \tag{30}$$

$$\underline{\nu}'_o = - E^{-1} \underline{\nu}' \tag{31}$$

Letting $\Delta\theta := \theta - \theta^*$, the tracking problem is then reduced to a regulator problem based on the reduced slew dynamics given by Eqn. (28), that is, to the choice of a stabilizing regulator $H_o(s)$ for the block diagram in Figure 1:

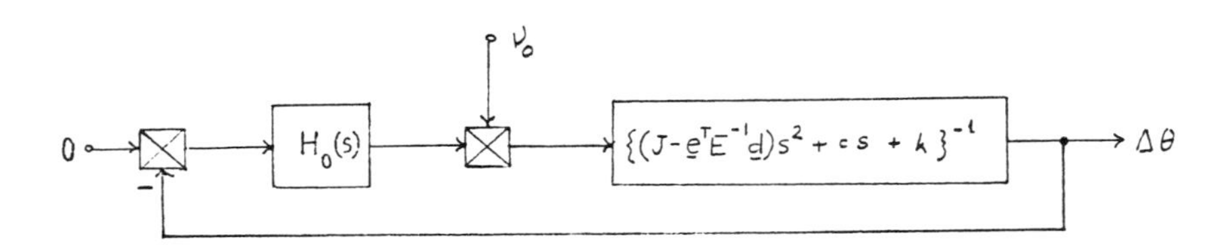

Figure 1

The slew torque is then given by Eqn. (32):

$$\tau = \tau_o = (J - \underline{e}^T E^{-1} \underline{d}) s^2 \theta^* - H_o(s)\Delta\theta + cs\theta^* + k\theta^* \tag{32}$$

For Eqn. (28) to be a valid slew dynamics model, it is also necessary to apply a deformation suppressing control $\underline{f}$, which is found from Eqn. (29) by insertion therein of the closed loop pointing angle θ found by inversion of Eqn. (28):

$$\underline{f} = \underline{f}_o = E^{-1}\underline{d}s^2\{(J - \underline{e}^T E^{-1}\underline{d})s^2 + cs + k\}^{-1}\tau_o = E^{-1}\underline{d}s^2\theta^*$$

$$- E^{-1}\underline{d}s^2\{(J - \underline{e}^T E^{-1}\underline{d})s^2 + cs + k\}^{-1}H_o(s)\Delta\theta \tag{33}$$

The "fast" term ν_o brought into Eq. (29), hence into Eqn. (33) by solving for θ is found to cancel with $\underline{\nu}_o'$ already there, so that $\underline{f}_o$ will have the bandwidth of τ_o.

In particular, if a proportional and derivative regulator $H_o(s) = a_1 s + a_o$ is selected, the closed loop pointing error dynamics becomes as shown in Eqn. (33):

$$(J - \underline{e}^T E^{-1}\underline{d})\Delta\ddot{\theta} + (c + a_1)\Delta\dot{\theta} + (k + a_o)\Delta\theta = \nu_o \tag{34}$$

The driving term ν_o is a transient, and thus, unlike Eqn. (5), no high frequency correction term is needed in the slew control law of Eqn. (32), at the cost of only a transient ripple on the pointing error. The need for structural sensors and high bandwidth actuators in the implementation of the control law given by Eqn. (5) is now replaced by the need for full order structural control. This is alleviated next.

First Order Deformation Shaping

If ε_o in Eqn. (25) is not invertible, the system cannot be "rigidfied", but the coupling term $\underline{n}^T\ddot{\underline{y}}$ can still be replaced by the fast transient plus higher derivative terms in the pointing variable angle. The simplest choice of such a slow manifold is $\underline{z} = z_o$, which exists if and only if it is possible to solve Eqn. (35) below for $\tau_o, \tau_1, \underline{f}_o$ and $\underline{f}_1$,

$$\underline{\phi}_1(s)\ \theta = \varepsilon_1(s)(\tau_o, \underline{f}_o^T, \varepsilon\tau_1, \varepsilon\underline{f}_1^T)^T + \underline{\psi}_{-1}(s)\nu \tag{35}$$

where

$$\underline{\psi}_1(s)^T = (1, -\underline{n}^T K^{-1} \underline{d} J^{-1} s, - \{[Ms + C]K^{-1} \underline{d} J^{-1} s\}^T) \tag{36}$$

$$\underline{\phi}_1^T(s) = (Js^2 + cs + k, - \underline{n}^T K^{-1} \underline{d} s^4 , - \{[Ms + C]K^{-1} \underline{d} s^3\}^T) \tag{37}$$

and

$$\varepsilon_1(s) = \begin{bmatrix} 1 & \underline{e}^T & 0 & \underline{0}^T \\ 0 & - \underline{n}^T K^{-1} E s^2 & 1 & \underline{e}^T \\ \underline{0} & - [Ms + C]K^{-1} E s & \underline{0} & E \end{bmatrix} \tag{38}$$

Equations (35) - (38) are obtained by solving Eqn. (20) for $\underline{z}_o$, to yield

$$\underline{z}_o = \overline{K}^{-1}\{E \underline{f}_o - \underline{d} s^2 \theta + \underline{\nu}'\} \tag{39}$$

in the frequency domain, then inserting Eqn. (39) into Eqns. (23) and (24) with p=1 (hence $\underline{z}_1 = \underline{0}$).

Noting that dim $\varepsilon_1(s) = (2+n) \times (2+2m)$, it is seen that $\varepsilon_1(s)$ is square if and only if m=n/2. In this case, if $\varepsilon_1(s)$ is generically invertible one may set

$$\varepsilon_1^{-1}(s)\underline{\phi}_1(s) =: (\gamma_o^\tau(s), \underline{\gamma}_o^f(s)^T, \gamma_1^\tau(s), \underline{\gamma}_1^f(s)^T)^T \tag{40}$$

$$\varepsilon_1^{-1}(s)\underline{\psi}_1(s) =: (\delta_o^\tau(s), \underline{\delta}_o^f(s)^T, \delta_1^\tau(s), \underline{\delta}_1^f(s)^T)^T \tag{41}$$

(which is independent of ε), and then define the slow transfer functions for $\tau = \tau_o + \varepsilon\tau_1$ and $\underline{f} = \underline{f}_o + \varepsilon\underline{f}_1$ as follows:

$$\gamma^\tau(s) := \gamma_o^\tau(s) + \gamma_1^\tau(s) \tag{42}$$

$$\underline{\gamma}^{\underline{f}}(s) := \underline{\gamma}_o^f(s) + \underline{\gamma}_1^f(s) \tag{43}$$

Equation (35) then yields

$$\gamma^\tau(s)\theta = \tau + \nu_1 \tag{44}$$

$$\underline{\gamma}^f(s)\theta = \underline{f} + \underline{\nu}_1' \tag{45}$$

where the transient disturbances ν_1 and $\underline{\nu}_1'$ are given in terms of ν given by Eqn. (14),

$$\nu_1 = \delta^{\tau}(s)\nu \qquad (46)$$

$$\underline{\nu}_1' = \underline{\delta}^{f}(s)\nu \qquad (47)$$

where

$$\delta^{\tau}(s) := \delta_o^{\tau}(s) + \delta_1^{\tau}(s) \qquad (48)$$
$$\underline{\delta}^{f}(s) := \underline{\delta}_o^{f}(s) + \underline{\delta}_1^{f}(s) \qquad (49)$$

The tracking problem is now reduced to a regulator problem based on the reduced slew dynamics given by Eqn. (44), that is, to the choice of a stabilizing regulator $H_1(s)$ for the block diagram of Figure 2:

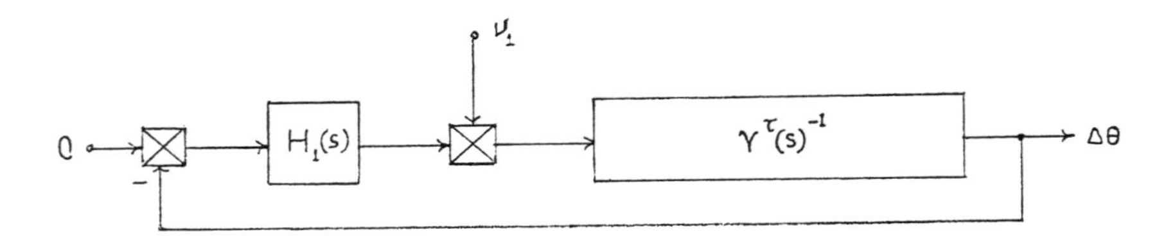

Figure 2

The low bandwidth slew torque control is then given by Eqn. (50):

$$\tau = \tau_o + \varepsilon\tau_1 = \gamma^{\tau}(s)\theta^* - H_1(s)\Delta\theta \qquad (50)$$

For Eqn. (44) to be a valid slew dynamics model, it is again necessary to apply a deformation shaping control $\underline{f}$, which is now found from Eqn. (45) by insertion therein of the <u>closed loop</u> pointing angle θ found by inversion of Eqn. (44):

$$\underline{f} = \underline{f}_o + \varepsilon\underline{f}_1 = \underline{\gamma}^{f}(s)\gamma^{\tau}(s)^{-1}\tau = \underline{\gamma}^{f}(s)\theta^* - \underline{\gamma}^{f}(s)\gamma^{\tau}(s)^{-1}H_1(s)\Delta\theta \qquad (51)$$

Again, ν_1 in Eqn. (44) is found to cancel with $\underline{\nu}_1'$ brought in from Eqn. (45) by θ, so that $\underline{f}$ will have the bandwidth of τ. The closed loop pointing error dynamics now becomes as shown in Eqn. (52):

$$\{\gamma^{\tau}(s) + H_1(s)\}\ \Delta\theta = \nu_1 \tag{52}$$

It turns out that <u>each scalar component</u> $\gamma_i^f(s)$ <u>of</u> $\underline{\gamma}^f(s)$ <u>has the same number of poles and the same number of zeros as</u> $\gamma^{\tau}(s)$. In particular, the poles of $\gamma^{\tau}(s)$ and of each $\gamma_i^f(s)$ are given by det $\varepsilon_1(s)$.

Thus, only the equalizer $H_1(s)$ determines the number of derivatives of the pointing error $\Delta\theta$ that are needed to generate $\underline{f}$ from Eqn. (51), as well as τ from Eqn. (50). This is an exact generalization of the "independent modal space control" case given by Eqns. (32) and (33). In particular, the high frequency disturbance to the slew dynamics is again reduced to a transient (now ν_1), and no high frequency correction is needed for the pointing control torque τ.

P - Th Order Deformation Shaping

More generally, a slow manifold with the representation of Eqn. (11) as a (p-1)-th order polynomial in ε exists if and only if Eqn. (53) below can be solved for τ_j, $\underline{f}_j$, $j = 0, \ldots, p$:

$$\underline{\phi}_p(s)\theta = \varepsilon_p(s)(\tau_o, \underline{f}_o^T, \ldots, \varepsilon^p\tau_p, \varepsilon^p\underline{f}_p^T)^T + \underline{\psi}_p(s)\nu \tag{53}$$

where $\underline{\phi}_p(s)$, $\underline{\psi}_p(s)$ and $\varepsilon_p(s)$ are found by successive elimination of $\underline{z}_o, \ldots, \underline{z}_{p-1}$ in Eqns. (20) - (24). Now $\underline{\phi}_p(s)$ is (p+1+n) - dimensional and dim $\varepsilon_p(s) = \{p+1+n\} \times \{(p+1)(1+m)\}$. It follows that $\varepsilon_p(s)$ is square if and only if the following relation holds true of $n = \dim \underline{y}$, $m = \dim \underline{f}$ and p:

$$p = \frac{n - m}{m} \tag{54}$$

2(p+1) derivatives of θ then determine the slow part $\varepsilon\underline{z}$ of $\underline{y}$.

For an invertible $\varepsilon_p(s)$ one again sets

$$\varepsilon_p(s)^{-1}\underline{\phi}_p(s) =: (\gamma_o^{\tau}(s), \underline{\gamma}_o^f(s)^T, \ldots, \gamma_p^{\tau}(s), \underline{\gamma}_p^f(s)^T)^T \tag{55}$$

$$\varepsilon_p(s)^{-1}\underline{\psi}_p(s) =: (\delta_o^{\tau}(s), \underline{\delta}_o^f(s)^T, \ldots, \delta_p^{\tau}(s), \underline{\delta}_p^f(s)^T)^T \tag{56}$$

to construct the slow transfer functions for $\tau = \tau_o + \ldots + \varepsilon^p \tau_p$ and for $\underline{f} = \underline{f}_o + \ldots + \varepsilon^p \underline{f}_p$, namely

$$\gamma^{\tau}(s)\theta = \tau + \nu_p \tag{57}$$

$$\underline{\gamma}^{f}(s)\theta = \underline{f} + \underline{\nu}'_p \tag{58}$$

where

$$\gamma^{\tau}(s) = \sum_{j=0}^{p} \gamma_j^{\tau}(s) \tag{59}$$

$$\underline{\gamma}^{f}(s) = \sum_{j=0}^{p} \underline{\gamma}_j^{f}(s) \tag{60}$$

and the transients ν_p, $\underline{\nu}'_p$ are again found from ν as in Eqns. (46) and (47), but now with the following replacement for Eqns. (48) and (49):

$$\delta^{\tau}(s) := \sum_{j=0}^{p} \delta^{\tau}(s) \tag{61}$$

$$\underline{\delta}^{\tau}(s) := \sum_{j=0}^{p} \underline{\delta}_j^{f}(s) \tag{62}$$

Instead of directly inverting $\varepsilon_p(s)$, it is also possible to proceed as follows:

(a) Solve recursively for z_{p-1} in terms of $\underline{f}_j$ for $j=0,\ldots,p$ (and $2p + 2$ derivatives of θ) from Eqns. (20) and (22);

(b) Insert $\underline{z}_{p-1}$ in terms of $\underline{f}_j$'s into Eqn. (24), which is decomposable into n equations in (p+1) m unknowns (the m components of $\underline{f}_j$ for $j=0,\ldots,p$), where here $n = (p+1)\,m$ according to Eqn. (54);

(c) Solve recursively for τ_o in terms of $\underline{f}_o$ from Eqn. (19) , then for τ_{j+1} in terms of $\underline{f}_{j+1}$ and $\underline{f}_j,\ldots,\underline{f}_o$ (through $\underline{z}_j$) from Eqn. (21).

It is from (b) that Eqn. (58) arises, and from (c) that Eqn. (57) arises. This quasi-triangular algorithm is a consequence of retaining the redundant terms $\ddot{\underline{y}}$ in Eqn. (1) and $\ddot{\theta}$ in Eqn. (2). The use of Eqns. (3) and (4) would require the direct inversion of $\varepsilon_p(s)$, since τ and $\underline{f}$ appear in both equations.

Again the tracking problem is reduced to the design of a stabilizing regulator $H_p(s)$ replacing $H_1(s)$ in the block diagram of Figure 2, where the high frequency transient ν_p replaces ν_1, with Eqn. (59) now giving the transfer function $\gamma^\tau(s)^{-1}$ instead of Eqn. (42).

The consequent low bandwidth slew torque τ is then given by the counterpart of Eqn. (50), and the shaping control $\underline{f}$ by the counterpart of Eqn. (51), with $H_1(s)$ replaced by $H_p(s)$ and with $\gamma^\tau(s)^{-1}$ and $\underline{\gamma}^f(s)$ found from Eqns. (59) and (60). It is again found that the "fast" term $\underline{\nu}'_p$ in Eqn. (58) cancels with ν_p brought to it when solving Eqn. (57) for θ, so that f will have the bandwidth of τ.

Fast Control

If there is insufficient or non-existent "a priori" damping C in the deformation dynamics for adequate time scale separation, the shaping control can be augmented by a "fast" term $\underline{f}'$, as advocated in the singular perturbation literature, e.g., [6] and references therein. That is, Eqn. (51) for $\underline{f}$ (or its $H_o(s)$ counterpart) can be replaced by Eqns. (63) and (64) below, with no change to Eqn. (50) for τ (or to its $H_p(s)$ counterpart),

$$\underline{f} = \underline{\gamma}^f(s)\ \gamma^\tau(s)^{-1}\tau - \underline{f}' \tag{63}$$

$$\underline{f}' = A_1(\hat{\dot{\underline{y}}} - \varepsilon\ \dot{\underline{z}}) + A_o(\hat{\underline{y}} - \varepsilon\ \underline{z}) \tag{64}$$

where $\hat{\underline{y}}$ and $\hat{\dot{\underline{y}}}$ now must be obtained from deformation sensors, but $\varepsilon\underline{z}$, $\varepsilon\dot{\underline{z}}$ from Eqn. (11), hence in terms of the $\underline{f}_j$'s, i.e.,. of $\underline{\gamma}_j^f(s)\gamma^\tau(s)^{-1}\tau$ for j=0,...,p.

Since the "fast" off-manifold term $\underline{y}'$ in Eqn. (10) is modeled by $\hat{\underline{y}} - \varepsilon\underline{z}$, it follows that $\underline{f}'$ is nothing but proportional plus derivative control of the fast dynamics,"transversal" to the manifold on which τ acts. In particular, $\underline{f}' \to \underline{0}$ as required in [6]: indeed, the fast dynamics of Eqn. (16) is now replaced by that of Eqn. (65),

$$M'\ddot{\underline{y}}' + [C + EA_1]\dot{\underline{y}}' + [K + EA_o]\underline{y}' = \underline{0} \tag{65}$$

hence $\underline{y}' \to \underline{0}$ by designer-selected choice of the matrix gains A_1 and A_o.

In the above approach, no change is needed in the slow dynamics and corresponding control laws, where the original K and C are retained, provided ν is replaced by $\nu - \underline{e}^T\underline{f}'$ ($\to$ 0) wherever appearing. An alternative is to design the fast control first, i.e., omit $\varepsilon\underline{z}$, $\varepsilon\dot{\underline{z}}$ from Eqn. (64), with consequently greater simplicity in the implementation of $\underline{f}'$. But then C must be replaced by $C + EA_1$ and K by $K + EA_o$ in all the slow control algorithms.

Example

Let $n=2$ and $m=1$, hence $p=1$, so that $\underline{z} = \underline{z}_o$ for appropriate choices of $\tau = \tau_o + \varepsilon\tau_1$ and of $\underline{f} = f = f_o + \varepsilon f_1$. Let also $E = e_1$ $e_2)^T$, $\underline{e} = e$, $\underline{d} = (d_1, d_2)^T$, $\underline{n} = \underline{d}$, $M = I$, $C = \text{diag}\ (c_1,\ c_2)$ and $K = \text{diag}\ (k_1,$ $k_2)$. One then finds

$$\underline{\phi}_1(s) = (\phi_{11}(s),\ \phi_{12}(s),\ \phi_{13}(s),\ \phi_{14}(s))^T \tag{66}$$

where

$$\phi_{11}(s) = Js^2 + cs + k \tag{67}$$

$$\phi_{12}(s) = (d_1{}^2/k_1 - d_2{}^2/k_2)s^4 \tag{68}$$

$$\phi_{13}(s) = (s + c_1)(d_1/k_1)s^3 \tag{69}$$

$$\phi_{14}(s) = (s + c_2)\ (d_2/k_2)s^3 \tag{70}$$

and

$$\varepsilon_1(s) = [\underline{\varepsilon}_{11}(s), \underline{\varepsilon}_{12}(s), \underline{\varepsilon}_{13}(s), \underline{\varepsilon}_{14}(s)] \tag{71}$$

where

$$\underline{\varepsilon}_{11}(s)^T = (1,0,0,0) \tag{72}$$

$$\underline{\varepsilon}_{12}(s)^T = (e, \{d_1(e_1/k_1) + d_2(e_2/k_2)\}s^2\ ,$$

$$(s + c_1)(e_1/k_1)s, (s + c_2)(e_2/k_2)s) \tag{73}$$

$$\underline{\varepsilon}_{13}(s)^T = (0,-1,0,0) \tag{74}$$

$$\underline{\varepsilon}_{14}(s)^T = (0,-e,-e_1,-e_2) \tag{75}$$

so that

$$\det\ \varepsilon_1(s) = e_1e_2\{(s + c_1)/k_1 - (s + c_2)/k_2\}s =: \delta_2(s) \tag{76}$$

Inversion of $\varepsilon_1(s)$ yields the following forms of Eqns. (42) and (43),

$$\gamma^T(s) = \frac{p_6^\tau(s)}{\delta_2(s)} \tag{77}$$

$$\gamma^f(s) = \frac{p_6^f(s)}{\delta_2(s)} \tag{78}$$

where the denominator is the quadratic polynomial given by Eqn. (76), while the numerators are 6-th degree polynomials.

In the idealized case of a perfect slew actuator, i.e., c=0 and k=0 in Eqn. (1), as well as no structural damping, i.e., C=0 in Eqn. (2), so that $c_1 = c_2 = 0$, the numerator and denominator of $\gamma^\tau(s)$ and $\gamma^f(s)$ take the forms below,

$$p_6^\tau = s^4(a_2 s^2 + a_o)/k_1 k_2 \tag{79}$$

$$p_6^f(s) = s^4(b_2 s^2 + b_o)/k_1 k_2 \tag{80}$$

$$\delta_2(s) = s^2 c_o/k_1 k_2 \tag{81}$$

where

$$a_o = e_1 e_2(k_2 - k_1) + e_1 d_2 k_1 - e_2 d_1 k_2 \tag{82}$$

$$a_2 = e_1 e_2(d_1^2 - d_2^2) + d_1 d_2(e_2^2 - e_1^2) + d_2 e_1 - d_1 e_2 \tag{83}$$

$$b_o = e_2 d_1 k_2 - e_1 d_2 k_1 \tag{84}$$

$$b_2 = d_1 e_2 - d_2 e_1 \tag{85}$$

$$c_o = e_1 e_2(k_2 - k_1) \tag{86}$$

It follows that the transfer function to be regulated in the block diagram of Figure 2 becomes the "type 2" system shown in Eqn. (87),

$$\gamma^{\tau}(s)^{-1} = \frac{c_o}{s^2(a_2 s^2 + a_o)} \tag{87}$$

which can be stabilized by a "damped PID" controller of the form given in Eqn. (88),

$$H_1(s) = K(s^2 + p_1 s + p_o)/(s + q) \tag{88}$$

or more generally,

$$H_1(s)\ K(s^3 + p_2 s^2 + p_1 s + p_o)/(s^2 + q_1 s + q_o) \tag{89}$$

In particular, by matching coefficients of the closed loop transfer function with those of $(s + \lambda)^5$, the filter parameters for Eqn. (88) can be chosen to be $q = 5\lambda$, $K = 10\lambda^3 a_2 - a_o q$, $p_1 = 5\lambda^4/c_o K$ and $p_o = 1/c_o K$, provided one can select $\lambda = \sqrt{a_o/10a_2}$ (but see below).

To have $(s + \lambda)^6$ for characteristic polynomial with Eqn. (89) one needs then only set $K = (20\lambda^3 a_2 - 6\lambda a_o)/c_o$, $p_2 = (15\lambda^4 a_2 - a_o q_o)/Kc_o$, $p_1 = 6\lambda^5 a_2/Kc_o$, $p_o = \lambda^6 a_2/Kc_o$, $q_1 = 6\lambda$ and $q_o = 15\lambda^2 a_o/a_2$, for free assignment of the slow control frequency λ. More generally, any control design method, be it LQR, H^{∞} or co-prime factorization, can be used.

The "slow" part of the shaping control is then given by Eqn. (51), where the feedback part consists of the compensator $H_1(s)$ selected as above for use in Eqn. (50), cascaded with the two stage lead-lag filter given below:

$$\gamma^{f}(s)\gamma^{\tau}(s)^{-1} = (b_2 s^2 + b_o)/(a_2 s^2 + a_o) \tag{90}$$

In the absence of damping as presumed here, a "fast" correction f' to f must be added, in accordance to Eqn. (64), where A_o and A_1 are freely chosen to impose time scale separation. The required "slow" term $\varepsilon\underline{z}$ in f' is found from Eqn. (39), yielding the expression below in terms of the independently designed slow slewing torque τ (where $\underline{\nu}'$ from Eqn. (39) is found to cancel with $\gamma^{\tau}(s)^{-1}\nu$ brought by θ from Eqn. (44)):

$$\underline{y} - \underline{y}' = \varepsilon \underline{z}_o = K^{-1}\{E\underline{Y}_o^f(s) - \underline{d}s^2\}Y^\tau(s)^{-1}\tau$$

$$= \{\begin{pmatrix} e_1/k_1 \\ e_2/k_2 \end{pmatrix} \frac{b_o}{a_2 s^2 + a_o} - \begin{pmatrix} d_1/k_1 \\ d_2/k_2 \end{pmatrix} \frac{c_o}{(a_2 s^2 + a_o)s^2}\}\tau \tag{91}$$

Alternatively, by omitting the corrections εz_o and $\varepsilon \dot{z}_o$ in the fast control $\underline{f}$, of Eqn. (64), so that K everywhere is replaced by $K + EA_o$ and C (here otherwise zero) by $C + EA_1$ (here EA_1), not only in Eqn. (65), the coefficients a_o and a_2 include dependence on a_{01}, a_{02} if $A_o = (a_{01}, a_{02})^T$ and on a_{11}, a_{12}, where $A_1 = (a_{11}, a_{12})^T$, permitting the adjustment of the slow time constant λ and retain the simpler PID control given by Eqn. (88). The lower order of the compensator given by Eqn. (88) is now obtained at the cost of a fixed relationship between fast and slow time constants, but no longer between the time constant λ and the structural parameters.

Extensions to Robust, Multiaxial and High Slew Rate Control

It is possible to extend the design methodology proposed herein to multiple slew axes as well as to high slew rates, in which case stiffness and damping coefficients in Eqn. (2) acquire angular rate and angular acceleration - dependent terms, while the rotational dynamics of Eqn. (1) acquire "vector product" terms: cf. [4], [5]. In this situation, the reduced order slewing dynamics obtained by inversion of Eqns. (19) - (24) become nonlinear. It is then possible to first globally feedback-linearize the nonlinear counterparts of Eqns. (1) and (2) with respect to pointing variables, such as in [9], [10] and elsewhere. Alternatively, it may be possible to first invert the nonlinear counterparts of the recursive Eqns. (19) - (24), to obtain nonlinear versions of the decoupled Eqns. (57) and (58) for the slow dynamics, and then seek to globally feedback-linearize Eqn. (57) (Eqn. (58)'s counterpart can be directly used to generate the shaping control $\underline{f}$ from τ without linearization).

Finally, the question of sensitivity to modeling errors may yield to variable structure control implementation: the presence only of the shaping control $\underline{f}$ in Eqn. (2) can be exploited to design an "interpolated switching control" following [11], with the selected slow manifolds regarded as the switching surfaces. The pointing control τ can then be designed for the reduced order dynamics on the switching surface, by the "hierarchical" method. These extensions will be discussed elsewhere.

References

[1] J. L. Junkins and J. D. Turner, Optimal Control of Spacecraft Rotational Maneuvers, Elsevier, 1986.

[2] E. Barbieri and U. Özgüner, "Unconstrained and Constrained Mode Expansions for a Flexible Slewing Link," Proc. American Control Conf. (Atlanta, GA, June 15-17, 1988) pp. 83-88.

[3] J. Baillieul and M. Levi, "Rotational Elastic Dynamics," Physica D, Vol. 27 (1987), pp. 43-62.

[4] T.A.W. Dwyer, III and F. Karray, "Nonlinear Modeling and Estimation of Slew-Induced Deformations," Proc. 1988 ASME Winter Annual Meeting (Chicago, IL, Nov. 27 - Dec. 2, 1988).

[5] W. H. Bennett, H. G. Kwatny and G. L. Blankenship, "Nonlinear Dynamics and Control Issues for Space-Based Directed Energy Weapons," Proc. 27th IEEE Conf. on Decision and Control (Austin, TX, December 12-14, 1988).

[6] M. W. Spong, K. Khorasani and P. V. Kokotovic, "An Integral Manifold Approach for the Feedback Control of Flexible Joint Robots," IEEE J. of Robotics and Automation, Vol. RA-3, No. 4 (August 1987), pp. 291-300.

[7] T.A.W. Dwyer, III, "Tracking and Pointing Maneuvers with Slew-Excited Deformation Shaping," AIAA Paper 87-2599, Proc. AIAA Guidance, Navigation and Control Conf. (Monterey, CA, August 17-19, 1987) Vol. 2, pp. 1503-1511.

[8] T.A.W. Dwyer, III, "Automatically Reconfigurable Control for Rapid Retargeting of Flexible Pointing Systems," SPIE Proceedings, Vol. 851, Space Station Automation III (1987), pp. 75-82.

[9] T.A.W. Dwyer, III, H. Sira-Ramirez, S. Monaco and S. Stornelli, "Variable Structure Control of Globally Feedback-Decoupled Deformable Vehicle Maneuvers," Proc. 26th IEEE Conf. on Decision and Control (Los Angeles, CA, December 9-11, 1987), pp. 1281-1287.

[10] A. Iyer and S. N. Singh, "Variable Structure Control of Decoupleable Systems and Attitude Control of Spacecraft in Presence of Uncertainty," Proc. American Control Conf. (Atlanta, GA, June 15-17, 1988), pp. 2238-2243.

[11] J. J. Slotine, "Sliding Controller Design for Nonlinear Systems," Internat. J. of Control, Vol. 40 (1984), pp. 421-434.

Department of Aeronautical and Astronautical Engineering
University of Illinois
Urbana, IL 61801

Contemporary Mathematics
Volume **97**, 1989

Feedback Equivalence and Symmetries of Brunowski Normal Forms

R. B. Gardner[1], W. F. Shadwick[2] and G. R. Wilkens

ABSTRACT The structure equations and differential invariants for nonlinear feedback equivalence of control systems are used to characterize Brunowski normal forms in terms of their symmetry groups. This provides a new solution to the linearization problem and constructs the differential invariants necessary for the classification of nonlinear systems.

1 INTRODUCTION

In this paper we indicate the way in which the structure equations and differential invariants for the equivalence method of E. Cartan [5][6] may be used to provide a new solution to the problem of local feedback linearization for control systems (see e.g. [1][12][13][14][15][17]). The point of view adopted here is that the *geometric content* of the control systems

$$\frac{dx}{dt} = f(x,u) \quad x \in \mathbb{R}^n, \ u \in \mathbb{R}^p \tag{1.1}$$

under feedback transformations

$$\Psi\colon (x,u) \mapsto (\varphi(x), \psi(x,u)) \tag{1.2}$$

is precisely the information and structure in (1.1) which is invariant under (1.2). This is the approach taken by F. Klein in his Erlangen Program: the geometry is the information which is invariant under a given transformation group. This point of view is particularly natural in applications to problems in Physics and Engineering, where one always has some freedom to change coordinates and the precise nature of the allowable changes of variable is dictated by the process or phenomenon which is being modeled. In the case of control systems, the coordinate freedom is given by (1.2) and is less than the full diffeomorphism group freedom because there is a physical difference between the states and the controls in the system.

We also take the point of view that one should, in general, use the largest class of changes of

---

[1]Supported by NSF grant DMS 8802918
[2]Supported by NSERC grant A7895

variable which preserve the interesting features of a problem and therefore we consider fully nonlinear feedback and do not make the assumption that (1.1) is control-linear. It is not clear *a priori* that this class of transformations will indeed preserve all of the structures of interest in control theory: it is possible for example that distinct Brunowski forms [2] could be equivalent under such transformations. As we demonstrate in this paper, however, *the Brunowski normal forms are characterized by differential invariants for fully nonlinear feedback. Moreover, these normal forms may also be characterized completely by their Lie pseudogroups of symmetries*. It is a feature of Cartan's method that the equivalence problem produces, as a by-product, the structure equations of the pseudogroup of symmetries which a given model admits. In fact these structure equations are always encoded in the structure equations of the equivalence problem and coincide with the latter in the case of a *transitive* symmetry group. This case is identified by the fact that the structure equations contain no *non-constant* invariants and the transitive action is used to obtain normal forms for models which have the given symmetry.

It is not difficult to construct the symmetries of the Brunowski normal forms under feedback. For example, if we consider controllable linear systems with four states and two controls, there are two Brunowski forms

$$\frac{dx^1}{dt} = x^2 , \frac{dx^2}{dt} = u^1 , \frac{dx^3}{dt} = x^2 , \frac{dx^4}{dt} = u^2 , \tag{1.3}$$

and

$$\frac{dx^1}{dt} = x^2 , \frac{dx^2}{dt} = x^3 , \frac{dx^3}{dt} = u^1 , \frac{dx^4}{dt} = u^2 . \tag{1.4}$$

The symmetry pseudogroups are generated by

$$\mathbf{x}^1 = \varphi(x^1, x^3) \quad \text{and} \quad \mathbf{x}^3 = \rho(x^1, x^3) \tag{1.5}$$

and by

$$\mathbf{x}^1 = \varphi(x^1) \quad \text{and} \quad x^4 = \rho(x^1, x^2, x^3, x^4) \tag{1.6}$$

respectively.

One may describe the system (1.3) in terms of the contact system for the one jets of maps from the line to the plane while (1.4) is given by a partial prolongation of the same system. The corresponding symmetries are the general group on the plane (1.5) and a subgroup of the general group in four variables (1.6). The system (1.3), or rather its equivalence class under (1.2), may be characterized as the class of systems which admit precisely the pseudogroup (1.5) as symmetries and similarly the system (1.4) is characterized by the pseudogroup (1.6).

In general, the Brunowski normal forms for a system in n states and p controls may all be recognized as partial prolongations of the contact system for maps from the line to p dimensional Euclidean space. The symmetry pseudogroups of such systems characterize them completely. This is a result which is striking in its own right, however, it has implications for the implementation of feedback linearization. In order to *find* the linearizing feedback for a linearizable system, one must solve differential equations. We show that the integration of these equations is naturally given by differential systems determined by the symmetry pseudogroup. As one cannot expect to solve these in closed form in general, numerical techniques will be essential. The fact that the geometry of the system is precisely the geometry of its symmetry pseudogroup means that numerical routines which are adapted to this structure can be expected to give the best results. This aspect of the problem will be treated elsewhere.

It is a pleasure to dedicate this paper to Roger Brockett, whose seminal paper on feedback equivalence of control systems [1] inspired much further work and who was responsible for introducing one of us (R.B.G.) to the study of the geometry of control problems.

2 THE EQUIVALENCE PROBLEM

We begin by describing the equivalence problem in the form which is adapted to the use of Cartan's method [5][6]. This problem falls into the class of overdetermined equivalence problems which are discussed in [8]. In that paper the conversion of the present problem to one in standard form is treated as an example. We indicate the result here and refer the reader to [8] for the details.

Two control systems of the form (1.1) will be called feedback equivalent iff there is a map

$$\Psi\colon \mathbb{R}^{n+p+1} \rightarrow \mathbb{R}^{n+p+1}$$

with the property that

$$\Psi^*(d\mathbf{x} - \mathbf{f}(\mathbf{x}, \mathbf{u})dt) = A(dx - f(x, u)dt)$$

and

$$\Psi(\mathbf{t}, \mathbf{x}, \mathbf{u}) = (\varphi(x), \psi(x, u)) .$$

It is not difficult to verify that these conditions may be formulated as follows. We work on open sets $\mathbf{U}$ and U on which $\mathbf{f}$ and f are nowhere zero. Let $\mathbf{L}$ and L be matrices on $\mathbf{U}$ and U respectively such that

$$\mathbf{L}f = \mathrm{L}f = t\,(1\ 0\ \ldots\ 0).$$

Then Ψ is an equivalence iff its Jacobian satisfies

$$\Psi^* \begin{bmatrix} \mathbf{L}\ \mathbf{dx} \\ \mathbf{du} \end{bmatrix} = \begin{bmatrix} 1 & A & 0 \\ 0 & B & 0 \\ C & D & E \end{bmatrix} \begin{bmatrix} \mathrm{L}\ \mathrm{dx} \\ \mathrm{du} \end{bmatrix} \tag{2.1}$$

where $B \in GL(n\text{-}1, \mathbb{R})$ and $E \in GL(p, \mathbb{R})$ and

$$\mathbf{t} \circ \Psi = t\,. \tag{2.2}$$

Following [5][6] we now construct 1 forms $\omega := {}^t(\eta\mu)$ on $U \times G$ where G is the Lie group of matrices of the form

$$S = \begin{bmatrix} 1 & A & 0 \\ 0 & B & 0 \\ C & D & E \end{bmatrix}$$

with $B \in GL(n-1, \mathbb{R})$ and $E \in GL(p, \mathbb{R})$. The 1-forms ω are defined by $\omega := S\begin{bmatrix} \mathbf{L}\ \mathbf{dx} \\ \mathbf{du} \end{bmatrix}$ for $S \in G$.

We note that the *control system is* described by the differential system $\Sigma = \{\eta^2,\ldots,\eta^n\}$ with independence condition given by η^1, i.e. the solutions of (1.1) are the curves on which

$$\eta^2 = \eta^3 = \ldots = \eta^n = 0 \quad \text{and } \eta^1 \neq 0\,.$$

We also note that the variational problem for the functional $\int \eta^1$ defined on integral curves of Σ is the *time optimal control problem* for the control system (1.1)[10]. The G-structure to which we are led by the study of feedback equivalence is therefore also that of a Pfaffian differential system with independence condition and of a general variational problem of the sort studied by Bryant and Griffiths [3][4][11][16]. The feedback problem corresponds to special classes of the latter problems which are characterized by the fact that the differential system $\{\eta^1,\Sigma\}$ is *completely integrable.* As we indicated in [10], the rich geometry of the variational problem has immediate implications for control systems and likewise all of the invariants of a Pfaffian system with independence condition will also have interpretations in the control system picture. The most basic invariant of a Pfaffian system is its derived structure defined as follows: the *first derived system* is the differential system

$$\Sigma^{(1)} := \{\eta \in \Sigma \mid d\eta \equiv 0 \mod \Sigma\}.$$

The *second derived system* of Σ is the first derived system of $\Sigma^{(1)}$ and so on inductively, to produce the sequence of subsystems called the derived flag for Σ. This process stabilizes when $\Sigma^{(N+1)} = \Sigma^{(N)}$ and the integer N is called the *derived length* of Σ. The derived structure is a key ingredient in the identification of Brunowski normal forms as Hermann has shown [12] in the (dual) vector fields approach. Of course, in the case of nonlinear (even bi-linear) systems the information in the derived flag is not sufficient and more refined invariants are essential.

We now proceed to the equivalence problem calculation following [6]. For convenience we omit the wedge symbol in products of differential forms and we use the notation Σ to indicate both the differential system $\{\eta^2, ..., \eta^n\}$ and the vector of 1-forms ${}^t(\eta^2, ..., \eta^n)$. With the latter convention, ω is given by $\omega = {}^t(\eta^1, \Sigma, \mu)$. The *structure equations* are now obtained by calculating the exterior derivative of ω and carrying out absorption of torsion as described in [6]. This leads to

$$d\eta^1 = \eta^1 k\mu + {}^t\alpha\Sigma \tag{2.3}$$

$$d\Sigma = \eta^1 K\mu + \beta\Sigma$$

$$d\mu = \gamma\eta^1 + \delta\Sigma + \varepsilon\mu .$$

The use of the identity $d^2 = 0$ in the structure equations yields the infinitesimal action on the invariants k and K:

$$\begin{aligned} dk + k\varepsilon - {}^t\alpha K &\equiv 0 \\ dK + K\varepsilon - \beta K &\equiv 0 \qquad \mod \eta^1, \Sigma, \mu. \end{aligned} \tag{2.4}$$

It follows from (2.4) that the matrix $\mathcal{K} := \begin{bmatrix} k \\ K \end{bmatrix}$ is acted on by the group G by right and left multiplication and that the ranks of $\mathcal{K}$ and of K are the only invariants. It is easy to verify that the rank of $\mathcal{K}$ is the number of controls, so we may assume that we have the full rank (i.e. rank p) case. There are now two possibilities to consider because K may have rank $p - 1$ or rank p. From the explicit form of the matrix $\mathcal{K}$ one may check that the condition that K has rank $p - 1$ is that there are controls for which the system has the form

$$\frac{dx^A}{dt} = u^A \qquad 1 \leq A \leq n - p \tag{2.5}$$

$$\frac{dx^\rho}{dt} = g^\rho(x,u) \quad p+1 \le \rho \le n$$

where each of the functions g^ρ is *homogeneous of degree one* in the controls. Thus there are no *linear* controllable systems found in this branch of the problem and we may consider the *generic case* in which K has rank p. The group action may be used to normalize K so that its last p rows are the $p \times p$ identity matrix. The first derived system for Σ is $\Sigma^{(1)} = {}^t(\eta^2, \ldots, \eta^{n-p})$. If we re-label the 1-forms η as ${}^t(\eta^1, \Sigma^{(1)}, \mu^{(1)})$, the 1-forms $\mu^{(1)}$ generate a complement to the first derived system and play the role of the *controls* for the states $\{\eta^1, \Sigma^{(1)}\}$. Thus they are analogous to the 1-forms μ which appear in the first structure equations. The structure equations for the reduced problem are

$$\begin{aligned} d\eta^1 &= {}^t\mu h^{(1)} \mu^{(1)} + {}^t\mu^{(1)} j^{(1)} \mu^{(1)} + \eta^1 k^{(1)} \mu^{(1)} + {}^t\alpha^{(1)} \Sigma^{(1)} \\ d\Sigma^{(1)} &= {}^t\mu H^{(1)} \mu^{(1)} + {}^t\mu^{(1)} J^{(1)} \mu^{(1)} + \eta^1 K^{(1)} \mu^{(1)} + \beta^{(1)} \Sigma^{(1)} \\ d\mu^{(1)} &= \eta^1 \mu + \delta^{(1)} \Sigma^{(1)} + \varepsilon^{(1)} \mu^{(1)} \end{aligned} \tag{2.6}$$

where $h^{(1)}$ and $H^{(1)}$ are symmetric and $j^{(1)}$ and $J^{(1)}$ are antisymmetric. (The equations for $d\mu$ are omitted as they are just those given by taking the exterior derivative of the equations for $d\mu^{(1)}$.)

The new torsion torsion terms $\mathcal{H}^{(1)} := \begin{bmatrix} h^{(1)} \\ H^{(1)} \end{bmatrix}$ and $\mathcal{J}^{(1)} := \begin{bmatrix} j^{(1)} \\ J^{(1)} \end{bmatrix}$ measure the *nonlinearity* of the system while $\mathcal{K}^{(1)} := \begin{bmatrix} k^{(1)} \\ K^{(1)} \end{bmatrix}$ measures the dependence of the states given by $\{\eta^1, \Sigma^{(1)}\}$ on the "current controls" $\mu^{(1)}$. It follows from the action of G on $h^{(1)}$ and $j^{(1)}$ which is given by

$$dh^{(1)} + {}^t\varepsilon^{(1)} h^{(1)} + h^{(1)} \varepsilon^{(1)} - {}^t\alpha^{(1)} H^{(1)} \equiv 0 \tag{2.7a}$$

$$dj^{(1)} + {}^t\varepsilon^{(1)} j^{(1)} + j^{(1)} \varepsilon^{(1)} - {}^t\alpha^{(1)} J^{(1)} \equiv 0 \mod \eta^1, \Sigma^{(1)}, \mu^{(1)}, \mu$$

that the vanishing of $h^{(1)}$ and $j^{(1)}$ imply the vanishing of $H^{(1)}$ and $J^{(1)}$ respectively. The vanishing of $h^{(1)}$ has the following significance.

Proposition 2.1 *For a control system which satisfies* rank $K = p$, *there is a feedback transformation which puts the system in control-linear form iff* $h^{(1)}$ *is identically zero.*

Proof One may calculate $\mathcal{H}^{(1)}$ directly in the same manner as in [10] where the case of n states and n-1 controls is treated. Intuitively, the first reductions are accomplished by assigning the group parameters as functions of the first derivatives of the functions $f^i(x,u)$. At this stage these functions are being differentiated twice with respect to the controls or more precisely twice in the direction of the 1-forms μ.

As may be verified from the form of $h^{(1)}$ it need not be readily apparent by inspection of the functions f^i that a control linear form is available by a feedback transformation. This is a feature of the use of fully nonlinear feedback - without which it is impossible to distinguish truly nonlinear models from ones which are feedback equivalent to models with a control-linear presentation. By a similar calculation one may also verify that for a linear system the torsion term $j^{(1)}$ vanishes and this establishes

Proposition 2.2 *The vanishing of* $h^{(1)}$ *and* $j^{(1)}$ *are necessary conditions for feedback equivalence to a linear system. Thus the complete integrability of the differential system* $\{\eta^1, \Sigma^{(1)}\}$ *is a necessary condition for equivalence to a linear system.*

We assume that the invariants $\mathcal{H}^{(1)}$ and $\mathcal{J}^{(1)}$ are identically zero so the remaining cases depend on the properties of $\mathcal{K}^{(1)}$. There may be several possibilities depending on the relative sizes of p and n. The ranks involved indicate how many of the "current controls" $\mu^{(1)}$ enter the system at this level of derivatives. Thus the cases where $\mathcal{K}^{(1)}$ has rank q with $1 \leq q \leq \min\{p, n-p\}$ may all occur, however, the rank 0 case may be discarded as this corresponds to a uncontrollable system. It is apparent from (2.6) that if $\mathcal{H}^{(1)}$, $\mathcal{J}^{(1)}$ and $\mathcal{K}^{(1)}$ are all zero, then the first derived system $\Sigma^{(1)}$ is *completely integrable*. It is clear that the control system cannot be driven to a point off of the integral manifolds of $\Sigma^{(1)}$ and is therefore not controllable. This means that we have an additional condition for equivalence to a controllable linear system:

Proposition 2.3 *The rank of* $\mathcal{K}^{(1)}$ *must be greater than zero for equivalence to a controllable linear system.*

As one should expect, the controllability condition appears as a genericity condition (requiring non-vanishing of an invariant) while the linearizability condition is non-generic (requiring the vanishing of invariants).

When $\mathcal{H}^{(1)}$ and $\mathcal{J}^{(1)}$ are zero the action on $\mathcal{K}^{(1)}$ is exactly as in the first normalization:

$$dk^{(1)} + k^{(1)}\,\varepsilon^{(1)} - {}^t\alpha^{(1)}\,K^{(1)} \equiv 0$$

$$(2.7b)$$

$$dK^{(1)} + K^{(1)}\,\varepsilon^{(1)} - \beta^{(1)}\,K^{(1)} \equiv 0 \quad \text{mod } \eta^1, \Sigma^{(1)}, \mu^{(1)}.$$

The reduction now proceeds by normalizing the invariant $\mathcal{K}^{(1)}$. This leads to new structure equations of exactly the same form:

$$d\eta^1 = {}^t\mu^{(1)} h^{(2)} \mu^{(2)} + {}^t\mu^{(2)} j^{(2)} \mu^{(2)} + \eta^1 k^{(2)} \mu^{(2)} + {}^t\alpha^{(2)} \Sigma^{(2)} \tag{2.8}$$

$$d\Sigma^{(2)} = {}^t\mu^{(1)} H^{(2)} \mu^{(2)} + {}^t\mu^{(2)} J^{(2)} \mu^{(2)} + \eta^1 K^{(2)} \mu^{(2)} + \beta^{(2)} \Sigma^{(2)}$$

$$d\mu^{(2)} = \eta^1 \mu^{(1)} + \delta^{(2)} \Sigma^{(2)} + \varepsilon^{(2)} \mu^{(2)}$$

where $\Sigma^{(2)}$ is the second derived system and $\mu^{(2)}$ is a complement to $\Sigma^{(2)}$ in $\Sigma^{(1)}$. As before, we may omit the equations for $d\mu^{(1)}$. The iteration of this process continues through the derived flag for the system Σ and leads to the following necessary and sufficient conditions for equivalence to a controllable linear system:

Theorem 2.4 *Let* rank $\mathcal{K}$ = rank K = p *and let* N *be the derived length of* Σ. *Necessary and sufficient conditions for equivalence to a controllable linear system are:*

a) $\mathcal{H}^{(i)} = \mathcal{J}^{(i)} = 0$ *for* $i = 1, \ldots, N$

b) $1 \leq$ rank $K^{(i)}$ *for* $i = 1, \ldots, N-1$

c) rank $K^{(i)}$ = rank $\mathcal{K}^{(i)}$ *for* $i = 1, \ldots, N-2$

d) *either* rank $K^{(N-1)}$ = rank $\mathcal{K}^{(N-1)}$ *and* rank $k^{(N)} = 1$

or rank $K^{(N-1)}$ = rank $\mathcal{K}^{(N-1)} - 1$.

Thus, given the ranks of $\mathcal{K}^{(i)}$ for $i = 1, \ldots, N$, the Brunowski normal forms are the *generic* systems with those ranks. Furthermore, after all reductions have been carried out, the structure equations for any system equivalent to a Brunowski normal form are the structure equations of the Lie pseudogroup of symmetries of that normal form.

Two remarks are in order here:

First, the complete integrability of the sequence of differential systems $\{\eta^1, \Sigma^{(i)}\}$ is of course the same as the vanishing of the invariants $\mathcal{H}^{(i)}$ and $\mathcal{J}^{(i)}$ however these invariants carry information which is important not just for the linearization problem but for the general feedback equivalence problem. In particular they determine the non-degeneracy conditions [16] which are necessary for the provision of closed loop time critical controls as solutions to the Euler Lagrange equations for the functional given by $\int \eta^1$ [10]. (In the case of linearizable systems the functional is totally degenerate.) They are also crucial to the identification of bilinear normal forms [18].

Second, after the reductions have been carried out, coordinates in which the system takes its normal form may be found in practice by inspection of the structure equations, in which the strings of single input systems which make up the Brunowski forms are readily apparent.

We illustrate with the case of systems with 6 states and 2 controls for which there are 3 Brunowski normal forms. These are given by

B(1): $\frac{dx^1}{dt} = x^2,\ \frac{dx^2}{dt} = x^3,\ \frac{dx^3}{dt} = x^4,\ \frac{dx^4}{dt} = x^5,\ \frac{dx^5}{dt} = u^1,\ \frac{dx^6}{dt} = u^2$

B(2): $\frac{dx^1}{dt} = x^2,\ \frac{dx^2}{dt} = x^3,\ \frac{dx^3}{dt} = x^4,\ \frac{dx^4}{dt} = u^1,\ \frac{dx^5}{dt} = x^6,\ \frac{dx^6}{dt} = u^2$

and

B(3): $\frac{dx^1}{dt} = x^2,\ \frac{dx^2}{dt} = x^3,\ \frac{dx^3}{dt} = u^1,\ \frac{dx^4}{dt} = x^5,\ \frac{dx^5}{dt} = x^6,\ \frac{dx^6}{dt} = u^2.$

These correspond to the following data:

B(1): The derived flag is $\Sigma = \{\eta^2, \eta^3, \eta^4, \eta^5, \eta^6\}$, $\Sigma^{(1)} = \{\eta^2, \eta^3, \eta^4\}$
$\Sigma^{(2)} = \{\eta^2, \eta^3\}$, $\Sigma^{(3)} = \{\eta^2\}$, $\Sigma^{(4)} = \{0\}$
$N = 4$, rank $K = 2$, rank $K^{(i)} = 1$ for $i = 1, 2, 3$, rank $k^{(4)} = 1$

B(2): The derived flag is $\Sigma = \{\eta^2, \eta^3, \eta^4, \eta^5, \eta^6\}$, $\Sigma^{(1)} = \{\eta^2, \eta^3, \eta^4\}$
$\Sigma^{(2)} = \{\eta^2\}$, $\Sigma^{(3)} = \{0\}$
$N = 3$, rank $K = 2$, rank $K^{(1)} = 2$, rank $K^{(2)} = 1$, rank $k^{(3)} = 1$

B(3): The derived flag is $\Sigma = \{\eta^2, \eta^3, \eta^4, \eta^5, \eta^6\}$, $\Sigma^{(1)} = \{\eta^2, \eta^3, \eta^4\}$
$\Sigma^{(2)} = \{\eta^2\}$, $\Sigma^{(3)} = \{0\}$
$N = 3$, rank $K = 2$, rank $K^{(1)} = 2$, rank $K^{(2)} = 1$, rank $k^{(2)} = 1$ and $k^{(3)}$ is identically zero.

We note that the systems **B(2)** and **B(3)** have the same derived flag and that the ranks of K and $K^{(i)}$ are the *type numbers* for the differential system Σ. Thus these numbers alone do not distinguish the Brunowski forms but must be supplemented by the rank of the matrix $\mathfrak{K}^{(i)}$ which yields the last reduction. These integers together with the vanishing of the invariants $\mathcal{H}$ and $\mathfrak{J}$, provide a complete characterization of the equivalence classes of the systems **B(1)**, **B(2)**, and **B(3)** under fully nonlinear feedback. The structure equations corresponding to these three cases are obtained as follows:

In each case we have rank $K = 2$ so the first normalization results in the structure equations

$$d\eta^1 = {}^t\mu h^{(1)} \mu^{(1)} + {}^t\mu^{(1)} j^{(1)} \mu^{(1)} + \eta^1 k^{(1)} \mu^{(1)} + {}^t\alpha^{(1)} \Sigma^{(1)} \tag{2.9}$$

$$d\Sigma^{(1)} = \mu H^{(1)} \mu^{(1)} + {}^t\mu^{(1)} J^{(1)} \mu^{(1)} + \eta^1 K^{(1)} \mu^{(1)} + \beta^{(1)} \Sigma^{(1)}$$

$$d\mu^{(1)} = \eta^1 \mu + \delta^{(1)} \Sigma^{(1)} + \varepsilon^{(1)} \mu^{(1)},$$

where $\Sigma^{(1)} = {}^t(\eta^2, \eta^3, \eta^4)$ and $\mu^{(1)} = {}^t(\eta^5, \eta^6)$. We assume that $\mathcal{H}^{(1)}$ and $\mathcal{J}^{(1)}$ are both zero. In this case, the generic branch is the one in which both rank $\mathcal{K}^{(1)}$ and rank $K^{(1)}$ are equal to 2 and this contains both **B(2)** and **B(3)**. The other branch is the one in which rank $\mathcal{K}^{(1)}$ = 1 and we consider it first. Here the only possibility for a controllable system is that $K^{(1)}$ also has rank 1 as otherwise the subsystem corresponding to $\Sigma^{(1)}$ is completely integrable. The reduction now proceeds by setting $\mathcal{K}^{(1)} = \begin{bmatrix} 0 & 0 \\ 1 & 0 \end{bmatrix}$ which reduces the group. The torsion term $\mathcal{K}^{(2)}$ is now a vector in the plane. The rank $K^{(1)} = 1$ case is the only controllable one in this branch and the reduction is obtained by normalizing $K^{(1)}$ to ${}^t(0\ 1)$. After two more iterations the derived flag has been exhausted and the matrix $\mathcal{K}^{(4)}$ reduces to $k^{(4)}$. The final reduction leads to the structure equations

$$\begin{aligned} d\eta^1 &= \eta^1\eta^2 \\ d\eta^2 &= \eta^1\eta^3 \\ d\eta^3 &= \eta^1\eta^4 + \eta^2\eta^3 \\ d\eta^4 &= \eta^1\eta^5 + 2\eta^2\eta^4 \\ d\eta^5 &= \eta^1\mu^1 + 2\eta^3\eta^4 + 3\eta^2\eta^5 \end{aligned} \tag{2.10}$$

and

$$d\eta^6 = \eta^1\mu^2 + \beta^2\eta^2 + \beta^3\eta^3 + \beta^4\eta^4 + \beta^5\eta^5 + \beta^6\eta^6 .$$

These are involutive and contain no non-constant invariants. Thus the structure equations for the system B(1) are the structure equations of a *transitive* Lie pseudogroup. This means that the Brunowski form B(1) is a *normal form* for any system whose symmetry group has these structure equations. The Cartan characters are $s_1 = 2$ and $s_2 = s_3 = s_4 = s_5 = s_6 = 1$ so there is $s_6 = 1$ arbitrary function of 6 variables. The group has the structure of the group on the line,

$$x^1 \mapsto \varphi(x^1) \tag{2.11a}$$

(whose prolongations are described by the structure equations for $d\eta^1$ through $d\eta^5$) coupled to the transformations given by

$$x^6 \mapsto \rho(x^1, x^2, x^3, x^4, x^5, x^6) \tag{2.11b}$$

It is easily verified that this is precisely the symmetry group of **B(1)**.

In order to obtain the usual form **B(1)** one *integrates* the structure equations (2.10) as follows. The structure equation for $d\eta^1$ yields coordinates x^1 and x^2 such that

$$\eta^1 = \frac{dx^1}{x^2}.$$

This determines η^2 up to a multiple of η^1 and we have a new coordinate x^3 such that

$$\eta^2 = \frac{dx^2}{x^2} - x^3\eta^1 .$$

This in turn determines η^3 up to a multiple of η^1 as

$$\eta^3 = dx^3 - x^3\eta^2 - x^4\eta^1$$

and so on to determine the coordinate x^4, and after two more iterations, x^5 and u^1.

Next we make use of the fact that the Lie pseudogroup given by (2.10) has a subgroup determined by setting $\beta^2 = \beta^3 = \beta^4 = \beta^5 = 0$. This means that, modulo η^2, η^3, η^4 and η^5, the structure equation for $d\eta^6$ is that of a *contact structure* on a three dimensional space and we obtain coordinates x^6 and u^2 in which

$$\eta^6 = \lambda(dx^6 - u^2\eta^1) .$$

The ambiguity in the choice of such coordinates is of course indicated by the fact that we are working modulo η^2, η^3, η^4 and η^5 and that η^6 is only given up to the multiple λ, i.e,. we have x^6 determined up to changes of the form (2.11) with the induced change in u^2.

Next we consider the generic branch of the problem in which we normalize to $\mathcal{K}^{(1)} = \begin{bmatrix} 1 & 0 \\ 0 & 1 \end{bmatrix}$ to obtain the structure equations

$$d\eta^1 = {}^t\mu^{(1)} h^{(2)} \mu^{(2)} + {}^t\mu^{(2)} j^{(2)} \mu^{(2)} + \eta^1 k^{(2)} \mu^{(2)} + {}^t\alpha^{(2)} \Sigma^{(2)}$$

$$d\Sigma^{(2)} = {}^t\mu^{(1)} H^{(2)} \mu^{(2)} + {}^t\mu^{(2)} J^{(2)} \mu^{(2)} + \eta^1 K^{(2)} \mu^{(2)} + \beta^{(2)} \Sigma^{(2)} \tag{2.12}$$

$$d\mu^{(2)} = \eta^1\mu^{(1)} + \delta^{(2)}\Sigma^{(2)} + \varepsilon^{(2)}\mu^{(2)}$$

with $\Sigma^{(2)} = (\eta^2)$, $\mu^{(1)} = (\eta^5, \eta^6)$ and $\mu^{(2)} = (\eta^3, \eta^4)$. It is a necessary condition for linearizability that $\mathcal{H}^{(2)}$ and $\mathcal{J}^{(2)}$ vanish. There are two branches here in which rank $\mathcal{K}^{(2)} = 1$ and 2 respectively. The generic cases in both branches correspond to Brunowski normal forms. The first branch contains the equivalence class of **B(2)** as is easily verified. This branch is controllable only if $K^{(2)}$ has rank 1 so we have the normalization $\mathcal{K}^{(2)} = \begin{bmatrix} 0 & 0 \\ 0 & 1 \end{bmatrix}$. This leads to the structure equations

$$d\eta^1 = {}^t\mu^{(2)}\, h^{(3)}\, \mu^{(3)} + \eta^1\, k^{(3)}\mu^{(3)}. \tag{2.13}$$

with $\mu^{(1)} = (\eta^5, \eta^6)$, $\mu^{(2)} = (\eta^3, \eta^4)$ and $\mu^{(3)} = (\eta^2)$. It is a necessary condition for linearizability that $h^{(3)}$ should vanish. The torsion term $k^{(3)}$ is now a single function and the controllability condition is satisfied if and only if $\mathcal{K}^{(3)}$ is non-zero. The final reductions yield the structure equations

$$d\eta^1 = \eta^1\eta^2 \tag{2.14}$$

$$d\eta^2 = \eta^1\eta^3$$

$$d\eta^3 = \eta^1\eta^5 + \eta^2\eta^3$$

$$d\eta^4 = \eta^1\eta^6 + \beta^2\eta^2 + \beta^3\eta^3 + \beta^4\eta^4$$

$$d\eta^5 = \eta^1\mu^1 + 2\eta^2\eta^5$$

and

$$d\eta^6 = \eta^1\mu^2 + \beta^5\eta^2 + \beta^6\eta^3 + \beta^7\eta^4 + \beta^3\eta^5 + \beta^4\eta^6.$$

These equations are involutive with Cartan characters $s_1 = 3$, $s_2 = s_3 = 2$, $s_4 = 1$ and there are no non-constant invariants. As in the case of **B(1)** this means that we may characterize the Brunowski normal form **B(2)** as the equivalence class of systems which have the Lie pseudogroup of symmetries given by (2.14). This group has the structure of the group on the line,

$$x^1 \mapsto \varphi(x^1) \tag{2.15a}$$

(which is described by the structure equations for $d\eta^1$, $d\eta^2$, $d\eta^3$ and $d\eta^5$) coupled to the transformations given by

$$x^4 \mapsto \rho(x^1, x^2, x^3, x^4) \tag{2.15b}$$

(described by the equations for $d\eta^4$ and $d\eta^6$). One may easily check directly that this is the symmetry group of **B(2)**.

To obtain the usual form **B(2)** one integrates the structure equations (2.14) as above. The structure equation for $d\eta^1$ yields coordinates x^1 and x^2 such that

$$\eta^1 = \frac{dx^1}{x^2}.$$

This determines η^2 up to a multiple of η^1 and we have a new coordinate x^3 such that

$$\eta^2 = \frac{dx^2}{x^2} - x^3\eta^1 .$$

This determines η^3 up to a multiple of η^1 as

$$\eta^3 = dx^3 - x^3\eta^2 - x^5\eta^1$$

and so on to determine the coordinates x^5, and after one more iteration u^1.

The symmetry group described by (2.14) has a subgroup whose structure equations are obtained by setting $\beta^2 = \beta^3 = 0$. If this is done the equation for η^4 becomes the structure equation of a contact structure and we obtain coordinates x^4 and u^2 in which

$$\eta^4 = \lambda(dx^4 - u^2\eta^1) .$$

The ambiguity in the choice of such coordinates is of course indicated by the fact that we are working modulo η^2 and η^3 and that is only given up to the multiple λ, i.e., we have x^4 determined up to changes of the form (2.15) with the induced change in u^2.

The final branch of the problem to be considered is the one in which rank $\mathcal{K}^{(2)} = 2$ and the second normalization is given by $\mathcal{K}^{(2)} = \begin{bmatrix} 1 & 0 \\ 0 & 1 \end{bmatrix}$. In this case the structure equations become

$$d\eta^1 = \eta^1\eta^3 + \beta^1\eta^2$$

$$d\eta^2 = \eta^1\eta^4 + \beta^2\eta^2$$

$$d\eta^3 = \eta^1\eta^5 + \beta^3\eta^2 + \beta^1\eta^4 \tag{2.16}$$

$$d\eta^4 = \eta^1\eta^6 + \eta^3\eta^4 + \beta^4\eta^2 + \beta^2\eta^4$$

$$d\eta^5 = \eta^1\mu^1 + \eta^3\eta^5 + \beta^5\eta^2 + \beta^6\eta^4 + \beta^1\eta^6$$

and

$$d\eta^6 = \eta^1\mu^2 + 2\eta^3\eta^6 + \beta^7\eta^2 + \beta^8\eta^4 + \beta^2\eta^6 .$$

This is just the partial prolongation (neglecting the equations for $d\beta^i$) of the structure equations of the group in the plane:

$$(x^1, x^2) \mapsto (\varphi(x^1, x^2), \rho(x^1, x^2)) . \tag{2.17a}$$

In this case the subgroup on the line

$$x^1 \mapsto \varphi(x^1) \tag{2.17b}$$

is described by $\beta^1 = 0$ and this allows us to obtain coordinates x^1 and x^2, as before, in which

$$\eta^1 = \frac{dx^1}{x^2}.$$

From this point the integration proceeds exactly as above.

References

[1] R. Brockett, Feedback invariants for nonlinear systems, *IFAC Congress Helsinki*, **2** (1978) 1115-1120.

[2] P. Brunowski, A classification of linear controllable systems, *Kibernetica cislo 3,* **6** (1970) 173-187.

[3] R. Bryant, On notions of equivalence for variational problems with one independent variable, in *Differential Geometry: The Interface Between Pure and Applied Mathematics*, M. Lucksic, C. Martin and W. Shadwick (Eds.) AMS Providence (1987) 65-76.

[4] R. Bryant and P. Griffiths, Reduction for constrained variational problems and $\int \frac{\kappa^2}{2}\, ds$, *American J. Math.* **108** (1986) 525-570.

[5] E. Cartan, Les sous-groupes des groupes continus de transformations, *Ann. Ec. Normale* **25**

(1908) 719-856.

[6] R.B. Gardner, Differential geometric techniques interfacing control theory, in *Differential Geometric Control Theory*, R. Brockett, R.Millman and H. Sussmann (Eds.) Birkhäuser, Boston (1983) 117-180.

[7] R.B. Gardner, W.F. Shadwick, Feedback equivalence of control systems, *Systems and Control Lett.* **8** (1987) 463-465.

[8] R.B. Gardner, W.F. Shadwick, Overdetermined equivalence problems with an application to feedback equivalence, in *Differential Geometry: The Interface Between Pure and Applied Mathematics*, M. Lucksic, C. Martin and W. Shadwick (Eds.) AMS, Providence (1987) 111-119.

[9] R.B. Gardner, W.F. Shadwick, Feedback equivalence of two-state systems with scalar control, (1988) (preprint).

[10] R.B. Gardner, W.F. Shadwick, G.R. Wilkens, A geometric isomorphism with applications to closed loop controls (1988) (preprint to appear in *SIAM J. Control and Opt.*).

[11] P. Griffiths, *Exterior differential systems and the calculus of variations*, Birkhäuser, Boston (1983).

[12] R. Hermann, The theory of equivalence of Pfaffian systems and input systems under feedback, *Math. Systems Theory* **15** (1982) 343-356.

[13] L.R. Hunt and R. Su, Linear equivalents of nonlinear time-varying systems, in *Proc. International Symposium on the Mathematical Theory of Networks and Systems*, Santa Monica, CA (1981) 119-123.

[14] B. Jakubczyk, W. Respondek, On linearization of control systems, *Bull. Acad. Polon. Sci.* **28** (1980) 517-522

[15] A. Krener, On the equivalence of control systems and the linearization of nonlinear systems, *SIAM J. Control & Opt.* **11** (1973) 670-676.

[16] W.F. Shadwick, The notion of strong nondegeneracy for variational problems with one independent variable (1988) (preprint).

[17] A. van der Schaft, Linearization and input-output decoupling for general nonlinear systems, *Systems and Control Lett.* **5** (1984) 27-33.

[18] G.R. Wilkens, Local feedback equivalence of control systems with 3 state and 2 control variables, University of North Carolina, Ph.D. Thesis (1987).

R.B. Gardner
Department of Mathematics
Phillips Hall 039A
University of North Carolina
Chapel Hill, NC 27514

W.F. Shadwick
Department of Pure Mathematics
University of Waterloo
Waterloo, Ontario N2L 3G1
CANADA

G.R. Wilkens
Department of Mathematics
University of Hawaii
Honolulu, Hawaii 96822

Contemporary Mathematics
Volume **97**, 1989

The Application of Total Energy as a Lyapunov Function for Mechanical Control Systems *

Daniel E. Koditschek †

February 14, 1989

Abstract

Examination of total energy shows that the global limit behavior of a dissipative mechanical system is essentially equivalent to that of its constituent gradient vector field. The class of "navigation functions" is introduced and shown to result in "almost global" asymptotic stability for closed loop mechanical control systems upon which a navigation function has been imposed as an artificial potential energy. Two examples from the engineering literature — satellite attitude tracking and robot obstacle avoidance — are provided to demonstrate the utility of these observations.

1 Introduction

It has been known for at least a century that the decrease in total energy of a dissipative mechanical system implies the local convergence of generalized position toward a minimum of the potential function. Lagrange demonstrated the stability of motion around the equilibrium state of a conservative system using total energy in 1788 [26]. Asymptotic stability of the potential field minima resulting from the introduction of dissipative forces to a conservative system was discussed by Lord Kelvin in 1886 [44]. These ideas were generalized by Lyapunov in his 1892 doctoral dissertation [28]. The use of total energy for control applications has been rediscovered many times in the engineering community since Lyapunov's ideas were first introduced by Kalman and Bertram [14]. For example, a two decade old paper by Pringle [36] paper concerns the applications of total energy to satellite control. Credit for first noting utility of artificial potential energy in robotics applications a decade ago would appear to be due Khatib [15]. Arimoto and colleagues [32, 42] contributed a precise demonstration that every minimum of the potential field is a local attractor of a dissipative mechanical system by application of LaSalle's Invariance Principle. Similar independent work of Van der Schaft [38] and this author [21] appeared subsequently. Of course, the familiar idea of energy dissipation and its consequences for local stability behavior is to be found in physics texts as well [1, 8].

In contrast, the extent to which global conclusions about the phase portrait of a mechanical system may be drawn from analysis of the total energy function appears not to have been addressed in the previous literature. Yet every mechanical system includes a lifted gradient vector field arising from its potential energy. Since the global limit set of a nondegenerate gradient vector field is trivial

* **Keywords:** Lyapunov functions, mechanical systems, gradient systems, Hamiltonian systems, robotics.
†This work was supported in part by the National Science Foundation under grant no. DMC-8505160

(consists of a finite number of isolated points) it seems reasonable to inquire whether such simple limit behavior lifts as well. This paper answers that question affirmatively and demonstrates the utility of the observation in the design of controllers for mechanical systems resulting in "almost global" asymptotically stable closed loops.

The mechanical systems form a large and important class of highly nonlinear finite dimensional dynamical plants which include, for example, all rigid link robots and satellites. Generally, the construction of feedback controllers for nonlinear systems which ensure that global convergence properties hold for the closed loop system is a hopeless task. Indeed, of the large and ever growing number of investigators who have exploited total energy as a Lyapunov function either for purposes of robotic manipulation [32, 12, 16, 35] or for satellite attitude control [36, 29, 13], there seem to be none who systematically have considered the global aspects of the problem. Here, via elementary arguments from the theory of dynamical systems, we show that the limit properties of a dissipative mechanical system on the phase space reduce to the trivial case of gradient dynamics on the configuration space. Both, of course, are constrained to respect the configuration space topology, and the selection of useful classes of gradient systems requires some care. In the end, ensuring the "strongest" global convergence behavior allowed by the configuration space topology from a reasonably large set of initial phase space states is guaranteed by the construction of scalar cost functions endowed with certain "navigation properties."

The paper is organized as follows. The mechanical control system, Σ, is defined in Section 2 with some care paid to its attendant global geometric features. The dissipative mechanical system, Δ, is shown to result from a state feedback law determined by a gradient system, Γ, along with a suitable dissipative vector field. The section ends with a recitation of Lord Kelvin's century old result (Theorem 1). Attention shifts to the global limit properties of the dissipative mechanical system, Δ, in Section 3. The "navigation function" is introduced as a central tool of feedback controller design yielding a "congruence" between the limit sets of Δ and Γ (Theorem 2). The section ends with the summary of our earlier result showing that smooth navigation functions always exist on smooth manifolds with boundary (Theorem 3). Application of these results to satellite attitude tracking, summarized by Theorem 4, and to robot obstacle avoidance, summarized by Theorem 5 and Theorem 6, is the concern of Section 4.

2 Dissipative Mechanical Systems

The geometry of classical physics has been extensively studied for decades, and recent years have witnessed the publication of numerous expository texts containing the background material required for the present paper. This section reviews the relevant ideas, following very closely the presentation in the excellent text of Abraham and Marsden [1].

The setting for all the results presented in this paper is a *configuration space*, $\mathcal{J}$, a compact Riemannian manifold with boundary. Mechanical systems may be approached from the point of view of symplectic geometry on the cotangent bundle, $\sigma : T^*\mathcal{J} \to \mathcal{J}$, or Riemannian geometry on the tangent bundle, $\tau : T\mathcal{J} \to \mathcal{J}$: it is the second point of view which will be most useful in this paper. A *Lagrangian dynamical system* is defined by the extrema of an integral cost functional applied to motion on the tangent bundle, $T\mathcal{J}$. Section 2.1 particularizes this idea to the instance of a *mechanical system* when the cost functional is defined in terms of a Riemannian metric, and there obtains the notion of kinetic energy. Adding a pointwise cost functional to the original configuration space introduces *potential and total energy,* discussed in Section 2.2. The *dissipative vector fields* are introduced in Section 2.3. Finally, the chief object of present study, the *dissipative mechanical system,* is defined in terms of all these ingredients.

2.1 Mechanical Systems

Given a Lagrangian function, $\lambda : T\mathcal{J} \to \mathbb{R}$, the *Variational Principle of Hamilton* [1], states that of all the curves, b, between two points, $q_1, q_2 \in \mathcal{J}$, the one whose tangent curve, $c \triangleq Tb$, minimizes

$$\int \lambda \circ c,$$

also satisfies the classical Euler-Lagrange coordinate equations,

$$\frac{d}{dt} D_{\dot{q}}\lambda - D_q\lambda = 0. \tag{1}$$

These coordinate equations define a *Lagrangian vector field* on the tangent bundle,

$$f_\lambda : T\mathcal{J} \to TT\mathcal{J},$$

which constitutes a *second order equation* on $\mathcal{J}$ in the sense that $T\tau\, f = 1_{T\mathcal{J}}$ [1].

2.1.1 Kinetic Energy as a Riemannian Metric

A particular Riemannian metric results from the choice of a *morphism* [10, Ch. 4.1], $M \in \mathcal{M}[T\mathcal{J}, T^*\mathcal{J}]$, from the tangent bundle to the co-tangent bundle which is *symmetric positive definite* (i.e. $(Mw, v) = (Mv, w)$ and $(Mv, v) > 0$ for all $v, w \in T\mathcal{J}$ assuming that $\tau(v) = \tau(w)$). We now suppose that a distinguished symmetric positive definite morphism, M, has been chosen and denote the resulting inner product

$$\langle\, v \mid w \,\rangle \triangleq (Mv, w).$$

This metric is understood to arise from the *kinetic energy* of the mechanical system, [1]

$$\kappa : T\mathcal{J} \to \mathbb{R}; \qquad v \mapsto \frac{1}{2}\langle\, v \mid v \,\rangle,$$

defined by the dynamical parameters of rigid bodies as determined by a kinematic relationship. For example, the author has presented a quick derivation of the Riemannian metric resulting from the sort of physical system which motivates the paper — the "kinematic chains" encountered in robotics — in a recent encyclopedia article [23]. Examples will be given in Section 4.

2.1.2 The Mechanical Control System

Define a *mechanical system* to be the Lagrangian dynamical system resulting from a cost functional $\lambda \triangleq \kappa$ specified by a kinetic energy — an analytic section of the tangent to the tangent bundle, $f_\kappa : T\mathcal{J} \to TT\mathcal{J}$. We now present a formal model expressing the manner in which control inputs may be applied to this system.

Suppose that the physical system is "actuated" by external forces which are under our control. In coordinates, the generalized forces, u, affect Lagrangian dynamics according to the relation

$$\frac{d}{dt} D_{\dot{q}}\lambda - D_q\lambda = u. \tag{2}$$

[1]Kinetic Energy is formally defined as a scalar valued map on $T^*\mathcal{J}$ [1], however it will simplify the discussion and do no technical harm in the present paper to speak of κ as a map on $T\mathcal{J}$.

The generalized forces may be modeled globally as points in the cotangent bundle, $\sigma : T^*\mathcal{J} \to \mathcal{J}$. However there is a formal problem in taking the space of combined inputs-states to be the simple cross product, $T^*\mathcal{J} \times T\mathcal{J}$, since a force, $u = (q_1, f) \in T^*_{q_1}\mathcal{J}$, cannot be applied at a phase, $v = (q_2, x) \in T_{q_2}\mathcal{J}$, unless both are vectors over the same configuration — $\sigma(u) = q_1 = \tau(v) = qs_2 \in \mathcal{J}$. [2] To resolve this problem it seems most natural to adopt the point of view introduced by Brockett [3] and model the input-state space as the *total space* , $\mathcal{T} \triangleq \sigma^*T\mathcal{J}$, obtained by pulling back $T\mathcal{J}$ over σ, according to the following commutative diagram,

$$\begin{array}{ccc} \sigma^*T\mathcal{J} & \xrightarrow{\psi} & T\mathcal{J} \\ \sigma^*\tau \downarrow & & \tau \downarrow \\ T^*\mathcal{J} & \xrightarrow{\sigma} & \mathcal{J} \end{array}$$

resulting in a vector bundle over the state space via the *natural vector bundle map* [10, Ch. 4.2] $\psi : \mathcal{T} \to T\mathcal{J}$. This simply means that $e \in \mathcal{T}$ is a valid input-state point if and only if its base, $u = \sigma^*\tau e$, and vector component, $v = \psi(e)$, are, in turn, vectors over the same configuration, $q = \sigma(u) = \tau(v)$.

Now, following Brockett, we are in a position to define the mechanical control system as a smooth section of the pullback of $TT\mathcal{J}$ through ψ,

$$\begin{array}{ccc} \psi^*TT\mathcal{J} & & TT\mathcal{J} \\ \psi^*\tau \downarrow & & \tau \downarrow \\ \mathcal{T} & \xrightarrow{\psi} & T\mathcal{J} \end{array}$$

as follows. Equation (2) implies that the generalized forces, $u \in \mathcal{U}$, affect the Lagrangian vector field on $TT\mathcal{J}$ through *vertical lift* , $V_v : T_{\tau v}\mathcal{J} \to T_v(T\mathcal{J})$ [1, Def. 3.7.5], of the vector field associated with $u \in T^*\mathcal{J}$ by the Riemannian metric, $M^{-1}u$. Thus, a *mechanical control system,* is the section $f_\Sigma : \mathcal{T} \to \psi^*TT\mathcal{J}$ determined by the ordered pair, consisting of a configuration space and a choice of kinetic energy,

$$\Sigma \triangleq (\mathcal{J}, \kappa), \tag{3}$$

defined by

$$f_\Sigma(v_u) \triangleq f_\kappa(v) - V_v \circ M_{\tau v}^{-1} u. \tag{4}$$

Assuming, as we do throughout this paper, that M is analytic, and that $\mathcal{J}$ is an analytic manifold, it follows that a mechanical control system (4) falls within the class of linear analytic control systems.

2.2 Gradient Vector Fields in Mechanical Systems

Consider the class of twice differentiable real valued functions $\varphi \in C^2[\mathcal{J}, \mathbb{R}]$ on a compact Riemannian manifold, $\mathcal{J}$. The *co-vector field,* $d\varphi$, is related to the *gradient vector field* , *grad* φ , of φ via the Riemannian metric inverse morphism,

$$grad\ \varphi \triangleq M^{-1}d\varphi.$$

Thus, a gradient system is determined by the triple consisting of a configuration space, a Riemannian metric, and a scalar valued function,

$$\Gamma \triangleq (\mathcal{J}, M, \varphi), \tag{5}$$

yielding the vector field $grad\ \varphi : \mathcal{J} \to T\mathcal{J}$.

[2] This problem may be resolved, of course, if $\mathcal{J}$ is parallelizable: the product notion will go through after the extra step of identifying $T\mathcal{J} \approx \mathcal{J} \times \mathbb{R}^n$ and $T^*\mathcal{J} \approx \mathcal{J} \times \mathbb{R}^n$ with the penalty of an extra coordinate transformation in all future computation.

2.2.1 Limit Behavior of Gradient Systems

One calls φ a *Morse function* if its hessian (matrix of second derivatives) is non-singular at every critical point [10]. Gradient vector fields resulting from Morse functions give rise to flows with simple limit behavior which may be summarized as follows.

Proposition 2.1 *Let φ be a twice continuously differentiable Morse function on a compact Riemannian manifold, $\mathcal{J}$. Suppose that* grad φ *is transverse and directed away from the interior of $\mathcal{J}$ on any boundary of that set. Then the negative gradient flow has the following properties:*

1. *$\mathcal{J}$ is a positive invariant set;*
2. *the positive limit set of $\mathcal{J}$ consists of the critical points of φ;*
3. *there is a dense open set $\tilde{\mathcal{J}} \subset \mathcal{J}$ whose limit set consists of the local minima of φ.*

Proof: Since the vector field is directed toward the interior of $\mathcal{J}$ on its boundaries by hypothesis, it follows that this set is positive invariant. The limit set for any trajectory of a gradient system is an equilibrium state [11], hence, in this case, a minimum, maximum, or saddle of φ in the interior of $\mathcal{J}$. It remains to demonstrate the third property.

Since φ is Morse, it has only isolated critical points and the number of local maxima and saddles is countable. Each of these has a stable manifold of lower dimension than $\mathcal{J}$, whose closure in $\mathcal{J}$ is consequently nowhere dense. But the domain of attraction of any maximum or saddle is contained in its stable manifold. It follows that the domain of attraction of the minima of φ, $\tilde{\mathcal{J}}$, is the complement of the countable union of nowhere dense sets in $\mathcal{J}$.

□

It is worth commenting on two technical assumptions in the hypothesis of this proposition. The first, the assumption of compactness, is a relatively harmless measure taken to guarantee that $\mathcal{J}$ is positive invariant with respect to the negative gradient flow. For consider the scalar function, $\varphi(q) \triangleq q^3$, defined on $\mathbb{R}$, yielding the the negative gradient dynamics,

$$\dot{q} = -3q^2.$$

Since this system has finite escape trajectories, it is not true that $\mathbb{R}$ is positive invariant under the flow. However, there is no possibility of finite escape from a compact set without boundary (and the tranversality assumption precludes finite escape otherwise) as the proposition requires. In the case that the configuration space is homeomorphic to Euclidean n-space, a "radial unboundedness" condition on φ [9] will yield equivalent results. However, for most mechanical systems of interest the configuration space (but, of course, not the phase space) is compact.

The more onerous requirement, that φ be a Morse function, guarantees that the limit set of every motion is a single critical point. For, absent such a restriction, connected limit sets are known to be possible [34][Example 1.1.3]. Yet isolated critical points might still be degenerate: a degenerate saddle of φ might well include an open set in its domain of attraction. Indeed, the previous example demonstrates the existence of a dynamical system whose unique isolated critical point is unstable, yet which has a domain of attraction which is open in the configuration space, $\mathbb{R}$. This possibility would invalidate the argument in the previous proof that any open set has a limit set comprised of minima. The Morse property guarantees that no co-dimension one set of saddle connections can separate $\mathcal{J}$. While this condition incurs an undesirable loss of generality [3] the technical problems which result in its relaxation require more attention than worthwhile in this paper.

2.2.2 Lifting Gradient Vector Fields

Any smooth real valued function, $\varphi : \mathcal{J} \to \mathbb{R}$ may be "pulled back" to $T\mathcal{J}$ in a natural fashion by defining $\tilde{\varphi} \triangleq \varphi \circ \tau$. This defines a new Lagrangian function,

$$\lambda \triangleq \kappa - \tilde{\varphi}, \tag{6}$$

whose Lagrangian vector field, $f_{\kappa-\tilde{\varphi}}$, can be shown to include a "lift" of the gradient vector field of φ

$$f_{\kappa-\tilde{\varphi}}(x) = f_{\kappa}(x) - V_x\,(grad\,\varphi\,) \circ \tau(x), \tag{7}$$

where V is the vertical lift. It is in this sense that we are justified in claiming that mechanical systems provide a "natural analog computer" for gradient vector fields. In the absence of dissipative forces, this "integration" occurs, through the conservation of a *total energy* function,

$$\eta \triangleq \kappa + \tilde{\varphi}, \tag{8}$$

in which sense φ may be regarded as a *potential energy* function. [4]

Proposition 2.2 ([1]) *The total energy, η is invariant under the flow of the lagrangian vector field, that is,*

$$d\eta f_{\kappa-\tilde{\varphi}} \equiv 0.$$

2.3 Dissipative Mechanical Systems

We now introduce the chief object of study in this paper, the dissipative mechanical system,

$$\Delta \triangleq (\mathcal{J}, \eta, f_d), \tag{9}$$

defined by the vector field, $f_\Delta : T\mathcal{J} \to TT\mathcal{J}$,

$$f_\Delta \triangleq f_{\kappa-\tilde{\varphi}} + f_d, \tag{10}$$

where $f_{\kappa-\tilde{\varphi}}$ is the Lagrangian vector field introduced in Section 2.2.2 and f_d is a dissipative vector field to be defined below. The aim here is to show how Δ arises from the closed loop of a certain class of feedback controllers for Σ.

---

[3] Redundant degrees of freedom in robotics will result in non-immersive output (kinematic) maps. Thus, the pullback of a nondegenerate "task specification" vector field in the output space, $\mathbb{R}^3 \times SO(3)$ will necessarily have degenerate equilibrium states in the configuration space.

[4] It is interesting to note that the base integral curves of $f_{\kappa-\tilde{\varphi}}$ are the geodesics of the *Jacobi metric* , $M_\varphi \triangleq (\varphi_0 - \varphi)M$, where M is the original metric, and φ_0 is any real value outside the range of φ (possibly restricted to a suitably defined compact positive invariant submanifold of $\mathcal{J}$) [1][Thm. 3.7.7]. However, this point of view does not seem to lead to any new understanding in the present situation.

2.3.1 Dissipative Vector Fields and Feedback

The last ingredient we require is the notion of a *dissipative* vector field: a vector field on the phase space, $f_d : T\mathcal{J} \to TT\mathcal{J}$, which is "vertical" (i.e. $T\tau\, f_d = 0$), and acts negatively on kinetic energy, $d\kappa f_d \leq 0$ [1]. Since f_d is vertical it follows that there exists a morphism, $G_d \in \mathcal{M}^\infty[T\mathcal{J}, T^*\mathcal{J}]$, such that

$$f_d(v) = V_v \circ M_{\tau v}^{-1} \circ G_d(v).$$

Thus, considering the following analytic section of the total space $\psi : \mathcal{T} \to T\mathcal{J}$,

$$g(v) \triangleq G_d(v) + d\varphi \circ \tau(v), \tag{11}$$

as a feedback law relating states to inputs, it is clear that the dissipative mechanical system on the state space, f_Δ, is the closed loop system resulting from the composition

$$f_\Delta = f_\Sigma \circ g.$$

In coordinates, $p : T\mathcal{J} \to \mathcal{O}$, for some convex open subset of $\mathbb{R}^n$, Δ gives rise to the differential equation

$$\begin{aligned} \dot{p}_1 &= p_2 \\ \dot{p}_2 &= -\underline{M}^{-1}(p_1)\left([C(p_1,p_2) + \underline{K}_2(p_1,p_2)]p_2 + \underline{K}_1(p_1)p_1\right), \end{aligned} \tag{12}$$

where C denotes the coriolis forces, $\underline{K}_2$ the damping law, and $\underline{K}_1$ the spring law.

2.3.2 Local Limit Behavior

In the presence of dissipative forces, integration of dissipative mechanical system, Δ, produces local limit behavior in the phase space, $T\mathcal{J}$ which is analogous to that of the original gradient system, Γ, in the configuration space, $\mathcal{J}$. This notion is made precise by the following familiar result.

Theorem 1 (Lord Kelvin (1886) [44, §345]) *If q_0 is a local minimum of φ in $\mathcal{J}$, then $(q_0, 0)$ is a stable equilibrium state of the dissipative mechanical system, Δ (9), in $T\mathcal{J}$. In particular, η (8) is a Lyapunov function for f_Δ, with with Lie derivative*

$$L_{f_\Delta}(\eta) = d\kappa f_d.$$

Since η is not a strict Lyapunov function — that is, $d\kappa f_d$ must vanish on the entire zero section of $T\mathcal{J}$ and the Lie derivative, $L_f(\eta)$ is negative semi-definite — rigorous conclusions about asymptotic properties of the flow near $(q_0, 0)$ require the application of LaSalle's Invariance Principle [27] as shown, for example, by Arimoto and colleagues [42].

3 Global Limit Properties of Dissipative Mechanical Systems

The global phase portrait of a regular gradient system, Γ (5), is as simple as one could imagine arising from a nonlinear differential equation. The feedback synthesis procedure introduced (11) for the class of mechanical control systems, Σ (3), results in a closed loop dissipative mechanical system, Δ (9), containing a "lift" of a constituent gradient vector field, *grad* φ . This Section will demonstrate that an appropriately chosen potential function, φ, called a "navigation function," induces limit behavior of Δ on a useful subset of the phase space, $T\mathcal{J}$, which may be identified with that of Γ on the configuration space, $\mathcal{J}$. Thus, at least with regard to steady state behavior, our synthesis procedure reduces the control designer's task from the consideration of complex hamiltonian dynamics to the manipulation of comparatively trivial gradient dynamics.

This Section is organized roughly congruently to Proposition 2.1 which we take as our model of desired steady state behavior. We first show that f_Δ fails to be transverse on the boundary of phase space, and find a reasonable substitute for this condition in Section 3.1. We next show in Section 3.2 that the global limit set of Δ is a lift of that of Γ into the phase space. This yields a global version of Lord Kelvin's result in the form of Proposition 3.6. We finally introduce the notion of a navigation function in Section 3.3. This restores to Δ a positive invariant set which includes all of the (lifted) configuration space and yields convergence properties which are as strong as the topology of $\mathcal{J}$ allows.

3.1 Positive Invariance

To begin with, we must preclude the possibility of finite escape: the transversality of $grad\ \varphi$ on $\partial\mathcal{J}$ is no longer enough. Intuitively, it is helpful to think of a marble rolling along a hilly terrain in the earth's gravitational field through a viscous atmosphere. Tranversality simply implies that the terrain slopes away from any forbidden region (the boundary of configuration space): a marble traveling with sufficient kinetic energy could, of course, roll uphill and crash through the boundary. We first give an example, and then establish a reasonably large subset of the phase space which is positive invariant under the flow induced by Δ.

3.1.1 Rolling Uphill

As an example, take $\mathcal{J}$ to be the closed real interval, $[-\epsilon, 1]$, with the Euclidean metric as kinetic energy,$\kappa \triangleq \frac{1}{2}\dot{q}^2$, and define the potential energy to be $\varphi \triangleq \frac{1}{2}q^2$. The gradient vector field, $grad\varphi = q$, is transverse and exterior directed on $\partial\mathcal{J} = \{-\epsilon, 1\}$ as required by Proposition 2.1 as long as $\epsilon > 0$. Now, using the coordinate system,

$$p = [p_1, p_2]^{\mathrm{T}} \in T\mathcal{J} = [-\epsilon, 1] \times \mathrm{I\!R},$$

choose a damping law $G(p) \triangleq \delta p_2$ corresponding to a valid dissipative vector field, $f_d(v) = [0, -\delta v_2]^{\mathrm{T}}$, as long as $\delta > 0$.

The associated dissipative mechanical system, Δ, is expressed in coordinates as as

$$\dot{p} = f_\Delta(p) = \begin{bmatrix} 0 & 1 \\ -1 & -\delta \end{bmatrix} p.$$

Now the boundary of the phase space is the union of the two parallel lines

$$\partial(T\mathcal{J}) = T_{\partial\mathcal{J}}\mathcal{J} = (\{-\epsilon\} \times \mathrm{I\!R}) \cup (\{1\} \times \mathrm{I\!R}),$$

on which f_Δ fails to be transverse at the points $[-\epsilon, 0], [1, 0]$. In fact, f_Δ is directed *away* from the interior of the phase space — an unfortunate circumstance — on the upper half of the line through $[1, 0]$ and the lower half of the line through $[-\epsilon, 0]$. Consequently, it may be observed that the trajectory of f_Δ through every initial condition in a neighborhood of these open half line segments must escape from $T\mathcal{J}$ in finite time. Thus, the transversality of the gradient field on $\partial\mathcal{J}$ fails to guarantee that a dissipative mechanical system gives rise to a complete dynamical system on $T\mathcal{J}$. This example holds true in general, for exactly the same simple reason, as the following result demonstrates.

Lemma 3.1 *If $\partial\mathcal{J} \neq \emptyset$ then $T\mathcal{J}$ contains an open set of points whose trajectories under Δ exhibit finite escape.*

Proof: It will suffice to show that there is a regular point of f_Δ in $\partial(T\mathcal{J})$ whose forward trajectory leaves $T\mathcal{J}$ in finite time — the open set is then contained within the local flow box [11, Ch. 11.2] around that point.

No point away from the zero section of $T\mathcal{J}$ can be an equilibrium state of f_Δ, so consider the curve $c_+(t)$, the forward trajectory of f_Δ through $v \in \partial(T\mathcal{J})$. We may assume with no loss of generality that v is not tangent to $\partial\mathcal{J}$ — that is $v \in T_{\tau v}\mathcal{J} - T_{\tau v}\partial\mathcal{J}$. Its base curve, $b_+(t) \triangleq \tau c(t)$, has the property that $b_+(0) \in \partial\mathcal{J}$. We have $\dot{b}_+(0) = T\tau\, f_\Delta(v) = v$ since f_Δ is second order as shown in Section 2. On the other hand, if $b_-(t)$ is the base curve of the forward trajectory through $-v$ then $\dot{b}_-(0) = -v$ in exactly the same way. Thus, either the curve b_- or b_+ leaves $\mathcal{J}$ after $t = 0$ from which it follows that either c_- or c_+ leaves $T\mathcal{J}$ after $t = 0$.

□

3.1.2 Invariance of the Lowest Boundary Energy Set

It is clear that f_Δ makes the boundary of $T\mathcal{J}$ "dangerous" away from the zero section. We now provide a simple means of characterizing positive invariant subsets of the phase space which intersect its boundary at most at points in the zero section.

Proposition 3.2 *Let $\varphi \in C^2[\mathcal{J}, \mathbb{R}]$ and suppose that*

$$b_1 \triangleq \inf_{q\in\partial\mathcal{J}} \varphi(q),$$

is a regular value of φ (and is understood to be $+\infty$ if $\partial\mathcal{J} = \emptyset$) and that

$$b_1 > b_0 \triangleq \inf_{q\in\mathcal{J}} \varphi(q).$$

Then for all constants, $b \in (b_0, b_1]$, the set of "bounded total energy" states of η (8),

$$\mathcal{E}^b \triangleq \{v \in T\mathcal{J} : \eta(v) \leq b\},$$

is a positive invariant set of the dissipative mechanical system (9), with non-empty interior.

Proof: Choosing $b > b_0$ guarantees that $\mathcal{E}^b$ has a non-empty interior. For $\eta^{-1}(b_0, b) \subseteq \mathcal{E}^b$ is both non-empty (it contains the set identified with $\emptyset \neq \varphi^{-1}(b_0, b)$ in the zero section) and open (η is continuous).

To demonstrate the positive invariant property, it suffices to show that $\partial\mathcal{E}^b \subseteq \eta^{-1}[c]$, for $d\eta f_\Delta \leq 0$ according to Theorem 1. To see that the boundary [5] is contained in the level set,

[5] Strictly speaking, the boundary of $\mathcal{E}^b$, considered as a point set, has no intersection with the boundary of the manifold in which it lies, $T\mathcal{J}$. To avoid the technicalities involved in distinguishing the two different notions of boundary, we assume that the original space $T\mathcal{J}$ is collared [10], thus both $T\mathcal{J}$ and $\mathcal{E}^b$ are closed subsets within the open collared set, $\widetilde{T\mathcal{J}}$ and the boundary operations are taken in the sense of point set topology so that the two subsets do indeed share a common boundary component as described.

as claimed, examine its partition,

$$\partial \mathcal{E}^b = \left(\partial (T\mathcal{J}) \cap \partial \mathcal{E}^b\right) \cup \left(\overset{\circ}{T\mathcal{J}} \cap \partial \mathcal{E}^b\right).$$

The first component is equivalent to $\partial (T\mathcal{J}) \cap \mathcal{E}^b$ which, in turn, is equivalent to $T_{\partial \mathcal{J}}\mathcal{J} \cap \mathcal{E}^b$. Now this set is entirely contained in the subset of the zero section of $T\mathcal{J}$ identified with $\varphi^{-1}[b] \cap \partial \mathcal{J}$. For $v \in T_{\partial \mathcal{J}}\mathcal{J}$ implies $\varphi \circ \tau(v) \geq b_1 \geq b$ since b_1 is the infimal value of φ on $\mathcal{J}$ by hypothesis. But $\eta(v) > \varphi \circ \tau(v)$ unless $v \in \kappa^{-1}[0]$, in which case $\eta(v) = \varphi \circ \tau(v)$ and we have $\tau(v) \in \varphi^{-1}[b]$. This demonstrates that the first component is equivalent to $\partial T\mathcal{J} \cap \eta^{-1}[b]$ (and is empty when $b < b_1$).

It remains to show that $\overset{\circ}{T\mathcal{J}} \cap \partial \mathcal{E}^b \subseteq \overset{\circ}{T\mathcal{J}} \cap \eta^{-1}[b]$: specifically, notice that any point not in the second set cannot be in the first. This is so, for if $v \in \overset{\circ}{T\mathcal{J}} \cap \left(\mathcal{E}^b - \eta^{-1}[b]\right)$ then it is necessarily in the inverse image of some open interval, $v \in \eta^{-1}[\eta(v) - \epsilon, \eta(v) + \epsilon]$, which is an open set in $\overset{\circ}{T\mathcal{J}} \cap \mathcal{E}^b$ since η is continuous.

□

With respect to the example of Section 3.1.1, note that $b_1 = \frac{1}{2}\epsilon^2$, and $b_0 = 0$, so that the positive invariant sets, $\mathcal{E}^b$ defined by the proposition are simply the closed disks around the origin of $\mathbb{R}^2$ whose radius is less than or equal to ϵ.

3.2 Limit Behavior

The hypothesis of Proposition 3.2 in some sense strengthens that required for conclusion (1) of Proposition 2.1: if b_1 is a regular value and a minimal value on the boundary, then *grad* φ must be transverse to $\partial \mathcal{J}$ at points near $\varphi^{-1}(b_1)$. In contrast, once sufficient conditions for positive invariance have established, a conclusion analogous to (2) for Proposition 2.1 follows quite easily: the critical points constitute the entire positive limit set, and their local stability properties are inherited directly from that of φ.

3.2.1 The Limit Set Consists of Equilibrium States

In general, Hamiltonian systems will have extremely complicated limit behavior. The addition of the damping term, f_d, makes the global limit behavior of the dissipative mechanical system Δ as trivial as that of its gradient model, Γ. The formal argument below is adapted from the discussion of gradient vector fields in [34, Example 1.1.3]. Somewhat surprisingly, the reasoning there applies to the present situation almost without change.

Proposition 3.3 *The positive limit set of any positive invariant subset of $T\mathcal{J}$ under the flow of a dissipative mechanical system, Δ, consists of the critical points of φ identified with their image in the zero section of $T\mathcal{J}$.*

Proof: It is clear that the equilibrium states of f_Δ correspond exactly to the critical points of φ, hence, form a totally isolated set in the zero section of $T\mathcal{J}$. Now suppose some initial condition, v_0, has a limit set which includes a state, v_1, whereon f_Δ does not vanish. It is clear that v_1 must be a regular point of η since the critical points of the latter correspond exactly to the critical points of φ (lifted into the zero section). Therefore, according to the implicit function theorem, the intersection of $\eta^{-1}[\eta(v_1)]$ with a sufficiently small neighborhood of v_1 is a codimension 1 submanifold of $T\mathcal{J}$. The computation of $\dot{\eta}$ in Theorem 1 demonstrates that the flow of f_Δ is transverse to this submanifold in a small enough neighborhood of v_1. Yet, since that point is in the limit set of v_0, the trajectory through v_0 must intersect this submanifold in more than one point (in fact, an infinity of points). This would contradict the monotone property of η on the flow.

□

3.2.2 Local Stability of the Equilibrium States

Lord Kelvin's result, Theorem 1, demonstrates that the minima of φ (suitably identified with points in the zero section of $T\mathcal{J}$) are local attractors. To gain a conclusion analogous to (3) in Proposition 2.1 it must be verified as well that local behavior of the negative gradient flow near a critical point other than a minimum is a model for local behavior of the dissipative mechanical system near the identified equilibrium state. To see this, first observe that the local linearized vector field at any equilibrium state of the dissipative mechanical system, Δ (9), is the linear time invariant dissipative mechanical system determined by the hessian of the potential function at that point.

Lemma 3.4 *Let q_0 be a critical point of φ. Then the linearized vector field at the equilibrium state, $(q_0, 0)$, of the dissipative mechanical field, (12), is given in local coordinates as*

$$[Df_\Delta](q_0,0) = \begin{bmatrix} 0 & I \\ -\underline{M}(q_0)^{-1}\underline{K}_1(q_0) & -\underline{M}(q_0)^{-1}\underline{K}_2(q_0,0) \end{bmatrix},$$

where $\underline{K}_1 = D^2\varphi$ is the hessian matrix of φ, $\underline{K}_2$ is the local matrix representation of the linearized dissipative field, and $\underline{M}$ is the matrix representation of the kinetic energy metric, as introduced in (10).

If M is a metric, then $\underline{M}(q_0)$ is a positive definite symmetric matrix. Similarly, if f_d is strict, then $\underline{K}_2(0,0)$ has a positive definite symmetric part by assumption. Thus the local stability properties of equilibria of f are determined by the nature of the Hessian at the identified critical point of φ as the following result indicates: that is, the stability of the origin of a linear time invariant dissipative mechanical system in $\mathbb{R}^{2n}$ is governed by the stability properties of the associated gradient system in $\mathbb{R}^n$.

Lemma 3.5 (Chetaev [4]) *Let M, K_2, K_1 be symmetric matrices in $\mathbb{R}^{n \times n}$ and let M, K_2 be positive definite. The origin of the linear time invariant system in $\mathbb{R}^{2n}$,*

$$\begin{bmatrix} \dot{x}_1 \\ \dot{x}_1 \end{bmatrix} = \begin{bmatrix} 0 & I \\ -M^{-1}K_1 & -M^{-1}K_2 \end{bmatrix} \begin{bmatrix} x_1 \\ x_2 \end{bmatrix} \tag{13}$$

has one of the following stability properties: (i) asymptotically stable; (ii) stable but not attractive; (iii) unstable; if and only if the origin of the linear time invariant system in $\mathbb{R}^n$,

$$\dot{x}_1 = -K_1 x_1. \tag{14}$$

has the corresponding property.

Putting these results together yields a global version of Lord Kelvin's result which compares favorably in strength to Proposition 2.1: the significant difference is in the stronger boundary condition required by the present version as motivated by the discusion surrounding Proposition 3.2.

Proposition 3.6 *Let Δ be a dissipative mechanical system (9). Suppose that*

$$b_1 \triangleq \inf_{q\in\partial\mathcal{J}} \varphi(q),$$

is a regular value of φ (and is understood to be infinite if $\partial\mathcal{J} = \emptyset$) and that

$$b_1 > b_0 \triangleq \inf_{q\in\mathcal{J}} \varphi(q).$$

Then for all constants, $c \in (b_0, b_1]$, the set of "bounded total energy" states of η (8),

$$\mathcal{E}^b \triangleq \{v \in T\mathcal{J} : \eta \leq c\},$$

has the following properties with respect to the flow of Δ:

1. *$\mathcal{E}^b$ is a positive invariant set with non-empty interior;*
2. *the positive limit set of $\mathcal{E}^b$ consists of those points in the zero section of $T\mathcal{J} \cap \mathcal{E}^b$ identified with a critical point of φ;*
3. *there is a dense open set in $\mathcal{E}^b$ whose limit set consists of those points in the zero section of $T\mathcal{J} \cap \mathcal{E}^b$ identified with a local minimum of φ;*

3.3 Navigation Functions

Proposition 3.6 suggests that the major effort in the construction of feedback controllers for mechanical systems will be spent finding suitable scalar valued functions, φ. In this section we address the following loosely posed control problem. Given a mechanical system, $\Sigma = (\mathcal{J}, \kappa)$ and a desired configuration, $q_d \in \mathcal{J}$ (identified, as usual, as a point in the zero section of $T\mathcal{J}$), to which we want to bring the system, what is a suitable choice of φ?

From the point of view of dynamical systems theory, linear feedback controllers for linear time invariant plants provide the means by which a desired equilibrium state of the closed loop system is made to attract the entire state space. The preceding results suggest two important relative deficiencies in the case of the mechanical control systems, Σ. First, no smooth vector field can have a global attractor unless the state space on which it is defined is homeomorphic to $\mathbb{R}^n$ [2]. Thus, as is generally the case in robotics and other applications of mechanical systems theory, when the configuration space, $\mathcal{J}$, (and, hence, the state space, $T\mathcal{J}$) is not a homeomorph of a Euclidean vector space, there can be no hope of global asymptotic stability. Second, when the configuration space has a boundary, Lemma 3.1 demonstrates that there does not even seem to be the possibility of avoiding finite escape from the entire phase space (without recourse to unbounded vector fields which we must discard as unrealizable).

We now offer a definition placing conditions on φ that result in what is arguably the "best possible" convergence behavior: we will ensure that the enirety of $\mathcal{J}$ (identified, as usual, with the zero section of $T\mathcal{J}$) is contained in a positive invariant subset of the phase space and will guarantee "almost global" asymptotic stability within that set.

Following Hirsch [10, Ch. 6.3], say that a smooth Morse function from a compact manifold, $\mathcal{F}$, is *admissible over the interval* $[a,b]$ if $\partial\mathcal{F} = \varphi^{-1}[a] \cup \varphi^{-1}[b]$, and a, b are regular values. This implies that each of $\varphi^{-1}[a], \varphi^{-1}[b]$ is a union of components of $\partial\mathcal{F}$, a regular set on which *grad* φ is transverse. Note that if either level set is empty then the (corresponding) minimum or maximum value of φ is taken only at critical points in the interior of $\mathcal{F}$. Adapting the terminology of Morse [33], say that φ is *polar* if it has a unique minimum.

Definition 1 *Let $\mathcal{F} \subset \mathbf{E}^n$ be a smooth compact connected manifold with boundary, and $q_d \in \mathring{\mathcal{F}}$, be a distinguished point in its interior. A polar Morse function, $\varphi \in C^2[\mathcal{F}, [0,1]]$, which is admissible over the interval $[-\epsilon, 1], \epsilon \in \mathbb{R}^+$, and takes its unique minimum at $\varphi(q_d) = 0$, is called a* navigation function .

As a direct consequence of Proposition 3.6 we have the following result.

Theorem 2 *Let $\Delta = (\mathcal{J}, \eta, f_d)$ be a dissipative mechanical system (9), and suppose that φ is a navigation function for $\mathcal{J}$. Then the entire zero section of $T\mathcal{J}$ is included in the closed subset, $\eta^{-1}[0,1] \subseteq T\mathcal{J}$, which is positive invariant with respect to the flow induced by Δ (and includes the entirety of $T\mathcal{J}$ if $\mathcal{J}$ has no boundary). Moreover, there is an open dense set in $\eta^{-1}[0,1]$ whose limit set is exactly the desired configuration, q_d, at zero velocity.*

3.3.1 Existence

When does a compact manifold with boundary admit a navigation function? In a recent paper [25] we have been able to answer this question quite unequivocally: we show that smooth navigation functions exist on any such smooth manifold for any desired interior point, q_d. We now review this result.

Smale proved the generalized "Poincaré's Conjecture" in higher dimensions roughly three decades ago. In so doing, he was led to develop a number of results concerning gradient systems of which the most important to us is the existence of "nice" functions. Suppose $\mathcal{M}$ is a smooth compact n-dimensional manifold whose boundary is the disjoint union of two closed components, $\partial\mathcal{M} = \mathcal{V}_1 \sqcup \mathcal{V}_2$. Smale calls a smooth Morse function, $\varphi \in C^\infty[\mathcal{M}, \mathbb{R}]$, *nice* if $\varphi(\mathcal{V}_1) = -\frac{1}{2}$, $\varphi(\mathcal{V}_2) = n + \frac{1}{2}$, and at a critical point p of φ, $\varphi(p) = \text{index } p$. He obtains a number of important results with this construction, including a generalization of the somewhat earlier result of Morse which demonstrates that every smooth manifold with no boundary admits a smooth polar non-degenerate function [33]. For our purposes, this result is important if it can be extended to the general case with boundary.

The desired extension obtains by applying the notion of "cancellation" of adjacent (in index) critical points that Morse and Smale developed in the course of their independent investigations. A (reasonably) self-contained exegesis upon these techniques is provided by Milnor [30], whose version may be rendered as follows. Suppose that φ is a smooth Morse function on $\mathcal{M}$ with two distinct interior critical points, p_1 and p_2, with indices λ_1, λ_2, respectively, possessing the properties $\lambda_1 \neq \lambda_2$ and $\varphi(p_1) \neq \varphi(p_2)$. These two points may be *cancelled* if there exists another smooth Morse function, φ', on $\mathcal{M}$, which agrees with φ everywhere away from a neighborhood of $\varphi^{-1}[\varphi(p_1)]$ and $\varphi^{-1}[\varphi(p_2)]$ in $\mathcal{M}$, yet which has two fewer critical points — one less critical point of index λ_1; one less critical point of index λ_2. It turns out that pairs of index 0 and index 1 critical points may be cancelled if the "lower boundary" has the right homology type [30, Thm. 8.1]. Moreover, there are always "enough" index 1 critical points to cancel all the minima if the manifold is connected: a proof may be found in [25] and was suggested to us by W. Massey. The application of these results in the present setting was suggested by M. Hirsch.

Theorem 3 ([25]) *For every smooth compact connected manifold with boundary, $\mathcal{M}$, and any point, $x_0 \in \mathring{\mathcal{M}}$, there exists a C^∞* navigation function.

3.3.2 Construction

Although questions of existence presumably have independent mathematical interest, the engineer is only concerned to know of a negative outcome: in the present case, since we are guaranteed that what we seek is available, attention shifts to the question of construction. Here, we are led to impose some additional constraints.

We have already observed in Section 2.1.2 that the mechanical control system, Σ (3), falls within the class of linear analytic systems. Since Δ results from the application of feedback, g (11), to f_Σ, as described in Section 2.3, it seemts only natural to demand that g be analytic as well. This, of course, implies that φ be not merely smooth, but analytic as well. Moreover, in their pioneering solution to the geometric robot navigation problem, Schwartz and Sharir [39] have argued persuasively that the class of real algebraic functions on real semi-algebraic varieties provides a practicable notion of "effective computation." We will adopt this point of view and insist that our constructions remain within the class of real algebraic functions: that is to say, the level set of a navigation function must be real semi-algebraic. Thus, the standard appeal to smooth functions "patched together" through a partition of unity will not avail. We are confronted, instead, with the harder task of building analytic algebraic navigation functions.

4 Applications of Total Energy as a Lyapunov Function

This section presents two engineering applications of the preceding results. In both cases the configuration space fails to be a homeomorph of $\mathbb{R}^n$, so that global asymptotic stability is impossible: we display analytic algebraic navigation functions resulting in closed loop dynamics which are dissipative mechanical systems whose limit behavior accomplishes the specified task. In the first case, the problem of satellite attitude tracking discussed in Section 4.1 (and taken from the longer treatment of [18]), the configuration space is a Lie group — the entirety of $SO(3)$. Since the boundary is empty, convergence is guaranteed from almost every initial phase. To the best of the author's knowledge, Theorem 4 represents the first feedback controller for a fully actuated satellite which is well defined on the entirety of $SO(3)$ and which achieves asymptotically exact attitude tracking around an arbitrary reference trajectory with probability one. In the second case, the problem of robot obstacle avoidance discussed in Section 4.2 (and taken from our ealier publications [25, 37]), the configuration space is a a subset of $\mathbb{R}^n$ (a Lie group as well) but this time it is bounded by a finite number of disjoint spheres. Here, the construction of navigation functions results in a robot which is guaranteed to approach a desired destination point in a cluttered space without hitting any of the clutter from every zero velocity initial condition excepting a set of zero measure. Again, to the best of the author's knowledge, Theorem 6 represents the first feedback controller which solves the global robot obstacle avoidance problem on nontrivial spaces of arbitary dimension.

4.1 Satellite Attitude Tracking

Now consider the application of the previous results to a classical control problem — asymptotically exact tracking — in a non-classical setting — the group of spatial rotations, $SO(3)$. Suppose there is a single rigid body actuated by three independent gas jets operating outside of the earth's gravitational field: the only forces operating on the body are the controlled inputs from the actuators which are capable of delivering any desired force in the "wrench space" of the body, $TSO(3)$. Both the position and the velocity of the body are available from sensors. It is desired to force the body to track an arbitrary but entirely known reference trajectory. Since the system is completely actuated, there is perfect state information, and all derivatives of the reference trajectory are known, the velocity tracking problem is trivial. Namely, all nonlinearities due to the kinetic energy may be exactly cancelled, leaving a completely decoupled linear time invariant system. This procedure may be recognized as a trivial implementation of the global exact linearization techniques which have become so popular in the nonlinear control literature.

Consider, instead, the problem of attitude tracking. Namely, given a desired motion, $D \in C^2[\mathbb{R}, SO(3)]$, construct a time invariant controller which causes the actual attitude to asymptotically approach $D(t)$ from any initial configuration, $A \in SO(3)$. So different is this from the trivial linear problem to which velocity tracking reduces that it is unsolvable as posed in that context. For, consider the particular case that $D(t) = D^*$ is some constant point. We seek a controller which makes that point (at zero angular velocity) a global attractor of the closed loop dynamics. Now, as was mentioned above, the domain of attraction of an attracting point is homeomorphic to some Euclidean vector space [2]. But the state space of our mechanical system — the tangent bundle over the rotation group — is clearly not homeomorphic to any Euclidean vector space. Thus, it would be impossible for our closed loop system to bring *all* initial conditions to the desired attitude. Evidently, the control system arising from a single rigid body is *not* globally linearizable by *any* technique since its state space is not a vector space. Our problem statement must be refined.

Since $SO(3)$ is a compact odd dimensional manifold without boundary, its Euler characteristic is zero [10]. It follows from the Theorem of Hopf [31] that any nondegenerate vector field on $SO(3)$ with an attracting equilibrium state has at least one other singularity which, if it is the only additional equilibrium state, must be totally unstable. Excepting the complement of some open dense set — in this case, the repelling point — trajectories of such a vector field are guaranteed to asymptotically approach the attracting point. Thus, although topological obstructions preclude a globally asymptotically stable system, a practically equivalent formulation which respects the underlying topology of the problem may be attainable. Say that a dynamical system is *almost globally* asymptotically stable if all trajectories starting in some open dense subset of the state space tend asymptotically to a specified stable equilibrium state. This we take as the criterion of convergence for our tracking algorithms on $SO(3)$: it is the best possible result. Moreover, Theorem 2 assures us that we may achieve this result by recourse to feedback of the form (11) if we find a navigation function for $SO(3)$. That is the task we now undertake.

4.1.1 The Mechanical Control System on $SO(3)$

The configuration space of a rigid body is the group of rigid transformations, $SO(3) \times \mathbb{R}^3$. If we are concerned only with the attitude of a rigid body, then it suffices to treat $SO(3)$ alone, which we now identitfy with a subset of $\mathbb{R}^9$,

$$SO(3) \triangleq \left\{ R \in \mathbb{R}^{3\times 3} : R^{\mathrm{T}} R = I \text{ and } |R| = 1 \right\}.$$

This Lie group has as its Lie algebra the set of skew symmetric matrices,

$$so(3) = \mathsf{skew}(3) \triangleq \left\{ J \in \mathbb{R}^{3\times 3} : J + J^{\mathrm{T}} = 0 \right\},$$

which is isomorphic to $\mathbb{R}^3$ according to the linear bijection

$$J : \begin{bmatrix} w_1 \\ w_2 \\ w_3 \end{bmatrix} \mapsto \begin{bmatrix} 0 & -w_3 & w_2 \\ w_3 & 0 & -w_1 \\ -w_2 & w_1 & 0 \end{bmatrix}.$$

The vector space of three by three matrices is the direct sum,

$$\mathbb{R}^{3\times 3} = \mathsf{sym}(3) \oplus \mathsf{skew}(3),$$

of the symmetric and skew-symmetric matrices. Thus, we may define a unique "pseudo-inverse" for J whose domain is extended to all of $\mathbb{R}^{3\times 3}$ by projection onto the linear subspace, $\mathsf{skew}(3)$.

$$J^{\dagger}(A) \triangleq J^{-1}(A - A^{\mathrm{T}}).$$

The maps J^{-1} and $J^{\dagger}$ have distinct domains, and must not be confused. On the other hand, we will be sloppy and not distinguish between the version of the linear map J whose range is $\mathsf{skew}(3)$ and the version whose range is $\mathbb{R}^{3\times 3}$.

The natural inner product on the vector space $\mathbb{R}^{3\times 3}$ is

$$(\, R_1 \mid R_2 \,) \triangleq \frac{1}{2}\, tr\, \left\{ R_1 R_2^{\mathrm{T}} \right\}.$$

Direct computation reveals that J is an isometry between $\mathbb{R}^3$ with its Euclidean norm, $\|w\|^2 = w^{\mathrm{T}} w$, and $\mathsf{skew}(3)$ with the norm corresponding to this inner product. Note, as well, that $\mathsf{sym}(3)$ is the orthogonal complement of $\mathsf{skew}(3)$ with respect to this inner product. Finally, the norm associated with $(\,\cdot \mid \cdot\,)$ defines a metric on the group $SO(3)$,

$$\rho(R_1, R_2) \triangleq \xi^{-1} \circ (\, R_1 - R_2 \mid R_1 - R_2 \,),$$

after composition with a suitable comparison function [18], [6]

$$\xi \in \mathcal{K}_{\infty}[[0,\pi],[0,4]] : \chi \mapsto 2(1 - \cos\chi).$$

On any Lie group, we may take the differential of left (or right) inverse translation and this is the canonical means of identifying left (or right) invariant vector fields with the Lie algebra. Thus,

$$T_R SO(3) = \left\{ RJ(w) \in \mathbb{R}^{3\times 3} : w \in \mathbb{R}^3 \right\},$$

is identified once and for all with $so(3) = \mathsf{skew}(3) \approx \mathbb{R}^3$, and we may take the tangent bundle to be the cross product

$$TSO(3) \approx SO(3) \times \mathbb{R}^3.$$

As sketched in Section 2.1, a mechanical control system arises from the choice of a kinetic energy. If $M \in \mathsf{sym}^{+}(3)$, a positive definite symmetric matrix, is the moment of inertia matrix of the rigid body then the kinetic energy at a phase, $v = (R, r) \in TSO(3)$, is

$$\kappa(v) = \langle\, v \mid v \,\rangle \triangleq (\, J(r)R \mid J(r)RM \,).$$

[6] The comparison functions, the group $\mathcal{K}_r(\mathcal{I}_1, \mathcal{I}_2)$, of monotone increasing C^r diffeomorphisms between two real intervals appears extensively in the engineering stability literature [9]. Some properties are reviewed in [18, 24].

This leads to a mechanical control system, Σ, whose internal dynamics may be expressed in body coordinates as

$$f_\Sigma(v_u) \triangleq \begin{bmatrix} RJ(r) \\ M^{-1}[u - J(r)Mr] \end{bmatrix}.$$

For the present application, we have assumed the à priori designation of a desired "reference trajectory", $v_d : \mathbb{R} \to T\mathcal{J}$, which is "second order." That is to say, if $v_d(t) = (D, d)(t)$, then $\dot{D} \equiv DJ(d)$. Now if $v = (A, a)$ denotes the actual trajectory of the rigid body, we will find it useful to consider the "error coordinate system" obtained via left translation by v_d,

$$v_e = (E, e) \triangleq L_{v_d} v = (D^{\mathrm{T}} A, a - A^{\mathrm{T}} Dd),$$

preserving the second order property, $\dot{E}(t) = EJ(e(t))$.

4.1.2 A Navigation Function on $SO(3)$

We are now ready to search for a potential function on $SO(3)$ whose lift into the physical Lagrangian system of the actuated rigid body will define almost globally asymptotically stable error dynamics. According to Theorem 2, we need merely ensure that φ is a navigation function. Since the configuration space has an empty boundary, this amounts to finding a Morse function with a unique minimum on $SO(3)$.

In a very nice report, Meyer [29] attempted to generalize PD techniques to the global control of spacecraft attitude. His point of view is very close to the spirit of this paper, and, in some sense, this application might be seen as a continuation and extension of that earlier work. Meyer chose for his potential law on $SO(3)$ the distance from a reference point measured by the natural metric, ρ, itself. This will not suffice for the present purposes since, the gradient of the distance function is necessarily undefined on its "cut locus" (the set of points whose minimal geodesic to a reference point is not unique) — an embedding of real projective two-space in $SO(3)$ in the present case. Intuitively, this is clear since ρ is a composition with the trace function on **skew**(3). The latter is unfortunately not a Morse function since it has a critical point at every symmetric rotation: the symmetric rotations — an embedding of real projective two-space, $\mathbb{RP}^2$, in $SO(3)$ [7] — comprise a connected set; the critical points are not isolated. Instead we will use a "modified trace" function according to the following result of Marsden and collaborators.

Lemma 4.1 (Chillingworth, Marsden, and Wan [5]) *If $P \in$ sym(3) has distinct eigenvalues, π_1, π_2, π_3, and*

$$(\pi_1 + \pi_2)(\pi_1 + \pi_3)(\pi_3 + \pi_2) \neq 0,$$

then there are exactly four rotations, $R \in SO(3)$ at which $PR \in$ sym(3). These are exactly the critical points of

$$(\, P \mid R \,) = tr\ \{PR\}$$

with Morse index specified by the number of positive eigenvalues of

$$tr\ \{RP\} I - PR$$

We are thus led to define as a navigation function on $SO(3)$

$$\varphi(R) \triangleq \frac{1}{\pi'}\, tr\ \{P(I - R)\} = \frac{1}{\pi'}(\, P \mid I - R \,), \tag{15}$$

where the factor involving $\pi' \triangleq \pi_2 + \pi_3 - \pi_1$ is added to keep the image in the interval $[0, 1]$ (assuming that $\pi_1 < \pi_2 < \pi_3$). If $P \in$ sym$^+(3)$, a positive definite symmetric matrix, which we now further assume, then the eigenvalue assumptions of the previous Lemma are assured. Since $d\varphi$ is a scalar multiple of Marsden's modified trace, φ is also a Morse function with four critical points specified in the same fashion. Moreover, φ takes its values on $\mathbb{R}^+$, vanishing only at $R = I$. Thus, φ is indeed a navigation function. In fact, it is the best we can find since any Morse function on $SO(3)$ must have at least four critical points according to the Lusternik-Schnirelmann theorem [41, p. 92]. Moreover, it is certainly algebraic so that its differential one-form — the critical ingredient of the feedback law, g (11) — may be readily computed [18] as

$$d\varphi = 2\, J^{\dagger}(PR).$$

According to Proposition 2.1 together with Lemma 4.1 the negative flow of the gradient vector field resulting from the kinetic energy metric,

$$grad\ \varphi \ = 2\, M^{-1} J^{\dagger}(PR),$$

takes all points of $SO(3)$ to one of four symmetric rotations — the identity, and the three orientations which are "180° away" along the x, y, z axes — and all points excepting a nowhere dense set to the identity.

4.1.3 Inverse Dynamics

In the linear time invariant setting, inverse dynamics amounts to the use of a precompensator to make the errors between the plant state and reference derivatives satisfy an asymptotically stable linear time invariant dynamical system. Entirely analogously in this setting, we will use the navigation function presented above to achieve an "almost global" asymptotically stable dynamical error system via the feedback law, g (11), applied to the error coordinates, v_e. We then pre-filter the reference signal, $(D, d)(t)$, so that the error dynamics fall within the class of dissipative mechanical systems, f_{Δ}, and apply Theorem 2 directly.

For example, in one degree of freedom, $\mathcal{J} = \mathbb{R}$, a point with unit mass gives rise to a kinetic energy metric given by the identification map, $M = 1$, between the tangent and cotangent space over each configuration. If an arbitrary bidirectional force, u, can be imposed upon the point mass then our mechanical control system is the familiar double integrator,

$$f_{\Sigma}(v_u) \triangleq \begin{bmatrix} 0 \\ u \end{bmatrix}.$$

Any Hook's Law spring potential, $\varphi \triangleq \frac{1}{2} K_1 q^2$, (where $K_1 \in \mathbb{R}^+$) in conjunction with a Rayleigh damper, $G_d(q, \dot{q}) \triangleq K_2 \dot{q}$, (where $K_2 \in \mathbb{R}^+$ as well) defines a globally asymptotically stable closed loop system on $T\mathcal{J} = \mathbb{R}^2$,

$$f_{\Delta}(v) \triangleq \begin{bmatrix} 0 & 1 \\ -K_1 & -K_2 \end{bmatrix} v,$$

resulting from the feedback law

$$g(q, \dot{q}) \triangleq K_2 \dot{q} + K_1 q$$

given by (11). Given a particular reference trajectory, $v_d = (r, \dot{r})$, and an error coordinate system again defined via left translation (identical, of course, to right translation since $\mathbb{R}^2$ is an Abelian Lie group)

$$v_e = (e, \dot{e}) \triangleq L_{v_d} v = (q - r, \dot{q} - \dot{r}),$$

we may now servo on the error, $g(v_e) = K_2\dot{e} + K_1 e$ and pre-filtering through the inverse of the closed loop plant,

$$u \triangleq g(v_e) + \ddot{r} - K_1 r - K_2 \dot{r}.$$

This results in globally asymptotically stable error dynamics,

$$\dot{v_e} = f_\Delta(v_e),$$

on $T\mathcal{J}$. We now implement a nonlinear version of this scheme in the present setting.

In place of the familiar Hooks's Law potential, we will substitute the navigation function, φ (15). Taking an arbitrary positive definite symmetric matrix, $K_2 \in Sp(3)$, we will use Rayleigh damping, $G_d(R, r) \triangleq K_2 r$, for the sake of simplicity. The feedback law (11) applied to the error coordinates is now computed as

$$g(E, e) = K_2 e + 2J^\dagger(PE).$$

Note that

$$\begin{aligned} \dot{e} &= \dot{a} - E^{\mathrm{T}}\dot{d} - \left[EJ(a - E^{\mathrm{T}}d)\right]^{\mathrm{T}} d \\ &= \dot{a} - E^{\mathrm{T}}\dot{d} + J(a)E^{\mathrm{T}}d. \end{aligned}$$

Thus, if we apply error feedback and build an appropriate pre-filter for the reference, $v_d = (D, d)$,

$$\begin{aligned} u &\triangleq M\left(E^{\mathrm{T}}\dot{d} - J(a)E^{\mathrm{T}}d\right) \\ &\quad + J(a)ME^{\mathrm{T}}d + J(E^{\mathrm{T}}d)Me + g(v_e) \end{aligned} \tag{16}$$

the closed loop angular aceleration will be given as

$$\dot{a} = E^{\mathrm{T}}\dot{d} - J(a)E^{\mathrm{T}}d - M^{-1}\left[J(e)Me + K_2 e + grad\,\varphi(E)\right],$$

or, in the "error coordinate system" of phase space, $TSO(3) = SO(3) \times \mathrm{IR}^3$,

$$\begin{aligned} \dot{E} &= EJ(e) \\ \dot{e} &= -M^{-1}\left[J(e)Me + K_2 e + grad\,\varphi(E)\right]. \end{aligned} \tag{17}$$

Equation (17) is a dissipative mechanical system, Δ based upon the navigation function (15). Theorem 2 applies directly using information about the critical points of φ supplied in Lemma 4.1.

Theorem 4 *All trajectories of (17) tend toward one of the four critical points of φ. A dense open set of initial conditions has its limit set at the desired point, $(E, e) = (I, 0)$.*

The satellite asymptotically attains the desired attitude trajectory, $D(t)$, except from a set of initial conditions of zero measure in the phase space.

4.2 Robot Navigation

Consider the following problem in robotics. A kinematic chain — a sequence of mutually constrained actuated rigid bodies — is allowed to move in a cluttered workspace. Contained within the *joint space* — an analytic manifold which forms the configuration space of the kinematic chain — is the *free space*, $\mathcal{F}$ — the set of all configurations which do not involve intersection with any of the "obstacles" cluttering the workspace. Given any interior "destination point," $q_d \in \mathring{\mathcal{F}}$, to which it is desired to move the robot, find a curve in $\mathcal{F}$ from an arbitrary initial point to the desired destination.

The purely geometric problem of constructing a path between two points in a space obstructed by sets with arbitrary polynomial boundary (given perfect information) has already been completely solved [40]. Moreover, a much more efficient solution has recently been offered for this class of problems as well [?]. The motivation for the present direction of inquiry (beyond its apparent academic interest) is the desire to incorporate explicitly aspects of the control problem — the construction of feedback compensators for a well characterized class of dynamical systems in the presence of well characterized constraints — in the planning phase of robot navigation problems. That is, the geometrical "find path" problem is generalized to the search for a family of paths in $\mathcal{F}$ (the one–parameter group of the gradient flow), which provides a feedback control law for the physical robot as well. It is clear from Theorem 2 that the construction of a navigation function on the freespace provides a solution to this problem.

4.2.1 The Mechanical Control System

The configuration space of a rigid robot with n moving joints is generally taken to lie within the cross product of a torus and a cartesian space $\mathcal{J} \subset T^{n-k} \times \mathbb{R}^k$, where k is number of "sliding joints" and $n-k$ is the number of "revolute joints" [23, 6], its boundary generally arising from physical limits on the range of motion of each joint. There is a "kinematic map," $k_i : \mathcal{J} \to \mathbb{R}^3 \times SO(3)$, which expresses the physical location and orientation of a distinguished frame of reference in the ith constituent rigid body of the robot robot as a function of the position in configuration space — k_i is a polynomial in transcendental functions of the generalized coordinates of $\mathcal{J}$. The "workspace," $\mathcal{W}$, lies within the n-fold cross product of this Euclidean group with itself, and represents the placements of the robot which do not intersect physical obstacles. The *freespace,* $\mathcal{F} \subset \mathcal{J}$ results after removing from $\mathcal{J}$ those configurations which involve any self-interesection or intersection with the obstacles.

The kinetic energy is determined by summing up the contribution of each of the robot's constituent rigid bodies. Let $\{M_i\}_{i=1} n$ be the moment of inertia matrices of these constituent rigid bodies. In the appropriate local coordinates at some point $q \in \mathcal{J}$, the kinetic energy morphism, is given by [23]

$$M(q) \triangleq \sum_{i=1}^{n} [[Dk_i](q)]^{\mathrm{T}} M_i [Dk_i](q),$$

so that given any phase, $v = (q, x)$,

$$\kappa(q, x) = \frac{1}{2} x^{\mathrm{T}} M(q) x \triangleq \frac{1}{2} \langle v \mid v \rangle.$$

Perhaps the most difficult aspect of the "generalized piano mover's problem" [40] is a precise determination of the freespace, $\mathcal{F}$, from information about the robot's consituent rigid bodies and the obstacles in the world. We will presume that this information has been furnished in the form of an implicit representation for each obstacle boundary. In particular, we will consider progressively more complicated versions of the freespace, and present analytic algebraic navigation functions for arbitrary destinations. These constructions solve the piano mover's "findpath" problem (for almost every initial configuration) via the resulting gradient, Γ on $\mathcal{F}$ according to Proposition 2.1. Moreover, they solve the piano mover's "find controller" problem (for almost every initial configuration in the bounded energy set, $\mathcal{E}^1 \subseteq T\mathcal{F}$ according to Theorem 2.

4.2.2 Navigation Functions on Euclidean Sphere Worlds

A "Euclidean sphere world" is a compact connected subset of $\mathbf{E}^n$ whose boundary is the disjoint union of a finite number, say $M+1$, of $(n-1)$-spheres. We suppose that perfect information about this space has been furnished in the form of $M+1$ center points $\{q_i\}_{i=0}^{M}$ and radii $\{\rho_i\}_{i=0}^{M}$ for each of the bounding spheres. In our previous work [25], we have shown how to use this information to build a navigation function on the particular sphere world, $\mathcal{M}$, considered as a simple freespace.

The proposed navigation function, $\varphi : \mathcal{M} \to [0,1]$, is a composition of three functions:

$$\varphi \triangleq \sigma_d \circ \sigma \circ \hat{\varphi}.$$

The function $\hat{\varphi}$ is polar, almost everywhere Morse and analytic, it attains a uniform height on $\partial\mathcal{M}$ by blowing up there. Its image is "squashed" by the diffeomorphism, σ, of $[0,\infty)$ into $[0,1]$, where

$$\sigma(x) \triangleq \frac{x}{1+x},$$

resulting in a polar, admissible, and analytic function which is non-degenerate on $\mathcal{M}$ except at one point — the destination. This last flaw is repaired by σ_d.

We distinguish between "good" and "bad" subsets of $\mathcal{M}$. When a point belongs to the "good" set, we expect the negative gradient lines to lead to it (here it is just $\{q_d\}$). The "bad" subset includes all the boundary points of the free space, and we expect the cost at such a point to be high. Let γ and β denote analytic real valued maps whose zero-levels, i.e. $\gamma^{-1}(0), \beta^{-1}(0)$, are respectively, the "good" and "bad" sets. We define $\hat{\varphi}$ to be,

$$\hat{\varphi} \triangleq \frac{\gamma}{\beta},$$

where $\gamma : \mathcal{M} \to [0,\infty)$ is

$$\gamma \triangleq \gamma_d^k \quad k \in \mathbb{N}; \qquad \gamma_d \triangleq \|q - q_d\|^2,$$

and $\beta : \mathcal{M} \to [0,\infty)$ is,

$$\beta \triangleq \Pi_{i=0}^{M} \beta_i,$$

where

$$\beta_0 \triangleq \rho_0^2 - \|q\|^2 \quad ; \quad \beta_j \triangleq \|q - q_j\|^2 - \rho_j^2 \;\; j = 1 \ldots M.$$

Due to the parameter k in $\hat{\varphi}$, the destination point is a degenerate critical point. To counteract this effect, the "distortion" $\sigma_d : [0,1] \to [0,1]$,

$$\sigma_d(x) \triangleq (x)^{\frac{1}{k}} \quad k \in \mathbb{N},$$

is introduced, to change q_d to a non-degenerate critical point.

Theorem 5 ([25]) *If the free space, $\mathcal{M}$, is a Euclidean sphere world then there exists a positive integer N such that for every $k \geq N$, for any finite number of obstacles, and for any destination point in the interior of $\mathcal{M}$,*

$$\varphi = \sigma_d \circ \sigma \circ \hat{\varphi} = \left(\frac{\gamma^k}{\gamma^k + \beta}\right)^{\frac{1}{k}}, \tag{18}$$

is a navigation function on $\mathcal{M}$.

In the proof of this theorem (which comprises the central contribution of [25]) a constructive formula for N is given with the "schematic" form,

$$N = \max_{\{\mathcal{O}_i\}_{i=0}^M} N_i\left(q_d, \begin{bmatrix} q_0 \\ \rho_0 \end{bmatrix}, \ldots \begin{bmatrix} q_M \\ \rho_M \end{bmatrix}\right),$$

where $\mathcal{O}_i$ is the ith obstacle. The functions N_i are given explicitly in an appendix of [25].

4.2.3 Navigation Functions Induced by Diffeomorphism

The Euclidean sphere world, of course, corresponds to a rather simplistic view of freespace. Fortunately, the navigation properties defined in Section 3.3 are invariant diffeomorphism, as the following result makes clear.

Proposition 4.2 ([25]) *Let $\varphi : \mathcal{M} \to [0,1]$ be a navigation function on $\mathcal{M}$, and $h : \mathcal{F} \to \mathcal{M}$ be a diffeomorphism. Then*

$$\tilde{\varphi} \triangleq \varphi \circ h,$$

is a navigation function on $\mathcal{F}$.

This result suggests that we might consider the Euclidean sphere world as a a "model space" used to induce navigation functions on more interesting "real spaces" in its analytic diffeomorphism class. The problem of constructing a navigation function on a member of this class reduces to the construction of an analytic diffeomorphism from this space onto its model.

Our constructive results to date encompass the class of "star worlds." A *star shaped set* is a diffeomorph of a Euclidean n-disk, $\mathcal{D}^n$ possessed of a distinguished interior *center point* from which all rays intersect its boundary in a unique point. A *star world* is a compact connected subset of $\mathbf{E}^n$ whose boundary is the disjoint union of a finite number of star shaped set boundaries. We now suppose the availability of an implicit representation for each boundary component, $\{\beta_j\}_{j=0}^M$, where $\beta_j \in C^\omega[\mathcal{F}, \mathrm{I\!R}]$ and

$$\partial\mathcal{F} \subseteq \bigcup_{j=0}^{M} \beta_j^{-1}[0],$$

as well as the obstacle center points, $\{q_j\}_{j=0}^M$. Further geometric information required in the construction to follow is detailed in the chief reference for this work [37]. A suitable Euclidean sphere world model, $\mathcal{M}$, is explicitly constructed from this data. That is, we determine (p_j, ρ_j), the center and radius of a model j^{th} sphere, according to the center and minimum "radius" (the minimal distance from q_j to the jth obstacle) of the jth star shaped obstacle. This in turn determines the model space "obstacle functions", $\left\{\hat{\beta}_j\right\}$ as well as the navigation function on $\mathcal{M}, \hat{\varphi}$, as described above.

A transformation, $h : \mathcal{M} \to \mathcal{F}$, may now be constructed in terms of the given star world and the derived model sphere world geometrical parameters as follows. Denote the "j^{th} omitted product", $\Pi_{j=0}^M \beta_j$ as $\bar{\beta}_j$. The "j^{th} analytic switch", $\sigma_j \in C^\omega[\mathcal{F}, \mathrm{I\!R}]$,

$$\sigma_j(q, \lambda) \triangleq \frac{x}{x+\lambda} \circ \frac{\gamma_d \bar{\beta}_j}{\beta_j} = \frac{\gamma_d \bar{\beta}_j b}{\gamma_d \bar{\beta}_j b + \lambda \bar{\beta}_j},$$

(where λ is a positive constant) attains the value one on the jth boundary and the value zero on every other boundary component of $\mathcal{F}$. The "j^{th} star set deforming factor", $\nu_j \in C^\omega[\mathcal{F}, \mathbb{R}]$,

$$\nu_j(q) \triangleq \rho_j \frac{1+\bar{\beta}_j(q)}{\|q-q_j\|},$$

scales the ray starting at the center point of the jth obstacle, q_j, through its unique intersection with that obstacle's boundary in such a way that q is mapped to the corresponding point on the jth model obstacle — a suitable sphere. The overall effect is that the complicated star shaped obstacle is is "deformed along the rays" originating at its center point onto the corresponding sphere in model space.

Definition 2 *The* star world transformation, h_λ, *is a member of the one-parameter family of analytic maps from an open neighborhood, $\tilde{\mathcal{F}} \subset \mathbf{E}^n$, containing $\mathcal{F}$, into $\mathbf{E}^n$, defined by*

$$h_\lambda(q) \triangleq \sum_{j=0}^{M} \sigma_j(q,\lambda)\,[\,\nu_j(q)\cdot(q-q_j)+p_j\,] + \sigma_d(q,\lambda)\,[\,(q-q_d)+p_d\,], \tag{19}$$

where σ_j is the jth analytic switch, σ_d is defined by

$$\sigma_d \triangleq 1 - \sum_{j=0}^{M} \sigma_j, \tag{20}$$

and ν_j is the jth star set deforming factor.

The "switches", make h look like the jth deforming factor in the vicinity of the jth obstacle, and like the identity map away from all the obstacle boundaries. With some further geometric computation we are able to prove the following.

Theorem 6 ([37]) *For any valid star world, $\mathcal{F}$, there exists a suitable model sphere world $\mathcal{M}$, and a positive constant Λ, such that if $\lambda \geq \Lambda$, then*

$$h_\lambda : \mathcal{F} \rightarrow \mathcal{M},$$

is an analytic diffeomorphism.

Thus, if φ is a navigation function on $\mathcal{M}$, the construction of h_λ automatically induces a navigation function on $\mathcal{F}$ via composition, $\tilde{\varphi} \triangleq \varphi \circ h_\lambda$, according to Proposition 4.2.

This family of transformations, mapping any star world onto the corresponding sphere world, induces navigation functions on a much larger class than the original sphere worlds, thus advancing our program of research toward the goal of developing "geometric expressiveness" rich enough for navigation amidst real world obstacles.

5 Conclusion

In recent years there has been a general resurgence of interest in the solution of abstract goals which are expressed by means of a constrained optimization problem. Extending well beyond the increasing body of robotics research [32, 16, 35], this point of view informs the recent activity in neural network research [43], and simulated annealing methods of VLSI design [17], as well. The appeal of cost functions, of course, is that they lead to gradient vector fields: the optimization problem is "solved by integrating" the gradient dynamics on a "network" of digital computers. Gradient vector fields, in contrast to most other classes of nonlinear dynamical systems, are known to possess simple limit sets — the extrema of the cost function. The central result of this paper, Theorem 2, shows that the dynamics arising from the natural motion of appropriately compensated mechanical systems are capable of "integrating out" the limit set of a gradient system as well. Thus the mechanical plant may itself be used as a "second order" analog computer to solve problems encoded via cost functions.

Although Theorem 1 has been known for more than a century and the present extension, Theorem 2, involves the most elementary application of qualitative dynamical systems theory, this work represents (to the best of the author's knowledge) the first systematic use of of total energy for synthesizing desirable global properties in closed loop mechanical control systems. In particular, the notion of navigation functions and demonstration, Theorem 3, that they exist on any reasonable configuration space is entirely new. Their utility is demonstrated by the new global results for satellite attitude tracking summarized in Theorem 4, and for our incipient program of research in robot osbtacle avoidance, summarized by Theorem 5 and Theorem 6.

There are at least two important failings of the theory presented here. The first arises from the "flawed" nature of total energy, η, when considered as a Lyapunov function: its derivative along the motions of a dissipative mechanical system, Δ, vanishes on the entire zero section of the phase space, $T\mathcal{J}$. Recent work [24] has resulted in a family of modified total energy functions whose derivatives vanish only on the equilibrium states of Δ: thus the full power of Lyapunov theory as a tool for studying robustness [20], convergence rates [22], and adaptive capabilities [19] of dissipative mechanical systems is now available. More problematic, Lyapunov analysis, itself, reveals little concerning transient properties of dynamical systems beyond (often crude) convergence rate estimates. While the phase portrait of a gradient system, Γ, is closely related to the level curves of the defining scalar map, φ, it is not at all clear how to "tune" the dissipative field, f_d, to obtain analogous behavior from the dissipative mechanical system, Δ. Preliminary results on a theory of "damping" for Δ have been presented in [19]: more comprehensive research addressing this question is presently in progress.

Acknowledgements

A number of conversations with R. Brockett, E. Rimon, and E. Sontag have been extremely helpful in preparing this manuscript.

References

[1] Ralph Abraham and Jerrold E. Marsden. *Foundations of Mechanics.* Benjamin/Cummings, Reading, MA, 1978.

[2] N. P. Bhatia and G. P. Szegö . *Dynamical Systems: Stability Theory and Applications.* Springer-Verlag, Berlin, 1967.

[3] R. W. Brockett. Control theory and analytical mechanics. In C. Martin and R. Hermann, editors, *Proc. 1976 NASA Ames Research Center Conference on Geometric Control Theory,* pages 1–49, Mathsci Press, 1977.

[4] N. G. Chetaev. *The Stability of Motion*. Pergammon, New York, 1961.

[5] D. R. J. Chillingworth, J. E. Marsden, and Y.H. Wan. Symmetry and bifurcation in three-dimensional elasticity, Part I. *Arch. Rat. Mech. Anal.*, 80(4):295–331, 1982.

[6] John J. Craig. *Introduction to Robotics Mechanics and Control*. Addison Wesley, Reading, MA, 1986.

[7] Theodore Frankel. Critical submanifolds and stiefel manifolds. In Stewart S. Cairns, editor, *Differential and Combinatorial Topology*, pages 37–54, Princeton University, Princeton, 1965.

[8] H. Goldstein. *Classical Mechanics*. Addison-Wesley, Reading, MA, 1980.

[9] Wolfgang Hahn. *Stability of Motion*. Springer-Verlag, New York, 1967.

[10] Morris W. Hirsch. *Differential Topology*. Springer-Verlag, NY, 1976.

[11] Morris W. Hirsch and Stephen Smale. *Differential Equations, Dynamical Systems, and Linear Algebra*. Academic Press, Inc., Orlando, Fla., 1974.

[12] Neville Hogan. Impedance control: an approach to manipulation. *ASME Journal of Dynamics Systems,Measurement, and Control*, 107:1–7, Mar 1985.

[13] T.A.W. Dwyer III. Exact nonlinear control of large angle rotational maneuvers. *IEEE Transactions on Automatic Control*, AC-29(9):769–774, 1984.

[14] R.E. Kalman and J.E. Bertram. "Control systems analysis and design via the 'second method' of Lyapunov". *Journal of Basic Engineering*, 371–392, June 1960.

[15] O. Khatib and J.-F. Le Maitre. Dynamic control of manipulators operating in a complex environment. In *Proceedings Third International CISM-IFToMM Symposium*, pages 267–282, Udine, Italy, Sep 1978.

[16] Oussama Khatib. Real time obstacle avoidance for manipulators and mobile robots. *The International Journal of Robotics Research*, 5(1):90–99, Spring 1986.

[17] S. Kirkpatrick, C. D. Gelatt, and M. P. Vecchi. *Optimization by Simulated Annealing*. Research Report RC 9355 (#41093), IBM Thomas J. Watson Research Center, Yorktown Heights, NY, Apr 1982.

[18] D. E. Koditschek. Application of a new Lyapunov function to global adaptive attitude tracking. In *Proc. 27th IEEE Conference on Decision and Control*, pages 63–68, Austin, TX, Dec 1988.

[19] Daniel E. Koditschek. Adaptive techniques for mechanical systems. In *Fifth Yale Workshop on Applications of Adaptive Systems Theory*, pages 259–265, Center for Systems Science, Yale University, New Haven, CT, May 1987.

[20] Daniel E. Koditschek. High gain feedback and telerobotic tracking. In *Workshop on Space Telerobotics*, pages 355–363, Jet Propulsion Laboratory, California Institute of Technology, Pasadena, CA, Jan 1987.

[21] Daniel E. Koditschek. Natural motion for robot arms. In *IEEE Proceedings 23rd Conference on Decision and Control*, pages 733–735, Las Vegas, Dec 1984.

[22] Daniel E. Koditschek. *Quadratic Lyapunov Functions for Mechanical Systems*. Technical Report 8703, Center for Systems Science, Yale University, Mar 1987 .

[23] Daniel E. Koditschek. Robot control systems. In Stuart Shapiro, editor, *Encyclopedia of Artificial Intelligence*, pages 902–923, John Wiley and Sons, Inc., 1987.

[24] Daniel E. Koditschek. Strict global lyapunov functions for mechanical systems. In *Proc. American Control Conference*, pages 1770–1775, American Automatic Control Council, Atlanta, GA., Jun 1988.

[25] Daniel E. Koditschek and Elon Rimon. Robot navigation functions on manifolds with boundary. *Advances in Applied Mathematics*, (to appear).

[26] J. L. LaGrange. *Méchanique Analytique.* Gauthier-Villars, Paris, 1788.

[27] J. P. Lasalle. *The Stability of Dynamical Systems.* Volume 25 of *Regional Conference Series in Applied Mathematics*, SIAM, Philadelphia, PA, 1976.

[28] A. M. Lyapunov. *Problème Gén éral de la Stabilité du Mouvement.* Princeton University, Princeton, NJ, 1949.

[29] George Meyer. *Design and Global Analysis of Spacecraft Attitude Control Systems.* NASA Technical Report TR R-361, Ames Research Center, Moffett Field, CA, Mar 1971.

[30] J. Milnor. *Lectures on the h-Cobordism Theorem.* Princeton University Press, Princeton, NJ, 1965.

[31] John W. Milnor. *Topology from the Differentiable Viewpoint.* The University Press of Virginia, Charlottesville, Va., 1965.

[32] Fumio Miyazaki and S. Arimoto. Sensory feedback based on the artificial potential for robots. In *Proceedings 9th IFAC*, Budapest, Hungary, 1984.

[33] Marston Morse. The existence of polar non-degenerate functions on differentiable manifolds. *Annals of Mathematics*, 71(2):352–383, Mar 1959.

[34] Jacob Palis, Jr. and Welington de Melo. *Geometric Theory of Dynamical Systems.* Springer-Verlag, New York, 1982.

[35] V. V. Pavlov and A. N. Voronin. The method of potential functions for coding constraints of the external space in an intelligent mobile robot. *Soviet Automatic Control*, (6), 1984.

[36] Ralph Pringle, Jr. On the stability of a body with connected moving parts. *AIAA*, 4(8):1395–1404, Aug 1966.

[37] E. Rimon and D. E. Koditschek. The construction of analytic diffeomorphisms for exact robot navigation on sphere worlds. In *Proc. IEEE International Conference on Robotics and Automation*, (to appear), Arizona, May 1989.

[38] A. J. Van Der Schaft. *Stabilization of Hamiltonian Systems.* Memo 470, Technische Hogeschool Twente, Twente, Netherlands, Jan 1985.

[39] Jacob T. Schwartz and Micha Sharir. On the 'Piano Movers' problem. II. general techniques for computing topological properties of real algebraic manifolds. *Advances in Applied Mathematics*, 4:298–351, 1983.

[40] Jacob T. Schwartz and Micha Sharir. *On the "Piano Movers" Problem I. The Case of a Two-Dimensional Rigid Polygonal Body Moving Amidst Polygonal Barriers.* Technical Report 39, N.Y.U. Courant Institute Department of Computer Science, New York, 1981.

[41] H. Seifert and W. Threfall. *Variationsrechnung im Grossen (Theorie von Marston Morse).* Chelsea, New York, 1948.

[42] Morikazu Takegaki and Suguru Arimoto. A new feedback method for dynamic control of manipulators. *ASME Journal of Dynamics Systems,Measurement, and Control*, 102:119–125, 1981.

[43] David W. Tank and John J. Hopfield. Simple "Neural Optimization Networks": an a/d converter, signal decision circuit, and a linear programming circuit. *IEEE Transactions on Circuits and Systems*, CAS-33(5):533–541, May 1986.

[44] Sir W. Thompson and P. G. Tait. *Treatise on Natural Philosophy.* University of Cambridge Press, 1886, Cambridge.

Center for Systems Science
Yale University, Department of Electrical Engineering
New Haven, CT 06520-2157

Contemporary Mathematics
Volume 97, 1989

CLASSICAL ADIABATIC ANGLES FOR SLOWLY MOVING MECHANICAL SYSTEMS

Jair Koiller[1]

ABSTRACT. Hannay's classical adiabatic angles are computed on several examples with and without symmetries: the Foucault pendulum, isotropic planar oscillators, the rotating elliptical billiard, and the rigid body with slowly varying inertia matrix. The nonintegrable case and the extension to coupled "slow-fast" systems are briefly discussed.

0. HISTORICAL REMARKS. An "adiabatic principle" was invoked by Ehrenfest to justify the old quantum mechanics recipies |1|. Kato made it into a theorem about slowly varying self-adjoint operators on a Hilbert space |2|. He remarked that the case in which the spectrum of each frozen Hamiltonian is discrete and nondegenerate was already treated by Born and Fock, and that the adiabatic transformation is "determined up to a phase factor for each eigenfunction".

Berry |3| was the first to recognize the geometric nature of the phase. Furthermore, Simon showed that the phase is the holonomy of Chern - Bott connection for a bundle of subspaces of the Hilbert space |4|. These papers inspired a great number of theoretical as well as experimental works |5|.

Einstein and Lorentz observed that a classical counterpart should exist for the adiabatic principle, and considered, as an example, a pendulum with slowly varying length |6|. Strangely enough, it is very hard to make the adiabatic principle into a theorem of classical mechanics, when there are two or more degrees of freedom. Landau and Lifschitz |7| outline an argument to show that for analytic Hamiltonians the variation of actions is exponentially small.

Even less attention has been given to the behavior of angle variables, until Hannay found one classical analogue of the quantum phase factor |8|.

1980 Mathematics Subject Classification (1985 Revision): 58F05, 70Q05.
[1]Research supported by a CAPES fellowship. Local expenses at Bowdoin supported by NSF; travel grant by IBM-Brasil.

Hannay's classical adiabatic angles were defined only assuming that each frozen Hamiltonian is integrable. As expected, Hannay's angles had also a geometric character.

1. CLASSICAL ADIABATIC ANGLES. Berry |9| also found a beautiful formula for Hannay's angles. Here is the gist of his argument: Let $H^o(z,x(t))$ a time-dependent Hamiltonian on a symplectic manifold Z; x(t) describes a path on a parameter space X. Suppose that each time frozen Hamiltonian $H^o(\cdot,x)$ is integrable, so that the phase space is foliated by invariant tori. Let $z=(p,q)$ be fixed canonical coordinates on Z and let $p=p(I,\theta,x)$, $q=q(I,\theta,x)$ describe this foliation. If $pdq-Id\theta=d_{(I,\theta)}S(I,\theta,x)$, then it follows that

$$pdq - \overset{o}{H}dt = Id\theta - (H^o-(pq_t-S_t))dt \quad . \tag{1.1}$$

Thus in the coordinates (I,θ) the motion is governed by the Hamiltonian

$$K(I,\theta,t) = H^o(I,x(t)) + H^1(I,\theta,t) \quad , \tag{1.2}$$

where $H^1=-(pq_t-S_t)$. Introduce a "slowness parameter" ε such that C: $x=x(\varepsilon t)$ moves slow relative to the frequencies $\omega=\omega(I,x)=\partial H^o/\partial I$.

Since H^1 is $O(\varepsilon)$, the adiabatic principle can be invoked. The actions I are adiabatically invariant. Berry claims that $\theta(t)$ contains, besides the <u>dynamical term</u> $\theta_o + \int_0^t \omega(I,x(\varepsilon t))dt$, a <u>geometrical term</u> coming from the perturbation:

$$(\Delta\theta)_{geo} = \partial/\partial I \int_C \lambda \qquad \lambda(I) = -<p\cdot q_x-S_x> dx \quad . \tag{1.3}$$

Here < > means averaging with respect to θ, so $\lambda(I)$ is a 1-form on parameter space. If C is closed and X is simply connected, Stokes theorem yields a 2-form which does not depend on the generating function:

$$(\Delta\theta)_{geo} = -\partial/\partial I \iint_R <d_xp \wedge d_xq> d^2x, \partial C = R \quad . \tag{1.4}$$

Notice that Hannay's angles depend only on the variation of the Lagrangian foliation, not on the Hamiltonians H^o. Let $\tau=\varepsilon t$, C: $x=x(\tau)$, $0\leq\tau\leq T$. The final time is T/ε, so $(\Delta\theta)_{dyn}=O(1/\varepsilon)$, $(\Delta\theta)_{geo}=O(1)$. What can be said about the higher order corrections?

We claim that these depend on H^o: indeed, applying normal forms techniques (see §21), based on KBM theory, one can confirm the correctness of Berry's insight. Further, one can show that the $O(\varepsilon)$ term is of the form

$$(\Delta\theta)_1 = \varepsilon\Phi(I,T,\int_0^{T/\varepsilon} \omega(I,\varepsilon t)dt) \tag{1.3}$$

where $\Phi(I,\tau,\theta)$ has zero-average with respect to θ. Not only the third argument but Φ itself depends on H^o through its frequencies.

In |10| we have posed a few questions relative to the general subject of Berry's phase and Hannay's angles. Among these, we argued that one should be able to give an intrinsic formulation for (1.3) or (1.4). This was done by Montgomery |11| and independently by Golin et al. |12|.

2. HANNAY-BERRY PARALLEL TRANSPORT. We present here a souped-down version of Montgomery's intrinsic characterization for the connection whose holonomy are the classical adiabatic angles. We use the canonical formalism as presented in Arnold |13|. Denote F^x: $R^n x T^n \to Z$, $z=z(I,\theta,x)$. The time-dependent family of symplectomorphisms $\phi^t = F^{x(t)}{}_o\ (F^{x(0)})^{-1}$ carries the initial Lagrangian foliation to the foliation corresponding to $x(t)$.

PROPOSITION 2.1. ϕ^t is the flow of the time-dependent Hamiltonian on Z given by $K(p,q,t)=p(I,\theta,t)q_t(I,\theta,t)-S_t$, where the right hand side is computed at $(I,\theta)=(F^{x(t)})^{-1}(p,q)=(I(z,t),\theta(z,t))$.

PROOF. Consider in the extended phase space (p,q,t) the Poincaré-Cartan form $pDq - Kdt$ (we employ now the notation $Dq=d_{(I,\theta)}q+q_t dt$). Pulling back, we get

$$pDq - Kdt = Id\theta + DS .$$

Thus the flow of K is given by $\dot{I}=0$, $\dot{\theta}=0$, i.e., ϕ^t.

PROPOSITION 2.2. The Hannay-Berry parallel transport along $x(t)$ is given by the flow of the Hamiltonian $K(z,t) - \langle K(z,t)\rangle$ where $\langle\ \rangle$ denotes the average with respect to the tori.

PROOF. Represent $K - \langle K\rangle$ in the (I,θ,t) coordinates: you get Berry's formula.

REMARK. In |9|, Berry introduced the two-form $\Lambda=d\lambda$ in order to get rid of the generating function. Observe, however, that $\partial/\partial I\ (-pq_t+S_t)=p_t q_I-p_I q_t$ so that

$$\text{(2.1)} \qquad \mathrm{grad}_I \lambda \ = -\ \langle p_t q_I - p_I q_t\rangle$$

i.e., the average of the <u>Lagrange brackets</u> (I,t). The generating function can be removed already at the level of the 1-form λ.

<u>PROBLEM</u>. The astute reader may object against the local formula $pdq-Id\theta=dS$. What if there is not such a coordinate system (p,q) on (Z,ω), together with this family of globally defined generating functions? We suspect that if F^x is not <u>free</u>, then I is not in general an adiabatic invariant.

Not so trivial are the consequences of a remark by Weinstein that the expression $-\int_R d_x p \wedge d_x q$ is equal to $-\int_D \omega$ where D: $z=z(I_o,\theta_o,x)$, $x \in R$, is a

2-disk inside Z bounded by the curve $z(I_o,\theta_o,x(t))$. Berry's two-form is obtained by averaging the pairing $-(\omega,D)$ over $D=D(I_o,\theta_o)$ with I_o fixed and θ_o describing the torus for $t=0$.

Averaging over the whole Z (for Z compact) gives rise to a new invariant on loops of symplectomorphisms (Weinstein |14|). Can there ideas be useful in the case where Z admits torus actions? Are there non evident relationships with the convexity properties of the momentum mappings?

3. SLOW-FAST SYSTEMS. The geometric constructions, used in the subject of "Berry's phase", can also be applied to the situation where the parameter space is replaced by a set of slow variables, whose dynamics is coupled to the fast ones. Quantum-mechanical examples go back to the Born-Oppenheimer method in nuclear physics |15|. Classical slow-fast systems are also widespread in natural phenomena; for these, analytical methods were extensively developed by the Soviet school, since Kapitza's famous experiments |16,17|. Recent technologies in micromechanics, involving energy transfer from fast micromotion to slow macromotion were described by Prof. R. Brockett at this meeting.

In many cases the slow dynamics can be decoupled (approximately) from the fast. The effect of the averaged fast motion is to introduce an extra, "magnetic-like" force in the slow phase space. In QM this was described as "anomalous commutators" |18|; through semiclassical "dequantization" the classical case was described by Gozzi and Thacker |19|.

We will give a direct approach for a slow-fast natural mechanical system

$$H = p^2/2\mu + p^2/2 + V(r,R) \tag{3.1}$$

where μ is the small parameter, so that (p,r) is the fast motion (see (15)). Assume that for frozen R the subsystem

$$h = p^2/2\mu + V(r,R) \tag{3.2}$$

is completely integrable. So there are action-angle variables (I,θ) and partial canonical transformations (here R is thought as a parameter)

$$p = p(I,\theta,R,\mu) \quad , \qquad q = q(I,\theta,R,\mu) \tag{3.3}$$

which action-anglize the (p,r) subsystems: $h=K(I,R,\mu)$. The mapping $(I,\theta,P,R)\to$ $\to(p,r,P,R)$ is no longer symplectic, since the symplectic form $\omega=dP\wedge dR+dp\wedge dr$ pulls back as

$$\omega = dP\wedge dR + dI\wedge d\theta + (I,R)dI\wedge dR + (R,\theta)dR\wedge d\theta + \Sigma(R_i,R_j)dR_i\wedge dR_j \ , \tag{3.4}$$

where $(u,v)=p_u q_v - p_v q_u$ denote Lagrange brackets.

PROPOSITION 3.1. The (unaveraged) equations of motion are given by

$$(3.5)\qquad \dot{I} = (\theta,R)P \qquad \dot{\theta} = K_I + (R,I)P$$
$$\dot{P} = -K_R + (R,R)P + (R,\theta)K_I \qquad \dot{R} = P$$

with hamiltonian $H=P^2/2+K(I,R,\theta)$ and pulled back symplectic form (3.4).

PROOF. We have $A(\dot{P},\dot{R},\dot{I},\dot{\theta})=\mathrm{grad}_{(P,R,I,\theta)}\, H$, where A is the matrix of the pulled back symplectic form. So it is enough to compute A^{-1}. But a simple calculation gives A=J+C where

$$(3.6)\qquad J = \begin{pmatrix} 0 & 1 & & \\ -1 & 0 & & \\ & & 0 & 1 \\ & & -1 & 0 \end{pmatrix} \qquad C = \begin{pmatrix} 0 & 0 & 0 & 0 \\ 0 & (R,R) & (R,I) & (R,\theta) \\ 0 & (I,R) & 0 & 0 \\ 0 & (\theta,R) & 0 & 0 \end{pmatrix}$$

and one checks that $C^2=0$. Thus $A^{-1}=-J-JCJ$ and the proposition follows.

Since the terms with Lagrange brackets are of smaller order, we can average with respect to θ. Assuming that (3.3) admits a global generating function $pdq-Id\theta=dS(I,\theta,R,\mu)$, it follows that $(\theta,R)=\partial/\partial\theta(pq_t-S_t)$, so I is an adiabatic invariant.

PROPOSITION 3.2. The averaged equations on the slow subsystem are

$$(3.7)\qquad \dot{P} = -K_R + \langle R,R\rangle\, P \quad , \qquad \dot{R} = P$$

where $\langle R;R\rangle$ P is a "magnetic force" (exerts no work). The hamiltonian is $H=P^2/2+K(I,R,\mu)$, with I acting as a parameter, and the symplectic form is

$$(3.8)\qquad \langle\omega\rangle = dP\wedge dR + \Sigma_{i<j}\langle R_i,R_j\rangle\, dR_i\wedge dR_j \quad .$$

PROPOSITION 3.3. Given a solution curve (P(t),R(r)) of the decoupled slow subsystem, the motion of the fast variables (p,r) can be approximately obtained via (3.3) with I=const. and

$$(3.9)\qquad \theta(t) = \theta_0 + (\Delta\theta)_{dyn} + (\Delta\theta)_{geo}$$

$$(3.10)\qquad (\Delta\theta)_{dyn} = \int_0^t K_I(I,R(t),\mu)dt \qquad (\Delta\theta)_{geo} = \int_0^t \langle R,I\rangle\, P(t)dt \quad .$$

It would be interesting to find a rigorous averaging scheme so that terms of higher order could be computed. (μ will be the expansion parameter). Also, one could study further reductions under Lie symmetries.

4. MOVING, STRONGLY CONSTRAINED NATURAL MECHANICAL SYSTEMS. Recall that given a time-independent natural mechanical system L=T-V on a Riemannian manifold M,

a <u>constraining potential</u> to a submanifold $N \subset M$ is a function W_N: $M \to R$ such that $W_N|N \equiv 0$ and the second derivative normal to N is positive definite. Rubin and Ungar, and later Takens |20,21|, characterized the conditions under which the limits, as $\lambda \to \infty$, of the motions in TM for the system $L = T - V - \lambda W_N$, converge to the motions of $L_\infty = T_{|N} - V_{|N}$. Here L_∞ is a Lagrangian <u>on N</u> where $T_{|N}$ is the induced metric.

In the time-dependent context, $W(\cdot,t)$ constraints to a "moving submanifold" N_t. Let $n = n(q,t)$, $q \in R^n$, a parametrization of N_t. Although we will neglect this issue here, one should be able to obtain similar conditions in order that the limit motions will be given by

$$(4.1) \qquad L(q,\dot{q},t) = (1/2)\ |d/dt(n(q(t),t)|^2 - V(n(q,t),t)\ .$$

<u>We specialize to the case where M is acted by a Lie group G_o of isometries and $W(m,t) = W_N(g(t)^{-1}m)$, so that $N_t = g(t)N$. Concerning the potential V, we assume either that (i) $V: M \to R$ is G_o equivariant, or (ii) $V: N_t \to R$ is defined by $V(gn) = V_o(n)$ where $V_o: N \to R$. In the former V is an "external" potential on M; in the latter V_o describes a physical experiment in the "laboratory" N, while N moves "noninertially" inside M.</u>

The Lagrangian (4.1) contains linear terms on the generalized velocities q, which are $O(\varepsilon)$ if $W = W(\cdot,\varepsilon t)$. The standard viewpoint is to study the effects of the fictitious "kinematical forces" on frame N; such is the approach usually taken in studying examples like the Foucault pendulum. We have chosen to follow a different viewpoint, namely: the dynamics is thought as occurring in T*M all along via $H(P_m,t) = (1/2)\| P_m\|^2 + V + \lambda W(m,t)$ with $\lambda \to \infty$.

LEMMA 4.1. The mapping $T^*(M\times R) \to T^*(M\times R)$ given by

$$(4.2) \qquad (p_m,E,t) \to (g(t)p_m,\ E-(J_o(p_m),g^{-1}\dot{g}),t) = (P_{gm},F,t)$$

is exact symplectic, where J_o: $T^*M \to g_o^*$ is the momentum mapping of the G_o action on M lifted to T*M.

COROLLARY 4.2. Let $\tilde{H} = H(P_m,t) + F$ the augmented hamiltonian on T*M with coordinates (P_m,F,t). Via (4.2) it pulls back to

$$(4.3) \qquad \tilde{H}(p_m,E,t) = (1/2)\|p_m\|^2 + V(m) + E - J_o^{g^{-1}\dot{g}}(p_m) + W_N(m)\ .$$

PROOFS. We leave (4.2) to the reader as an exercise on symplectic algebra. Now (4.3) follows from the equivariance assumptions stated above. It is perhaps worth recalling that $g\cdot p_m = (L_g - 1)^* p_m$ and $J_o(p_m) = R_m^*\ p_m$, where L_g: $M \to M$, $m \to gm$ and R_m: $G \to M$, $g \to gm$.

As $\lambda\to\infty$ the system constraints to $N_t \subset M$. Seen by an observer in N, via the mapping (4.2), the system is governed by

$$(4.3)' \qquad \tilde{H}(p_n,E,t) = H^0(p_n) + E + H^1(p_n,t) \quad , \qquad p_n \in T^*N ,$$

with

$$(4.4) \qquad H^0(p_n) = (1/2)\| p_n \|^2 + V(n) \ , \quad H^1(p_n,t) = -J_0^{g^{-1}\dot{g}}(p_n) \text{ (see (5.3))} .$$

REMARK. If the potential $V\colon M\to R$ is not G_0 equivariant (see (i) above) then $V(n)$ in (4.4) must be replaced by $V(gn)$. In case (ii) V_0 can be a time dependent potential. The term H^0 describes the dynamics if $\dot{g}=0$; the correction term H^1 incorporates the motion of N inside M, keeping in mind that we use the pulled-back coordinates (4.2).

5. EQUATIONS OF MOTION. Suppose now that $H^0\colon T^*N\to R$ is invariant under a subgroup $G\subset G_0$, i.e., G fixes N (not necessarily pointwise) and $V\colon N\to R$ is G equivariant. Notice that this second condition is not automatic in case (ii). Assume that the action of G on N is principal; since all we will need are local formulas, we may assume that there is a trivialization $G\times B\equiv N$, which is then lifted to $T^*(G\times B)\equiv T^*N$. The canonical 1-form of T^*N pulls back to $pdq+\alpha_G$, where α_G is the canonical 1-form of T^*G, and (p,q) are coordinates on T^*B. We leave to the reader the proof of the following

LEMMA 5.1. Identifying $T^*G\equiv G\times g^*$ via left translations, $p_h\to(h,L_h{}^*p_h)$, the simplectic form $d\alpha_G$ pulls back to

$$(5.1) \qquad (d\alpha_G)_{h,\mu}(X_1,z_1;X_2,z_2) = z_1(L_h-1)_* X_2 - z_2(L_h-1)_* X_1 - \mu(L_h-1)_* [X_1,X_2]$$

Given a Hamiltonian $H(p_h)=\tilde{H}(\mu,h)$, $\mu=L_h^*p_h$ (not necessarily left invariant), the equations of motion are

$$(5.2) \qquad \dot{\mu} = \mu[\delta\tilde{H}/\delta\mu,\cdot] - L_h^*(\partial\tilde{H}/\partial h) \quad , \quad \dot{h} = (L_h)_* \, \delta\tilde{H}/\delta\mu \quad .$$

Here $\partial\tilde{H}/\partial\mu$ is the Lie-algebra gradient given by $d_\mu\tilde{H}\cdot\eta=(\eta,\delta\tilde{H}/\delta\mu)$.

REMARK. If $H(p_h)$ is left invariant, then $\tilde{H}$ does not depend on h and (5.2) becomes the familiar Euler equations (|13|, appendix 2).

COROLLARY 5.2. Consider a Hamiltonian $H(p_n,E,t)$ on $T^*(N\times R)$. Using the coordinates (p,q,μ,h) on T^*N, $H=\tilde{H}(p,q,\mu,h)$ and the equations of motion are (5.2) coupled with the traditional $\dot{p}=-\partial\tilde{H}/\partial q$, $\dot{q}=\partial\tilde{H}/\partial p$.

CAVEAT. Before writing the equations of motion for (4.3)' it is important to understand carefully the term $H^1(p_n,t)=-p_n\cdot d/dt|_{t=0}\ e^{tX}\cdot n$, $X=g^{-1}\dot{g}$. Since

$X \in g_o$, in general the tangent vector points out of TN. How can we apply p_n to it? Recall that the Legendre transformation on TN $\subset$TM allows us to consider T*N $\subset$ T*M. In other words, (4.4) is actually:

(5.3) $$H^1(p_n,t) = -p_n \cdot \mathrm{proj}_N (R_n)_* X \ .$$

NOTATION. Let $\mathfrak{g}$ be the Lie algebra of G and let $\mathfrak{g}^1$ any complement in the Lie algebra $\mathfrak{g}_o$ (just as vector spaces). Write X=Y+Z with $Y \in \mathfrak{g}$ and $Z \in \mathfrak{g}^1$. Then $H^1(p_n,t)=-J^Y(p_n)-K(p_n,Z)$ where

(5.4) $$K(p_n,Z) = p_n \cdot \mathrm{proj}_N (R_n)_* Z$$

and J: T*N$\to$g* is the momentum mapping of the action of G on T*N.

REMARK. Using the coordinates (p,q,μ,h) on T*N, J writes simply as

(5.5) $$J(p,q,\mu,h) = Ad_h{-1} * \mu$$

However, there is not such a simple expression for K.

6. CASE M=N, G_o=G. Two typical examples are the circular hoop |8| with $T^*N=T^*S^1$, $G=S^1$, and the rigid body whose center of mass is moved arbitrarily. In the latter N=SO(3)=G (think of a gyroscope inside an airplane: it must keep the orientation in spite of the pilot's loops).

Here $J=J_o$ and $K\equiv 0$. From (5.5) it follows immediately that

(6.1) $$\delta J^X/\delta\mu = Ad_h{-1}\ X \quad , \quad L_h^*\ \partial J^X/\partial h = \mu[Ad_h{-1}\ X,\cdot] \ .$$

Applying (6.1) to the equations from Corollary 5.2, a nice cancellation occurs in the equation for μ, and we will get

PROPOSITION 6.1. Under the notations introduced in §5 with M=N, $G=G_o$,

(6.2) $$\dot q = H_p^o(p,q,\mu) \quad , \quad \dot p = -H_q^o(p,q,\mu)$$

(6.3) $$\dot\mu = \mu[\delta H^o/\delta\mu,\cdot] \quad , \quad \dot h = (L_h)_*(\delta H^o/\delta\mu - Ad_h{-1}\ g^{-1}\dot g) \ .$$

Thus the dynamics for q,p,μ is the same as if $g(t)\equiv id$; the influence of the "noninertial motion" is reflected only in the correction term $-Ad_h{-1}\ X$, where X describes the instantaneous change of attitude: $X=g^{-1}\dot g$.

COROLLARY 6.2. The equation of motion for $J=Ad_h{-1}{*}\ \mu$ is

(6.4) $$\dot J = (J, [X,\cdot]) \ .$$

Thus the momentum moves inside its coadjoint orbit; $J(t) \in Orb(J(0))$.

REMARK. Seen in the underlying inertial frame the momentum should be conserved. Indeed, let $\eta(t)=Ad^*_{(g(t)h(t))^{-1}}\,\mu(t)$. Exercise: show that $\dot{\eta}\equiv 0$. Notice also that it is not necessary that g(t) moves slowly.

7. GENERAL CASE: EQUATIONS OF MOTION. Again, under the notations of §5, we decompose X=Y+Z with $Y \in \mathfrak{g}$ and $Z \in \mathfrak{g}^{\perp}$.

PROPOSITION 7.1. The constrained equations on $N_t=g(t)N$, pulled back to T*N, are

$$\begin{aligned} \dot{p} &= -H^o_q(p,q,\mu) + \partial/\partial q\; K^Z(p,q,\mu,h) \\ \dot{q} &= H^o_p(p,q,\mu) - \partial/\partial p\; K^Z(p,q,\mu,h) \\ \dot{\mu} &= \mu[\delta H^o/\delta\mu,\cdot] - \mu[\delta K^Z/\delta\mu,\cdot] + L^*_h\; \partial K^Z/\partial h \\ \dot{h} &= (L_h)_* \,(\delta H^o/\delta\mu - Ad_{h^{-1}} Y - \delta/\delta\mu\; K^Z) \end{aligned} \tag{7.1}$$

Notice that if $g=g(\varepsilon t)$ then X (hence Y and Z) are $O(\varepsilon)$.

PROOF. One just applies Corollary 5.2, noticing that for Z≡0 one should recover (6.2) and (6.3). Also, a short calculation gives

COROLLARY 7.2. The equation of motion for $J=Ad^*_{h^{-1}}\,\mu$ is

$$\dot{J} = (J,\; [Y,\cdot]) + R^*_h\; \partial K^Z/\partial h \tag{7.2}$$

QUERY. What may happen if the second term is transversal to Orb(J(t))? (§15)

8. INTEGRABILITY VIA AN ABELIAN GROUP OF SYMMETRIES. The following situation is often found in applications: G is a maximal torus T^n of a compact Lie group G_o, and dim N=n or n+1. Clearly any G-invariant Hamiltonian H^o on T*N will be integrable. Coordinates θ on G are also coordinates on N via the action; hence at most one more coordinate is needed. The momentum mapping of the induced action on T*N gives n action coordinates, and the last pair of conjugate canonical coordinates can be action-anglized by quadratures.

We adapt the notations of §7 to this setting: $(p,q) \in R^2$ (or do not exist) and $(\mu,h)=(I,\theta) \in R^n\times T^n$. Equations (7.1) become

$$\begin{aligned} \dot{p} &= -H^o_q(p,q,I) + \partial/\partial q\; K^Z(p,q,I,\theta) \\ \dot{q} &= H^o_p(p,q,I) - \partial/\partial p\; K^Z(p,q,I,\theta) \\ \dot{I} &= \partial K^Z/\partial\theta \quad , \qquad \dot{\theta} = \partial H^o/\partial I - Y - \partial K^Z/\partial I \;. \end{aligned} \tag{8.1}$$

Here is natural to take $\mathfrak{g}^{\perp}$ as the Killing orthogonal to $\mathfrak{g}$ inside g^o. Now let (L,ϕ) action-angle variables such that $p=p(L,\phi;I)$, $q=q(L,\phi;I)$ make $H^o(p,q,I)=\tilde{H}(L,I)$. The argument below will hold even if $(p,q) \in R^{2k}$; all that matters is that $H^o(p,q;I)$ should be integrable for each value of I. In general the angle variables ϕ do not have a group theoretical character (see §16).

The frequencies of the unperturbed system are $\tilde{H}_L(L,I)$ for ϕ and for θ we get the expression $\partial H^o/\partial I$ computed at I, and p,q as functions of L,I and a time dependent $\phi(t)=\phi_o+t\,\tilde{H}_L(L,I)$. Thus $\dot{\theta}$ is not uniform; nevertheless, the measure $d\phi d^n\theta$ is preserved. In fact, observe that $dp\wedge dq=dL\wedge d\phi$ + terms involving dI, so that Liouville's form is $dpdqd^nId^n\theta=(dL\wedge d^nI)\wedge(d\phi\wedge d^n\theta)$.

Use of the average principle requires in principle to average with respect both of ϕ and θ. Thanks however to a trick by Atiyah, we are home free:

LEMMA 8.1. The average $\langle K^Z(p,q,I,\theta)\rangle_\theta\equiv 0$.

THEOREM 8.2. There is no holonomy for ϕ. The holonomy for θ is the hodograph of g(t) over the toral algebra t^n

$$(8.2)\qquad (\Delta\theta)_{geo} = -\mathrm{proj}_{t^n}\int_0^t X(u)du \quad , \qquad X = g^{-1}\dot{g} \ .$$

PROOF OF LEMMA. Take $p_n \in T^*N$, Z in the Killing orthogonal to t^n. By the equivariance property of the momentum, $J^Z(g\cdot p_n)=(J(p_n),Ad_gZ)$. Thus it sufficies to show that $\int_G Ad_gZ\ dg=0$. This follows from basic Lie group theory (see e.g., |22|, §4.10): the situation is reduced to the case $G=S^1$ acting on the plane (horizontal components of the angular momentum).

9. A REMARK ON THE NONINTEGRABLE CASE. The situation becomes more complicated if $H^o(p,q,I)$ is not integrable. Imagine, however, that the time average along a particular solution curve $(p(t),q(t),\theta(t))$ can be replaced by a space average with respect to some measure. If this measure is a product containing $d^n\theta$, then (8.2) will still be true. This happens for instance if the systems $H^o(p,q,I)$ are ergodic on a compact region. It would be interesting to find an example so biased, that K^Z will give an important contribution to the holonomy.

10. INTEGRABILITY WITHOUT SYMMETRIES. Here G=id, i.e., the Hamiltonian H^o on T*N does not have Lie symmetries. Nonetheless, H^o may yet be integrable due to some miracle (coming, say, from a Lax trick, or from algebraic geometry). So in (7.1) μ,h disappear and $p=p(I,\theta)$, $q=q(I,\theta)$, where we stress again the non-geometric nature of I,θ. Moreover, $H^o=H^o(I)$. The domain of (4.3)' can be pulled back to I,θ,E,t and the system is governed by

$$(10.1)\qquad \tilde{H}(I,\theta,E,t) = H^o(I) + E - J^X(I,\theta) \quad , \qquad X = g^{-1}\dot{g} \ .$$

PROPOSITION 10.1. Berry's 1-form is left-invariant in parameter space G_o. Its value at the identity element is (see the notation of §1)

$$(10.2)\qquad \lambda(I) = -\langle J(I,\theta)\rangle \ .$$

PROPOSITION 10.2. For a natural mechanical system being transported inside an Euclidian space, G_o is the group of rigid motions $T(R^3)_{s.d.} \oplus SO(3)$. Then λ vanishes on the translations. If J_{ang} is the total angular momentum (under the usual identifications $sO(3)=sO(3)^*=R^3$), then $\lambda(I)=-\langle J_{ang}(I,\theta)\rangle$.

PROOF. A uniform motion does not change Newton's equations, so produces no holonomy (for an algebraic proof, see my technical report |23|, available upon request).

EXERCISE. Suppose that H^o is time-dependent (each frozen one integrable). Thus not only $N_t=g(t)N$ is moving, but the dynamics inside N changes adiabatically. Calculate Berry's 1-form.

HINT. Compose the mapping (4.2) with another symplectic mapping sending (I,θ,E_1,t) to a point in $T^*(N\times R)$. Caveat: the energy value will change.

11. EXAMPLES. Before reading any further, the reader should try to assess the degrees of difficulty of the following:

(i) isotropic oscillators transported along space or surface curves.

(ii) a 3-dimensional top inside R^4 such that the axis e_4 is adiabatically moved.

(iii) the Foucault pendulum; what happens if gravity is turned "off"?

(iv) the rotating hoop (Hannay, Berry) and the rotating elliptical billiard.

(v) the rigid body with slowly varying matrix of inertia.

The reader should amuse him(her)self by cooking up other examples.

12. ISOTROPIC OSCILLATOR TRANSPORTED ALONG A SAPCE CURVE. Let a planar isotropic harmonic oscillator be adiabatically transported along a curve C in R^3. The Frenet frame $R=(\vec{t},\vec{n},\vec{b})$ satisfies the Frenet-Serret equations |24, §1.6|

$$R^{-1}dR = (-\tau,\ 0\ ,\ -\kappa)\ ds \quad . \tag{12.1}$$

Use polar coordinates (r,θ) on the plane N. Applying (8.2) we get:

(i) The oscillator is transported along the oscullating plane $N_t=\{\vec{t},\vec{n}\}$. Here the Lie algebra of $G(\cong S^1)$ is the third axis, and

$$(\Delta\theta)_{geo} = \int_C \kappa\ ds \tag{12.2}$$

(ii) $N_t=\{\vec{n},\vec{b}\}$ (normal plane). Then the Lie algebra is the first axis, so

$$(\Delta\theta)_{geo} = \int_C \tau\ ds \tag{12.3}$$

(iii) $N_t=\{\vec{t},\vec{b}\}$. There is no holonomy.

REMARK. The holonomy (12.3) has been verified in optical fibers. See |25,26|.

REMARK. The reader can verify that a <u>nonisotropic</u> planar oscillator satisfies $\langle J\rangle \equiv 0$. Berry's 1-form is zero, but "purely topological" holonomies can arise from the nontriviality of the torus bundle over parameter space. This is discussed in |10, §3.6| and in more detail by Montgomery |11, §6|. We illustrate the phenomenon with a simple example. Consider a parametrized family of 2-degrees of freedom oscillators. The parameter space is here the neighborhood of an umbilical point on a two dimensional surface, like an ellipsoid of three unequal axis |24, §2,6|. Suppose that the kinetic energy is given by the first fundamental form, and the potential energy by the second.

Then the normal modes are tangent to the lines of curvature, and the oscillator at the umbilic point is isotropic. If we perform an adiabatic excursion around the umbilic point, then the phases are changed by 180º! However, if we compute Berry's 1-form, the answer is identically zero. The problem here is <u>not</u> caused by the parameter space, since it is simply connected. Our interpretation for the trouble is that the action-angle variables cannot be defined globally. This is similar to Duistermaat's monodromy |27|. The quantum mechanical analogue is the behaviour of Hermitian matrices near double eigenvalues, which was already discussed by Berry in his seminal paper |1|.

13. ISOTROPIC OSCILLATOR TRANSPORTED ALONG A SURFACE CURVE. The adequate frame here is Darboux's $R=(\vec{t},\vec{n},\vec{N})$ where $\vec{t}$ is tangent to the curve C and $\vec{N}$ is normal to the surface. The structure equations are |24, §4.1| $R^{-1}dR=(\tau_g,-\kappa_n,\kappa_g)ds$ where κ_n is the normal curvature, and κ_g (resp. τ_g) the geodetic curvature and torsion. It follows from (8.2) that $(\Delta\theta)_{geo}$ is <u>minus</u> the total geodetic curvature along C. This was known already by J. Radon |28| who found the "mechanical implementation of Levi-Civita's parallel transport". Radon's result has been rediscovered recently, |29|; compare theirs with the approach we use.

14. HANNAY'S TOP. In |8| Hannay considers a "spinning symmetrical top with fixed base point whose axis is pulled around a closed circuit (in S^2) enclosing a solid angle Ω". Just by a physical reasoning he is able to conclude that the holonomy is the solid angle Ω itself. On the other hand, our previous section will predict, by the Gauss-Bonnet theorem,

$$(14.1) \qquad (\Delta\theta)_{geo} = -\int\kappa_g = -2\pi + \iint K dA = -2\pi + \Omega \quad .$$

The discrepancy comes from the way we are measuring the angle variable θ, namely, from the tangent vector $\vec{t}$ to the curve $C \subset S^2$ described by the axis. $\vec{t}$ makes a complete 2π turn around C (Hopf's "umlaufsatz") so an extra 2π must be added to (14.1) to get the correct result. The same remark holds for §13 so there we will get

(14.2) $(\Delta\theta)_{geo}$ = total curvature of the region enclosed by C.

We cannot refrain ourselves from presenting yet another argument for the symmetrical top. It sufficies to consider curves $C \subset S^2$ bounding a region that omits two antipodal points, $\pm p$. Geographic coordinates, with $\pm p$ as poles, yield a map $S^2-\{\pm p\} \to SO(3)$, $e_3 \to (e_1(e_3), e_2(e_3), e_3) = R$. Clearly $R^{-1}dR = (-e_2 de_3, e_1 de_3, -e_1 \cdot de_2)$. Berry's 1-form $e_1 \cdot de_2$ can be pulled back to S^2, and Stokes theorem gives the 2-form $de_1 \cdot de_2$ which is easily recognized as the area element of the sphere.

15. GENERALIZED HANNAY'S TOP. Consider inside R^4 a rigid body contained in 3-space e_1, e_2, e_3 (with different moments of inertia I_1, I_2, I_3). One would like to know what happens when e_4 is pulled around a closed circuit in S^3.

In the setting of §4, we have here N=SO(3), M=SO(4) (extend the left invariant metric T of SO(3) to M in an arbitrary fashion). SO(4) has also the standard bi-invariant metric I. The Lie groups of the problem are also G=SO(3) and G_0=SO(4). The Lie algebras are $\mathfrak{g} = so(3) \cong R^3$ and $so(4) \cong R^6$. Let $\mathfrak{g}^\perp$ be the complement of $\mathfrak{g}$ with respect to the bi-invariant metric.

The manifold B from §5 is here just a point, so the coordinates (p,q) do not exist here. The function K^Z is given by

$$(15.1) \qquad K^Z(\mu,h) = (\mu, \mathrm{proj}_{\mathfrak{g}}^T h^{-1} Z h)$$

where we project over $\mathfrak{g}$ according to metric T.

The third equation in (7.1) is here

$$(15.2) \qquad \dot{\mu} = \mu[\delta T/\delta\mu, \cdot] + (\mu, \mathrm{proj}_{\mathfrak{g}}^T [\mathrm{proj}_{\mathfrak{g}^\perp}^T h^{-1} Zh, \cdot])$$

which is therefore coupled to the fourth. The analysis of this problem will be the subject of a future work. We antecipate that one could be able to pump momentum to the rigid body via an adiabatic excursion out of SO(3). This is because we believe that, in general, the second term is transversal to the coadjoint orbit.

16. THE FOUCAULT PENDULUM. The spherical pendulum in the regime of small oscillations can be approximated by an isotropic harmonic oscillator. Let C a parallel of colatitude α on the earth (α measured from the north pole). In the notation of §13, the tangent vector $\vec{t}$ points "east", $\vec{n}$ points "north" and $\vec{N}$ to the "sky". This is the natural frame for an observer moving with the earth.

Thus equation (14.1) must be applied without correction. Hence

$$(16.1) \qquad (\Delta\theta)_{geo} = -2\pi + \iint K dA = -2\pi + 2\pi(1-\cos\alpha) = -2\pi\cos\alpha \quad .$$

Given $I \neq 0$, corollary 8.2 guarantees that (16.1) is valid even for large oscillations (big energies), as long as the two frequencies of motion are much greater than $2\pi/(1 \text{ day})$. However, the usual Foucault experiment is done with a pendulum starting on a vertical plane, with $\dot{\theta}=0$. We cannot apply lemma 8.1!

Let's see why the result is still true. Suppose that the system is being transported (along a smooth road) inside a box whose center describes a curve $C(t)$ and whose sides are parallel to moving axis $R=(e_1,e_2,e_3)$, with $-e_3$ pointing to the earth center. Use as coordinates for the unperturbed system, (I,θ,p,q) where q is the angle from the negative vertical axis $-e_3$ to the pendulum bob. We have the familiar Hamiltonian

$$(16.2) \qquad H^0 = (1/2m\ell^2)(p^2+I^2/\sin^2 q) - mg\ell\cos q, \quad I = m\ell^2\sin^2 q\dot{\theta} \ , \ p = m\ell^2\dot{q} \ .$$

A short calculation gives for the total angular momentum

$$(16.3) \qquad J = (\sin\theta\cdot p+\cos\theta\cdot I \operatorname{cotg} q, -\cos\theta\cdot p+\sin\theta\cdot I \operatorname{cotg} q, I)$$

For the full Hamiltonian (4.3)' we also need the linear momentum

$$(16.4) \qquad P = (1/\ell)(-I\sin\theta/\sin q+p\cos q\cos\theta, I\cos\theta/\sin q+ +p\cos q\sin\theta, p\sin q)$$

The perturbative term (5.3) is here

$$(16.5) \qquad H^1(p,q,I,\theta) = -P\cdot C'(t) - J\cdot X(t)$$

where $X(t)$ is the vector associated to the antisymmetric matrix $R^{-1}\dot{R}$,

$$(16.6) \qquad X = (-e_2\cdot\dot{e}_3, e_1\dot{e}_3, -e_1\dot{e}_2)$$

Let's obtain the full-blown Hamiltonian for the Foucault case $C(t)$, $\vec{t},\vec{n},\vec{N}$. Let ρ be the earth radius, Ω its angular velocity. Some elementary differential geometry gives

$$(16.7) \qquad dC/dt = (ds/dt)\vec{t} \quad , \qquad ds/dt = \Omega\rho\sin\alpha$$

$$(16.8) \qquad X = (\tau_g, -\kappa_n, \kappa_g)ds/dt \quad , \quad \tau_g = 0 \quad , \quad \kappa_n = -1/\rho \quad , \quad \kappa_g = \operatorname{cotg}\alpha/\rho$$

We have written all the ingredients for the complete Hamiltonian $H=H^0+H^1$. It would be interesting to study it numerically, or via perturbation theory; the latter requires passing (I,θ,p,q) in (16.2) to action-angle variables (I,θ are not true action-angle variables for this problem (see §22)!). The aim of this study would be to determine the range of values I,J such that the adiabatic approximation yielding (16.1) is good.

We proceed with heuristic reasonings. We want to justify why the Foucault experiment works with I=0. The unperturbed Hamiltonian is the vertical pendulum $H^o=(1/2m\ell^2)p^2-\ell mg\cos q$. From (16.3) and (16.4) we get

(16.9) $\quad \partial P/\partial I = (-\sin\theta/\sin q, \cos\theta/\sin q, 0)\ell$

(16.10) $\quad \partial J/\partial I = (\cos\theta\,\mathrm{cotg}\,q, \sin\theta\,\mathrm{cotg}\,q, 1)$.

All functions appearing in these expressions (except the constant 1) will average zero for any pendulum motion, be it a libration or a circulation, just by symmetry properties. It seems that (16.1) holds because, in the notation of §8, $<\partial/\partial I\ K^Z>_\phi=0$. The failure of lemma 8.1 is avenged, provided we overlook the fact that these functions blow up at $q=0,\pi$ (averaging out a big vectorfield is dangerous). We may want also to look at

(16.11) $\quad \partial J/\partial\theta_{I=0} = (p\cos\theta, p\sin\theta, 0)$

(16.12) $\quad \partial P/\partial\theta_{I=0} = (-p\cos q\sin\theta, p\cos q\cos\theta, 0)$.

Again, all these functions average zero for the libration regime, by symmetry. However, p and p cos q do not average zero for the circulation regime. One hopes that in this case I will go to a nonzero value such that averaging over θ is again justified. At any rate, the adiabatic approximation should be poorer for small I and energies near and bigger that of the unstable equilibrium.

Another way to think about the last assertion is to reduce the value of gravity g. When $g\equiv 0$ we will be back to the setting of §6.

Essentially all of the above conclusions were stated by Montgomery |11, §4|.

17. HANNAY'S SLOWLY ROTATING PLANAR HOOP |8,9|. A bead of unit mass moves on a curve C: $x=x(s)$, $y=y(s)$ parametrized by arc length s. Let L be the total length of C. Action-angle coordinates are given by $\theta=(2\pi/L)s$ and $I=(L/2\pi)\dot{s}$, since $dI\wedge d\theta=d\dot{s}\wedge ds$. We are in the context of §10, so

(17.1) $\quad (\Delta\theta)_{geo} = (2\pi)\ d/dI\{(1/2\pi)\int_0^{2\pi} -J_3\ d\theta\}$.

The first factor 2π arises from the integral over parameter space S^1 (the hoop is tied to a slowly rotating plane). The expression in brackets is the average value of minus the angular momentum $J_3 = x\dot{y} - y\dot{x}$. Now,

$$(17.2)\qquad -J_3\, d\theta = (y\dot{x} - x\dot{y})d\theta = (ydx - xdy)\dot{\theta} = (ydx - xdy)(2\pi/L)\dot{s}$$
$$= (2\pi/L)^2\, I(ydx - xdy)\quad .$$

Therefore by Green's theorem we have

$$(17.3)\qquad (\Delta\theta)_{geo} = 4\pi^2/L^2 \int_C ydx - xdy = -8\pi^2 A/L^2 \quad (A = \text{area bounded by } C).$$

18. THE SLOWLY ROTATING ELLIPTICAL BILLIARD. Esoteric as it may seem, in reality this system appears as a model in optical fiber theory |30,31| and in nuclear physics |32|. The elliptical billiard has been treated recently by several authors |33,34|; actually it can be seen as a degenerate case of the celebrated problem of Jacobi's geodesics on a triaxial ellipsoid |35|.

Jacobi's treatment is based on the confocal quadrics coordinates λ_i

$$(18.1)\qquad \begin{aligned} &1 = \Sigma_{i=1}^3\, x_i^2/(a_i+\lambda)\quad , \quad a_3 > a_2 > a_1 > 0 \\ &-a_3 < \lambda_3 < -a_2 < \lambda_2 < -a_1 < \lambda_1 \quad . \end{aligned}$$

We specialize to $x_1 = a_1 = \lambda_1 \equiv 0$. The ellipsoid becomes a two-sided elliptical table, parametrized by $-a_3 < \lambda_3 < -a_2 < \lambda_2 < 0$. The curves $\lambda_2 = ct.$ are ellipses, and the curves $\lambda_3 = ct.$ are hyperbola. The range of parameters are given in the figure below. There are 8 copies of the fundamental region in the first quadrant.

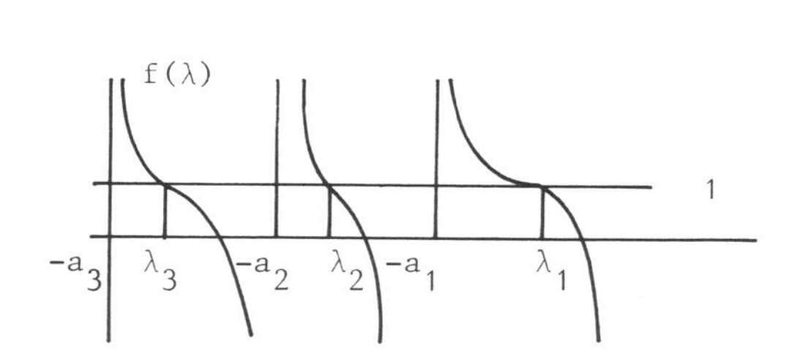

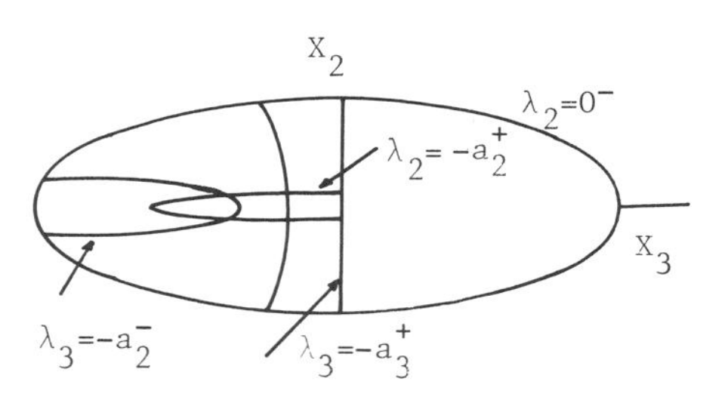

The kinetic energy of a unit mass particle is

$$(18.2)\qquad \begin{aligned} &H = (1/8)(M_2\, \dot{\lambda}_2^2 + M_3\, \dot{\lambda}_3^2) = 2(p_2^2/M_2 + p_3^2/M_3) \\ &M_2 = (\lambda_2-\lambda_3)/(a_2+\lambda_2)(a_3+\lambda_2)\quad , \quad M_3 = (\lambda_3-\lambda_2)/(a_2+\lambda_3)(a_3+\lambda_3)\quad . \end{aligned}$$

Let $S = S_2(\lambda_2, h, \beta) + S_3(\lambda_3, h, \beta)$ a trial generating function for a canonical transformation $(q_h, q_\beta, h, \beta) \to (\lambda_2, \lambda_3, p_2, p_3)$ with

$$(18.3)\qquad p_2 = \partial S/\partial\lambda_2\quad , \quad p_3 = \partial S/\partial\lambda_3\quad , \quad q_h = \partial S/\partial h\quad , \quad q_\beta = \partial S/\partial\beta\ .$$

The Hamilton-Jacobi equation $h=2(S^2_{\lambda_2}/M_2+S^2_{\lambda_3}/M_3)$ separates:

$$(a_2+\lambda_2)(a_3+\lambda_2)S^2_{\lambda_2} - h\lambda_2/2 = (a_2+\lambda_3)(a_3+\lambda_3)S^2_{\lambda_3} - h\lambda_3/2 = h\beta/2 \tag{18.4}$$

where the second equality defines the common value.

PROPOSITION 18.1. The generating function can be obtained by elliptic integrals, and (18.1) gives the inequalities $-\lambda_2<\beta<-\lambda_3$,

$$\begin{aligned}(a_2+\lambda_2)(a_3+\lambda_2)S^2_{\lambda_2} &= h(\beta+\lambda_2)/2 > 0\\ (a_2+\lambda_3)(a_3+\lambda_3)S^2_{\lambda_3} &= h(\beta+\lambda_3)/2 < 0\end{aligned} \tag{18.5}$$

PROPOSITION 18.2. The product J_1J_2 of the angular momenta with respect to the foci is a second constant of motion. In fact, $J_1J_2=2h(a_2-\beta)$.

This can be proven by a direct calculation. We will consider only the regime of motion where $J_1J_2>0$. Hence $0<\beta<a_2$ and (18.5) implies that $\lambda_2 \in(-\beta,0)$. The motion is therefore confined to the region bounded by the confocal ellipses $\lambda_2=0$, $\lambda_2=-\beta$. This region must be duplicated, because the table has two sides; the copies are sewn along the outer ellipses. Moreover, this folded ribbon itself is the projection of the Lagrangian torus

$$\begin{aligned}p_2^2 &= (h/2)(\beta+\lambda_2)/\{(a_2+\lambda_2)(a_3+\lambda_2)\}\\ p_3^2 &= (h/2)(-\beta-\lambda_3)/\{(-a_2-\lambda_3)(a_3+\lambda_3)\}\end{aligned} \tag{18.6}$$

so two copies of the folded ribbon are to be sewn along the inner ellipses. In conclusion, each torus is built out of 16 copies of a fundamental region in the first quadrant $x_1>0$, $x_2>0$.

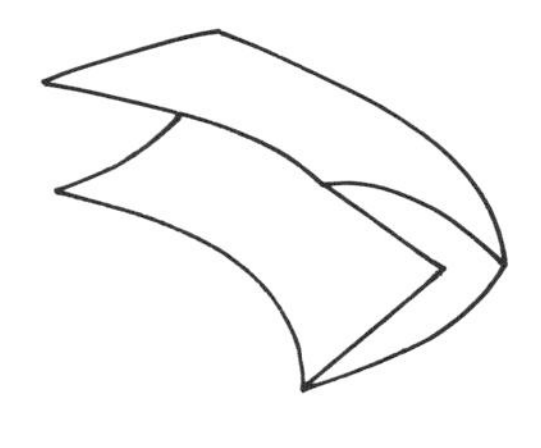

a. 1/4 of a torus

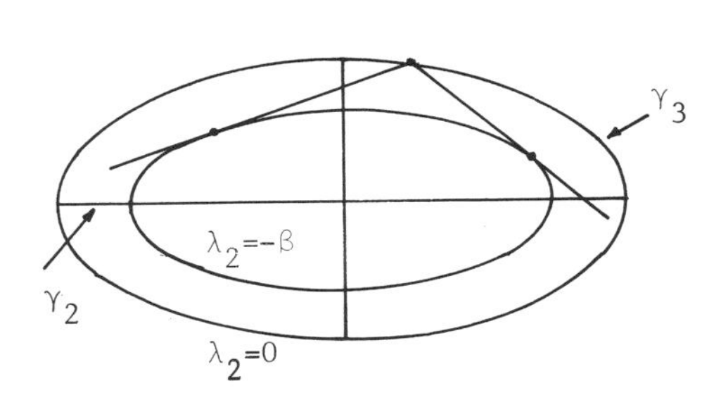

b. the regime $J_1J_2>0$

The generators for the homology group of the torus (h,β) are the curves γ_2: $\lambda_3=-a_3$ and γ_3: $\lambda_2=0$. The actions $2\pi I_j=\int_{\gamma_j} p_{\lambda_2}d\lambda_2+p_{\lambda_3}d\lambda_3$ are given by the elliptic integrals

$$2\pi I_2 = (h/2)^{1/2} f_2(\beta) \quad , \qquad 2\pi I_3 = (h/2)^{1/2} f_3(\beta)$$

$$f_2(\beta) = 4\int_{-\beta}^{0} \{(\lambda_2+\beta)/[(a_2+\lambda_2)(a_3+\lambda_2)]\}^{1/2} d\lambda_2 \tag{18.7}$$

$$f_3(\beta) = 4\int_{-a_3}^{-a_2} \{(-\lambda_3-\beta)/[(-a_2-\lambda_3)(a_3+\lambda_3)]\}^{1/2} d\lambda_3$$

The angle variable θ_i just rescales the corresponding λ_i coordinate; this depends only on β:

$$d\theta_2/d\lambda_2 = (2\pi/f_2(\beta))\ \{(\lambda_2+\beta)/[(a_2+\lambda_2)(a_3+\lambda_2)]\}^{1/2} \tag{18.8}$$

$$d\theta_3/d\lambda_3 = (2\pi/f_3(\beta))\ \{(-\lambda_3-\beta)/[(-a_2-\lambda_3)(a_3+\lambda_3)]\}^{1/2}$$

Fix the direction of motion so that the angular momentum $J>0$. One gets, after a lengthy calculation,

$$J = \{(2h)^{1/2}/(\lambda_2-\lambda_3)\}\ \{Y(\lambda_2,\lambda_3) \pm Z(\lambda_2,\lambda_3)\} \tag{18.9}$$

$$Y = (a_2+\lambda_2)(a_3+\lambda_2)(-\beta-\lambda_3) \quad , \qquad Z = (-a_2-\lambda_3)(a_3+\lambda_3)(\beta+\lambda_2) \quad .$$

Each factor in Y is greater than the corresponding factor in Z. Moreover, the plus and minus signs correspond, respectively, to regions where $\dot{\lambda}_2$ and $\dot{\lambda}_3$ have opposite (resp. equal) signs.

Now, according to Proposition 10.1, Berry's 1-form is minus the average of J with respect to the θ_i's. This can be done just on the fundamental region on the first quadrant, averaging first the two choices of signs. We get:

PROPOSITION 18.3.

$$\langle J\rangle = \{4(2h)^{1/2}/\pi^2\} \int_0^{\pi/2}\int_0^{\pi/2} d\theta_2\ d\theta_3\ f(\lambda_2,\lambda_3) \tag{18.10}$$

$$f(\lambda_2,\lambda_3) = \{(a_2+\lambda_2)(a_3+\lambda_2)(-\beta-\lambda_3)\}^{1/2}/(\lambda_2-\lambda_3)$$

As a check, let's make $\beta\to 0^+$, so that the billiard ball moves tangentially to the outer ellipse. The variable λ_2 is squeezed to 0^-, the corresponding average over θ_2 disappears, and we must compute "only"

$$\langle J\rangle = (2ha_2a_3)^{1/2} R/S$$

$$R = \int_{-a_3}^{-a_2} d\lambda_3/\{(-a_2-\lambda_3)(a_3+\lambda_3)\}^{1/2} = \pi \tag{18.11}$$

$$S = \int_{-a_3}^{-a_2} d\lambda_3 \cdot \{-\lambda_3/(-a_2-\lambda_3)(a_3+\lambda_3)\}^{1/2} = L/2$$

where L is the length of the outer ellipse. Hence $<J> = 2(2h)^{1/2}A/L$ where A is the area.

(The first integral can be obtained by residues, while the second is recognized as an arc length by a reparametrization $x=(-a_2)^{1/2}\cos\phi$, $y=(-a_3)^{1/2}\sin\phi$. The residue calculation can be summarized as follows: make a cut on the Riemann sphere along the segment $-a_3,-a_2$. The integrand can be made analytic on C-{cut}, so 2R, the integral over a loop around the cut, is equal the integral around the point at infinity. It is readily seen that the residue there is 1.).

Finally, taking into account that $\dot{s}=(2h)^{1/2}=2\pi I/L$ for any hoop, we get

$$<J> = 4\pi\ IA/L$$

$$(\Delta\theta)_{geo} = -8\pi^2 A/L^2 \qquad \text{(as already known by Berry and Hannay)}.$$

What is the point of going through all this algebra, just to obtain the same result (17.3)? It turns out that (18.10) can be explicitly calculated:

19. EVALUATION OF $<J>$ FOR $0 < \beta < a_2$. Substituting (18.8) into (18.10) we get immediately

$$<J> = (2h)^{1/2}\ C/AB$$

$$A = \int_{-\beta}^{0} \{(\lambda_2+\beta)/[(a_2+\lambda_2)(a_3+\lambda_2)]\}^{1/2} d\lambda_2$$

$$\text{(19.1)} \qquad B = \int_{-a_3}^{-a_2} (-\lambda_3-\beta)/[(-a_2-\lambda_3)(a_3+\lambda_3)]\}^{1/2} d\lambda_3$$

$$C = \int_{-a_3}^{-a_2} d\lambda_3 (-\beta-\lambda_3) g(\lambda_3,\beta)/\{(-a_2-\lambda_3)(a_3+\lambda_3)\}^{1/2}$$

where

$$\text{(19.2)} \qquad g(\lambda_3,\beta) = \int_{\lambda_2=-\beta}^{0} \{(\lambda_2+\beta)^{1/2}/(\lambda_2-\lambda_3)\} d\lambda_2 \quad .$$

Gradshtein-Ryzhik's table |36| gives (p. 71)

$$\text{(19.3)} \qquad g = 2\sqrt{\beta} - 2(-\beta-\lambda_3)^{1/2}\ \mathrm{arctg}\{\beta/(-\beta-\lambda_3)\}^{1/2} \quad .$$

For $0<\beta<a_2/2$ we can expand

$$\text{(19.4)} \qquad \mathrm{arctg}\{\beta/(-\beta-\lambda_3)\}^{1/2} = \sum_0^{\infty} \{(-1)^n/(2n+1)\}\{\beta/(-\beta-\lambda_3)\}^{(2n+1)/2}$$

so that

$$(19.5)\qquad C = \frac{2\pi\ \beta^{3/2}}{3} + \frac{\beta^{(n+1/2)}}{2n+1}\int_{-a_3}^{-a_2} \frac{d\lambda_3}{\{(-a_2-\lambda_3)(a_3+\lambda_3)\}^{1/2}(\beta+\lambda_3)^{n-1}}\ .$$

These integrals can be evaluated by residues, and we obtain:

$$(19.6)\qquad C = 2\pi\{\beta^{3/2} + \sum_{n\geq 2} \frac{\beta^{n+(1/2)}}{2n+1}\ h_{n-2}(\beta,a_2,a_3)$$

where we have the generating series

$$(19.7)\qquad \sum_{k\geq 0} h_k(z+\beta)^k = \{(z+a_2)(z+a_3)\}^{-1/2}\ .$$

Furthermore, from Byrd-Fridman's table |37| we get

$$(19.8)\qquad A = 2(a_2-\beta)\{\mathrm{dn}u\mathrm{tn}u - E(\phi,\kappa)\}/[\kappa'^2(a_3-\beta)^{1/2}]\}$$

$$(19.9)\qquad B = 2(a_2-\beta)E(\kappa)/\{\kappa'^2(a_3-\beta)^{1/2}\}$$

where

$$(19.10)\qquad \kappa^2 = (a_3-a_2)/(a_3-\beta)\quad ,\quad \mathrm{sn}u = \sin\phi = (\beta/a_2)^{1/2}\ .$$

It would be interesting to make an expansion in powers of β near $\beta=0$, in order to get the corrections to the holonomy of the 1-degree of freedom hoop. In the future we plan to pursue an in depth study of pool tables (theory and practice).

20. THE RIGID BODY WITH SLOWLY VARYING MATRIX OF INERTIA. Although this seems a contraction in terms, we have in mind a system such as a space structure, consisting of N massive points m_i connected by very light rods on universal joints. The lengths ℓ_{ij} of the rods can be adjusted by internal control mechanisms. These distances form a 3N-6 dimensional parameter space.

Except for a lower dimensional singularity set, there is a mapping F

$$(20.1)\qquad (\ell_{ij};\ m_i) \to (Q_1,\ \ldots,\ Q_N) \in R^{3N}$$

such that (i) the center of mass of the Q_i is at the origin; (ii) the mutual distances are precisely the ℓ_{ij}; (iii) the ellipsoid of inertia is in the standard position (the principal axis are aligned with the vectors $\vec{i},\vec{j},\vec{k}$ and $I_1>I_2>I_3$; these are also functions of ℓ_{ij},m_i).

REMARK. The bifurcation set consists of values of m_i, ℓ_{ij} such that two or three principal moments of inertia coincide. We will assume that this will not happen during the course of a variation $\ell_{ij}(t)$.

REMARK. Obtaining the map F seems to be a not so easy exercise on implicit functions. Old advanced calculus problems are back in fashion, due to the advances in computer graphics and its applications to topology |38|.

REMARK. Using the Jacobian matrix of F one can express $\dot{Q}_i$ as a linear function of the $\dot{\ell}_{ij}$. This trivial observation will be valuable soon.

The positions of masses m_i relative to the center of mass can now be described by the attitude matrix $R \in SO(3)$ such that $q_i = RQ_i$. We are following Arnold's notation, according to which the capital letters designate entities in the body's frame. Differentiating, we get $\dot{q}_i = \dot{R}Q_i + R\dot{Q}_i$, so that

$$(20.2) \qquad \vec{m} = \Sigma\, m_i\, q_i \times \dot{q}_i = \Sigma\, m_i\, RQ_i \times (\dot{R}Q_i + R\dot{Q}_i) = RN \quad ,$$

where the modified body angular momentum vector N is given by

$$(20.3) \qquad N = A(t)\Omega + f(\ell_{ij}, \dot{\ell}_{ij}) \quad , \quad \Omega = R^{-1}\dot{R} \quad , \quad A = \mathrm{diag}(I_1, I_2, I_3) \ .$$

Here the vector f is given by

$$(20.4) \qquad f(\ell_{ij}, \dot{\ell}_{ij}) = \Sigma\, m_i Q_i \times \dot{Q}_i$$

which in principle can be explicitly "computed" via the mapping (20.1).

PROPOSITION 20.1. The body angular momentum N obeys Euler's equations

$$(20.5) \qquad \dot{N} = N \times \mathrm{grad}_N H \quad , \quad H(N,t) = (1/2)(A^{-1}(t)N,N) - (A^{-1}(t)f(t),N) \quad .$$

PROOF. The total angular momentum vector $\vec{m}$ is conserved, $0 = \dot{R}N + R\dot{N}$, so that $\dot{N} = N \times \Omega$. Solve for Ω in (20.3) and plug it back.

REMARK. f is a linear function of the $\dot{\ell}$'s.

If the variations $\ell_{ij} = \ell_{ij}(\varepsilon t)$ are slow compared with the frequency of the frozen 1-degree of freedom systems $H^o = (1/2)(A^{-1}(\varepsilon t)N,N)$ (on the coadjoint orbits $|N| = \mathrm{const.}$), we can apply perturbation methods.

We outline how to obtain the classical adiabatic angle, yielding a good approximation for N(t).

The solution of the rigid body problem in terms of elliptic functions is a an old and honorable result; see eg. |39, chapt. 4|. Via the equiareal maps

$$\text{sphere} \to \text{cylinder} \to \text{plane} \quad ,$$

one can write the perturbation problem in more familiar terms. Berry's 1-form will therefore have two terms: one comes from the family of action angle variables reducing the unperturbed system H^o; the other comes from the average of the perturbative term $(A^{-1}f,N)$ in (20.5). Both are manifestly 1-forms on parameter space $\{\ell_{ij}\}$.

Once N(t) is obtained to the desired order of approximation, the angular velocity Ω is recovered via $\Omega=A(t)^{-1}(N(t)-f)$ (see (20.3) and the Euler angles of the attitude matrix via (|7, §37|)

$$\cos\theta = I_3\Omega_3/|\vec{m}| \quad , \quad \mathrm{tg}\,\psi = I_1\Omega_1/(I_2\Omega_2) \tag{20.6}$$
$$d\phi/dt = |\vec{m}|\ (I_1\Omega_1^2+I_2\Omega_2^2)/(I_1^2\Omega_1^2+I_2^2\Omega_2^2)$$

Here we choose the coordinate axis so that m is in the direction of $\vec{k}$. We plan to attempt these calculations futurely. As in §19, the key difficulty is to average certain elliptic functions.

One possible application of this, and other related problems, is the design of strategies to recover defective orbiting objects. Classical adiabatic angles could be used to "fine tune" the attitude of a remote controlled manipulator, whose task is to capture a certain fixture in the satellite.

21. MORRISON'S AVERAGING TECHNIQUE |40,41|. Consider a system of the form $\dot{y}=Y(y,\theta,\varepsilon)$, $\dot{\theta}=\omega(y)+Z(y,\theta,\varepsilon)$, $Y,Z=O(\varepsilon)$, $y \in R^p$ and $\theta \in T^m$. One expands $Y=\Sigma_1\varepsilon^n Y_n$, $Z=\Sigma_1\varepsilon^n Z_n$ and one looks for a near identity transformation $y=y^*+F(y^*,\theta^*,\varepsilon)= =y^*+\Sigma_1\varepsilon^n F_n$, with $\langle F\rangle=0$, $\theta=\theta^*+\Phi(y^*,\theta^*,\varepsilon)=\theta^*+\Sigma_1\varepsilon^n\Phi_n$, with $\langle\Phi\rangle=0$.

This coordinate change must yield averaged equations so that $\dot{y}^*$ decouples: $\dot{y}^*=A(y^*,\varepsilon)$, $\dot{\theta}^*=\omega(y^*)+\Omega(y^*,\varepsilon)$. One gets recursive expressions for $F_n, A_n;\ \Phi_n,\ \Omega_n$, where $A=\Sigma_1\varepsilon^n A_n$, $\Omega=\Sigma_1\varepsilon^n\Omega_n$.

The slowly varying Hamiltonian of §1 is precisely in this setting, with $y=(I,\tau)$, where $\tau=\varepsilon t$ is the "slow time". Assuming that the deviation of I is exponentially small, the series can be constructed so that all $A_n\equiv 0$.

22. ADDENDUM TO §8. We have overlooked the fact that θ_i are not "true angle coordinates", as it is obvious from the fact that the frequencies $\dot{\theta}_i = \frac{\partial H^o}{\partial I}(p,q,I)$ are not uniform. We must show that (8.2) is still true. Let $p=p(\phi,L;I)$, $q=q(\phi,L;I)$ be obtained via the generating function $S=S(q,L;I)$ depending on I as a parameter. The complete generating function is $\tilde{S}=S+I\theta$, so that

$$\alpha = \partial\tilde{S}/\partial I = \theta + \partial S(q,L;I)/\partial I \tag{22.1}$$

This angle coordinate α, truly conjugate to I, changes uniformly,

$$\dot{\alpha} = \partial\tilde{H}(L,I)/\partial I \tag{22.2}$$

so that for the unperturbed system

$$\theta(t) = (\partial\tilde{H}/\partial I)t - \partial S/\partial I\,|(q(\phi_o+(\partial\tilde{H}/\partial I)t),L,I) + \text{const.} \tag{22.3}$$

This formula shows that $\theta(t)$, although not varying uniformly, generically covers the torus preserving the measure $d^n\theta$. Thus we can perform the space average with respect to θ and (8.2) follows.

23. FINAL COMMENTS AND TOPICS FOR FURTHER RESEARCH.

(i) Higher order corrections to Hannay's angles (§1, 21). The technique used in |42| seems to be congenial to Landau-Lifschitz's approach |7|. The exponentially small deviation of the actions for analytic, adiabatically varying Hamiltonians should follow from the analysis of the poles of $\int_{-\infty}^{\infty} \frac{\partial H}{\partial \theta}\, dt$. Other averaging schemes, and counterexamples for non-analytic Hamiltonians are given in |43,44|.

(ii) Rigidity of the image of the momentum mapping (§2). For symplectic torus actions on compact symplectic manifolds, the image of the momentum mapping is a convex polytope |45|. By the adiabatic principle, changing the action together with a suitably normalized change of symplectic structure should not change the image.

(iii) Slow-fast systems with symmetries (§3). Consider a symplectic fibration $M \subset P \xrightarrow{\pi} B$ (all ingredients are symplectic manifolds), see |46|. M is the phase space for a "fast" and B for a "slow" dynamics, coupled in P. Suppose G_o acts on P, $G_b \subset G_o$ acts on $\pi^{-1}(b)$. Can the averaged dynamics in B be further reduced?

(iv) Strongly constrained time-dependent systems (§4); nonintegrable case (§9). More sophisticated results, based on Kummer's construction |47| are expected to come out soon |48|. For the nonintegrable case, see also |49|.

(v) Generalized Hannay's top (§15). Perturbations transversal to coadjoint orbits, related to averaging methods, were considered in |50|.

(vi) The Foucault pendulum (§16). Strictly speaking, the spherical pendulum has no circulation regime, as shown in |27|. As $I \to 0$, $\dot\theta = I/\sin^2 q$ is near zero except when q is close to 0 or π; at these moments θ changes by near 180° in a very short interval. The analysis by perturbation theory is further complicated by the fact that there is not a single family of action angle coordinates near I=0, H=energy of the unstable equilibrium.

(vii) Separatrix crossings. The adiabatic approximation fails as one of the frequencies of the frozen systems approaches zero. This may happen either by a resonance mechanism (eg. frequency crossings) or by a degenerescence of the foliation by tori (homoclinic surfaces). These phenomena are very important in celestial mechanics |51-55| and in plasma physics |56-57|. The process of diffusion, or accuracy loss for the actions only recently began to be better understood |58-65|.

24. ACKNOWLEDGEMENTS. I would like to thank the organizers for the invitation to give a plenary talk (it came as a big surprise!); Prof. Vincent Moncrief for sponsoring a visit to Yale where this research started. Correspondence with Richard Montgomery about our common interests was very stimulating; Susana Falcão Bodin de Morais has been a very thoughtful symplectic listener. Thanks to Maria Cristina Lima Raymundo for a superb typing job.

BIBLIOGRAPHY

1. P. Ehrenfest, "Adiabatic invariants and the theory of quanta", Phil. Mag. 33 (1917), 500-513.

2. T. Kato, "On the adiabatic theorem of quantum mechanics", J. Phys. Soc. Japan 5 (1950), 435-439.

3. M.V. Berry, "Quantal phase factors accompanying adiabatic changes", Proc. Roy. Soc. London A 392 (1984), 45-57.

4. B. Simon, "Holonomy, the quantum adiabatic theorem, and Berry's phase", Phys. Rev. Lett. 51 (1983), 2167-2170.

5. R. Jackiw, "Berry's phase - topological ideas from atomic, molecular and optical physics, to appear in Comments Atomic and Molecular Phys.

6. J.E. Littlewood, "Lorentz pendulum problem", Ann. Phys. 21 (1063), 233-242.

7. L.D. Landau, E.M. Lifschitz, Mechanics, Pergamon Press, 3rd. ed., N.Y., 1976.

8. J.H. Hannay, "Angle variable holonomy in adiabatic excursion of an integrable Hamiltonian, J. Phys. A: Math. Gen. 18 (1985), 221-230.

9. M.V. Berry, "Classical adiabatic angles and quantal adiabatic phase", J. Phys. A: Math. Gen. 18 (1985) 15-27.

10. J. Koiller, "Some remarks on Berry's phase", 26º Sem. Bras. Anal. (1987), 325-333.

11. R. Montgomery, "The connection whose holonomy is the classical adiabatic angles of Hannay and Berry and its generalization to the non-integrable case", to appear in Commun. Math. Phys.

12. S. Golin, A. Knauf, S. Marmi, "The Hannay angles: geometry, adiabaticity, and an example", preprint, 1988.

13. V.I. Arnol'd, Mathematical Methods of Classical Mechanics, Springer-Verlag, Berlin, Heildelberg, New York, 1978.

14. A. Weinstein, "Cohomology of symplectomorphism groups and critical values of Hamiltonians", preprint, Berkeley, 1988.

15. R. Jackiw, "Three elaborations on Berry's connection, curvature and phase", preprint CTP-MIT 1529, 1987.

16. P.L. Kapitza, "Dynamical stability of a pendulum when its point of suspension vibrates", "Pendulum with a vibrating suspension", Collected papers, vol. II, 714-737, Pergamon Press, N.Y., 1965.

17. I.I. Blekhman, "The development of the concept of direct separation of motions in nonlinear mechanics", in A. Yu. Ishlinsky, F. Chernousko, eds., Theoretical and Applied Mechanics, MIR, Moscou, 1981.

18. S. Iida, H. Kuratsuji, "Adiabatic theorem and anomalous commutators", Phys. Lett. B 184: 2,3 (1987), 242-246.

19. E. Gozzi, W.D. Thacker, "Classical adiabatic holonomy and its canonical structure", Phys. Rev. D 35 (1987), 2398-2406.

20. H. Rubin, P. Ungar, "Motion under a strong constraining force", Commun. Pure Appl. Math. X (1957), 65-87.

21. F. Takens, "Motion under the influence of a strong constraining force", in Springer Lect. Notes Math. 819, Springer, Berlin, Heildelberg, New York, 1979.

22. J.F. Adams, Lectures on Lie Groups, Benjamin, N.Y., 1969.

23. J. Koiller, "Classical adiabatic angles for slowly moving mechanical systems, preprint LNCC-CNPq, 1988.

24. D.J. Struik, Lectures on Classical Differential Geometry, Add. Wesley, Reading, 1961.

25. A. Tomita, R.Y. Chiao, "Observation of Berry's topological phase by use of an optical fiber", Phys. Rev. Lett. 57 (1986), 937-940.

26. M.V. Berry, "Interpreting the anholonomy of coiled light", preprint.

27. J.J. Duistermaat, "On global action-angle coordinates", Commun. Pure Appl. Math 33 (1980), 687-706.

28. F. Klein, Vorlesungen uber Hohere Geometry, Chelsea, London, 1949.

29. J.B. Hart, R.E. Miller, R.L. Mills, "A simple geometric model for visualizing the motion of a Foucault pendulum", Am. J. Phys. 55 (1987), 67-70.

30. R.J. Black, A. Ankiewicz, "Fiber-optic analogies with mechanics", Am. J. Phys. 53 (1985), 554-563.

31. A. Ankiewicz, C. Pask, "Regular and irregular motion: new mechanical results and fibre optics analogies", J. Phys. A: Math. Gen. 16 (1983), 3657-3673.

32. M. Sarraceno, personal communication.

33. S.I. Pidkuiko, A.M. Stepin, "On Hamiltonian systems with dynamical symmetries", in Springer Lect. Notes Math. 1108, Springer, Berlin, Heildelberg, New York, 1984 .

34. S. Chang, R. Friedberg, "Elliptical billiards and Poncelet's theorem", J. Math. Phys. 29 (1988), 1537-1550.

35. C. Jacobi, Dynamik, Werke, v. 8, Chelsea, London, 1969.

36. I.S. Gradshtein, I.M. Ryzhik, Tables of Integrals, Series, and Products, Ac. Press, N.Y., 1981.

37. P. Byrd, M. Friedman, Handbook of elliptic integrals for Engineers and Scientists, Springer- Verlag, Berlin, Heildelberg, New York, 1971.

38. E. Allgower, S. Gnutzman, "An algorithm for piecewise-linear approximation of implicitly defined two-dimensional surfaces", SIAM J. Num. Anal. 24 (1987), 452-469.

39. J. Wittenburg, Dynamics of Systems of Rigid Bodies, Teubner, Bonn, 1977.

40. J.A. Morrison, "Generalized method of averaging and the von Zeipel method", in V. Szebehely, R. Duncombe, eds. Methods in Astrodynamics and Celestial Mechanics, Progress in Astronautics and Aeronautics 17, Ac. Press, N.Y., 1966.

41. J.A. Morrison, "An averaging scheme for some nonlinear resonance problems", SIAM J. Appl. Math. 16 (1968), 1024-1047.

42. P. Holmes, J. Marsden, J. Scheurle, "Exponentially small splitting of separatrices", preprint PAM-364, Berkeley, 1987.

43. Bakhtin, V.I., Averaging in multifrequency systems, Funcl. Anal. Appl. 20 (1986), 83-88.

44. V.I. Arnol'd, V.V. Kozlov, A.I. Neishtadt, Mathematical aspects of classical and celestial mechanics, Encycl. Mathl. Sci., vol. 3, Dynamical Systems III, Springer-Verlag, Berlin, Heildelberg, New York, 1988.

45. F. Kirwan, "Convexity properties of the momentum mapping, III", Invent. Math. 77 (1984), 547-552.

46. M. Gotay, R. Lashof, J. Sniatycki, A. Weinstein, "Closed forms on symplectic fiber bundles", preprint CPAM, 111, 1980.

47. M. Kummer, "On the construction of the reduced phase space of a hamiltonian system with symmetry", Indiana U. Math. J. 30 (1981), 281-291.

48. J. Marsden, R. Montgomery, T. Ratiu, "Hannay's angles for nonintegrable and constrained systems", work in progress.

49. A. Weinstein, "Connections of Berry and Hannay type for moving Lagrangian submanifolds, preprint, Berkeley, 1988.

50. Yu. M. Vorobev, M.V. Karasev, "Poisson manifolds and the Schouten Bracket", Functl. Anal. Appl. 22 (1988), 1-9; "Corrections to classical dynamics and quantization conditions arising in the deformations of Poisson brackets", Sov. Math. Dokl. 36 (1988), 594-598.

51. J. Henrard, "Slow chaotic motion in Hamiltonian systems", Proc. First Belgian Cong. Th. Appl. Mech., 1987.

52. J. Henrard, A. Lemaitre, "Hamiltonian chaos in perturbed resonance problems", Proc. 10th Eur. Astr. Meet. IAU, 1987.

53. J. Henrard, "The adiabatic invariant in celestial mechanics", to appear in Dynamics Reported.

54. J. Wisdom, "A perturbative tratment of motion near the 3/1 commensurability", Icarus 63 (1985), 272-286.

55. J. Koiller, J.M. Balthazar, T. Yokoyama, "Relaxation chaos phenomena in celestial mechanics", Physica 26D (1987), 85-122.

56. B.V. Chirikov, "Particle dynamics in magnetic traps", in Reviews of Plasma physics, M.A. Leontovich, ed., Consultants Bureau, New York, 1987.

57. D.F. Escande, "Hamiltonian chaos and adiabaticity", Proc. Int. Workshop, Kiev, 1987.

58. L.D. Akulenko, "Averaging in a quasilinear system with a strongly varying frequency, PMM USSR 51 (1987), 189-196.

59. A.I. Neishtadt, "Persistence of stability loss for dynamical bifurcations", Diff. Eq. 23 (1988), 1385-1390.

60. A.N. Vasil'ev, M.A. Guzev, "Particle capture by a slowly varying periodic potential", Th. Math. Phys. 68 (1986), 907-916.

61. J.L. Tennyson, J.R. Cary, D.F. Escande, "Change of the adiabatic invariant due to separatrix crossing", Phys. Rev. Lett. 56 (1986), 2117-2120.

62. J.R. Cary, D.F. Escande, J.L. Tennyson, "Adiabatic-invariant change due to separatrix crossing", Phys. Rev. A 34 (1986), 4256-4274.

63. J.H. Hannay, "Accuracy loss of action invariance in adiabatic change of a one-freedom hamiltonian", J. Phys. A: Math. Gen. 19 (1986), L1067-L1072.

64. S. Wiggins, "On the detection and dynamical consequences of orbits homoclinic to hyperbolic periodic orbits and normally hyperbolic invariant tori in a class of ordinary differential equations", SIAM J. Appl. Math. 48 (1988), 262-285.

65. J. Murdock, "On the length of validity of averaging and the uniform approximation of elbow orbits with an application to delayed passage through resonance", ZAMP 39 (1988), 586-596.

LABORATÓRIO NACIONAL DE COMPUTAÇÃO CIENTÍFICA
Rua Lauro Muller, 455 - Urca
Rio de Janeiro, RJ, Brazil, 22250

INSTITUTO DE MATEMÁTICA DA UFRJ
C.P. 68530, Cidade Universitária
21945 Rio de Janeiro, RJ, Brazil

Contemporary Mathematics
Volume 97, 1989

EULERIAN MANY-BODY PROBLEMS

P.S. Krishnaprasad *

ABSTRACT. The hamiltonian dynamics of coupled structures is discussed. There are geometric parallels in earlier work on the Newtonian (gravitational) many- body problem. In the study of relative equilibria, a theorem due to Smale has a useful role. Relative stability modulo a group of symmetries can be determined using the energy-Casimir (or energy - momentum) method. For nongeneric values of momenta, the Poisson structure can affect stability.

1. INTRODUCTION. The central role of the Newtonian (gravitational) many-body problem in celestial mechanics has inspired major advances in mathematics and physics. For an exposition see (Abraham and Marsden [1]) and (Smale [31]). In recent years, engineering applications have brought to the forefront, questions concerning the dynamics of systems of kinematically coupled structures composed of rigid and flexible bodies. We refer to these as Eulerian many-body problems to emphasize the role of Euler forces (or frame forces) in determining the interactions. Eulerian many-body problems arise as models of robotic manipulators, high speed mechanical machinery, complex spacecraft with articulated components, space-based sensors etc. See (Wittenburg [36]) and [8], [12] for expositions of engineering aspects and basic formulations of underlying models.

In recent work, [13] [17] [18] [26] [27] [29] [33], we have explored the rich geometry of Eulerian many-body problems. We have used the geometry of symplectic manifolds, Poisson structures, and reduction by symmetry groups in creating a framework for the study of the dynamic behavior of certain classes of Eulerian many-body problems. Among the classes of problems we have investigated, we include rigid bodies carrying rotors, planar many- body systems, three dimensional systems coupled by ball and socket joints, and rigid bodies with flexible attachments modeled by geometrically exact formulations of elasticity. Our methods shed light on questions regarding relative equilibria, periodic orbits, stability,

---

1980 Mathematics Subject Classification (1985 Revision): 58F05, 58F10
* This work was supported in part by the AFOSR University Research Initiative Program under grant AFOSR- 87-0073 and by the National Science Foundation's Engineering Research Centers Program: NSFD CDR 8803012.

conservation laws (e.g. Casimir functions) and controllability on level sets of conservation laws.

The present paper simply highlights some key geometric aspects of these later developments.

ACKNOWLEDGEMENT

This is based on a stimulating collaboration with Robert Grossman, Jerrold Marsden, Yong-Geun Oh, Tom Posbergh, Juan Simo and N. Sreenath, and their contributions are gratefully acknowledged. We have also benefited from conversations with John Baillieul, Anthony Bloch, Mark Levi, Debra Lewis and Tudor Ratiu. A special thanks to Jerrold Marsden for his enthusiastic support over the years as friend, collaborator and teacher.

2. GEOMETRY

The abstract framework for Eulerian many-body problems is the one isolated by Smale in his study of the gravitational many-body problem. Let (M, K) be a Riemannian manifold and let G be a Lie group with associated action,

$$\begin{aligned} \Phi : G \times M &\to M \\ (g, q) &\mapsto \Phi_g(q) \end{aligned}$$

where Φ_g is an isometry for all $g \in G$. The Riemannian metric induces a vector bundle isomorphism

$$K^\flat : TM \to T^*M$$

defined by

$$K^\flat(v_q) \cdot w_q = K(v_q, w_q), \text{ forall } v_q, w_q \in TM_q .$$

The canonical symplectic structure $\omega = -d\theta_0$ on T^*M can be pulled back to

$$\Omega = (K^\flat)^*(\omega),$$

also an exact symplectic structure on TM. The action Φ lifts to symplectic actions $T\Phi$ and $T\Phi^*$ on TM and T^*M respectively.

Let $V : M \to \mathbb{R}$ be a G-invariant (potential) function on M. The hamiltonian $H : T^*M \to \mathbb{R}$, is defined by,

$$H(\alpha_q) = \frac{1}{2} K \left((K^{\flat-1})\,\alpha_q,\ (K^{\flat-1})\alpha_q\right) + V_\circ \tau_M^*(\alpha_q)$$

where $\tau_M : T^*M \to M$ is the canonical projection.

Associate to H a vector field X_H on T^*M by requiring that,

$$dH(Y) = \omega(X_H, Y)$$

for all vector fields Y on T^*M. The hamiltonian system $(T^*M,\ \omega,\ X_H)$ is a simple mechanical system with symmetry in the sense of Smale. It admits a momentum mapping in a natural way. To see this, let $\Im$ denote the Lie algebra of G and $\Im^*$ the dual space of $\Im$. The symplectic action $T\Phi^*$ on T^*M, defines a Lie algebra homomorphism, of $\Im$ into hamiltonian vector fields on T^*M; we denote this correspondence as $\xi \mapsto \xi_{T^*M}$. Then the map,

$$J : T^*M \to \Im^*$$

defined by,

$$J(\alpha_q)\cdot\xi = (i_{\xi_{T^*M}}\,\theta_0)(\alpha_q)\ , \quad \xi \in \Im$$

is an Ad^* - equivariant momentum mapping. Hence J is a conserved quantity of the system (T^*M, ω, X_H).

The framework sketched so far is the proper setting for Eulerian many-body problems in our sense (as it is for Smale's approach to the gravitational many-body problem).

EXAMPLE 1 (Planar two-body problem)

Imagine two rigid laminae connected by a pin joint, floating in a gravity-free planar universe (see figure 1). For an observer at the center of mass of the system of two bodies, the absolute orientations of the two bodies, determined say by attaching body-frames, are sufficient to determine the absolute configuration of the pair. The group S^1 of spatial rotations of the observer's frame is a symmetry group for the problem. Thus, $M = S^1 \times S^1$, $G = S^1$ acting on M via the diagonal action and the metric on M is given by

$$\begin{aligned} K(\dot\theta_1, \dot\theta_2) &= 2 \times \text{Kinetic energy} \\ &= \left\langle \begin{bmatrix} \tilde I_1 & \lambda(\theta) \\ \lambda(\theta) & \tilde I_2 \end{bmatrix} \begin{pmatrix} \dot\theta_1 \\ \dot\theta_2 \end{pmatrix} , \begin{pmatrix} \dot\theta_1 \\ \dot\theta_2 \end{pmatrix} \right\rangle \\ &= \langle \mathbf{I}_p\omega,\ \omega\rangle \end{aligned}$$

where, $\tilde I_i = I_i + \epsilon\, d_i^2$, $i = 1,2$, are augmented inertias of the bodies, $\epsilon = m_1 m_2 / (m_1 + m_2)$ is a reduced mass, and for the choice of body frames as in figure 1,

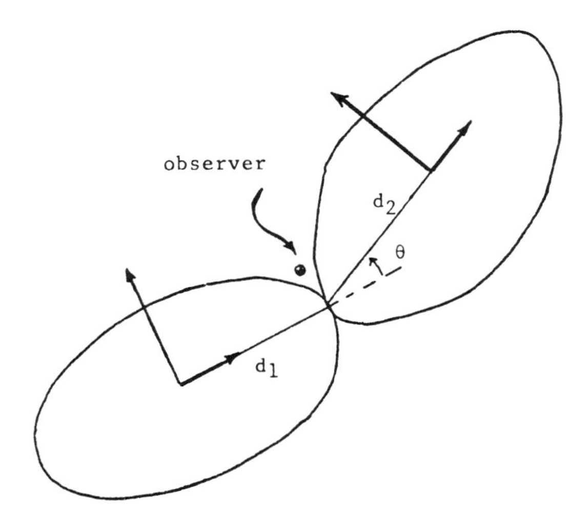

Figure 1. Planar Two-Body Problem

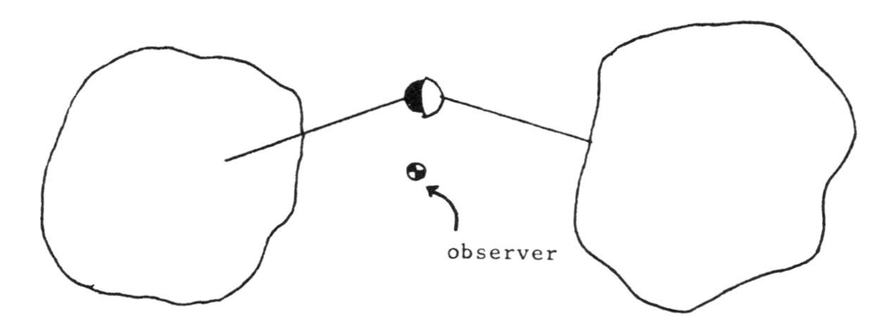

Figure 2. Rigid Bodies Coupled by a Ball and Socket Joint

$\lambda(\theta) = \epsilon\, d_1 d_2 \sin(\theta_1 - \theta_2)$ is a function of the joint angle. Since K depends only on the difference $\theta_1 - \theta_2$, it is invariant under the S^1 action $(\theta_1, \theta_2) \mapsto (\theta_1 + g, \theta_2 + g)$, $g \epsilon S^1$. The subscript in $\mathbf{I}_p$ is in reference to planarity. The vector bundle map K^b is given by

$$K^b(\omega) = \mu = \mathbf{I}_p \omega.$$

The momentum mapping for the S^1 action is then,

$$J_p : T^*(S^1 \times S^1) \to \mathbb{R}$$

$$(\theta_1, \theta_2, \mu_1, \mu_2) \mapsto \mu_1 + \mu_2$$

It is just the angular momentum of the system with respect to the observer at the center of mass.

EXAMPLE 2. (Rigid bodies coupled by a ball and socket joint)

This is a spatial analog of the previous planar example. The two bodies are free to move in three dimensions, subject to a (three degrees of freedom) ball and socket coupling. As before, the observer is at the center of mass of the system of two bodies. See figure 2 below for a representation.

In this case $M = SO(3) \times SO(3)$ and $G = SO(3)$ acts diagonally on M

$$\Phi : SO(3) \times M \to M$$
$$(P, A_1, A_2) \mapsto (PA_1, PA_2).$$

This is just the symmetry associated to the freedom of the observer to make arbitrary spatial rotations of his frame.

The action Φ leaves the kinetic energy metric invariant, the latter given by a 6×6 positive definite quadratic form $\mathbf{I}_s$ analogous to $\mathbf{I}_p$ in example 1, with only off-diagonal terms dependent on configurations. For $SO(3)$ invariance, these in fact depend only on $A_1^{-1} A_2$ the relative configuration of two bodies. Once again an Ad^*- equivariant moment mapping $J_s : T^*(SO(3) \times SO(3)) \to SO(3)^* \simeq \mathbb{R}^3$ can be written down. It is just the angular momentum of the system with respect to the observer at the center of mass.

In the thesis of Sreenath and in the papers by Oh, Sreenath, Marsden and Krishnaprasad, planar coupled systems such as that in example 1 are investigated.

In the paper of Grossman, Krishnaprasad and Marsden the example 2 is discussed.

For the most part, in these references the situations analyzed require that the potential $V \equiv 0$. However, in Sreenath's thesis, <u>control functions</u> at the joints of planar many-body systems are considered and the associated feedback laws may in certain cases be interpreted as arising from potential functions due to torsional springs at the joints.

Other interesting examples including flexible bodies (attachments) appear in [18], [27], [30], and in the papers of Baillieul and Levi [5] [6] [7].

Poisson structures are central to our point-of-view. A Poisson manifold P is simply a smooth manifold equipped with an $\mathbb{R}$ -bilinear map (Poisson structure),

$$\{\cdot,\cdot\}_P \ : C^\infty(P) \ \times \ C^\infty(P) \to C^\infty(P)$$

satisfying the axioms

(i) $\{f,g\}_P \ = \ - \ \{g,f\}_P$

(ii) $\{fg,h\}_P \ = \ g\{f,h\}_P \ + \ f\{g,h\}_P$

(iii) $\{f, \ \{g,h\}_P \ \}_P \ + \ \{g,\{h,f\}_P\}_P \ + \ \{h,\{f,g\}_P\}_P \ = 0.$

We outline the general theory a little before we specialize to the mechanical setting. First, associated to a Poisson structure, there is a unique, twice contravariant skew-symmetric, smooth tensor field Λ on P such that,

$$\{f,g\}_P \ = \ \Lambda \ (df, dg).$$

For a proof see p. 109 of [19]. The tensor Λ defines a vector-bundle morphism,

$$\begin{aligned} \Lambda^{\#} : T^*P \ &\to \ TP \\ \alpha_x \ &\mapsto \ \Lambda^{\#}(\alpha_x) \ \epsilon \ TP_x \end{aligned}$$

satisfying,

$$\beta_x \ (\Lambda^{\#}(\alpha_x)) \ = \ \Lambda \ (x) \ (\alpha_x, \beta_x) \text{ for all } \beta_x \ \epsilon \ TP_x.$$

The rank of the Poisson structure at $x \ \epsilon \ P$ is defined to be the rank of the Poisson tensor Λ at x . This is simply the rank of the (characteristic) distribution $C \ = \ \Lambda^{\#}(T^*M) \ \subset \ TM$ at the point x. The rank may vary on P. However, it is a theorem of Kirillov [16] that $\Lambda^{\#}(T^*M)$ defines a generalized foliation on P such that through each point $x \ \epsilon \ P$, passes a leaf carrying a unique symplectic structure that makes the injection map of that leaf a Poisson morphism. (See Weinstein [34] and Libermann-Marle [19]). Thus a Poisson manifold is a union of symplectic leaves.

A function $f \ \epsilon \ C^\infty(P)$ is called a <u>Casimir function</u> if

$$\{f,g\}_P = 0 \quad \forall\, g \in C^\infty(P).$$

Casimir functions are constant on symplectic leaves.

Let G be a Lie group and let $\Psi : G \times P \to P, (g,x) \mapsto \Psi_g(x)$, be a group action such that, $\Psi_g(\cdot)$ is a Poisson morphism for every $g \in G$. Further, suppose that the action is proper and free. Then there exists a good quotient P/G that carries a Poisson structure $\{\cdot,\cdot\}_{P/G}$ induced from the one on P satisfying,

$$\{f,g\}_{P/G} = \{f \circ \pi, g \circ \pi\}_P.$$

Here $\pi : P \to P/G$ is the canonical projection. By construction, it is a Poisson morphism.

G-equivariant dynamics on P induce dynamics on P/G. Suppose $h : P \to \mathbb{R}$ is a G-invariant hamiltonian function on P, i.e.,

$$h(\Psi_g(x)) = h(x) \ \forall\, g \in G.$$

Define a vector field X_h by

$$X_h f = \{f,h\}_P \ \forall\, f \in C^\infty(P).$$

The hamiltonian h descends to $\hat{h} : P/G \to \mathbb{R}$ and determines a <u>reduced</u> dynamics $\hat{X}_{\hat{h}}$ on P/G by

$$\hat{X}_{\hat{h}}(\hat{f}) = \{\hat{f},\hat{h}\}_{P/G} \ \forall \hat{f} \in C^\infty(P/G).$$

Here $\hat{h}([x]) = h(x)$ for any equivalence class $[x]$ in P/G. From, the definition of the characteristic distribution $C = \Lambda^{\#}(T^*M)$, it follows that the hamiltonian vector fields $\hat{X}_{\hat{h}}$ leave invariant the symplectic leaves. Thus any Casimir function is an integral of motion for $\hat{X}_{\hat{h}}$. The trajectories of X_h project under π to trajectories of $\hat{X}_{\hat{h}}$. The steps just outlined constitute the essence of Poisson reduction. See [22] for more details. We give some examples of Poisson structures.

EXAMPLE 3. (P,ω) is a connected symplectic manifold and $\{f,g\}_P := \omega(X_f, X_g)$.

Here the rank = dimension of P and there is just one symplectic leaf. Simple mechanical systems with symmetry yield interesting rank-degenerate cases. Referring to example 1 (the planar two-body problem), set $(P,\omega) = \left(T^*(S^1 \times S^1), \omega\right)$. The diagonal S^1 action, being symplectic, also leaves the Poisson structure on $T^*(S^1 \times S^1)$ invariant. The Poisson-reduced phase space $\left(T^*(S^1 \times S^1)\right)/S^1$ has a bracket structure

$$\{f,g\} = \left(\frac{\partial f}{\partial \theta}\frac{\partial g}{\partial \mu_1} - \frac{\partial f}{\partial \mu_1}\frac{\partial g}{\partial \theta_2}\right)$$
$$- \left(\frac{\partial f}{\partial \theta}\frac{\partial g}{\partial \mu_2} - \frac{\partial f}{\partial \mu_2}\frac{\partial g}{\partial \theta}\right)$$

where $\theta = \theta_1 - \theta_2 =$ joint angle.

Symplectic leaves on $(T^*(S^1 \times S^1))/S^1$ are cylinders (the corresponding characteristic distribution is of rank 2 everywhere) and are level sets of the Casimir function $\phi(\mu_1, \mu_2, \theta) = \mu_1 + \mu_2$.

EXAMPLE 4 (dual space of $\mathfrak{G}$)

$\mathfrak{G}^*$ carries the Lie-Berezin-Kirillov-Kostant-Souriau Poisson structure (s), defined by

$$\{f,g\}_{\mp}(\mu) = \mp\left\langle \mu, \left[\frac{\delta f}{\delta \mu}, \frac{\delta g}{\delta \mu}\right]\right\rangle$$

where $f, g \in C^\infty(\mathfrak{G}^*)$ and $\mu \in \mathfrak{G}^*$. The minus (plus) bracket is obtained by viewing $\mathfrak{G}^*$ as the left (right) Poisson reduction of T^*G by G.

The symplectic leaf through μ is $\mathcal{O}_\mu = \{\ell \in \mathfrak{G}^* : \ell = Ad^*_{g^{-1}}(\mu), g \in G\}$ the coadjoint orbit through μ.

When $\mathfrak{G} = so(3)$, the Poisson structure on $\mathfrak{G}^*$ is of rank 2 everywhere except at the origin where it is of rank 0.

We close this section with some remarks about dual pairs. Given a symplectic manifold S and Poisson manifolds P_1, P_2, suppose maps J_1 and J_2 can be found such that the following is a diagram of Poisson morphisms:

$$P_1 \xleftarrow{J_1} S \xrightarrow{J_2} P_2$$

The diagram is a <u>dual pair</u> in the sense of Marsden and Weinstein [24] [34] if the function algebras $\mathcal{F}_1 = J_1^*(C^\infty(P_1))$ and $\mathcal{F}_2 = J_2^*(C^\infty(P_2))$ are polar i.e.,

$$\{\mathcal{F}_1, \mathcal{F}_2\} = 0.$$

In that case the Casimir functions on P_1 and P_2 are in one-to-one correspondence and to the space $\mathcal{F}_1 \cap \mathcal{F}_2$.

Suppose the G action Φ on a simple mechanical system with symmetry is proper and free. Then there is an associated dual pair,

$$\mathfrak{G}^* \xleftarrow{J} T^*M \xrightarrow{\pi} T^*M/G.$$

Let $\mathcal{O}_\mu$ be the coadjoint orbit through $\mu \in \Im^*$ and let G_μ = isotropy subgroup of μ under the coadjoint action. Then $\mathcal{O}_\mu \simeq G/G_\mu$. Furthermore, the symplectic leaves in T^*M/G are the manifolds $\pi\ (J^{-1}(\mathcal{O}_\mu)) = J^{-1}(\mathcal{O}_\mu)/G$. They are isomorphic to the Marsden - Weinstein - Meyer spaces of symplectic reduction [23].

3. RELATIVE EQUILIBRIA. Much work on the gravitational many-body problem has concentrated on special uniformly rotating configurations (e.g. Moulton's theorem on collinear configurations). These are relative equilibria. The search for relative equilibria in Eulerian many-body problems has yielded some interesting results [33] [26].

Consider the dual pair

$$\Im^* \overset{J}{\leftarrow} (S,\omega) \overset{\pi}{\rightarrow} S/G$$

and $h : S \rightarrow \mathbb{R}$ a hamiltonian invariant under the action of G

DEFINITION. $z_e \in \mathcal{S}$ is a <u>relative equilibrium</u> (or the flow $F^t_{X_h}\ (z_e)$ is a <u>stationary motion</u>) if there exists $\xi \in \Im$ such that

$$F^t_{X_h}\ (z_e) = \Psi\ (exp\ (t\xi),\ X_e).$$

THEOREM (Relative Equilibrium)

The following are equivalent:

(i) z_e is a relative equilibrium;

(ii) z_e is a critical point of $h_\xi = h - < J, \xi >$, for some $\xi \in \Im$;

(iii) $\pi\ (z_e)$ is an equilibrium for the dynamics $\hat{X}_{\hat{h}}$ on S/G ■

REMARK. See Abraham & Marsden, chapter 4, for proofs. Part (ii) above is also a consequence of the Souriau-Smale-Robbin theorem.

For simple mechanical systems with symmetry, there is an elegant characterization of relative equilibria (due to Smale [31], although special versions have been known earlier).

THEOREM (Smale). Consider a simple mechanical system with symmetry (T^*M, ω, X_H) as defined in Section 2. Define,

$$V_\xi : M \rightarrow \mathbb{R}$$

$$q \mapsto V(q) - \frac{1}{2} K\ (\xi_M(q), \xi_M(q))$$

for each $\xi \in \Im$.

Then $z_e = (q_e, p_e) \in T^*M$ is a relative equilibrium iff, q_e is a critical point of V_ξ for some $\xi \in \mathfrak{I}$ and $p_e = K^\flat(\xi_M(q_e))$. ∎

For a proof of Smale's theorem see Smale [31] or Abraham & Marsden, pp 355. In his well-known paper [31], Smale uses this theorem to prove Moulton's theorem on the number of collinear configurations for the gravitational many-body problem.

Smale's theorem provides a convenient technique to compute relative equilibria. V_ξ is a G_ξ-invariant function on M the configuration space, where

$$G_\xi = \{g \in G : Ad_g(\xi) = \xi\}$$

and we are in the setting of equivariant Morse theory [3].

EXAMPLE 5. (planar 2-body problem continued). Returning to examples 1 and 3, we note, for $\xi \in \mathfrak{I} \equiv \mathbb{R}$, ξ_M is given by

$$\xi_M((\theta_1, \theta_2)) = \xi \left(\frac{\partial}{\partial\theta_1} + \frac{\partial}{\partial\theta_2}\right).$$

Thus, setting $V \equiv 0$,

$$\begin{aligned} V_\xi((\theta_1, \theta_2)) &= -\frac{\xi^2}{2}(1, 1)\mathbf{I}_p\left(\begin{pmatrix}1\\1\end{pmatrix}\right) \\ &= -\frac{\xi^2}{2}\left(\tilde{I}_1 + \tilde{I}_2 + 2\lambda(\theta_1 - \theta_2)\right). \end{aligned}$$

The S^1-equivalence classes of critical points of V_ξ are given by,

$$\frac{d\lambda}{d\theta} = 0 \leftrightarrow \theta = 0 \text{ or } \pi.$$

More generally, for a chain of n planar laminae, one expects at least $(1 + 1)^{n-1} = 2^{n-1}$ relative equilibrium classes since the Poincare$'$ polynomial of the $(n-1)$ torus is $(1+t)^{n-1}$.

We add that L-S. Wang, at the University of Maryland, has begun a numerical search for stable relative equilibria in the ball and socket problem of example 2 by numerical minimization of V_ξ.

4. RELATIVE STABILITY MODULO G. In the presence symmetries, a natural notion of stability is the following.

DEFINITION. Let $z_e \in S$ be a relative equilibrium for the dynamics X_h corresponding to a G-invariant hamiltonian h on (S, ω). We say that z_e is <u>relatively stable modulo</u>

G if $\pi\ (z_e)$ is a Lyapunov stable equilibrium for the Poisson reduced dynamics $\hat{X}_{\hat{h}}$ on S/G.

There is a sufficient condition for relative stability modulo G.

THEOREM (Relative Stability). $\pi\ (z_e)$ is an equilibrium point of $\hat{X}_{\hat{h}}$ *iff* it is a critical point of $\hat{h}_{|L}$ the restriction of $\hat{h}$ to the symplectic leaf L through $\pi\ (z_e)$. In that case, $\pi\ (z_e)$ is Lyapunov stable if,

(i) the Hessian $D^2\ (\hat{h}_{|L})\ (\pi\ (z_e))$ is definite.

(ii) the point $\pi\ (z_e)$ has a neighborhood W on which the rank of the Poisson structure $\{\cdot\ ,\ \cdot\}_{S/G}$ is constant.

REMARKS. In the form stated, the relative stability theorem appears to be due to Arnold. See also [19], Theorem 12.4 in chapter III. Points in S/G satisfying condition (ii) are called generic points. At generic points, nontrivial (local) Casimir functions C_ϕ exist. One can verify condition (i) by seeking a (local) Casimir C_ϕ such that $\pi\ (z_e)$ is an unconstrained critical point of $\hat{h} + C_\phi$ and $D^2\ (\hat{h} + C_\phi)$ at $\pi\ (z_e)$ is definite. This is the essence of the energy - Casimir method. Equivalently one can find $\xi \in \Im$ such that $dh_\xi\ (z_e) = 0$ and $D^2\ h_\xi\ (z_e)$ is definite in directions transversal to neutral directions associated to G_ξ. This is the essence of the energy - momentum method. For simple mechanical systems with symmetry $(S = T^*M)$, using Smale's theorem of section 3 and a splitting of $T(T^*M)$, this reduces to checking $D^2 V_\xi\ (q_e)$ is positive definite (see the paper of Marsden and Simo in this volume).

At nongeneric points (where condition (ii) above does not hold), one may, by ad hoc methods find conserved Lyapunov functions for $\hat{X}_{\hat{h}}$. But there exist examples due to Weinstein [35] and Libermann - Marle [19] indicating that at nongeneric points in S/G, definiteness of $D^2\ \hat{h}_{|L}$ does not imply stability. The Poisson bracket in $\{\cdot\ ,\ \cdot\}_{S/G}$ can affect relative stability modulo G. See the appendix to this paper for details of an example due Libermann and Marle.

We must add that we are aware of no "physically motivated" example that parallels the one in the appendix. It would be interesting to explore this further.

EXAMPLE 6. (Planar 2-body Problem)

By energy-Casimir the stretched out relative equilibrium $(\theta = 0)$ is relatively stable mod S^1 and the folded over relative equilibrium $(\theta = \pi)$ is unstable. This is true even at zero total angular momentum since the Poisson tensor is of constant rank 2.

5. HOLONOMY. In 1987, Jair Koiller introduced us to the concept of Berry's geometric phase. Inspired by his remarks, we worked out a formula for planar n-body chains that

admits interpretation via holonomy of a connection.

Consider a chain of planar rigid bodies floating in a planar gravity-free universe as in figure 3. Suppose each joint is actuated so as to permit free adjustment of joint angles. Assume that the whole assembly is at rest (angular momentum =0).

PROBLEM. Suppose the joint angles are varied continuously in a prescribed manner and brought back to their initial condition of rest. What will be the displacement of body 1 from its initial absolute orientation?

In geometric terms, a loop is traversed in T^{n-1} the joint angle space (or labelled shape space in the terminology of R. Montgomery) and we are interested in measuring the holonomy or extent to which it fails to lift to a loop in the absolute configuration space T^n. Such liftings require connections [20] and there is a natural one in the problem obtained by taking the orthogonal complement of the subspace spanned by vertical vector fields. Postponing the details to a future publication we would like to give a formula answering the problem above.

Let $\mathbf{I}_p^n$ denote the $n \times n$ quadratic form associated to the planar n-body system analogous to $\mathbf{I}_p$ in example 1. (see the thesis of Sreenath for explicit form of $\mathbf{I}_p^n$). Then the angular momentum relative to the observer at the center of mass is.

$$c = e \cdot \mathbf{I}_p^n \, \omega,$$

where $e = (1, 1, \cdots, 1)'$, and ω is the vector of angular velocities of the system. Admissible motions of the system leave,

$$e \cdot \mathbf{I}_p^n \, \omega = 0.$$

Then the phase shift of body 1 is given by

$$\Delta\theta_1 = - \int_\Gamma \frac{e \cdot \mathbf{I}_p^n M d\phi}{e \cdot \mathbf{I}_p^n e},$$

where $d\phi = (d\phi_1, \cdots, d\phi_{n-1})$ is the vector of joint differentials and M is an $n \times (n-1)$ matrix satisfying

$$M_{ij} = \begin{cases} 0 & i = 1 \\ 1 & i > j \geq 1 \\ 0 & \text{otherwise,} \end{cases}$$

and Γ is the loop traversed in joint-angle space.

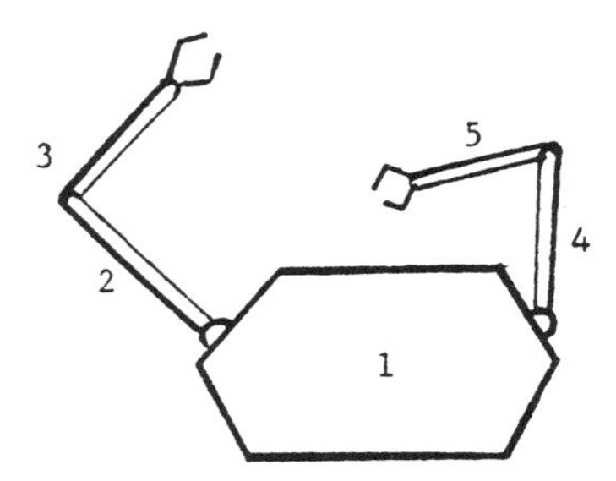

Figure 3. Planar n-Body System

The above formula can be useful in practical computations. Nonabelian analogs of this formula applicable to say the ball and socket problem can be derived from the theory of connections.

There is a related question of great interest in control theory.

PROBLEM. Among all possible parameterized paths Γ in joint angle space, find one that minimizes the action,

$$\int_\Gamma \omega \cdot \mathbf{I}_p^n \, \omega \, dt$$

and attains a prescribed phase shift $\triangle\theta_1$.

Control theoretic antecedents of this problem in the setting of Lie groups go back to the early papers of Brockett [9] and the Ph.D thesis of Baillieul [4]. The work of Brockett [10] [11] on singular Riemannian geometry and the recent results of Richard Montgomery [25] are directly applicable. We hope to report on this at a later date.

REFERENCES

1. R. Abraham and J.E. Marsden (1978), *Foundations of Mechanics*, Second Edition, revised, enlarged, reset. Benjamin/Cummings, Reading.

2. V.I. Arnold (1978), *Mathematical Methods of Classical Mechanics*, Springer-Verlag, New York.

3. M.F. Atiyah and R. Bott (1984), "The Moment Map and Equivariant Cohomology", *Topology*, vol 23, no.1, 1-28.

4. J. Baillieul (1978), "Geometric Methods for Nonlinear Optimal Control Problems", *Journal of Optimization Theory & Applications*, vol. 25, no. 4, 519-548.

5. J. Baillieul (1987), "Equilibrium Mechanics in Rotating Systems", *Proc. 26th IEEE Conf. Dec. Control,* IEEE, New York, 1429-1434.

6. J. Baillieul (1988). "Linearized Models for the Control of Rotating Beams", *Proc. 27th IEEE Conf. Dec. Control,* IEEE, New York. 1726-1731.

7. J. Baillieul and M. Levi (1987), "Rotational Elastic Dynamics", *Physica D*, 27D, 43-62.

8. G. Bianchi and W. Schielen (1985), eds. *Dynamics of Multibody Systems*, Proceedings of IUTAM/IFToMM Symposium Udine/Italy 1985, Springer Verlag, Berlin.

9. R. W. Brockett (1973), "Lie Theory and Control Systems Defined on Spheres", *SIAM Journal on Applied Mathematics*, vol. 25, no. 2, 1973.

10. R.W. Brockett (1981), "Control Theory and Singular Riemannian Geometry", in P.J. Hilton and G.S. Young eds., *New Directions in Applied Mathematics*, Springer-Verlag, Berlin, pp 11-27.

11. R.W. Brockett (1983), "Nonlinear Control Theory and Differential Geometry", in *Proc. Intl. Congr. Math.*, Warsaw, 1357-1368.

12. CIME (1972), *Stereodynamics*, edizione cremonese, Roma.

13. R. Grossman, P.S. Krishnaprasad and J.E. Marsden (1987) "The Dynamics of Two Coupled Three Dimensional Rigid Bodies" in F. Salam & M. Levi, eds. *Dynamical Systems Approaches to Nonlinear Problems in Systems and Circuits,* pp. 373-378, SIAM Publ., Philadelphia, 1988.

14. V. Guillemin and S. Sternberg (1984), *Symplectic Techniques in Physics*, Cambridge University Press, Cambridge.

15. D. Holm, J.E. Marsden, T. Ratiu and A. Weinstein (1984), "Stability of Rigid Body Motion using the Energy - Casimir Method", in J.E. Marsden ed. *Fluids and Plasmas: Geometry and Dynamics*, in series *Contemporary Mathematics*, vol. 28, 15-23, AMS, Providence.

16. A.A. Kirillov (1976), "Local Lie Algebras", *Russian Math. Surveys*, vol. 31, 56-75.

17. P.S. Krishnaprasad (1985), "Lie-Poisson Structures Dual-Spin Spacecraft and Asymptotic Stability", *Nonlinear Analysis : Theory, Methods and Applications*, vol. 9, no. 10, 1011-1035.

18. P.S. Krishnaprasad and J.E. Marsden (1987) "Hamiltonian Structures & Stability for Rigid Bodies with Flexible Attachments." *Arch. Rat. Mech. Anal. 98,* 71-93.

19. P. Libermann and C-M. Marle (1987), *Symplectic Geometry and Analytical Mechanics*, D. Reidel Publ., Dordrecht.

20. J.E. Marsden, R. Montgomery and T. Ratiu (1988), "Reduction, Symmetry and Berry's Phase in Mechanics", Preprint, Cornell University.

21. J.E. Marsden, T. Ratiu and A. Weinstein (1984), "Reduction and Hamiltonian Structures on Duals of Semidirect Product Lie Algebras", in J.E. Marsden ed. *Fluids and Plasmas: Geometry and Dynamics*, in series *Contemporary Mathematics*, vol. 28, 55-100, AMS, Providence.

22. J.E. Marsden and T. Ratiu (1986), "Reduction of Poisson Manifolds", *Letters in Math. Phys.*, vol 11, 161-169.

23. J. E. Marsden and A. Weinstein (1974), "Reduction of Symplectic Manifolds with Symmetry", *Reports in Math. Phys.*, vol. 5, 121-130.

24. J.E. Marsden and A. Weinstein (1983), "Coadjoint Orbits, Vortices, and Clebsch Variables for Incompresible Fluids", *Physica 7D,* 305-323.

25. R. Montgomery (1988), "Shortest Loops with a Fixed Holonomy", Preprint, Mathematical Sciences Resarch Institute, Berkeley, MSRI 01224-89.

26. Y.G. Oh, N. Sreenath, P.S. Krishnaprasad and J.E. Marsden (1988), "The Dynamics of Coupled Planar Rigid Bodies Part II: Bifurcation, Periodic Orbits, and Chaos", (in press) *J. Dynamics & Differential Equations.*

27. T. Posbergh, P.S. Krishnaprasad and J.E. Marsden (1987), "Stability Analysis of a Rigid Body with a Flexible Attachment using the Energy-Casimir Method" in M. Luksic, C. F. Martin, W. Shadwick eds. *Differential Geometry : The Interface between Pure and Applied Mathematics*, in series, *Contemporary Math.*, vol. 68, 253-273. AMS, Providence.

28. T. Posbergh (1988) Ph.D. Thesis, University of Maryland *"Modeling and Control of Mixed and Flexible Structures"*. Also, Systems Research Center Technical Report SRC TR 88-58.

29. J.C. Simo, J.E. Marsden and P.S. Krishnaprasad (1988), "The Hamiltonian Structure of Nonlinear Elasticity: The Material and Convective Representation of Rods, Plates and Shells", *Arch. Rat. Mech. & Anal.,* vol. 104, no. 2, 125-183.

30. J.C. Simo, T. Posbergh and J.E. Marsden (1988), "Nonlinear Stability of Geometrically Exact Rods by the Energy-Momentum Method". Preprint, Stanford University, Division of Applied Mechanics.

31. S. Smale (1970) "Topology and Mechanics, I, II", *Invent. Math.*, vol. 10, 305-331 and vol. 11, 45-64.

32. N. Sreenath (1987) Ph.D. Thesis, University of Maryland *"Modeling and Control of Multibody Systems"*. Also, Systems Research Center Technical Report SRC TR 87-163.

33. N. Sreenath, Y.G. Oh, P.S. Krishnaprasad and J.E. Marsden (1988), "The Dynamics of Coupled Planar Rigid Bodies Part I: Reduction, Equilibria & Stability *Dynamics & Stability of Systems,* vol. 3, no. 1&2, 25- 49.

34. A. Weinstein (1983), "The Local Structure of Poisson Manifolds", *J. Diff. Geom.*, vol. 18, 523-557 and vol. 22; (1985), 255.

35. A. Weinstein (1984), "Stability of Poisson-Hamilton Equilibria", in J.E. Marsden ed. *Fluids and Plasmas: Geometry and Dynamics*, in series *Comtemporary Mathematics*, vol 28, 3-13, AMS, Providence.

36. J. Wittenburg (1977), *Dynamics of Multibody Systems*, B.G. Teubner, Stuttgart.

APPENDIX. On An Example of P. Libermann and C.-M. Marle

(written with the assistance of L-S. Wang)

In this appendix, we work out an example suggested by Libermann & Marle (p. 274, [1]) to investigate the notion of relative stability modulo a group G of symmetries in the sense of Liapunov. The main purpose here is to show that for nongeneric momenta, stability may depend on the Poisson structure also.

First, the symplectic manifold in this example is

$$(M,\omega) = (\mathbb{R}^4, dq^1 \wedge dp_1 + dq^2 \wedge dp_2).$$

The Lie group here is

$$\begin{aligned} G &= Aff_+(\mathbb{R}) \\ &= \left\{ \begin{array}{ll} (a,b)| & a,b \in \mathbb{R}^2 \text{ with group law} \\ & (a,b)\cdot(a',b') \\ & = (a+a', b+e^a b') \end{array} \right\}. \end{aligned}$$

G acts on (M,ω) by the following rule.

$$\begin{aligned} \Phi : G \times M &\to M \\ (g,x) &\mapsto \Phi_g(x) \end{aligned}$$

$$\begin{aligned} ((a,b),(q^1,q^2,p_1,p_2)) &\mapsto (a+q^1, b+e^a q^2, p_1, e^{-a}p_2) \\ &= \Phi_g(x). \end{aligned} \tag{1}$$

It is easy to check that this is an action. Moreover, since

$$\begin{aligned} d(a + q^1) \wedge dp_1 + d(b + e^a q^2) \wedge d(e^{-a} p_2) \\ = dq^1 \wedge dp_1 + dq^2 \wedge dp_2, \end{aligned}$$

this action is actually symplectic (i.e. leaves ω invariant).

Let $\theta = p_1 dq^1 + p_2 dq^2$. Then $\omega = -d\theta$. The action of G also leaves θ invariant. Hence by theorem 4.2.10 (Abraham & Marsden [3]) there is an Ad^* equivariant momentum mapping J, defined by,

$$\begin{aligned} J : \mathbb{R}^4 \to \mathfrak{I}^* = \mathbb{R}^2 \\ J(x) \cdot \xi = (i_{\xi_M} \theta)(x). \end{aligned}$$

We now compute J explicitly.

First, the Lie algebra corresponding to G with the Lie bracket $[\cdot,\cdot]$ is

$$\begin{aligned} \mathfrak{I} = \{\xi = (\xi^1, \xi^2) \epsilon \mathbb{R}^2 \mid \\ [\xi, \eta] = (0, \xi^1 \eta^2 - \xi^2 \eta^1)\}. \end{aligned}$$

It follows that for $\xi \epsilon \mathfrak{I}$, the exponential map from $\mathfrak{I}$ to G is given by

$$\exp(t\xi) = \begin{cases} (0, t\xi^2) & \text{if } \xi^1 = 0 \\ (t\xi^1, (e^{t\xi^1} - 1)\frac{\xi^2}{\xi^1}) & \text{if } \xi^1 \neq 0. \end{cases} \tag{2}$$

The adjoint action of G on $\mathfrak{I}$ is given by

$$\begin{aligned} \text{Ad} : G \times \mathfrak{I} &\to \mathfrak{I} \\ (g, \xi) &\mapsto Ad_g(\xi) \\ &= T_e(R_{g-1} L_g)\xi. \end{aligned}$$

In our case, for $g = (a, b)$, $\xi = (\xi^1, \xi^2)$,

$$Ad_g \xi = (\xi^1, e^a \xi^2 - b\xi^1),$$

or

$$Ad_g \begin{pmatrix} \xi^1 \\ \xi^2 \end{pmatrix} = \begin{pmatrix} 1 & 0 \\ -b & e^a \end{pmatrix} \begin{pmatrix} \xi^1 \\ \xi^2 \end{pmatrix}. \tag{3}$$

The coadjoint action of G on $\Im^*$ is

$$Ad^* : G \times \Im^* \rightarrow \Im^*$$
$$(g, \ell) \mapsto Ad^*_{g-1}(\ell).$$

For $\ell = (\ell_1, \ell_2)$, $g = (a, b)$, we have

$$Ad^*_{g-1}\begin{pmatrix}\ell_1\\ \ell_2\end{pmatrix} = \begin{pmatrix}1 & e^{-a}b\\ 0 & e^{-a}\end{pmatrix}\begin{pmatrix}\ell_1\\ \ell_2\end{pmatrix}. \tag{4}$$

The infinitesimal generator of the action corresponding to $\xi = (\xi^1, \xi^2)$ can be obtained as, for $x = (q^1, q^2, p_1, p_2)$,

$$\begin{aligned}\xi_M(x) &= \frac{d}{dt}\Phi(exp\ (t\xi), x)|_{t=0}\\ &= (\xi^1, q^2\xi^1 + \xi^2, 0, -\xi^1 p_2).\end{aligned} \tag{5}$$

We are now ready to compute the momentum mapping J. The computation is as follows.

$$\begin{aligned}J(x)\ (\xi) &= (i_{\xi_M}\theta)(x)\\ &= (p_1 dq^1 + p_2 dq^2)\left(\xi^1\frac{\partial}{\partial q^1} + (q^2\xi^1 + \xi^2)\frac{\partial}{\partial q^2} - \xi^1 p_2\frac{\partial}{\partial p_2}\right)\\ &= p_1\xi^1 + p_2(q^2\xi^1 + \xi^2).\end{aligned}$$

We may then write J as

$$\begin{aligned}J : \mathbb{R}^4 &\rightarrow \mathbb{R}^2 = \Im^*\\ \begin{pmatrix}q^1\\ q^2\\ p_1\\ p_2\end{pmatrix} &\mapsto \begin{pmatrix}p_1 + p_2 q^2\\ p_2\end{pmatrix}\end{aligned} \tag{6}$$

We now carry out the Marsden-Weinstein (symplectic) reduction procedure, [2].

First, we choose $\mu = (0,0) \in \Im^*$. By (6), we know that, $J^{-1}(0) = \{(q^1, q^2, 0, 0,)|q^1, q^2 \in \mathbb{R}\}$.

Since the Jacobian matrix of J is

$$DJ = \begin{bmatrix}0 & p_2 & 1 & q^2\\ 0 & 0 & 0 & 1\end{bmatrix}$$

which has rank 2 for all points in $J^{-1}(0)$, it follows that $\mu = (0,0)$ is a regular value of J. Next, we find that the isotropy group at (0,0) is just the whole group G, which can be checked from (4).

$$G_0 = \{g = (a,b) | Ad^*_{g^{-1}}(0) = 0\}$$
$$= G.$$

The action of G_0 on $J^{-1}(0)$ is just the action G on $J^{-1}(0)$.

In order to have a good quotient space, we need to check if the action is free and proper (which implies the action is simple.) Obviously, the map,

$(a,b) \rightarrow (a + q^1, b + e^a q^2, 0, 0)$ is one-to-one. Thus, the action is free. Next, assume that, as $n \rightarrow \infty$,

$(q_n^1, q_n^2, 0, 0 \rightarrow (q^1, q^2, 0, 0)$ and $(a_n + q_n^1, b_n + e^{a_n} q_n^2, 0, 0) \rightarrow (\gamma^1, \gamma^2, 0, 0)$,

$$\text{Then} \qquad \begin{cases} a_n \rightarrow \gamma^1 - q^1 \\ b_n \rightarrow \gamma^2 - e^{\gamma^1 - q^1} q^2, \end{cases}$$

which shows the action is proper. Now we can apply Thm. 4.3.1. [Abraham and Marsden] to find the symplectic reduced manifold. In fact G_0 acts transitively on $J^{-1}(0)$, and hence P_0 is a 1 point manifold. From the reduction theorem, there exists a unique ω_0 (symplectic form) on P_0 such that

$$\pi_0^* \, \omega_0 = i_0^* \omega$$

$$\text{where} \qquad \begin{cases} \pi_0 : J^{-1}(0) \rightarrow P_0 & \text{is the canonical projection,} \\ i_0 : J^{-1}(0) \hookrightarrow P & \text{is the inclusion map.} \end{cases}$$

In our case, ω_0 degenerates to 0. Now we consider the dynamics on the manifold. If we define the Hamiltonian function on $\mathbb{R}^4$ by

$$H : \mathbb{R}^4 \rightarrow \mathbb{R}$$
$$(q^1, q^2, p_1, p_2) \mapsto p_2 e^{q^1},$$

It is easy to check that H is G invariant and,

$H = e^{q^1} \frac{\partial}{\partial q^2} - p_2 e^{q^1} \frac{\partial}{\partial p_1}$ is the Hamiltonian vector field on $(\mathbb{R}^4, \omega)$ associated to H.

The flow of X_H is given by,

$$F^t_{X_H}\left((q^1, q^2, p_1, p_2)\right) = (q^1, q^2 + te^{q^1}, p_1 - tp_2 e^{q^1}, p_2) \tag{7}$$

Now, we can apply Theorem 4.3.5. in [2] to find the reduced dynamics as

$$H_\mu = 0,$$
$$X_{H_\mu} = 0.$$

Obviously, the reduced dynamics is trivially Lyapunov stable.

Now we are ready to investigate the notion of relative stability. Before doing that, we note that any point in $J^{-1}(0)$ maps to one point in $P_0 = J^{-1}(0)/G_0$. Hence any point x in $J^{-1}(0)$ is a relative equilibrium, (i.e. the corresponding flow $F^t_{X_H}(x)$ is a stationary motion (in the sense of Libermann-Marle)).

The following computations confirm the argument we have made above.

For $x \in J^{-1}(0)$, we have to find $\xi \in \mathfrak{I}$ such that

$$F^t_{X_H|J^{-1}(0)}(x) = \Phi(exp\ (t\xi), x). \tag{8}$$

In coordinates,

$$F^t_{X_H|J^{-1}(0)}(x) = (q^1, q^2 + te^{q^1}, 0, 0)$$

$$\Phi(exp\ (t\xi), x) = \begin{cases} (q^1 + t\xi^1, e^{t\xi^1}q^2 + (e^{t\xi^1} - 1)\frac{\xi^2}{\xi^1}, 0, 0) & \text{if } \xi^1 \neq 0 \\ (q^1, q^2 + t\xi^2, 0, 0) & \text{if } \xi^1 = 0. \end{cases}$$

If we choose,

$$\xi^1 = 0 \ , \quad \xi^2 = e^{q^1},$$

then (8) is satisfied.

Remark

$M/G \approx \mathbb{R}^2$ is a good quotient.

The G invariant dynamics X_H descends to M/G. We denote the quotient dynamics as $\hat{X}$.

If we denote by $\pi : \ M \to M/G$ the canonical projection, then the following are equivalent characterizations of relative equilibria.

$x \epsilon J^{-1}(\mu)$ is a relative equilibrium

$\leftrightarrow$

$F^t_{X_H}(x) = \Phi(exp\ (t\xi), x)$

for some $\xi \epsilon \mathfrak{I}$

$\leftrightarrow$

$F^t_{X_H}(x)$ is a stationary motion

$\leftrightarrow$ $X_{H_\mu}(\pi_\mu(x)) = 0$

$\leftrightarrow$ $\hat{X}(\pi(x)) = 0$

Definition We say that $F^t_{X_H}(x)$ is relatively stable mod G if $\pi(x)$ is a Lyapunov stable equilibrium point of $\hat{X}$.

Now consider $x = (q^1, q^2, p_1, p_2) = (0,0,0,0) \epsilon \mathbb{R}^4$

$F^t_{X_H}(x) = (0,t,0,0)$ is a stationary motion since $F^t_{X_H}(x) \subset J^{-1}(0)\ \forall\ t \epsilon \mathbb{R}$

Is this motion relatively stable mod G? We can coordinatize M/G via

$$\pi(x) = \begin{pmatrix} p_1 \\ p_2 e^{q^1} \end{pmatrix}.$$

This is because

$$\begin{pmatrix} q^1 \\ q^2 \\ p_1 \\ p_2 \end{pmatrix}$$

is in the G orbit of

$$\begin{pmatrix} 0 \\ 0 \\ p_1 \\ e^{q_1} p_2 \end{pmatrix}.$$

$$\text{Clearly,} \quad \pi\{(0,t,0,0,)|t \epsilon \mathbb{R}\} = (0,0)$$
$$= \text{equilibrium of } \hat{X}$$

as it should be.

The question of relative stability mod G of the flow $(0,t,0,0)$ reduces to a question of Lypunov stability of the equilibrium $\pi(x) = (0,0) \epsilon M/G$.

Choose

$$\lambda = \begin{pmatrix} \lambda_1 \\ \lambda_2 \end{pmatrix} = \begin{pmatrix} p_1 \\ e^{+q^1} p_2 \end{pmatrix}$$

in a neighborhood of $\begin{pmatrix} 0 \\ 0 \end{pmatrix} \epsilon M/G$. Let $x_\lambda = (0,0,\lambda_1,\lambda_2) \epsilon \pi^{-1}(\lambda)$.

Then,

$$F^t_{X_H}(x_\lambda) = (0,t,\lambda_1 - t\lambda_2, \lambda_2)$$

$$\pi \circ F^t_{X_H}(x_\lambda) = \begin{pmatrix} \lambda_1 - t\lambda_2 \\ \lambda_2 \end{pmatrix}.$$

Clearly if $\lambda_2 \neq 0$, $\pi \circ F^t_{X_H}(x_\lambda)$ leaves any neighborhood of $\begin{pmatrix} 0 \\ 0 \end{pmatrix} \epsilon M/G$ in finite time!

Hence the stationary motion $(0,t,0,0)$ is not relatively stable mod G.

References to the Appendix

[1] P. Libermann, C.-M. Marle, "Symplectic Geometry and Analytical Mechanics", D. Reidel Publishing Company 1987.

[2] J.E. Marsden, A. Weinstein, "Reduction of Sympletic Manifolds with Symmetry". Reports on Mathematical Physics. 5. 1974. pp. 121-130.

[3] R. Abraham, J.E. Marsden, "Foundations of Mechanics" 2nd Edition., The Benjamin/Cummings Publishing Company 1978.

Department of Electrical Engineering and Systems Research Center
University of Maryland, College Park
College Park, MD 20742

Contemporary Mathematics
Volume 97, 1989

Morse theory for a model space structure.

Mark Levi*
Boston University

In this note we outline the results of qualitative analysis of a natural perturbation of the classical problem of dynamics of a free rigid body in space. Our aim here is to describe a bifurcation effect previously unobserved in flexible space structures. To that end we choose a maximally simple model to make the ideas as transparent as possible with the minimum of technicalities.

By coupling two classical and well-understood problems–a free rigid body in space and a damped oscillator we produce a dynamical system which can be viewed as the simplest meaningful model of a flexible body, such as a spacecraft with flexible attachments. In fact, our system can be viewed as a **one-mode truncation** of a flexible structure with infinitely many degrees of freedom. The phenomena described here are certain to occur in such infinite degree of freedom structures as well.

One such surprising phenomenon is the "symmetry–seeking" property of the model, manifesting itself in the existence of purely rotational asymptotic motions in which the effective ellipsoid of inertia of the rotating body becomes rotationally symmetric after the internal vibrations have dissipated, for a whole **interval** of angular momentum values. Moreover, the axis of rotation of the system will not coincide in these motions with any of the symmetry axes of the system. Finally, the deflections of the elastic part of the system will be shown to be *independent of the value of the angular momentum* in a certain range.

It seems to be a common belief that any elastic structure with internal dissipation tends, in the absence of all external forces, to a pure rotation with a constantt angular velocity around one of the axes of the effective ellipsoid of inertia (cf. Kaplan [5]). We show that, generally speaking, this is false, and propose a condition under which the asymptotic motions indeed are pure rotations.

In the presence of internal dissipation the flow in the reduced state space the system is almost always a Morse flow, that is, every motion tends to an equilibrium, with the energy playing the role of the Lyapunov function. Physical implications of this transparent geometrical picture may seem rather surprising; these are described in detail below.

1. Introduction

The problems of dynamics of non-rigid structures in space have been studied extensively, notably, by Poincaré, Lyapunov, Kovalevskaya, and more recently by Chandrasekhar [4] and others. These remarkable studies deal with the equilibrium shapes of self-gravitating fluids.

---

* Supported by AFOSR Grant #85-0144.

Since the early 1960's the new interest in related questions arose in connection with the development of artificial satellites, and more recently, in the context of large flexible space structures.

We mention in passing a very interesting problem of the motion of a rigid body with a fluid-filled cavity; for the ideal fluid with no bubbles, the problem has been solved completely over 100 years ago [8]. As it turns out, the motion of the fluid-filled body is indistinguishable from that of a rigid body with an appropriate tensor of inertia(!).

Recent ideas and methods of dynamical systems have not yet exercised their full power on this class of problems, and have only started penetrating this particular area. Very recently, various aspects of dynamics of flexible space structures have attracted attention of mathematicians. Among the new studies are stability proofs of the "least energy rotations" based on Arnold's ideas [2],[7], bifurcation analyses of rotating systems consisting of a rigid part with flexible attachments [3], [7], studies of the dynamics of coupled rigid bodies; the references to this subject can be found in the present volume.

The Hamiltonian approach of Arnold [2] has been carried over to some models [7] and the Lagrangian approach was developed [3].

We mention also a somewhat related work starting with Kolodner [1], [6], [9] on the equilibria and dynamics of moving chains.

Our understanding of the full dynamics of flexible space structures has been very limited. Almost all existing results on the qualitative aspects of dynamics fall into the following categories:

- The characterization of "rigid modes" in *some* models, i.e. the classification (in a very few special examples) of possible equilibrium motions and of their bifurcations [3].
- Proof (in a few examples) of stability of rigid rotations in the least energy case [7].
- Proof (in some special cases) of the trend to a pure rotational motion [3].

While analyzing the effects of infinite-dimensionality of the beam, the body–beam model [3] deliberately excluded the effects of spatial orientation. In this note, we give a complete description of a model problem possessing the full rotational freedom.

2. Description of the model

We consider a rigid body, Figure 2.1, whose tensor of inertia is I, and we orient the coordinate axes x, y, z along the eigendirections of I, denoting the corresponding eigenvalues, i.e. the moments of inertia, by $I_1 > I_2 > I_3$. Let a mass m be constrained to slide along the x–axis and let it be a subject to an elastic restoring force $k(x - x_0)$ with the spring constant k. Here x denotes the position of the mass along the x–axis and x_0 is the equilibrium position of the mass when the entire system is at rest. To reduce the technicalities to a minimum, we also assume the center of mass of the rigid body to be fixed in space–or alternatively, we could take another mass $m_1 = m$ positioned symmetrically with the first mass, with symmetric initial conditions. Let $c\dot{x}$ be the damping force acting on the mass. We emphasize that the linearity assumptions on the elasticity and the dissipation have nothing to do with the phenomena described below; these assumptions are chosen to bring out the essential features unobstructed by inessential technicalities. The equations of motion of the system are given by

$$\begin{cases} \dot{M} + \omega \times M = 0, \\ \ddot{x} + c\dot{x} + k(x - x_0) = (\omega_2^2 + \omega_3^2)x, \end{cases} \tag{2.1}$$

where $M = I\omega + mr \times (r_t + \omega \times r)$ is the total angular momentum of the system expressed in the body frame, ω is the angular velocity of the rigid body expressed in the body frame and r is the position vector of the mass in that frame. This form of equations follows at once from the general Lagrangian aproach [3]; one can also derive them directly. The first equation, for instance, expresses the conservation of the angular momentum but written in the non–inertial frame of the rigid body. Since the vectors r and r_t are parallel, we have $M = I\omega + mr \times (\omega \times r) \equiv I(x)\omega$, where $I(x) = diag(I_1, I_2 + x^2, I_3 + x^2)$ is the (variable) tensor of inertia of the whole system; we will refer to it as the effective tensor of inertia. The equations of can now be written as

$$\begin{cases} \frac{d}{dt}[I(x)\omega] + \omega \times I(x)\omega = 0 \\ \ddot{x} + c\dot{x} + k(x - x_0) = (\omega_2^2 + \omega_3^2)x, \end{cases} \tag{2.2}$$

where we took the mass $m = 1$, or more explicitly,

$$\begin{cases} I\dot{\omega} + \omega \times I\omega + 2x\dot{x}\omega^1 + x^2\dot{\omega}^1 + x^2\omega_1(0, -\omega_3, \omega_2)^T = 0 \\ \ddot{x} + c\dot{x} + k(x - x_0) = (\omega_2^2 + \omega_3^2)x, \end{cases} \tag{2.3}$$

where $\omega^1 = (0, \omega_2, \omega_3)^T$. Physical meaning of the terms in the first equation of (2.3) is as follows.

— the third term gives the Coriolis torque exerted by the mass upon the rigid body.
— the fourth term gives the torque created by the inertial force due to the angular acceleration of the rigid body
— the last term gives the torque created by the centrifugal (inertial) force.

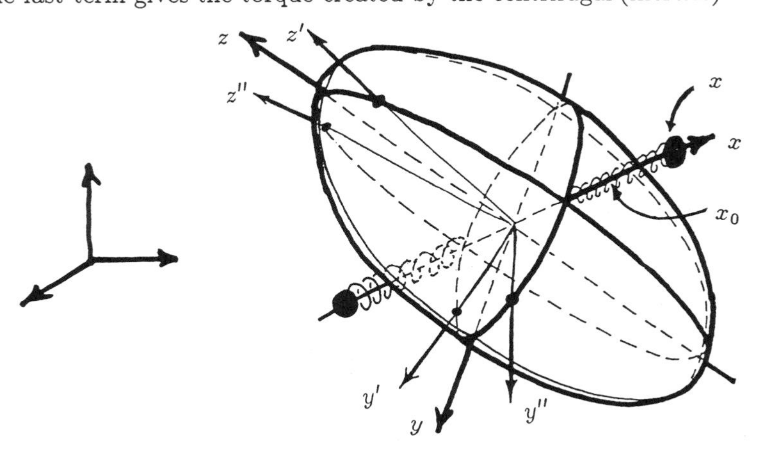

Figure 2.1. A model problem.

The right–hand side term in the second equation in (2.3) gives the centrifugal inertial force acting upon the mass exerted by the constraint of staying on the x–axis.

Energy of the system is given by

$$E = \frac{1}{2}(I\omega, \omega) + \frac{1}{2}[\dot{x}^2 + (\omega_2^2 + \omega_3^2)x^2] + \frac{1}{2}k(x - x_0)^2,$$

the expression in bracket giving the kinetic energy of the mass. The dissipation relation

$$\dot{E} = -c\dot{x}^2$$

can be checked either directly or from the general principle of Routh (cf. [3]). From the left-invariance of the Lagrangian, i.e. from the independence of the kinetic energy of the position of the body relative to the inertial frame, it follows that $|M|$ is a conserved quantity; in the presence of dissipation ($c \neq 0$) it is the only nontrivial conserved quantity.

3. Dynamics and Bifurcations

In this section we describe the asymptotic dynamics of the system and point out some interesting "symmetry–seeking" effects.

The first step in the direction was made in [3], although the first heuristic discussion is mentioned already in [5]. As it turns out, that heuristic discussion is valid only for moderate values of angular momentum.

3.1 Asymptotic behavior.

Theorem 3.1. *Every solution of eq. (2) tends to the zero dissipation set (the set of rigid configurations)*

$$\{(x, \dot{x}, \omega) : \dot{x} = 0\},$$

if the damping $c > 0$.

Proof. We have proved a similar theorem in [3] for an infinite– dimensional system but with only one rotational degree of freedom; the main difference here is that the system has a full rotational freedom.

We have to show that given any $\epsilon > 0$ and any initial condition $Z_0 = (x(0), \dot{x}(0), \omega_0)$, there exists $T = T(Z_0, \epsilon)$ such that for all $t \geq T$ the solution $Z(t, Z_0) = (x, \dot{x}, \omega)$ stays in the ϵ-strip around a (x, ω)-hyperplane in $\mathbf{R}^5 \equiv \{(x, \dot{x}, \omega)\}$. Assuming the contrary, we would have an $\epsilon > 0$ and Z_0 such that the solution starting at Z_0 would venture repeatedly outside the strip $|\dot{x}| \leq \epsilon$. For $|\dot{x}| > \epsilon$ we have $\dot{E} = -c\dot{x}^2 < -c\epsilon^2$, and thus we conclude that $Z(t)$ can spend only finite time outside the strip $|\dot{x}| \leq \epsilon$; since $\epsilon > 0$ was arbitrary, $x(t)$ will repeatedly come arbitrarily close to $\{\dot{x} = 0\}$. Consequently the strip $\frac{\epsilon}{2} \leq \dot{x} \leq \epsilon$ will be crossed infinitely many times. Since the amount of energy lost in each crossing is greater than some $\delta > 0$, we conclude: $E \to -\infty$, a contradiction.

We have used the same argument in an infinite-dimensional setting in [3]. The above result says that the pendulum stops oscillating, but leaves open the question on the limiting motion of the system as a whole. This question is answered in the next statement.

Theorem 3.2 *If $I_2 \neq I_3$ then for any initial condition in eq. (2.2) we have $\omega(t) \to \omega_\infty = const$, $x(t) \to x_\infty = const$, $\dot{x} \to 0$, with the limiting values satisfying*

$$I(x_\infty)\omega_\infty = \lambda\omega_\infty, \quad x_\infty = \frac{kx_0}{k - \omega_2^2 - \omega_3^2}.$$

We conclude that eq. (2.2) is a Morse system for $I_2 \neq I_3$. **Proof.** The proof is the same as for a similar theorem in [3].

Remark. It is important to note that the apparently reasonable conclusion $\omega \to const$ is false in general: if $I_2 = I_3$, a "typical" solution tends to $x \to const$, $\omega \to (\omega_1, a\cos b(t-t_0), a\sin b(t-t_0))$ with constant a, b, ω_1. Physically, this says that even in the presence of elasticity some structures need not tend to a pure rotation but rather may tend asymptotically to a precessing motion. One expects, however, that this is an exceptional phenomenon; in the case $I_2 = I_3$ at hand it is – in fact, the asymptotic precession is nongeneric and in fact, is of codimension 1, as theorem 3.2 shows. It should be pointed out that this phenomenon is due to a "residual rigidity" in the system: the motion of the mass is constrained to a line. Were we to relax this constraint (thus also increasing the dimension of the problem), we would have eliminated the possibility of a limiting motion with precession. It should also be pointed out that in the presence of the symmetry $I_2 = I_3$ one can carry out a further reduction; in the reduced phase space the system tends to an equilibrium and is a Morse system (which it is not in $\mathcal{M}_\mu$ in the symmetric case).

The model at hand suggests that, as a general principle, in order to eliminate the possibility of an asymptotic precession in a flexible space structure it suffices to assume that any change in the angular velocity of the structure leads to a deformation of some flexible part of the structure. To illustrate this point on the present model we note that the last condition fails precisely in the case $I_2 - I_3$; indeed, for a precession we have $\dot{\omega} = (0, -ab\sin bt, ab\cos bt) \neq 0$, which produces no change in the inertial forces acting upon the mass and thus no deformation; had the constraint of the mass to the x–axis been relaxed, the precessional motion would not have existed as an asymptotic state.

3.2 Global dynamics

Theorem 3.2 above describes the fate of an individual solution, but leaves open the question of global dynamics of the system and of its bifurcations – that is, it does not say what happens as one changes the initial consitions or the parameters. In this section we describe the effect by which the system "seeks out" a symmetry which persists for an open interval of parameter values.

Before stating the result, we note that the phase space R^5 is foliated by the invariant surfaces of constant angular momentum, and it suffices therefore to analyze the flow of eq. (2) restricted to the angular momentum surface

$$\mathcal{M}_\mu = \{(\omega, x, \dot{x}) : |M| = |I(x)\omega| = \mu\},$$

for different values of the constant μ. Topologically, $M_\mu = S^2 \times \mathbf{R}^2$. We fix the magnitude μ of the angular momentum and choose the Hooke's constant k of the spring as the parameter (the choice of μ as the parameter instead of k would amount simply to a rescaling of time).

Theorem 3.3 *Fix $|M| = \mu$, and assume that $I_1 > I_2 + x_0^2$, i.e. that the effective principal moments of inertia $I_1 > I_2 + x_0^2 > I_3 + x_0^2$ of the undeformed system are ordered in the same way as the moments of inertia $I_1 > I_2 > I_3$ of the rigid body. For any k in the interval*

$$0 < k < \frac{\mu}{I_1^2}\left(1 - \frac{x_0}{\sqrt{I_1 - I_2}}\right)^{-1} \equiv k_y \tag{3.1}$$

the system admits a rotation around two axes y', y'' in the (x, y)-plane not coinciding with any of the principal inertial axes x or y. Moreover, the position x_∞ of the mass in such a limiting motion is independent of k. This position is such as to make the effective ellipsoid of inertia rotaionally symmetric around the z-axis. The same statements hold when one replaces the y-axis with the z-axis, the xy-plane with the xz-plane, I_2 with I_3, k_y with $k_z = \frac{\mu}{I_1^2}(1 - \frac{x_0}{\sqrt{I_1 - I_3}})^{-1}$, etc..

Proof. We start by defining x_∞ by the symmetry condition $I_1 = I_2(x_\infty) \equiv I_2 + x_\infty^2$, i.e. by

$$x_\infty^2 = I_1 - I_2 > x_0^2.$$

Next we define the angular velocity ω_∞; in order that eq. (2.2b) hold for the constant $x(t) \equiv x_\infty$, it is necessary that

$$\omega_2^2 + \omega_3^2 = k\frac{x_\infty - x_0}{x_\infty} > 0;$$

the last inequality follows from the assumption made in the statement of the theorem. For eq. (2.2a) to hold one must have $\omega \times I(x_\infty)\omega = 0$, i.e. there must exist $\lambda > 0$ such that $I(x_\infty)\omega = \lambda\omega$, or $(I_1\omega_1, I_1\omega_2, (I_3 + x_\infty^2)\omega_3) = \lambda(\omega_1, \omega_2, \omega_3)$. Choosing $\omega_3 = 0$, $\lambda = I_1$, we satisfy eq. (2.2a); eq. (2.2) is thus satisfied by any $\omega = (\omega_1, \omega_2, 0)$ with

$$\omega_2^2 = k\frac{x_\infty - x_0}{x_\infty} > 0 \quad \text{and} \quad x = x_\infty \equiv I_1 - I_2. \tag{3.2}$$

Recalling the angular momentum constraint $|M| = \mu$, we obtain $\omega_1^2 + \omega_2^2 = \frac{\mu}{2I_1^2}$. Thus if

$$A \equiv \frac{\mu}{I_1^2} - k\frac{x_\infty - x_0}{x_\infty} > 0 \tag{3.3}$$

then $\omega_1 = \sqrt{A}$ and ω_2 given by eq. (3.2) satisfy the equations of motion (2.2). Condition (3.3) is in turn equivalent to (3.1), q.e.d. .

3.3. Bifurcations and stability.

It is not hard to show that for certain values of the parameters our system has **seven** distinct equilibrium rotations, among which three can be stable; all these are depicted in figure 3.1. (We identify, of course, the rotations around the same axis in the opposite direction.) This is to be contrasted with the well-known situation of only three rotations for the rigid bodies with distinct principal moments of inertia. For large k the spring is stiff and the system has three equilibrium rotations only, just as a rigid body. As k is decreased it passes through two bifurcation values $k_z > k_y > 0$ specified in Theorem 3.1. At k_z the rotational equilibrium around the z–axis undergoes a pitchfork bifurcation giving rise to two additional rotations around the z' and z''–axes which bifurcate off of the z–axis and which lie in the xz–plane. As k decreases further, these axes rotate towards the x–axis as shown in figure 3.1. As k decreases through the next value $k_y < k_z$, two axes of rotation y' and y'' bifurcate off of the y–axis and rotate in the xy–plane towards the x–axis. The stability picture is summarized in the following table. We note in particular, that the "saddle" rotation around the y–axis splits into two saddles and a sink – as it turns out, this sink is a sink since it (locally) minimizes the energy on the surface $\mathcal{M}_\mu$. Similarly, the z–rotation undergoes a pitchfork bifurcation and splits into a saddle and two unstable foci. The detailed analysis of this picture will appear elsewhere. From the more practical point of view one might want to change μ rather than k; it is not hard to see, in fact, that the bifurcation diagram with μ as a parameter is obtained from the diagram in figure 3.1 by relabeling the k–axis into the μ–axis and by reversing the direction of the parameter axis.

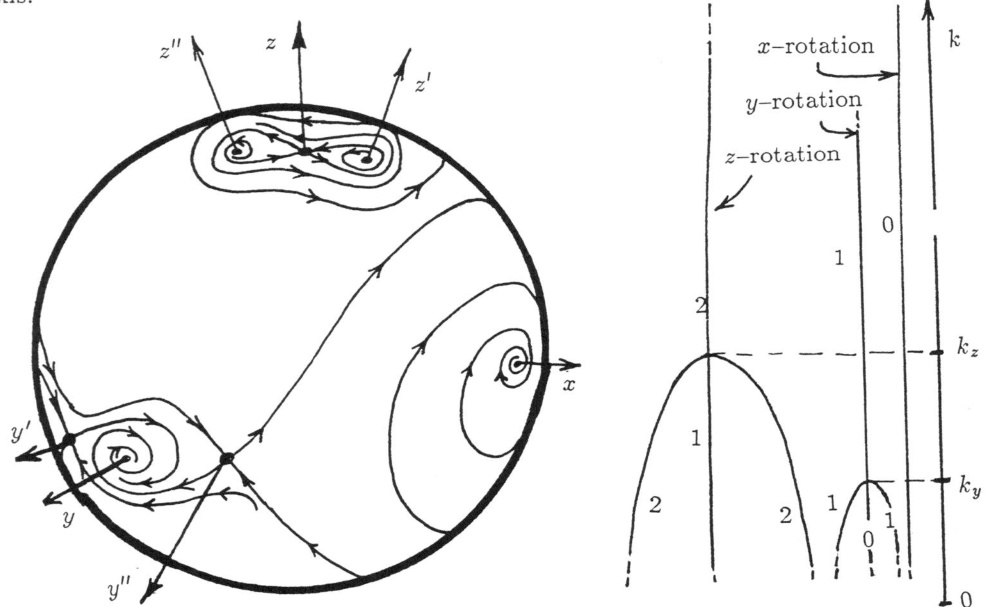

Figure 3.1. Equilibrium rotations of the system in $\mathcal{M}_\mu$ for $k < k_y$ and the bifurcation diagram.

<table>
<tr><th>axis \ k</th><th>$0 < k < k_y$</th><th>$k_y < k < k_z$</th><th>$k_z < k < \infty$</th></tr>
<tr><td>x</td><td colspan="3">0</td></tr>
<tr><td>y</td><td>0</td><td colspan="2">1</td></tr>
<tr><td>y', y''</td><td>1</td><td colspan="2">—</td></tr>
<tr><td>z</td><td colspan="2">1</td><td>2</td></tr>
<tr><td>z', z''</td><td colspan="2">2</td><td>—</td></tr>
</table>

Table 3.2. The entries in the table give the Morse index of the equilibrium points of the flow (2.2) constrained to $\mathcal{M}_\mu$.

References

[1] S.S. Antman, Large Lateral Buckling of Nonlinear Elastic Beams. Arch. Rat. Mech. Anal., 84(1984), pp. 293–305.

[2] V.I. Arnold, Mathematical Methods in Classical Mechanics. Springer–Verlag, NY 1983

[3] J. Baillieul and M. Levi, Rotational Elastic Dynamics. Physica 27D, pp. 43–62, 1987.

[4] S. Chandrasekhar, Hydrodynamic and Hydromagnetic Stability, Dover, NY, 1961.

[5] M.H. Kaplan, Modern Spacecraft Dynamics and Control. Wiley, New York 1976.

[6] I.I. Kolodner, Heavy Rotating Chain – a Nonlinear Eigenvalue Problem. Comm. Pure Appl. Math., 8(1955), pp. 395–408.

[7] P.S. Krishnaprasad and J.E. Marsden, Hamiltonian Structures and Stability for Rigid Bodies with Flexible Attachments, Arch. Rat. Math. Mech, 1987.

[8] N.N. Moiseev and V.V Rumyantsev, Dynamics of the Body With Fluid–filled Cavities, Moscow (1965) (in Russian).

[9] M. Reeken, Classical Solutions of the Chain Equations. II. Math. Z. 166, pp. 67–82(1979).

Contemporary Mathematics
Volume **97**, 1989

A Unified Approach for the Control of Multifingered Robot Hands

Zexiang Li * and Shankar Sastry *

September 9, 1988

Abstract

In the study of multifingered robot hands, the process of manipulating an object from one *grasp* configuration to another is called dexterous manipulation. In this paper, we study controls for dexterous manipulation by a multifingered robot hand. We use the basic building blocks developed by previous investigators to formulate the kinematics of a multifingered robot hand system. For finger contact with an object, we classify three useful types of constraints: fixed point of contact, rolling contact and sliding contact. Then, we propose control laws for dexterous manipulation of the object under these contact constraints. We show that the control laws realize both the desired position trajectory and the desired grasp force simultaneously. We also provide simulation results based on a planar manipulation system.

1 Introduction

A new avenue of progress in the area of robotics is the use of a multifingered robot hand for fine motion manipulation. The versatility of robot hands accrues from the fact that fine motion manipulation can be accomplished through relatively fast and small motions of the fingers and from the fact that they can be used on a wide variety of different objects (obviating the need for a large stockpile of custom end effectors). Several articulated hands such as the JPL/Stanford hand (Salisbury 1982 [12]), the Utah/MIT hand (Jacobsen et. al. 1985 [5]) have recently been developed to explore problems relating to manipulation of objects. It is of interest to note that the coordinated action of multiple robots in a single manufacturing cell may be treated in the same framework as a multifingered hand.

*Research supported in part by National Science Foundation PYI Grant DMC 8451129, and by the Defense Advanced Research Projects Agency (DoD), monitored by Space and Naval Warfare Systems and Command under Contract N00039-88-C-0292.

Manipulation of objects by a multifingered robot hand is more complicated than the manipulation of an object rigidly attached to the end of a six-axis robotic arm for two reasons: the kinematic relations between the finger joint motion and the object motion are complicated, and the hand has to firmly grasp the object during its motion.

The majority of the literature in multifingered hands has dealt with kinematic design of hands and the automatic generation of stable grasping configurations (see for example Salisbury 1982 [12], Kerr 1985 [6], Li and Sastry 1988 [8]). A few control schemes for the coordination of a multifingered robot hand or a multiple robotic system have been proposed by Nakamura et. al. [11], Zheng and Luh [13], Arimoto [1] and Hayati [4]. The most developed scheme is the master-slave methodology ([13] and [1]) for a two-manipulator system. The schemes developed so far all suffer from the drawback that they either assume rigid attachment of the fingertips to the object or are open loop. The schemes do not account for an appropriate contact model between the fingertips and the object.

In this paper, we study control laws for coordinated manipulation by a multifingered robot hand under the following contact constraints: *(1) Fixed points of contact and (2) rolling contacts.* In [7], a robot hand manipulating an object with fixed points of contact is called coordinated manipulation, and is called rolling motion with rolling contacts.

A brief outline of the paper is as follows: In Section 2, we review some basic concepts concerning rigid body motion and kinematics of contact. In Section 3, we formulate the kinematics of a multifingered robot hand system, and develop the force/velocity transformation relations. In Section 4, we propose the corresponding coordinated control laws. We also present simulation results on a planar example. In Section 5, we conclude this paper with several important remarks.

2. Preliminaries

In this section, we discuss concepts concerning rigid body motion and kinematics of contact. For further treatment of this subject see (Li, Hsu and Sastry [9]) and (Montana [10]).

Notation 1 *Let C_i and C_j be two coordinate frames of $\Re^3$, where i, j are arbitrary subscripts. Then, $r_{i,j}$ and $R_{i,j}$ denote the position and orientation of C_i relative to C_j. Furthermore, $g_{i,j} \triangleq (r_{i,j}, R_{i,j}) \in SE(3)$ denotes the configuration of C_i relative to C_j, where the Euclidean group, $SE(3)$, denotes the configuration space of an object.*

Definition 1 *The velocity of C_i relative to C_j is defined using left translation by*

$$\begin{bmatrix} v_{i,j} \\ w_{i,j} \end{bmatrix} = \begin{bmatrix} R_{i,j}^t \dot{r}_{i,j} \\ \mathcal{S}^{-1}(R_{i,j}^t \dot{R}_{i,j}) \end{bmatrix} \tag{1}$$

where the map $\mathcal{S}$

$$\mathcal{S} : \Re^3 \longrightarrow so(3) : \begin{bmatrix} w_1 \\ w_2 \\ w_3 \end{bmatrix} \longmapsto \begin{bmatrix} 0 & -w_3 & w_2 \\ w_3 & 0 & w_1 \\ -w_2 & w_1 & 0 \end{bmatrix}$$

identifies $\Re^3$ with $so(3)$, the space of 3 by 3 skey-symmetric matrices (or the Lie algebra of SO(3)).

Proposition 1 *Consider three coordinate frames C_1, C_2 and C_3. The following relation exists between their relative velocities:*

$$\begin{bmatrix} v_{3,1} \\ w_{3,1} \end{bmatrix} = Ad_{g_{3,2}^{-1}} \begin{bmatrix} v_{2,1} \\ w_{2,1} \end{bmatrix} + \begin{bmatrix} v_{3,2} \\ w_{3,2} \end{bmatrix} \tag{2}$$

where $Ad_{g_{3,2}^{-1}}$, the Adjoint map of SE(3), is a similarity transformation given by

$$Ad_{g_{3,2}^{-1}} = \begin{bmatrix} R_{3,2}^t & -R_{3,2}^t \mathcal{S}(r_{3,2}) \\ 0 & R_{3,2}^t \end{bmatrix}.$$

The proof is rather straightforward, see for example [9].

Corollary 1 *Consider three coordinate frames C_1, C_2 and C_3. Suppose that C_3 is fixed relative to C_2. Then, the velocity of C_3 relative to C_1 is related to that of C_2 by a constant transformation, given by*

$$\begin{bmatrix} v_{3,1} \\ w_{3,1} \end{bmatrix} = \begin{bmatrix} R_{3,2}^t & -R_{3,2}^t \mathcal{S}(r_{3,2}) \\ 0 & R_{3,2}^t \end{bmatrix} \begin{bmatrix} v_{2,1} \\ w_{2,1} \end{bmatrix} \tag{3}$$

In this paper, we will assume that the object in consideration is smooth and convex. This assumption also applies to the fingers of a robot hand.

Definition 2 *The boundary of a smooth rigid object is an embeded 2-dimensional manifold $S \subset \Re^3$, which can be expressed as the union of finitely many open sets $\{S\}_i$, such that S_i is the image of a diffeomorphism*

$$\varphi : U \subset \Re^2 \longrightarrow S_i \subset \Re^3.$$

The pair (φ, U) is called a coordinate system of S. The coordinates of a point $s \in S_i$ are $\mathbf{u} = (u, v) = \varphi^{-1}(s)$.

We assume that an object in consideration has an orthogonal coordinate system, in the sense that

$$\varphi_u(\mathbf{u}) \cdot \varphi_v(\mathbf{u}) = 0, \quad \forall \mathbf{u} \in U,$$

where $\varphi_u(\mathbf{u})$ denotes the partial derivative of φ with respect to u. When (φ, U) is orthogonal, we define the Gauss frame at a point $\mathbf{u} \in U$ as the coordinate frame with origin at $\varphi(\mathbf{u})$ and coordinate axes

$$\mathbf{x}(\mathbf{u}) = \varphi_u(\mathbf{u})/\|\varphi_u(\mathbf{u})\|, \ \mathbf{y}(\mathbf{u}) = \varphi_v(\mathbf{u})/\|\varphi_v(\mathbf{u})\| \text{ and } \mathbf{z}(\mathbf{u}) = \mathbf{x}(\mathbf{u}) \times \mathbf{y}(\mathbf{u}).$$

Definition 3 *Consider a manifold S with an orthogonal coordinate system (φ, U). At a point $s \in S_i$, the curvature form K is defined as the 2×2 matrix*

$$K = [\mathbf{x}(\mathbf{u}), \mathbf{y}(\mathbf{u})]^t [\mathbf{z}_u(\mathbf{u})/||\varphi_u(\mathbf{u})||, \mathbf{z}_v(\mathbf{u})/||\varphi_v(\mathbf{u})||],$$

the torsion form T as the 1×2 matrix

$$T = \mathbf{y}(\mathbf{u})^t [\mathbf{x}_u(\mathbf{u}) \backslash ||\varphi_u(\mathbf{u})||, \mathbf{x}_v(\mathbf{u}) \backslash ||\varphi_v(\mathbf{u})||],$$

and the metric as the 2×2 diagonal matrix

$$M = diag(||\varphi_u(\mathbf{u})||, ||\varphi_v(\mathbf{u})||).$$

Example 1. Consider the sphere of radius R, with the following coordinate system

$$U = \{(u, v) \in \Re^2 | -\pi/2 < u < \pi/2, -\pi < v < \pi\}$$

and the map

$$\varphi : U \longrightarrow \Re^3 : (u, v) \longmapsto (R \cos u \cos v, -R \cos u \sin v, R \sin u).$$

The coordinates u and v are known as the latitude and longitude, respectively. The Gauss frame is defined by

$$\mathbf{x}(\mathbf{u}) = \begin{bmatrix} -\sin u \cos v \\ \sin u \sin v \\ \cos u \end{bmatrix}, \mathbf{y}(\mathbf{u}) = \begin{bmatrix} -\sin v \\ -\cos v \\ 0 \end{bmatrix}, \mathbf{z}(\mathbf{u}) = \begin{bmatrix} \cos u \cos v \\ -\cos u \sin v \\ \sin u \end{bmatrix}.$$

The curvature form, torsion form and metric are, respectively,

$$K = \begin{bmatrix} 1/R & 0 \\ 0 & 1/R \end{bmatrix}, T = [0, \ -\tan u/R], M = \begin{bmatrix} R & 0 \\ 0 & R\cos u \end{bmatrix}.$$

We now consider two objects, called obj1 and obj2, that move while maintaining contact with each other (Figure 1). By the earlier convexity assumption the objects will make contact over isolated points. We wish to describe here motion of the contact points in response to a relative motion of the objects.

Let C_{r1} and C_{r2} be the coordinate frames fixed relative to obj1 and obj2, respectively. Let S_1 and S_2 be the embeded 2-manifolds representing the boundary of obj1 and obj2. S_1 and S_2 can be expressed as the union of open sets, $S_1 = \bigcup_j S_1^j$, and $S_2 = \bigcup S_2^j$, where S_1^j has orthogonal coordinate system (φ_1^j, U_1^j) and S_2^j has orthogonal coordinate system (φ_2^j, U_2^j). Let $c_1(t) \in S_1$, and $c_2(t) \in S_2$ be the positions at time t of the point of contact relative to C_{r1} and C_{r2}, respectively. We will restrict our attention to an interval I so that $c_i(t)$ belong to a single coordinate system of $S_i, i = 1, 2$.

The coordinate system $(\varphi_i^j, U_i^j), i = 1, 2$, induces a Gauss frame at all points of S_i, which will be denoted by $C_{ci}, i = 1, 2$. We also define a continuous family of coordinate frames, two for each $t \in I$, as follows. Let the local frames at time t, $C_{l1}(t)$ and $C_{l2}(t)$ be the coordinate frames fixed relative to C_{r1} and C_{r2}, respectively, that coincide at time t with the Gauss frames at $c_1(t)$ and $c_2(t)$.

The five parameters that describe the 5 degrees of freedom for the motion of the points of contact are: the coordinates of the point of contact relative to the coordinate system (φ_1^j, U_1^j) and (φ_2^j, U_2^j), given by, respectively,

$$\mathbf{u}_1(t) = (\varphi_1^j)^{-1}(c_1(t)) \in U_1^j, \text{and } \mathbf{u}_2(t) = (\varphi_2^j)^{-1}(c_2(t)) \in U_2^j.$$

and the angle of contact, $\phi(t)$, defined as the angle between the x-axes of C_{c1} and C_{c2}. We chose the sign of ϕ so that a rotation of C_{c1} through angle $-\phi$ around its z-axis aligns the x-axes.

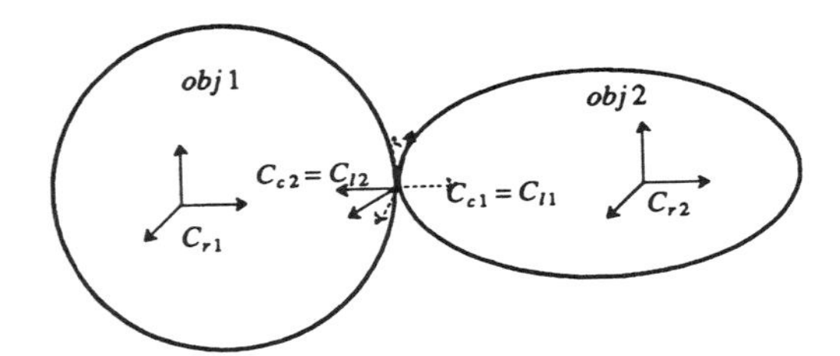

Figure 1: Coordinate frames for two objects which are in contact.

We describe the motion of obj1 relative to obj2 at time t, using the local coordinate frames $C_{l1}(t)$ and $C_{l2}(t)$. Let (v_x, v_y, v_z) and (w_x, w_y, w_z) be the translational and rotational velocities, respectively, of $C_{l1}(t)$ relative to $C_{l2}(t)$. These provide the 6 degrees of freedom for the relative motion between the objects.

The symbols K_1, T_1 and M_1 represent, respectively, the curvature form, torsion form and metric at time t at the point $c_1(t)$ relative to the coordinate system (φ_1^j, U_1^j). We can analogously define K_2, T_2 and M_2. We also let

$$R_\phi = \begin{bmatrix} \cos\phi & -\sin\phi & 0 \\ -\sin\phi & -\cos\phi & 0 \\ 0 & 0 & -1 \end{bmatrix} \triangleq \begin{bmatrix} \hat{R}_\phi & \begin{matrix} 0 \\ 0 \end{matrix} \\ 0\,0 & -1 \end{bmatrix}, \quad \tilde{K}_2 = \hat{R}_\phi K_2 \hat{R}_\phi.$$

Note that R_ϕ is the orientation matrix of C_{c1} relative to C_{c2}. Hence, $\tilde{K}_2$ is the curvature of obj2 seen from obj1. By the convexity assumption, the relative curvature form $K_1 + \tilde{K}_2$ is invertible. The following equations that relate motion of the points of contact to the relative velocity of the objects are due to Montana ([10]).

Theorem 1 *The point of contact and the angle of contact evolve according to*

$$\begin{cases} \dot{\mathbf{u}}_1 = M_1^{-1}(K_1+\tilde{K}_2)^{-1}\left(\begin{bmatrix} -w_y \\ w_x \end{bmatrix} - \tilde{K}_2\begin{bmatrix} v_x \\ v_y \end{bmatrix}\right), \\ \dot{\mathbf{u}}_2 = M_2^{-1}\hat{R}_\phi(K_1+\tilde{K}_2)^{-1}\left(\begin{bmatrix} -w_y \\ w_x \end{bmatrix} + K_1\begin{bmatrix} v_x \\ v_y \end{bmatrix}\right), \\ \dot{\phi} = w_z + T_1M_1\dot{\mathbf{u}}_1 + T_2M_2\dot{\mathbf{u}}_2, \\ 0 = v_z. \end{cases} \tag{4}$$

Montana ([10])calls the first three equations of (4) the kinematic equations of contact, and the last equation the constraint equation.

We define three special modes of contact in terms of the relative velocity components (v_x, v_y, v_z) and (w_x, w_y, w_z) by

(1) Fixed point of contact:

$$\begin{bmatrix} v_x \\ v_z \end{bmatrix} = 0, \ \textit{and} \begin{bmatrix} w_x \\ w_y \end{bmatrix} = 0; \tag{5}$$

(2) Rolling contact:

$$\begin{bmatrix} v_x \\ v_y \end{bmatrix} = 0, \quad \textit{and} \ w_z = 0; \tag{6}$$

(3) Sliding contact:

$$\begin{bmatrix} w_x \\ w_y \\ w_z \end{bmatrix} = 0. \tag{7}$$

We have from (4) that

Corollary 2 *The kinematic equations of contact correspond to each of the contact modes are*

$$\begin{cases} \dot{\mathbf{u}}_1 = 0, \\ \dot{\mathbf{u}}_2 = 0, \\ \dot{\phi} = w_z, \end{cases} \tag{8}$$

for fixed point of contact,

$$\begin{cases} \dot{\mathbf{u}}_1 = M_1^{-1}(K_1+\tilde{K}_2)^{-1}\begin{bmatrix} -w_y \\ w_x \end{bmatrix}, \\ \dot{\mathbf{u}}_2 = M_2^{-1}\hat{R}_\phi(K_1+\tilde{K}_2)^{-1}\begin{bmatrix} -w_y \\ w_x \end{bmatrix}, \\ \dot{\phi} = T_1M_1\dot{\mathbf{u}}_1 + T_2M_2\dot{\mathbf{u}}_2. \end{cases} \tag{9}$$

for rolling contact, and

$$\begin{cases} \dot{\mathbf{u}}_1 = -M_1^{-1}(K_1+\tilde{K}_2)^{-1}\tilde{K}_2\begin{bmatrix} v_x \\ v_y \end{bmatrix}, \\ \dot{\mathbf{u}}_2 = M_2^{-1}\hat{R}_\phi(K_1+\tilde{K}_2)^{-1}K_1\begin{bmatrix} v_x \\ v_y \end{bmatrix}, \\ \dot{\phi} = T_1M_1\dot{\mathbf{u}}_1 + T_2M_2\dot{\mathbf{u}}_2. \end{cases} \tag{10}$$

for sliding contact.

When a robot hand manipulates an object with its fingers contacting the object by one of the above contact modes, it defines, respectively, coordinated manipulation, rolling motion and sliding motion ([7]).

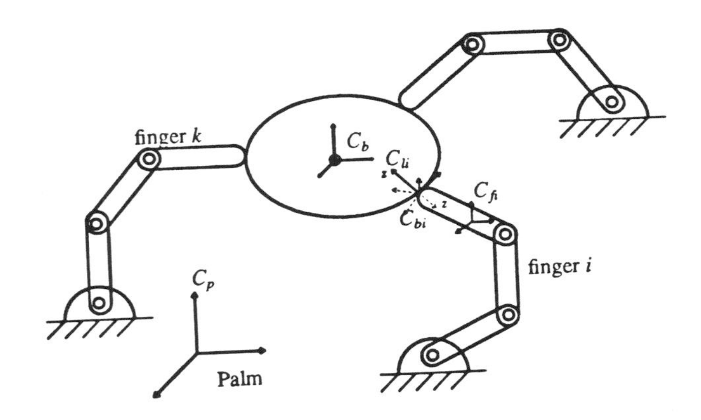

Figure 2: A hand manipulation system

Kinematics of a Multifingered Robot Hand

In this section, we derive the kinematics of a multifingered robot hand system and formulate the the velocity/force transformation relations so that control schemes for coordinated manipulation and for rolling motion can be studied easily in the section that follows. See [3] for the control of sliding motion.

Consider the hand manipulation system shown in Figure 2, which consists of an object and a $k-$fingered robot hand.

We denote by $m_i, i = 1, ...k$, the number of joints of finger i, and by $\theta_i, \tau_i \in \Re^{m_i}$, the vector of joint angles and the vector of joint torque, respectively, of finger i. We define a set of coordinate frames as follows. The reference frame of the system is C_p, which is fixed to the hand palm. The body coordinate frame is C_b, which is fixed to the mass center of the object. At time t the local frame of the object at the point of contact with finger i is $C_{bi}(t)$, which by our earlier convention is fixed relative to C_b. The reference frame of finger i is C_{fi}, which is fixed to the last link of finger i, and the local frame of finger i at time t is $C_{li}(t)$, which is fixed relative to C_{fi} and has origin at the point of contact with the object.

Let $c_{oi}(t) \in S_o$ and $c_{fi}(t) \in S_i$ be the positions at time t of the point of contact of the object with finger i relative to C_b and C_{fi}, respectively. Here, S_o denotes the boundary of the object and S_i the boundary of the last link of finger i. We assume that the contact point relative to finger i occurs only over the last link. We will restrict our attentions to a time interval I so that $c_{oi}(t)$ belongs to a single coordinate system (φ_o^j, U_o^j) of S_o, and $c_{fi}(t)$

belongs to a single coordinate system (φ_i^j, U_i^j) of S_i, $i = 1, \ldots k$. The coordinates of $c_{oi}(t)$ relative to (φ_o^j, U_o^j) will be denoted by $\mathbf{u}_{oi}(t) \in \Re^2$, and the coordinates of $c_{fi}(t)$ relative to (φ_i^j, U_i^j) by $\mathbf{u}_{fi}(t) \in \Re^2$. Let $\phi_i(t)$ be the angle of contact of the object at time t with respect to finger i.

Let (v_x^i, v_y^i, v_z^i) be the components of translational velocity of $C_{bi}(t)$ relative to $C_{li}(t)$, and (w_x^i, w_y^i, w_z^i) the components of rotational velocity (to apply the results of the previous section, let the object be objl, and finger i be obj2.). Since the local frames $C_{bi}(t)$ and $C_{li}(t)$ share a common origin (i.e., $r_{bi,li} = 0$), we obtain from Proposition 1 the following velocity constraint relation:

$$\begin{bmatrix} v_{bi,p} \\ \omega_{bi,p} \end{bmatrix} = \begin{bmatrix} R_{\phi_i} & 0 \\ 0 & R_{\phi_i} \end{bmatrix} \begin{bmatrix} v_{li,p} \\ \omega_{li,p} \end{bmatrix} + \begin{bmatrix} v_x^i \\ v_y^i \\ v_z^i \\ w_x^i \\ w_y^i \\ w_z^i \end{bmatrix} \tag{11}$$

where

$$R_{\phi_i} = \begin{bmatrix} \cos\phi_i & \sin\phi_i & 0 \\ -\sin\phi_i & -\cos\phi_i & 0 \\ 0 & 0 & -1 \end{bmatrix}$$

is the orientation matrix of C_{bi} relative to C_{li}. Note that by the constraint equation of (4), we have in (11) that $v_z^i = 0$.

On the other hand, by Corollary 1, the velocity of C_{bi} is related to the velocity of C_b by a constant transformation

$$\begin{bmatrix} v_{bi,p} \\ w_{bi,p} \end{bmatrix} = \begin{bmatrix} R_{bi,b}^t & -R_{bi,b}^t S(r_{bi,b}) \\ 0 & R_{bi,b}^t \end{bmatrix} \begin{bmatrix} v_{b,p} \\ w_{b,p} \end{bmatrix} = Ad_{g_{bi,b}^{-1}} \begin{bmatrix} v_{b,p} \\ w_{b,p} \end{bmatrix}, \tag{12}$$

and similarly for finger i one has

$$\begin{bmatrix} v_{li,p} \\ w_{li,p} \end{bmatrix} = \begin{bmatrix} R_{li,fi}^t & -R_{li,fi}^t S(r_{li,fi}) \\ 0 & R_{li,fi}^t \end{bmatrix} \begin{bmatrix} v_{fi,p} \\ w_{fi,p} \end{bmatrix} = Ad_{g_{li,fi}^{-1}} \begin{bmatrix} v_{fi,p} \\ w_{fi,p} \end{bmatrix}. \tag{13}$$

Moreover, the velocity of the finger reference frame, C_{fi}, is related to the velocity of the finger joints, $\dot{\theta}_i$, by the finger Jacobian,

$$\begin{bmatrix} v_{fi,p} \\ w_{fi,p} \end{bmatrix} = J_i(\theta_i)\dot{\theta}_i. \tag{14}$$

Substituting (12), (13) and (14) into (11) yields the constraint equation relating velocity of the object to the joint velocity of the fingers.

$$Ad_{g_{bi,b}^{-1}} \begin{bmatrix} v_{b,p} \\ w_{b,p} \end{bmatrix} = J_{fi}\dot{\theta}_i + \begin{bmatrix} v_x^i \\ v_y^i \\ 0 \\ w_x^i \\ w_y^i \\ w_z^i \end{bmatrix}, \tag{15}$$

where

$$J_{fi} \triangleq \begin{bmatrix} R_{\phi_i} & 0 \\ 0 & R_{\phi_i} \end{bmatrix} \cdot Ad_{g_{li,fi}^{-1}} \cdot J_i(\theta_i).$$

For a point contact with friction, a finger can exert linear forces upon the object about the point of contact. Thus, only contact constraints in the directions of the translational velocities can be re-enforced. When finger i contacts the object with a fixed point of contact, substituting (5) into (15) yields the following specialized constraint equation

$$B_i^t Ad_{g_{bi,b}^{-1}} \begin{bmatrix} v_{b,p} \\ w_{b,p} \end{bmatrix} = B_i^t J_{fi} \dot{\theta}_i, \tag{16}$$

where

$$B_i^t = \begin{bmatrix} 1 & 0 & 0 & 0 & 0 & 0 \\ 0 & 1 & 0 & 0 & 0 & 0 \\ 0 & 0 & 1 & 0 & 0 & 0 \end{bmatrix}.$$

Note that if finger i is a soft finger as in (Salisbury [12]), which enables the finger to exert an additional torque upon the object about the contact normal, or if finger i is rigidly attached to the object as in (Zheng [13]), which enables the finger to exert all six components of forces and torques upon the object, B_i^t can be modified accordingly as in (Li, Hsu and Sastry [9]), and (16) still holds.

When finger i contacts the object with rolling constraint, the corresponding constraint equation remains the same form as in (16), except that the contact coordinates $\mathbf{u}_{oi}(t)$ and $\mathbf{u}_{fi}(t)$ are no longer stationary and evolve according to (9).

When finger i slides across the object, the corresponding constraint equation becomes

$$B_i^t Ad_{g_{bi,b}^{-1}} \begin{bmatrix} v_{b,p} \\ w_{b,p} \end{bmatrix} = B_i^t J_{fi} \dot{\theta}_i + \begin{bmatrix} v_x^i \\ v_y^i \\ 0 \end{bmatrix}, \tag{17}$$

and the contact coordinates evolve according to (10).

We now examine briefly contact constraints in terms of contact wrenches. For a contact model, let n_i denote the total number of independent contact wrenches that finger i can apply to the object. For a point contact with friction, $n_i = 3$ (i.e., a force in the normal direction and two components of frictional forces in the tangent directions) but for a soft finger $n_i = 4$ (i.e., in addition to the three contact wrenches of a frictional point contact, a torque about the contact normal.). The resulting body wrench from applied contact wrenches of finger i can be expressed as

$$\begin{bmatrix} f_b \\ m_b \end{bmatrix} = Ad_{g_{bi,b}^{-1}}^t B_i x_i \tag{18}$$

where $f_b \in \Re^3$ is a linear force and $m_b \in \Re^3$ is a torque about the origin of C_b, and $x_i \in \Re^{n_i}$ is the magnitude vector of applied contact wrenches along the basis directions of B_i. For a frictional point contact, x_i is constrained to lie in the frictional cone K_i specified by

$$K_i = \left\{x_i \in \Re^{n_i}, x_{i,3} \leq 0, x_{i,1}^2 + x_{i,2}^2 \leq \mu^2 x_{i,3}^2\right\}$$

where μ is the coefficient of static Coulomb friction. Note that when finger i slides across the object, contact wrenches are restricted to the boundary ∂K_i of the friction cone, given by

$$\partial K_i = \left\{x_i \in \Re^{n_i}, x_{i,3} \leq 0, x_{i,1}^2 + x_{i,2}^2 = \mu^2 x_{i,3}^2\right\}.$$

By the *Principle of Virtual Work*, the joint torque required for maintaining static equilibrium in the presence of contact wrench $x_i \in \Re^{n_i}$, is given by

$$\tau_i = J_{fi}^t B_i x_i. \tag{19}$$

Finally, for the hand manipulation system, we define by $m = \sum_{i=1}^k m_i$, the total number of joints; $n = \sum_{i=1}^k n_i$, the total number of constraints; $\theta = (\theta_1^t, ...\theta_k^t)^t$, $\tau = (\tau_1^t, ...\tau_k^t)^t \in \Re^m$, respectively, the hand joint variable and the hand joint torque vectors; $B = \text{diag}(B_1, ...B_k)$ the basis matrix; $x = (x_1^t, ...x_k^t)^t \in \Re^n$ the magnitude vector of contact wrenches along the directions of B, and $K = K_1 \oplus ... \oplus K_k$ the force cone. Then, the contact constraint equation (16) for both fixed points of contact, and rolling contacts can be concatenated for $i = 1, ...k$ to give,

$$G^t \begin{bmatrix} v_{b,p} \\ w_{b,p} \end{bmatrix} = J_h \dot{\theta}, \tag{20}$$

where

$$G = [Ad_{g_{bi,b}^{-1}}^t, ...Ad_{g_{bi,b}^{-1}}^t]B \text{ and } J_h = B^t \text{diag}\{J_{f1}, ...J_{fk}\}$$

is called the grip Jacobian and the hand Jacobian, respectively.

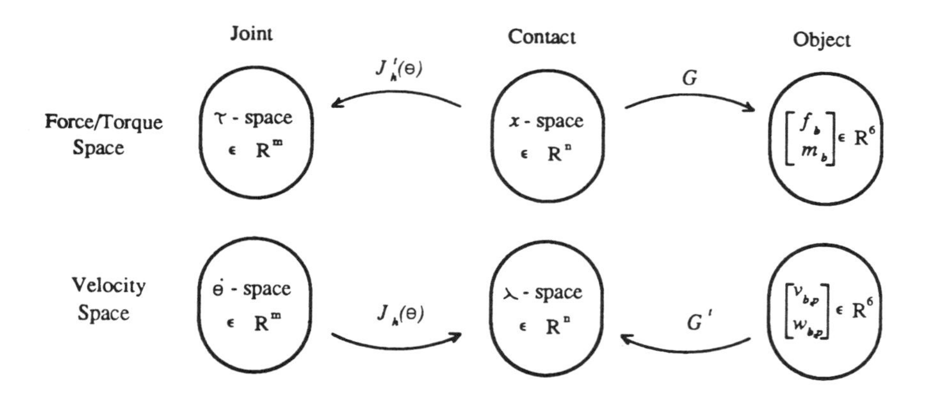

Figure 3: Force/velocity transformation for a hand manipulation system.

It is important to observe that *(1) for fixed points of contact G is constant, but J_h is not necessary constant unless $w_z^i = 0$. (2) For rolling contact both G and J_h depend on the contact coordinates, which are not stationary.*

The equation that relates the resulting body wrench to the applied contact wrenches is

$$\begin{bmatrix} f_b \\ m_b \end{bmatrix} = Gx \tag{21}$$

and the equation that relates contact wrenches to the required joint torques for maintaining static equilibrium is

$$\tau = J_h^t x. \tag{22}$$

The relations have been summarized in Table 1, while Figure 3 illustrates transformation of forces and motion in a hand manipulation system. Note that the vector $\lambda \in \Re^n$ is called the contact velocity.

	Force Torque Relations	Velocity Relations
Body to Fingertip	$\begin{bmatrix} f_b \\ m_b \end{bmatrix} = Gx$	$\lambda = G^t \begin{bmatrix} v_{b,p} \\ w_{b,p} \end{bmatrix}$
Fingertip to Joints	$\tau = J_h^t(\theta)x$	$J_h(\theta)\dot{\theta} = \lambda$

Table 1. Force/velocity transformation for a robot hand system.

Remarks: *(1) The null space of the grip Jacobian G, denoted as $\eta(G)$, is called the space of internal grasping forces (Kerr [6]). Any applied finger forces in $\eta(G)$ do not contribute to the motion of the object. However, during the course of manipulation a set of nonzero internal grasping forces is needed to assure that the grasp is maintained. Both Kerr and Roth ([6]); Nakamura et. al. ([11]) have presented detailed discussions on the optimal selection of internal grasping forces.*

The following dual definitions are now intuitive.

Definition 4 *(Stability and Manipulability of a Grasp) Define a grasp by a multifingered hand by $\Omega \triangleq (G, K, J_h)$ (see Figure 3). Then, for $K = R^n$ we have:*

1. *The grasp Ω is said to be stable if, for every body wrench $(f_b^t, m_b^t)^t$, there exists a choice of joint torque τ to balance it.*
2. *The grasp Ω is said to be manipulable if, for every body motion $(v_{b\ p}^t, w_{b,p})^t$, there exists a choice of joint velocity $\dot{\theta}$ to accommodate this motion without breaking contact.*

Remarks: *(1) A stable grasp has been called a force-closure grasp by Salisbury (1982). It is important to note that stability is not to be understood in the sense of Lyapunov since* *we are not discussing stability of a differential equation. (2) A manipulable grasp is called a grasp with full mobility by Salisbury (1982).*

Grasp stability and manipulability are now easily characterized for a given position of the fingers by

Proposition 2 *(1) A grasp is stable if and only if G is onto, i.e. the range space of G is the entire $\Re^6$. (2) A grasp is manipulable if and only if $R(J_h) \supset R(G^t)$, where $R(\cdot)$ denotes the range space.*

We remark that the conditions (1) and (2) superficially appear to be distinct, but they are related. In particular, a stable grasp which requires zero joint torque to balance a non-zero body wrench will be non-manipulable. Conversely, a manipulable grasp which requires zero joint motion to accommodate a non-zero body motion will be non-stable. Figure 4 (a) shows a planar two-fingered grasp, where each finger is one-jointed and contacts the object with a point contact with friction. Clearly the grasp is stable and a force f_y can be resisted with no joint torques. But the grasp is not manipulable, since a y-direction velocity on the body cannot be accommodated. Figure 4(b) shows a grasp of a body in

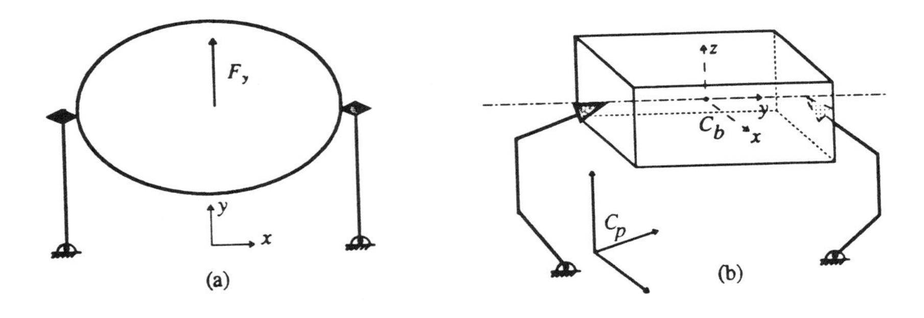

Figure 4: (a) A stable but not manipulable grasp, (b) a manipulable but not stable grasp.

$\Re^3$ by two three jointed fingers. The contacts are point contacts with friction. The grasp is manipulable, though the object can spin around the y-axis with zero joint velocities $\dot{\theta}$. However the grasp is not stable since a body torque τ_n about the y-axis cannot be resisted by any combination of joint torques.

In view of the preceding remarks, we will require a grasp to be both manipulable and stable, i.e.,

$$R(G) = \Re^6, \text{ and } R(J_h) \supset R(G^t). \tag{23}$$

Condition (1) suffers from the drawback that the force domain is left completely unconstrained. As we have seen earlier that the forces are constrained to lie in a convex cone K, taking into account the unidirectionality of the contact forces and finite friction al forces, in which case the image of $K \cap R(J_h)$ under G should cover all of $\Re^6$. Thus, we have

Corollary 3 *A grasp under unisense and finite frictional forces is both stable and manipulable if and only if*

$$G(K \cap R(J_h)) = \Re^6, \text{ and } R(J_h) \supset R(G^t). \tag{24}$$

4. Control Algorithms for Dexterous Manipulation

In this section, we develop control algorithms for dexterous manipulation. The objectives for each of the manipulation modes are:

1. *Coordinated Manipulation:* Control the fingers, *with fixed points of contact*, so that the object can be manipulated along a prescribed trajectory in $SE(3)$ while exerting possibly a set of desired contact forces on the environment.
2. *Rolling Motion:* Control the fingers, *with rolling contacts*, so that the object can be manipulated along a prescribed trajectory in $SE(3)$, while exerting a set of desired contact forces on the environment.

We will first propose the control scheme for coordinated manipulation, and then show that, after minor modifications, it also applies to rolling motion.

4.2 A Control Algorithm for Coordinated Manipulation

We assume that the desired trajectory of the object is

$$g^d_{b,p}(t) = (r^d_{b,p}(t), A^d_{b,p}(t)) \in SE(3), t \in [t_0, t_f]. \tag{25}$$

Since $SO(3)$ is a three dimensional manifold, we may choose pitch-roll-yaw variables, or the exponential coordinates to parameterize it. Let $\phi_{b,p} = (\phi_1, \phi_2, \phi_3)^t$ be one of these locally nonsingular parameterization. Then, we can express a trajectory of the object in terms of the parameterization variables

$$g_{b,p}(t) = (r_{b,p}(t), A_{b,p}(\phi_{b,p}(t))) \in SE(3),$$

and the body velocity in terms of the derivatives of the parameterization variables

$$\begin{bmatrix} v_{b,p} \\ w_{b,p} \end{bmatrix} = U(r_{b,p}, \phi_{b,p}) \begin{bmatrix} \dot{r}_{b,p} \\ \dot{\phi}_{b,p} \end{bmatrix} \tag{26}$$

Clearly, $U(r_{b,p}, \phi_{b,p})$ is a nonsingular 6×6 matrix. Differentiating (26) with respect to time t, yields the acceleration relation

$$\begin{bmatrix} \dot{v}_{b,p} \\ \dot{w}_{b,p} \end{bmatrix} = U(r_{b,p}, \phi_{b,p}) \begin{bmatrix} \ddot{r}_{b,p} \\ \ddot{\phi}_{b,p} \end{bmatrix} + \dot{U}(r_{b,p}, \phi_{b,p}) \begin{bmatrix} \dot{r}_{b,p} \\ \dot{\phi}_{b,p} \end{bmatrix}. \tag{27}$$

Note that to realize a desired trajectory of the object it is important that contact constraints be retained. On the other hand, validity of the contact constraints depends on if a proper set of internal grasp forces can be maintained. Thus, the second goal of the control algorithm is to regulate the internal grasp force to the following desired value

$$x_o^d(t) \in \eta(G), t \in [t_0, t_f], \tag{28}$$

where $\eta(G)$ is the null space of G. For fixed points of contact G is time independent, and a set of constant internal grasp forces suffices to guarantee contact constraints. Nevertheless, one may choose $x_o^d(t)$ to realize any other criteria.

We now proceed to formulate the control algorithms with (25) and (28) as our objectives.

From Section 3, the finger joint velocity $\dot{\theta}$ and the object velocity $(v_{b,p}^t, w_{b,p}^t)^t$ are related by the following constraint equation

$$J_h(\theta)\dot{\theta} = G^t \begin{bmatrix} v_{b,p} \\ w_{b,p} \end{bmatrix} \tag{29}$$

(29) is valid regardless of rolling or fixed points of contact. Differentiating (29) with respect to time t, yields the following acceleration constraint equation

$$J_h(\theta)\ddot{\theta} + \dot{J}_h(\theta)\dot{\theta} = G^t \begin{bmatrix} \dot{v}_{b,p} \\ \dot{w}_{b,p} \end{bmatrix} + \dot{G}^t \begin{bmatrix} v_{b,p} \\ w_{b,p} \end{bmatrix}. \tag{30}$$

Note that for fixed points of contact G is constant and the second term to the right hand side of (30) vanishes.

We will need the following assumption about the grasp.

A1. The grasp is both stable and manipulable.

Generation of object/finger trajectories so that assumption (A1) can be satisfied is discussed in (Li, Canny and Sastry [7]).

By Assumption (A1), we have that $R(J_h) \supset R(G^t)$ and we may express the joint acceleration $\ddot{\theta}$ in terms of the object acceleration by

$$\ddot{\theta} = J_h^+ G^t \begin{bmatrix} \dot{v}_{b,p} \\ \dot{w}_{b,p} \end{bmatrix} + J_h^+ \dot{G}^t \begin{bmatrix} v_{b,p} \\ w_{b,p} \end{bmatrix} - J_h^+ \dot{J}_h \dot{\theta} + \ddot{\theta}_o. \tag{31}$$

Here $J_h^+ = J_h^t (J_h J_h^t)^{-1}$ is the generalized inverse of J_h, and $\ddot{\theta}_0 \in \eta(J_h)$ is the internal motion of redundant joints not affecting the object motion.

Remarks: *(1) Using (31) will lead to a control algorithm in the task space. But if we express the object acceleration in terms of $\ddot{\theta}$ by*

$$\begin{bmatrix} \dot{v}_{b,p} \\ \dot{w}_{b,p} \end{bmatrix} = (GG^t)^{-1} G \left(J_h \ddot{\theta} + \dot{J}_h \dot{\theta} - \dot{G}^t \begin{bmatrix} v_{b,p} \\ w_{b,p} \end{bmatrix} \right)$$

a control algorithm in the joint space of the fingers can be developed. (2) When J_h is square, its generalized inverse J_h^+ is just the usual inverse, and $\ddot{\theta}_o$ disappears from (31). This also implies that the joint motion is determined uniquely by the motion of the object.

The dynamics of the object are given by the Newton-Euler equations

$$\begin{bmatrix} \hat{m} & 0 \\ 0 & \mathcal{I} \end{bmatrix} \begin{bmatrix} \dot{v}_{b,p} \\ \dot{w}_{b,p} \end{bmatrix} + \begin{bmatrix} w_{b,p} \times \hat{m} v_{b,p} \\ w_{b,p} \times \mathcal{I} w_{b,p} \end{bmatrix} = \begin{bmatrix} f_b \\ m_b \end{bmatrix}, \tag{32}$$

where $\hat{m} \in \Re^{3\times 3}$ is the diagonal matrix with the object mass in the diagonal, $\mathcal{I} \in \Re^{3\times 3}$ is the object inertia matrix with respect to the body coordinates, and $[f_b^t, m_b^t]^t$ is the applied body wrench in the body coordinates which is also related to the applied finger wrench $x \in \Re^n$ through

$$Gx = \begin{bmatrix} f_b \\ m_b \end{bmatrix}. \tag{33}$$

Notice that gravity and interaction forces from the environment can always be added to the right hand side of (32), and corresponding contact wrenches will be generated to counteract them.

Assumption A1 also implies that G is onto, and we can solve (33) as

$$x = G^+ \begin{bmatrix} f_b \\ m_b \end{bmatrix} + x_o, \tag{34}$$

where $G^+ = G^t(GG^t)^{-1}$ is the left inverse of G, and $x_o \in \eta(G)$ is the internal grasping force. The second control goal is to steer the internal grasping force x_o to the desired value $x_o^d(t) \in \eta(G)$.

Combining (32) and (34) yields

$$x = G^+ \left\{ \begin{bmatrix} \hat{m} & 0 \\ 0 & \mathcal{I} \end{bmatrix} \begin{bmatrix} \dot{v}_{b,p} \\ \dot{w}_{b,p} \end{bmatrix} + \begin{bmatrix} w_{b,p} \times \hat{m} v_{b,p} \\ w_{b,p} \times \mathcal{I} w_{b,p} \end{bmatrix} \right\} + x_o. \tag{35}$$

The dynamics of the ith finger manipulator is given by

$$M_i(\theta_i)\ddot{\theta}_i + N_i(\theta_i, \dot{\theta}_i) = \tau_i - J_i^t(\theta_i) B_i x_i. \tag{36}$$

Here, as is common in the literature, $M_i(\theta_i) \in \Re^{m_i \times m_i}$ is the moment of inertia matrix of the ith finger manipulator, $N_i(\theta_i, \dot{\theta}_i) \in \Re^{m_i}$ is the centrifugal, Coriolis and gravitational force terms, τ_i is the vector of joint torque inputs and $B_i x_i \in \Re^6$ the vector of applied finger wrenches. Define

$$M(\theta) = \begin{bmatrix} M_1(\theta_1) & \cdots & 0 \\ \vdots & \ddots & \vdots \\ 0 & \cdots & M_k(\theta_k) \end{bmatrix}, N(\theta, \dot{\theta}) = \begin{bmatrix} N_1(\theta_1, \dot{\theta}_1) \\ \vdots \\ N_k(\theta_k, \dot{\theta}_k) \end{bmatrix} \text{ and } \tau = \begin{bmatrix} \tau_1 \\ \vdots \\ \tau_k \end{bmatrix}. \tag{37}$$

Then, the finger dynamics can be grouped to yield

$$M(\theta)\ddot{\theta} + N(\theta, \dot{\theta}) = \tau - J_h^t(\theta) x. \tag{38}$$

The control objectives are to specify a set of joint torque inputs τ so that both the desired body trajectory (25) and the desired internal grasping force (28) can be realized.

The first main theorem of the paper is the following control algorithm for coordinated manipulation by robot hands with *non-redundant fingers* .

Theorem 2 *Assume that (A1) holds and the fingers are non-redundant, i.e., $m_i = n_i$, for $i = 1, \ldots k$. Denoting the position trajectory tracking error by $e_p \in \Re^6$, and the internal grasping force error by $e_f \in \Re^6$,*

$$e_p(t) = \begin{bmatrix} r_{b,p}(t) \\ \phi_{b,p}(t) \end{bmatrix} - \begin{bmatrix} r^d_{b,p}(t) \\ \phi^d_{b,p}(t) \end{bmatrix} \text{ and } e_f(t) = x_o(t) - x^d_o(t). \tag{39}$$

Then, the control law specified by (40) realizes, with fixed points of contact, not only the desired trajectory of the object but also the desired internal grasping force.

$$\begin{aligned} \tau &= N(\theta, \dot\theta) + J^t_h G^+ \begin{bmatrix} w_{b,p} \times \hat m v_{b,p} \\ w_{b,p} \times \mathcal{I} w_{b,p} \end{bmatrix} - M(\theta) J_h^{-1} \dot J_h \dot\theta + M_h \dot U \begin{bmatrix} \dot r_{b,p} \\ \dot\phi_{b,p} \end{bmatrix} \\ &+ J^t_h (x^d_o - K_I \int e_f) + M_h U \left\{ \begin{bmatrix} \ddot r^d_{b,p} \\ \ddot\phi^d_{b,p} \end{bmatrix} - K_v \dot e_p - K_p e_p \right\}, \end{aligned} \tag{40}$$

where

$$M_h = M(\theta) J_h^{-1} G^t + J^t_h G^+ \begin{bmatrix} \hat m & 0 \\ 0 & \mathcal{I} \end{bmatrix} \tag{41}$$

and K_I is a matrix such that the null space of G is K_I-invariant.

Note that

$$G J_h^{-t} M_h = G J_h^{-t} M(\theta) J_h^{-1} G^t + \begin{bmatrix} \hat m & 0 \\ 0 & \mathcal{I} \end{bmatrix}$$

is called the generalized inertia matrix of the hand system.

Remark: *The first four components in (40) are used for cancellation of Coriolis, gravitational and centrifugal forces. These terms behave exactly like the nonlinearity cancellation terms in the computed torque control for a single manipulator; the term $J^t_h(x^d_o - K_I \int e_f)$ is the compensation for the internal grasping force loop, and the last term is the compensation for the position loop. We will see in the proof that the dynamics of the internal grasping force loop and that of the position loop are mutually decoupled. Consequently, we can design the force error integral gain K_I independently from the position feedback gains K_v and K_p.*

Proof.

The proof is very procedural and straightforward. First, for nonredundant fingers with fixed points of contact, internal motion of the fingers reflected by $\ddot\theta_o$ in (31) disappear and G is constant. Thus, the acceleration constraint equation (31) simplifies to

$$\ddot\theta = J_h^{-1} G^t \begin{bmatrix} \dot v_{b,p} \\ \dot w_{b,p} \end{bmatrix} - J_h^{-1} \dot J_h \dot\theta. \tag{42}$$

Substitute (42) and (35) into (38) we have

$$M\left\{J_h^{-1}G^t\begin{bmatrix}\dot{v}_{b,p}\\ \dot{w}_{b,p}\end{bmatrix} - J_h^{-1}\dot{J}_h\dot{\theta}\right\}+N = \tau - J_h^t\left\{G^+\begin{bmatrix}\hat{m} & 0\\ 0 & \mathcal{I}\end{bmatrix}\begin{bmatrix}\dot{v}_{b,p}\\ \dot{w}_{b,p}\end{bmatrix} + G^+\begin{bmatrix}w_{b,p}\times\hat{m}v_{b,p}\\ w_{b,p}\times\mathcal{I}w_{b,p}\end{bmatrix}\right\} - J_h^t x_o. \tag{43}$$

Linearizing (43) with the following control

$$\tau = N(\theta,\dot{\theta}) + J_h^t G^+\begin{bmatrix}w_{b,p}\times\hat{m}v_{b,p}\\ w_{b,p}\times\mathcal{I}w_{b,p}\end{bmatrix} - M(\theta)J_h^{-1}\dot{J}_h\dot{\theta} + \tau_1 \tag{44}$$

where τ_1 is to be determined, we have that

$$\left\{M(\theta)J_h^{-1}G^t + J_h^t G^+\begin{bmatrix}\hat{m} & 0\\ 0 & \mathcal{I}\end{bmatrix}\right\}\begin{bmatrix}\dot{v}_{b,p}\\ \dot{w}_{b,p}\end{bmatrix} = \tau_1 - J_h^t x_o, \tag{45}$$

or

$$M_h\begin{bmatrix}\dot{v}_{b,p}\\ \dot{w}_{b,p}\end{bmatrix} = \tau_1 - J_h^t x_o.$$

Substitute (27) into the above equation, we have

$$M_h\left\{U\begin{bmatrix}\ddot{r}_{b,p}\\ \ddot{\phi}_{b,p}\end{bmatrix} + \dot{U}\begin{bmatrix}\dot{r}_{b,p}\\ \dot{\phi}_{b,p}\end{bmatrix}\right\} = \tau_1 - J_h^t x_o. \tag{46}$$

Further, let the control input τ_1 be

$$\tau_1 = M_h U\left\{\begin{bmatrix}\ddot{r}_{b,p}^d\\ \ddot{\phi}_{b,p}^d\end{bmatrix} - K_v\dot{e}_p - K_p e_p\right\} + M_h\dot{U}\begin{bmatrix}\dot{r}_{b,p}\\ \dot{\phi}_{b,p}\end{bmatrix} + J_h^t\left(x_o^d - K_I\int e_f\right) \tag{47}$$

and apply it to (46) to yield:

$$M_h U\{\ddot{e}_p + K_v\dot{e}_p + K_p e_p\} = -J_h^t(e_f + K_I\int e_f). \tag{48}$$

Multiply (48) by GJ_h^{-t}, we obtain the following equation.

$$GJ_h^{-t}M_h U\{\ddot{e}_p + K_v\dot{e}_p + K_p e_p\} = -G(e_f + K_I\int e_f) = 0 \tag{49}$$

where we have used the facts that $\eta(G)$ is constant and the internal grasping forces lie in the null space of G, i.e.,

$$G(e_f + K_I\int e_f) = 0. \tag{50}$$

Since $GJ_h^{-t}M_h = GJ_h^{-t}M(\theta)J_h^{-1}G^t + \begin{bmatrix}\hat{m} & 0\\ 0 & \mathcal{I}\end{bmatrix}$ is positive definite and U is non-singular, (49) implies that

$$\ddot{e}_p + K_v\dot{e}_p + K_p e_p = 0. \tag{51}$$

Thus, we have shown that the position trajectory tracking error e_p can be driven to zero with proper choice of the feedback gain matrices K_v and K_p.

The last step is to show that e_f also goes to zero. If we substitute (51) into (48) and notice that J_h is nonsingular, we have the following equation.

$$e_f + K_I \int e_f = 0. \tag{52}$$

With proper choice of K_I, the above equation implies that the internal grasping force error e_f converges to zero.

Q.E.D.

Theorem 2 provides a control law for coordinated manipulation by robot hands with non-redundant fingers. This implies that, for a point contact with friction, each finger of the hand has exactly three joints. But, in many industrial applications, several robots which often have more than three degrees of freedom are integrated to maneuver a massive load, or to perform a sophisticated task. Under the point contact model assumption the system has redundant degrees of freedom. It is desirable to have a control law that works for a robot hand with redundant degrees of freedom.

The control law of Theorem 2 can be modified for this purpose.

Corollary 4 *For a robot hand with redundant degrees of freedom, i.e., $m_i \geq n_i, i = 1, ...k$ assume that assumption A1 holds. Then, the control law given by (53) will realize both the desired object trajectory and the desired internal grasp force, with fixed points of contact.*

$$\begin{aligned}\tau = {} & N(\theta, \dot{\theta}) + J_h^t G^+ \begin{bmatrix} w_{b,p} \times \hat{m} v_{b,p} \\ w_{b,p} \times \mathcal{I} w_{b,p} \end{bmatrix} - M J_h^+ \dot{J}_h \dot{\theta} + M J_h^+ (J_h M^{-1} J_h^t) M_h \dot{U} \begin{bmatrix} \dot{r}_{b,p} \\ \dot{\phi}_{b,p} \end{bmatrix} + \\ & M J_h^+ (J_h M^{-1} J_h^t)(x_o^d - K_I \textstyle\int e_f) + M J_h^+ (J_h M^{-1} J_h^t) \hat{M}_h \left\{ \begin{bmatrix} \ddot{r}_{b,p}^d \\ \ddot{\phi}_{b,p}^d \end{bmatrix} - K_v \dot{e}_p - K_p e_p \right\},\end{aligned} \tag{53}$$

where

$$\hat{M}_h = (J_h M^{-1} J_h^t)^{-1} G^t + G^+ \begin{bmatrix} \hat{m} & 0 \\ 0 & \mathcal{I} \end{bmatrix}. \tag{54}$$

and $J_h^+ = J_h^t (J_h J_h^t)^{-1}$.

4.2 A Control Algorithm for Rolling Motion

Controls for dexterous manipulation with rolling constraints have been studied in (Cole, Hauser and Sastry [2]), and (Kerr [6]). Here, we modify Theorem 2 to give a control law for rolling motion. The main difference of rolling motion from coordinated manipulation is that the grip Jacobian G, which depends on the contact coordinates given by (9), is time varying. Consequently, the null space of $G(t)$, denoted by $V(t) \subset \Re^n$, is also time varying. In general $e_f(t) = x_o(t) - x_o^d(t) \in V(t)$ does not imply, unless $V(t)$ is time independent, that $\int_0^t e_f(\tau)d\tau \in V(t)$, nor $\dot{e}_f(t) \in V(t)$. Thus, we can not introduce dynamic feedback in the force loop, as we did in Theorem 2, to create linear force error equation. But,

$$G(t)e_f(t) = 0 \text{ implies that } G(t)\dot{e}_f(t) + \dot{G}(t)e_f(t) = 0. \tag{55}$$

Lemma 1 *Consider the following differential equation*

$$\dot{x}(t) = A(t)x(t). \tag{56}$$

Let $\mu(A(t)) = \lambda_{max}(A^*(t) + A(t))/2$ *be the matrix measure of* $A(t)$*, where* λ_{max} *stands for the maximum eigenvalue value. Then,*

$$\|x(t)\| \leq \|x(t_0)\| \exp \int_{t_0}^{t} \mu(A(\tau))d\tau.$$

In other words, if $\mu(A(t)) < 0, \forall t$, and $A(t)$ is sufficiently slow time-varying, then, the system (56) is exponentially stable.

Proposition 3 *Assume that Assumption (A1) holds for a robot hand with non-redundant degrees of freedom. Then, the following control law, along with the kinematic equations of contact given by (9), realizes both the desired position trajectory and the desired internal grasp force, for rolling contacts.*

$$\begin{aligned} \tau \;=\; & N(\theta,\dot{\theta}) + J_h^t G^+ \begin{bmatrix} w_{b,p} \times \hat{m} v_{b,p} \\ w_{b,p} \times \mathcal{I} w_{b,p} \end{bmatrix} - M(\theta) J_h^{-1} \dot{J}_h \dot{\theta} + M_h \dot{U} \begin{bmatrix} \dot{r}_{b,p} \\ \dot{\phi}_{b,p} \end{bmatrix} + \underline{M(\theta) J_h^{-1} \dot{G}^t \begin{bmatrix} v_{b,p} \\ w_{b,p} \end{bmatrix}} \\ & + J_h^t \Big(x_o^d - \underline{\dot{e}_f/\delta - G^+ \dot{G} e_f/\delta} \Big) + M_h U \left\{ \begin{bmatrix} \ddot{r}_{b,p}^d \\ \ddot{\phi}_{b,p}^d \end{bmatrix} - K_v \dot{e}_p - K_p e_p \right\}, \end{aligned} \tag{57}$$

where

$$M_h = M(\theta) J_h^{-1} G^t + J_h^t G^+ \begin{bmatrix} \hat{m} & 0 \\ 0 & \mathcal{I} \end{bmatrix} \tag{58}$$

and δ *is a sufficiently large number so that the force error equation can be made to be exponentially stable.*

Note that in order to compare the control law with that of Theorem 2, we have underlined the terms which are different here.

Proof.

The proof is very similar to that of Theorem 2 and an outline of it is given here. Under rolling constraints, (31) becomes

$$\ddot{\theta} = J_h^{-1} G^t \begin{bmatrix} \dot{v}_{b,p} \\ \dot{w}_{b,p} \end{bmatrix} + J_h^{-1} \dot{G}^t \begin{bmatrix} v_{b,p} \\ w_{b,p} \end{bmatrix} - J_h^{-1} \dot{J}_h \dot{\theta}. \tag{59}$$

Substitute (59) and (35) into the system dynamic equation (38) and linearize the resulting equation with the appropriate terms in the control inputs (57), we get

$$M_h \begin{bmatrix} \dot{v}_{b,p} \\ \dot{w}_{b,p} \end{bmatrix} = \tau_1 - J_h^t x_o. \tag{60}$$

Substituting (27) into the above equation and applying the rest of the control inputs yield

$$M_h\{\ddot{e}_p + K_v \dot{e}_p + K_p e_p\} = -J_h^t \Big\{ e_f + \dot{e}_f/\delta + G^+ \dot{G} e_f/\delta \Big\}. \tag{61}$$

Multiply (61) by GJ_h^{-t} and notice that because

$$Ge_f = 0, \text{ and } (G\dot{e}_f + \dot{G}e_f)/\delta = 0.$$

we have

$$G(e_f + \dot{e}_f/\delta + G^+\dot{G}e_f/\delta) = 0$$

which implies that

$$\ddot{e}_p + K_v \dot{e}_p + K_p e_p = 0. \tag{62}$$

This shows that the position error goes to zero. On the other hand, substituting (62) into (61), and using the fact that J_h is of full rank, we conclude that

$$(\delta I + G^+\dot{G})e_f + \dot{e}_f = 0. \tag{63}$$

Let $A(t) = -(\delta I + G^+\dot{G})$. It is easy to see that by choosing δ sufficiently large, $\mu(A(t))$ is negative for all $t \in [t_0, t_f]$. Consequently, by Lemma 1 force error e_f also goes to zero.

Q.E.D.

Combining Proposition 3 with Corollary 4 produces a control law for rolling motion by a robot hand with redundant degrees of freedom.

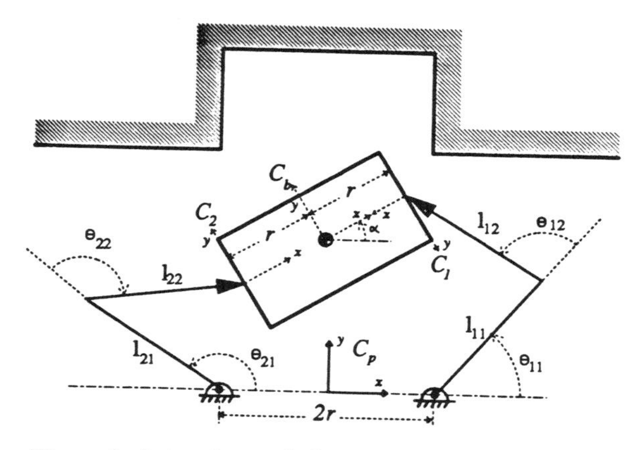

Figure 5: A two-fingered planar manipulation system.

4.3 Simulation

Consider the two-fingered planar manipulation system shown in Figure 5, where the two fingers are assumed to be identical. We model the contact to be a point contact with friction. Let the object width and the finger spacing be 2 units. The grip Jacobian and the hand Jacobian are

$$G = \begin{bmatrix} -1 & 0 & 1 & 0 \\ 0 & -1 & 0 & 1 \\ 0 & -1 & 0 & -1 \end{bmatrix}$$

and

$$J_h = \begin{bmatrix} J_1 & 0 \\ 0 & J_2 \end{bmatrix}$$

where

$$J_1 = \begin{bmatrix} \cos\alpha & -\sin\alpha \\ \sin\alpha & \cos\alpha \end{bmatrix} \begin{bmatrix} -\sin\theta_{11} - sin(\theta_{11}+\theta_{12}) & -\sin(\theta_{11}+\theta_{12}) \\ \cos\theta_{11} + \cos(\theta_{11}+\theta_{12}) & \cos(\theta_{11}+\theta_{12}) \end{bmatrix}$$

and

$$J_2 = \begin{bmatrix} -\cos\alpha & \sin\alpha \\ -\sin\alpha & -\cos\alpha \end{bmatrix} \begin{bmatrix} -\sin\theta_{21} - sin(\theta_{21}-\theta_{22}) & \sin(\theta_{21}-\theta_{22}) \\ \cos\theta_{21} + \cos(\theta_{21}-\theta_{22}) & -\cos(\theta_{22}+\theta_{22}) \end{bmatrix}.$$

The grasp will be stable and manipulable for the object along the following trajectory

$$x(t) = c_1 \sin(t),\ y(t) = c_2 + c_1 \cos(t), \alpha(t) = c_3 \sin(t).$$

With the control law of Theorem 2, we have simulated the system using a program designed to integrate differential equations with algebraic constraints. Figure 6 shows that the initial position error diminishes exponentially as predicted by (51).

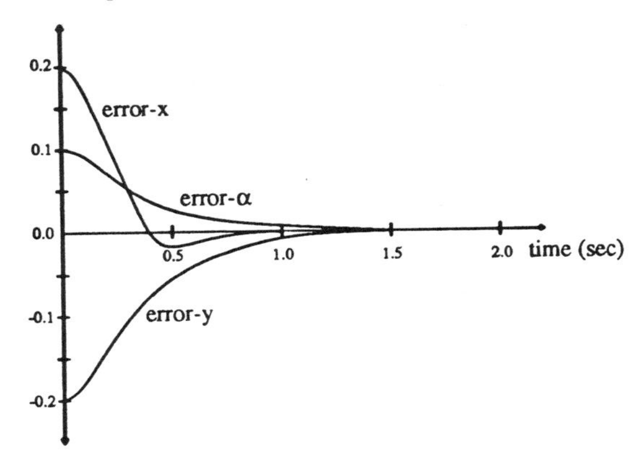

Figure 6: Position error from simulation.

5. Conclusions

In this paper, using the basic building blocks developed by previous investigators, we have formulated the kinematics of a multifingered robot hand system. For finger contact with an object, we have classified three types of useful contacts and developed control laws for dexterous manipulation by a robot hand under these contact constraints. We have shown that the control laws realize not only the desired position trajectory of the object, but also the desired internal grasp force.

References

[1] Arimoto, S. 1987. Cooperative motion control of multi-robot arms or fingers. *Proc. IEEE Int. Conf. on Robotics and Automation.* pp. 1407-1412.

[2] Cole, A., J. Hauser and S. Sastry. 1988. Kinematics and control of multifingered hands with rolling contact. *Proc. IEEE Int. Conf. on Robotics and Automation.* pp. 228-233.

[3] Cole, A., P. Hsu and Shankar Sastry. 1988. Dynamic regrasping by coordinated control of sliding for a multifingered robot hand . Memo. No. UCB/ERL M88/66, University of California, Berkeley.

[4] Hayati, S. 1986. Hybrid position/force control of multi-arm cooperating robots. *Proc. IEEE Int. Conf. on Robotics and Automation.* pp. 82-89.

[5] Jacobsen, S., J. Wood, K. Biggers and E. Iversen. 1985. The Utah/MIT hand: work in progress. *Int. J. of Robotics Res.* Vol. 4, No. 3.

[6] Kerr, J. and B. Roth. 1986. Analysis of multifingered hands. *Int. J. of Robotics Res.* Vol.4, No. 4.

[7] Li, Z.X., J. Canny and S. Sastry, 1988. On motion planning for dexterous manipulation, part I: The problem formulation. Submitted to *IEEE International Conference on Robotics and Automation.*

[8] Li, Z.X. and S. Sastry. 1988. Task oriented optimal grasping by multifingered robot hands. *IEEE J. of Robotics and Automation.* Vol. RA 2-14, No.1

[9] Li, Z.X., P. Hsu and S. Sastry. 1987. On grasping and coordinated coordination of multifingered robot hands. Memo. No. UCB/ERL M87/63, University of California, Berkeley. Also, to appear in International Journal of Robotics Research.

[10] Montana, D. 1986. Tactile sensing and kinematics of contact. Ph.D. Thesis, Division of Applied Sciences, Harvard University.

[11] Nakamura,Y., K. Nagai and T. Yoshikawa. 1987. Mechanics of coordinative manipulation by multiple robotic mechanisms. *Proc. IEEE Int. Conf. on Robotics and Automation.* pp. 991-998.

[12] Salisbury, J. 1982. Kinematic and force analysis of articulated hands. Ph.D. Thesis, Department of Mechanical Engineering, Stanford University.

[13] Zheng Y.F. and J.Y.S. Luh. 1985. Control of two coordinated robots in motion. *Proc. 24th Contr. and Dec. Conf.* pp. 1761-1766.

Department of Electrical Engineering and Computer Sciences
University of California at Berkeley, Berkeley, CA 94720

Department of Electrical Engineering and Computer Sciences
University of California at Berkeley, Berkeley, CA 94720

Contemporary Mathematics
Volume 97, 1989

Tethered Satellite System Stability

D.-C. Liaw and E.H. Abed

Department of Electrical Engineering
and the Systems Research Center
University of Maryland
College Park, MD 20742 USA

Abstract

Issues of stability of the Tethered Satellite System (TSS) during station-keeping, deployment and retrieval are considered. The basic nonlinear equations of motion of the TSS are derived using the system Lagrangian. Using the Hopf bifurcation theorem, tension control laws are established which guarantee the stability of the system during the station-keeping mode. A constant angle control method is hypothesized for subsatellite deployment and retrieval. It is proved that this control law results in stable deployment but unstable retrieval. An enhanced control law for deployment is also proposed, which entails use of the constant angle method followed by a station-keeping control law once the tether length is sufficiently near the desired value. Simulations are given to illustrate the conclusions.

I. Introduction

The Tethered Satellite System (TSS) [1]-[8] consists of a satellite and subsatellite connected by a tether, in orbit around the Earth. Many potential applications of the TSS have been proposed, including deployment of scientific instruments and study of the Earth's magnetic field [1], [12]. Station-keeping, deployment and retrieval of payloads are the three major modes of operation.

Arnold [2] proposed a constant angle method for deployment and retrieval of the subsatellite of the tethered satellite system. In [2], the satellite and subsatellite are

0271-4132/89 $1.00 + $.25 per page

modeled as point masses and the tether is assumed massless and of length small compared with the radius of the satellite's orbit. Based on these assumptions, Arnold obtained an approximate model of the TSS by applying the gravity-gradient method and argued that the constant angle scheme would result in stable deployment and unstable retrieval.

One goal of this paper is to give a proof of the validity of these conclusions. First, however, general dynamic equations for the TSS are derived by using the system Lagrangian. The applied tension control force is assumed to be the only external force acting on the TSS. Next, we consider stabilization of the TSS during the station-keeping mode, in which the tether's length is regulated to remain nearly fixed. We observe from linear analysis of the system equilibria the presence of two pairs of pure imaginary eigenvalues in the absence of feedback. This suggests the possibility of librations superimposed upon the orbital motion. It is observed that nonlinear stability analysis is needed to study stability and stabilization of the TSS during station-keeping. The program of [15], which considers Hopf bifurcation control algorithms, is employed to derive stabilizing control laws. Both linear and nonlinear feedback controls are achieved which guarantee asymptotic stability during station-keeping.

Viewing the tether length as an input variable for deployment and retrieval of the TSS, a constant in-plane angle control scheme is considered next. Within this setting, we prove stability of constant-angle deployment and instability of constant-angle retrieval. This is achieved through the construction of appropriate Liapunov-like functions and by appealing to the finite-time stability theory. A new control strategy for deployment of the subsatellite is also proposed. This control law consists of the constant angle scheme followed by the stabilizing station-keeping control.

Finally, simulation results are given to demonstrate the analytical conclusions of the paper.

Notation

E - Earth

S - Satellite

m - Subsatellite and its mass

G - Gravitational constant

M, m_s - Mass of the Earth, mass of the satellite

(x_m, y_m, z_m) - Earth-based rotating Cartesian coordinates of subsatellite, with z_m in the local outgoing vertical direction, and x_m in the direction of motion of the satellite in its orbit (see Figure 1)

$(\hat{x}_m, \hat{y}_m, \hat{z}_m)$ - Inertial coordinates of subsatellite

$(\hat{x}_s, \hat{y}_s, \hat{z}_s)$ - Inertial coordinates of satellite

Ω - Constant angular velocity of satellite in circular orbit

θ, ϕ - In-plane angle and out-of-plane angle of subsatellite relative to local vertical

$\omega_\phi := \dot{\phi}$, $\omega_\theta := \dot{\theta}$, ℓ - Tether length, $v := \dot{\ell}$

r_0, r_m - Radius of satellite orbit, radius of subsatellite orbit

$(\cdot)^*$ - Evaluation at equilibrium

τ_θ, τ_ϕ - Torques in directions θ, ϕ

F_ℓ - Force along tether (i.e., the applied tension force)

$\mathbf{F} := \frac{\tau_\theta}{\ell \cos\phi}\hat{\theta} + \frac{\tau_\phi}{\ell}\hat{\phi} + F_\ell \hat{\ell}$, where a hat indicates a unit vector in the given direction

II. System Dynamic Equations

The coordinate system of a typical tethered satellite system is depicted in Figure 1. Referring to this coordinate system, we make the following *simplifying assumptions*: The satellite and the subsatellite are point masses. Their masses are related as $m_s \gg m$, and hence the center of mass of the TSS may be taken to coincide with the satellite. The tether is massless and rigid, and the gravitational attraction between the subsatellite and the satellite is neglected. Finally, the TSS experiences no aerodynamic drag forces, and the satellite is in a circular orbit around the Earth.

From Figure 1, we note the following relationships (see the notation list in Section I):

$$x_m = \ell \cos\phi \sin\theta \tag{1}$$

$$y_m = \ell \sin\phi \tag{2}$$

$$z_m = r_0 + \ell \cos\phi \cos\theta \tag{3}$$

$$r_m^2 = r_0^2 + \ell^2 + 2r_0\ell \cos\phi \cos\theta. \tag{4}$$

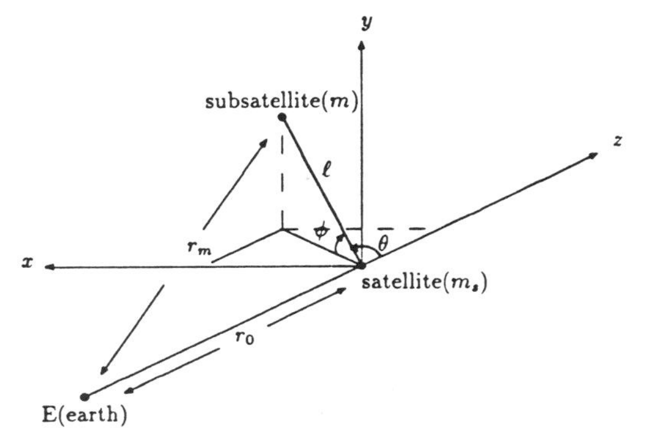

Figure 1. Coordinate System of the TSS

The inertial coordinates of the subsatellite and satellite are

$$\begin{pmatrix} \hat{x}_m \\ \hat{y}_m \\ \hat{z}_m \end{pmatrix} = \begin{pmatrix} \cos\Omega t & 0 & \sin\Omega t \\ 0 & 1 & 0 \\ -\sin\Omega t & 0 & \cos\Omega t \end{pmatrix} \begin{pmatrix} x_m \\ y_m \\ z_m \end{pmatrix}, \tag{5a}$$

$$\begin{pmatrix} \hat{x}_s \\ \hat{y}_s \\ \hat{z}_s \end{pmatrix} = \begin{pmatrix} \cos\Omega t & 0 & \sin\Omega t \\ 0 & 1 & 0 \\ -\sin\Omega t & 0 & \cos\Omega t \end{pmatrix} \begin{pmatrix} 0 \\ 0 \\ r_0 \end{pmatrix}, \tag{5b}$$

respectively, where an implicit choice of time reference is understood.

The kinetic energy of the TSS is

$$KE = \frac{1}{2} m_s (\dot{\hat{x}}_s^2 + \dot{\hat{y}}_s^2 + \dot{\hat{z}}_s^2) + \frac{1}{2} m (\dot{\hat{x}}_m^2 + \dot{\hat{y}}_m^2 + \dot{\hat{z}}_m^2). \tag{6}$$

Using Eq. (5), this may be rewritten as

$$KE = \frac{1}{2} m_s \Omega^2 r_0^2 + \frac{1}{2} m \{\dot{\ell}^2 + \ell^2 \dot{\phi}^2 + \ell^2 \cos^2\phi (\dot{\theta} + \Omega)^2 + \Omega^2 r_0^2$$

$$+ 2\Omega r_0 \dot{\ell} \cos\phi \sin\theta - 2\Omega r_0 \ell \sin\phi \sin\theta \dot{\phi} + 2\Omega r_0 \ell \cos\phi \cos\theta (\dot{\theta} + \Omega)\}. \tag{7}$$

The potential energy of the TSS is due to gravity and is given by

$$PE = -\frac{GMm_s}{r_0} - \frac{GMm}{r_m}. \tag{8}$$

The satellite is in a circular orbit, implying that it is in a zero-g orbit. Hence

$$\frac{GMm_s}{r_0^2} = m_s \Omega^2 r_0, \tag{9}$$

i.e., we can take $GM = \Omega^2 r_0^3$.

Using the system Lagrangian $L = KE - PE$ and invoking the Lagrangian formulation of the system's dynamics, the equations of motion of the TSS are obtained as

$$\tau_\theta = m\ell^2 \cos^2\phi \{\ddot{\theta} + 2\frac{\dot{\ell}}{\ell}(\dot{\theta} + \Omega) - 2\tan\phi(\dot{\theta} + \Omega)\dot{\phi}$$

$$+ \frac{\Omega^2 r_0 \sin\theta}{\ell \cos\phi}(1 - \frac{r_0^3}{r_m^3})\} \tag{10}$$

$$\tau_\phi = m\ell^2 \{\ddot{\phi} + 2\frac{\dot{\ell}}{\ell}\dot{\phi} + \cos\phi \sin\phi (\dot{\theta} + \Omega)^2$$

$$+ \frac{\Omega^2 r_0}{\ell} \cos\theta \sin\phi (1 - \frac{r_0^3}{r_m^3})\} \tag{11}$$

$$F_\ell = m\{\ddot{\ell} - \ell(\dot{\phi})^2 - \ell \cos^2\phi (\dot{\theta} + \Omega)^2$$

$$+ \frac{\Omega^2 r_0^3 \ell}{r_m^3} - \Omega^2 r_0 \cos\phi \cos\theta (1 - \frac{r_0^3}{r_m^3})\}. \tag{12}$$

For the case $r_0 \gg \ell$, we have $r_m \simeq r_0$. Moreover, in this case Eq. (4) implies

$$1 - \frac{r_0^3}{r_m^3} \simeq 3\cos\phi\cos\theta\frac{\ell}{r_0}. \tag{13}$$

Hence, the approximate equation of motion for the system for the case $r_0 \gg \ell$ is obtained as

$$\begin{aligned}\mathbf{F} &= m\hat{\ell}\{\ddot{\ell} - \ell\dot{\phi}^2 - \ell\cos^2\phi(\dot{\theta}+\Omega)^2 + \ell\Omega^2 - 3\Omega^2\ell\cos^2\phi\cos^2\theta\} \\ &+ m\hat{\theta}\{\ddot{\theta}\ell\cos\phi + 2(\dot{\theta}+\Omega)(\dot{\ell}\cos\phi - \ell\dot{\phi}\sin\phi) + 3\ell\Omega^2\cos\theta\cos\phi\sin\theta\} \\ &+ m\hat{\phi}\{\ell\ddot{\phi} + 2\dot{\ell}\dot{\phi} + \ell\cos\phi\sin\phi(\dot{\theta}+\Omega)^2 + 3\ell\Omega^2\cos^2\theta\cos\phi\sin\phi\}.\end{aligned}$$

This agrees with a result of Arnold [2]. Note, however, that we do *not* require $r_0 \gg \ell$ below.

III. Stabilization of System During Station-Keeping Mode

To facilitate design of stabilizing control laws for the TSS, we first rewrite the system of second order equations (10)-(12) in state space form. This is possible under the assumption $\cos\phi \neq 0$ (i.e., $\phi \neq \pm\frac{\pi}{2}$). In addition, we assume that the only external force acting on the TSS is the applied tension F_ℓ. Setting $\tau_\theta = \tau_\phi = 0$ in (10)-(12) and denoting $T := F_\ell$, the equations of motion in state space form are found to be

$$\dot{\phi} = \omega_\phi \tag{14}$$

$$\dot{\omega}_\phi = -\frac{2v}{\ell}\omega_\phi - \frac{1}{2}\sin(2\phi)(\omega_\theta+\Omega)^2 - \frac{\Omega^2 r_0}{\ell}\cos\theta\sin\phi(1 - \frac{r_0^3}{r_m^3}) \tag{15}$$

$$\dot{\theta} = \omega_\theta \tag{16}$$

$$\dot{\omega}_\theta = -\frac{2v}{\ell}(\omega_\theta+\Omega) + 2\tan\phi(\omega_\theta+\Omega)\omega_\phi - \frac{\Omega^2 r_0\sin\theta}{\ell\cos\phi}(1 - \frac{r_0^3}{r_m^3}) \tag{17}$$

$$\dot{\ell} = v \tag{18}$$

$$\begin{aligned}\dot{v} &= \ell\omega_\phi^2 + \ell\cos^2\phi(\omega_\theta+\Omega)^2 - \frac{\Omega^2 r_0^3\ell}{r_m^3} \\ &+ \Omega^2 r_0\cos\theta\cos\phi(1 - \frac{r_0^3}{r_m^3}) + \frac{T}{m}.\end{aligned} \tag{19}$$

With the tether length held constant ($\ell = \ell^*$), this system has exactly two equilibrium points: (0,0,0,0,ℓ^*,0) and (0,0,π,0,ℓ^*,0). Moreover, the linearized system of Eqs. (14)-(17)

at such an equilibrium is found to have two pairs of pure imaginary eigenvalues, given by

$$\lambda_{1,2} = \pm i\Omega\sqrt{1 + \frac{r_0}{\ell^*}(1 - \frac{r_0^3}{r_{m,0}^3})}, \tag{20}$$

$$\lambda_{3,4} = \pm i\Omega\sqrt{\frac{r_0}{\ell^*}(1 - \frac{r_0^3}{r_{m,0}^3})}, \tag{21}$$

where $r_{m,0} = r_0 + \ell^*$ at $\theta = 0$ and $r_{m,0} = r_0 - \ell^*$ at $\theta = \pi$. The pairs $\lambda_{1,2}$ and $\lambda_{3,4}$ of eigenvalues are associated with the out-of-plane and in-plane dynamics, respectively, suggesting that the TSS may undergo librations with two different frequencies superimposed upon the orbital motion, near either of the equilibria. In addition [9], the set $\phi = 0, \omega_\phi = 0$ is an invariant manifold for Eqs. (14)-(19), regardless of the tension control force T. Although this implies the uncontrollability of the system, a tension control law still can be designed to guarantee asymptotic stability.

Stabilizability will first be studied at the equilibrium point $x_0 = (0,0,0,0,\ell^*,0)^T$. Apply a linear tension control law $T = -m(U + k_1\theta + k_2\omega_\theta + k_3\tilde{\ell} + k_4 v)$, where $k_i, i = 1, \cdots, 6$ are constant control gains, $\tilde{\ell} := \ell - \ell^*$ and

$$U := \frac{(3r_0^2\ell^* + 3r_0\ell^{*2} + \ell^{*3})\Omega^2}{(r_0 + \ell^*)^2}. \tag{22}$$

The characteristic equation of the linearized closed-loop model of system (14)-(19) is found, after some manipulation, to be

$$(\lambda^2 + a_1^2)\,(\lambda^4 + k_4\lambda^3 + b_1\lambda^2 + b_2\lambda + b_3) = 0, \tag{23}$$

where

$$a_1 = (\frac{4r_0^3 + 6r_0^2\ell^* + 4r_0\ell^{*2} + \ell^{*3}}{(r_0 + \ell^*)^3})^{0.5}\Omega, \tag{24}$$

$$b_1 = k_3 - \frac{(2r_0\ell^{*2} + \ell^{*3})\Omega^2}{(r_0 + \ell^*)^3} - \frac{2\Omega k_2}{\ell^*} + 4\Omega^2, \tag{25}$$

$$b_2 = \frac{k_4(3r_0^3 + 3r_0^2\ell^* + r_0\ell^{*2})\Omega^2}{(r_0 + \ell^*)^3} - \frac{2\Omega k_1}{\ell^*}, \tag{26}$$

$$b_3 = \frac{(3r_0^3 + 3r_0^2\ell^* + r_0\ell^{*2})\Omega^2}{(r_0 + \ell^*)^3}(k_3 - \frac{(3r_0^3 + 3r_0^2\ell^* + 3r_0\ell^{*2} + \ell^{*3})\Omega^2}{(r_0 + \ell^*)^3}). \tag{27}$$

By applying the Routh-Hurwitz test to Eq. (23), we obtain the following preliminary result.

Lemma 1. If the tension control force is given as $T = m(-U - k_1\theta - k_2\omega_\theta - k_3\tilde{\ell} - k_4 v)$, then the linearized closed-loop system of Eqs. (14)-(19) at the equilibrium point x_0 has a pair of pure imaginary eigenvalues and four eigenvalues with negative real parts, provided that the gains $k_i, i = 1, \ldots, 4$ satisfy the following two conditions:

(i) $b_1, b_2, b_3, k_4 > 0$

(ii) $k_4 b_1 b_2 - b_2^2 - k_4^2 b_3 > 0$

Here, the $b_i, i = 1, 2, 3$ are as given in (25)-(27).

Note that, by Eq. (23), the system (14)-(19) has an uncontrollable pair of pure imaginary eigenvalues $\pm i a_1$, unaffected by linear state feedback. It is easily checked that this holds even with feedback of states ϕ and ω_ϕ. The stability of the closed-loop system cannot, therefore, be determined from the linearized model. That is, this is an example of a critical case in nonlinear stability. Stability results for Hopf bifurcation of one-parameter families of nonlinear systems can be employed in studying the stability of critical systems with a single pair of pure imaginary eigenvalues. For a discussion, see [15]. Two types of stabilizing tension control laws are given in the next two theorems for the station-keeping application. Theorem 1 involves the use of linear state feedbacks, while Theorem 2 accounts for a class of nonlinear feedbacks. Details, proofs and simpler results in the form of corollaries can be found in [13].

Theorem 1. If a linear state feedback controller T of the type specified in Lemma 1 is applied, with

$$(\frac{a_4}{2} + \frac{\ell^* a_1^2}{2} - \frac{\ell^*}{4}(k_3 - a_3))d_1 - \frac{1}{2}k_4 a_1 \ell^* d_2 < 0,$$

then the equilibrium point x_0 is rendered asymptotically stable for the system (14)-(19), where a_1 is as given in (24), and where

$$d_1 = k_1 + \frac{\ell^* k_4}{2\Omega}(a_2 + 4a_1^2), \tag{28}$$

$$d_2 = 2a_1(k_2 - 2\ell^*\Omega) + \frac{\ell^*(a_2 + 4a_1^2)}{4a_1\Omega}(a_3 + 4a_1^2 - k_3), \tag{29}$$

$$a_2 = -\frac{(3r_0^3 + 3r_0^2\ell^* + r_0\ell^{*2})\Omega^2}{(r_0 + \ell^*)^3}, \tag{30}$$

$$a_3 = \frac{(3r_0^3 + 3r_0^2\ell^* + 3r_0\ell^{*2} + \ell^{*3})\Omega^2}{(r_0 + \ell^*)^3}, \tag{31}$$

$$a_4 = -\frac{(8r_0^4\ell^* + 14r_0^3\ell^{*2} + 16r_0^2\ell^{*3} + 9r_0\ell^{*4} + 2\ell^{*5})\Omega^2}{2(r_0 + \ell^*)^4}. \tag{32}$$

Theorem 2. If the applied tension control force is given as $T = m(-U - k_1\theta - k_2\omega_\theta - k_3\tilde{\ell} - k_4 v - q_1\phi^2 - q_2\phi\omega_\phi - q_3\omega_\phi^2)$ with k_i, $i = 1, \ldots, 4$ satisfying the conditions of Lemma 1, then the system (14)-(19) is rendered asymptotically stable if the condition

$$(\frac{-q_1 + a_5}{2} + (\frac{\ell^* + q_3}{2})a_1^2 - (k_3 - a_3)\frac{\ell^*}{4})d_1 + \frac{a_1}{2}(-q_2 - \ell^* k_4)d_2 < 0$$

holds. Here, d_1, d_2 and $a_i, i = 1, \ldots, 4$ are as defined above.

Similar results have also been obtained for stabilizing the system at the equilibrium point $(0, 0, \pi, 0, \ell^*, 0)$. The details appear in [16].

IV. Constant In-Plane Angle Control

In this section, we consider deployment and retrieval of the subsatellite of the tethered satellite system. Viewing ℓ as an external control input, the state equations (10)-(12) of the system are

$$\dot{\theta} = \omega_\theta \tag{33}$$

$$\dot{\omega}_\theta = -\frac{2\dot{\ell}}{\ell}(\omega_\theta + \Omega) + 2\tan\phi(\omega_\theta + \Omega)\omega_\phi - \frac{\Omega^2 r_0 \sin\theta}{\ell\cos\phi}(1 - \frac{r_0^3}{r_m^3}) \tag{34}$$

$$\dot{\phi} = \omega_\phi \tag{35}$$

$$\dot{\omega}_\phi = -\frac{2\dot{\ell}}{\ell}\omega_\phi - \frac{1}{2}\sin(2\phi)(\omega_\theta + \Omega)^2 - \frac{\Omega^2 r_0}{\ell}\cos\theta\sin\phi(1 - \frac{r_0^3}{r_m^3}). \tag{36}$$

At an equilibrium point $(\theta^*, \omega_\theta^*, \phi^*, \omega_\phi^*)$ of (33)-(36), if one exists, we have $\omega_\theta^* = \omega_\phi^* = 0$, $\dot{\ell}$ must satisfy (from Eq. (34))

$$\dot{\ell} = -\frac{\Omega r_0}{2\cos\phi^*}\sin\theta^*(1 - \frac{r_0^3}{(r_m^*(\ell))^3}), \tag{37}$$

and ϕ^* must satisfy either

$$\sin\phi^* = 0, \quad \text{or} \tag{38a}$$

$$\cos\phi^* = -\frac{r_0}{\ell}(1 - \frac{r_0^3}{(r_m^*(\ell))^3})\cos\theta^*, \tag{38b}$$

where

$$r_m^*(\ell) := (r_0^2 + \ell^2 + 2r_0\ell\cos\theta^*\cos\phi^*)^{1/2}. \tag{39}$$

Remark. In fact, only the case $\sin\phi^* = 0$ is realistic. To see this, briefly consider the possibility (38*b*), which, using (37), would imply that at equilibrium $\dot{\ell}$ obeys

$$\dot{\ell} = \frac{\Omega\ell}{2}\tan\theta^*. \tag{40}$$

Since $-\frac{\pi}{2} \leq \phi^* \leq \frac{\pi}{2}$, we have $\cos\phi^* \geq 0$ (see Figure 1). Considering the possibilities $0 < \phi^* \leq \frac{\pi}{2}$ and $-\frac{\pi}{2} \leq \phi^* < 0$ separately, and referring to Figure 1 for the relative magnitudes of $r_m^*(\ell)$ and r_0, we find that the left and right sides of (38*b*) are then of opposite sign unless they both vanish. Hence, we obtain $\phi^* = \theta^* = \pm\frac{\pi}{2}$, implying $\dot{\ell}$ of (40) would be infinite.

In view of the Remark, we let $\phi^* = 0$. Eq. (37) now implies that, at equilibrium, ℓ satisfies

$$\dot{\ell} = -\frac{\Omega r_0}{2}(1 - \frac{r_0^3}{(\hat{r}_m^*(\ell))^3})\sin\theta^*, \tag{41}$$

where

$$\hat{r}_m^*(\ell) := (r_0^2 + \ell^2 + 2r_0\ell\cos\theta^*)^{1/2}. \tag{42}$$

This control law, which is a *constant in-plane angle control method*, has the feature that it results in the existence of an equilibrium point of (33)-(36). Moreover, the associated equilibrium point of system (33)-(36) will then be (θ^*, 0,0,0), where θ^* is the desired in-plane angle.

V. Stability Analysis of the TSS During Retrieval

Suppose for simplicity that $\dot{\ell} < 0$ throughout retrieval. From Eq. (41) we have

$$\dot{\ell} < 0 \iff -\frac{\Omega r_0}{2}(1 - \frac{r_0^3}{(\hat{r}_m^*(\ell))^3})\sin\theta^* < 0.$$

Denote by ℓ_i the initial (pre-retrieval) tether length. Then the condition for $\dot{\ell} < 0$ is that θ^* satisfies either $0 < \theta^* < \frac{\pi}{2}$ or $-\pi < \theta^* < \theta_1$ (see Figure 2), where $\theta_1 = \theta_1(\ell_i)$ is such that

$$\cos\theta_1 = -\frac{\ell_i}{2r_0}, \quad -\pi < \theta_1 < -\frac{\pi}{2}. \tag{43}$$

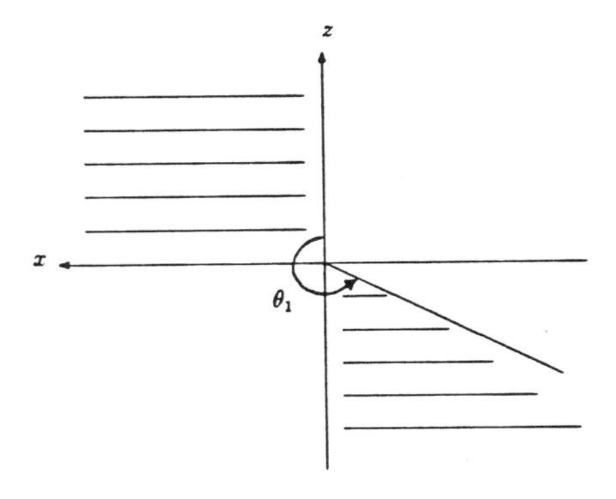

Figure 2. Retrieval Regions for θ^* with $\phi^* = 0$

From the discussion above, we have $\dot{\ell} < 0$ and $\ell > 0$ during retrieval. In addition, $\dot{\ell} = 0$ occurs only at $\ell = 0$. Hence, ℓ must approach 0 asymptotically.

Denoting $\tilde{\theta} := \theta - \theta^*$, the linearized system state equations at the equilibrium point $(\theta^*, 0, 0, 0)$ are found to be

$$\dot{\tilde{\theta}} = \omega_\theta \tag{44}$$

$$\dot{\omega}_\theta = (2\Omega\frac{\dot{\ell}}{\ell}\cot\theta^* + \frac{3\Omega^2 r_0^5}{(\hat{r}_m^*(\ell))^5}\sin^2\theta^*)\tilde{\theta} - 2\frac{\dot{\ell}}{\ell}\omega_\theta \tag{45}$$

$$\dot{\phi} = \omega_\phi \tag{46}$$

$$\dot{\omega}_\phi = (-\Omega^2 + 2\Omega\frac{\dot{\ell}}{\ell}\cot\theta^*)\phi - 2\frac{\dot{\ell}}{\ell}\omega_\phi \tag{47}$$

The *linearized* model of the system (33)-(36) is therefore seen to *decouple* into the subsystems (44)-(45) and (46)-(47). It follows from Lemma 2 and Theorem 3 below that these two decoupled subsystems are unstable for constant angle retrieval for any potential value of θ^* (see Figure 2), and that the original system (33)-(36) is then also unstable.

Lemma 2. If there is a constant $\delta > 0$ such that $b(t) > \delta$ for all $t \geq t_0$, then the origin is unstable for the system

$$\dot{x}_1 = x_2 \tag{48}$$

$$\dot{x}_2 = a(t)x_1 + b(t)x_2. \tag{49}$$

Proof: By the Abel-Jacobi-Liouville theorem (e.g., [10]), we have

$$\det(\Phi(t, t_0)) = \exp(\int_{t_0}^{t} b(s)ds),$$

where $\Phi(t, t_0)$ is the state transition matrix of the linear system (48)-(49). From the hypothesis, it now follows that $\det\Phi(t, t_0)$ approaches ∞ as $t \to \infty$. Hence, the origin is unstable.

□

Theorem 3. If $r_0 > \ell_i$, then the origin is unstable for the linearized system (44)-(47) during constant in-plane angle retrieval, and (θ^*,0,0,0) is an unstable equilibrium point of the original nonlinear system (33)-(36).

Proof: First, we show that there is a constant $\epsilon > 0$ such that $\frac{-\dot{\ell}}{\ell} > \epsilon$ for all $t \geq t_0$. By the discussion above, the tether length ℓ approaches 0 asymptotically. In addition, $\dot{\ell} = 0$ when $\ell = 0$. Applying L'Hôpital's Rule to (41)-(42), we obtain

$$\lim_{\ell \to 0} \frac{\dot{\ell}}{\ell} = \lim_{\ell \to 0} \frac{\partial \dot{\ell}}{\partial \ell} = -\frac{3}{2}\Omega \cos\theta^* \sin\theta^* \neq 0,$$

for any admissible θ^* (see Figure 2). Moreover, it is obvious that $-\frac{\dot{\ell}}{\ell} > 0$, for all $t \geq t_0$ during retrieval. Thus, $\frac{-\dot{\ell}}{\ell}$ is bounded and ϵ exists.

Next, we check the signs of the coefficients multiplying ω_θ and ω_ϕ in Eqs. (45) and (47), respectively. Since $0 < \epsilon < \frac{-\dot{\ell}}{\ell}$ for all $t \geq t_0$, Lemma 2 implies the origin is unstable for the two subsystems (44)-(45) and (46)-(47). This implies instability of the equilibrium point $(\theta^*, 0, 0, 0)$ of system (33)-(36) during constant in-plane angle retrieval.

□

VI. Stability Analysis of the TSS During Deployment

In this section, we consider application of the constant in-plane angle strategy of Section IV to subsatellite deployment. For simplicity, suppose that $\dot{\ell} > 0$ for all $t \geq t_0$. Since $\dot{\ell}$ is always positive in this consideration, one might expect that the tether length ℓ increases without bound. In reality, only a finite final tether length is meaningful for deployment. Stability of the TSS is hence only considered in a finite time interval, where standard Liapunov stability criteria cannot be employed. Results from finite-time stability shall be applied to study the behavior of the TSS during deployment.

Basic definitions and conditions for finite-time stability are given in Section VI.1. Then these finite-time stability criteria, especially the contractive stability criteria, are used to study the stability of the TSS during constant in-plane angle deployment. In addition to the proof of finite-time stability of deployment, a switching type control law combining constant angle deployment and station-keeping control is proposed to achieve asymptotic stability. Details of this are given in Section VI.2.

VI.1. Results on finite-time stability

Consider a system given by

$$\dot{x} = f(t, x), \tag{50}$$

where $f : \Gamma \times R^n \rightarrow R^n$ and $\Gamma := [t_0, t_0 + T)$ for some $t_0 \in R$, $T > 0$. Let x_0 denote the initial condition of (50) at t_0, and let $\phi(t; t_0, x_0)$ be the solution of (50) at time t satisfying the initial condition. We recall several relevant definitions [14].

Definition 1. System (50) is *finite-time stable* with respect to $(\alpha, \beta, \Gamma, ||\cdot||)$, $\alpha \leq \beta$, if for every trajectory $\phi(t; t_0, x_0)$ with $||x_0|| < \alpha$, we have $||\phi(t; t_0, x_0)|| < \beta$, $\forall\, t \in \Gamma$.

Definition 2. System (50) is *uniformly finite-time stable* with respect to $(\alpha, \beta, \Gamma, ||\cdot||)$, $\alpha \leq \beta$, if for any trajectory $\phi(t; s, x)$ with $||x|| < \alpha$, $\forall\, s \in \Gamma$, we have $||\phi(t; s, x)|| < \beta$, $\forall$ $t \in \Gamma$.

Definition 3. System (50) is *quasi-contractively stable* with respect to $(\alpha, \gamma, \Gamma, ||\cdot||)$, $\gamma < \alpha$, if for any trajectory $\phi(t; t_0, x_0)$ with $||x_0|| < \alpha$, there exists $t_1 \in \Gamma$ so that $||\phi(t; t_0, x_0)|| < \gamma$, $\forall\, t \in [t_1, t_0 + T)$.

Definition 4. System (50) is *contractively stable* with respect to $(\alpha, \beta, \gamma, \Gamma, ||\cdot||)$, $\gamma < \alpha \leq \beta$, if it is finite-time stable with respect to $(\alpha, \beta, \Gamma, ||\cdot||)$ and quasi-contractively stable with respect to $(\alpha, \gamma, \Gamma, ||\cdot||)$.

For given α, β, Γ, and $||\cdot||$, a necessary and sufficient condition for uniform finite-time stability follows.

Lemma 3 [14]. System (50) is uniformly finite-time stable with respect to $(\alpha, \beta, \Gamma, ||\cdot||)$, $\alpha < \beta$, if and only if there exists a continuous function $V(t, x)$ such that

$$\dot{V}(t, x) \leq 0, \quad \forall\ x \in \overline{B(\beta)}, \qquad t \in \Gamma, \tag{51}$$

$$V_M^\delta(t_1) < V_m^\beta(t_2), \quad \forall\ t_2 > t_1, \quad \forall\ \delta < \alpha, \quad t_1, t_2 \in \Gamma, \tag{52}$$

where

$$B(\beta) := \{x : ||x|| < \beta\}, \tag{53}$$

$||\cdot||$ denotes any norm on R^n, $\overline{B(\beta)}$ denotes the closure of $B(\beta)$, and

$$V_M^\alpha(t) := \sup_{||x||=\alpha} V(t, x), \tag{54}$$

$$V_m^\alpha(t) := \inf_{||x||=\alpha} V(t, x). \tag{55}$$

Here, $\dot{V}(t, x)$ is the time derivative of $V(t, x)$ along trajectories of system (50).

Stability properties of a system may be investigated without reference to the specific bounds on the states (i.e., α, β and γ). In the following lemma and theorem, two sufficient conditions are introduced for this type of stability. These provide a means for finding the associated bounds α, β, γ. Lemma 4 gives a sufficient condition for uniform finite-time stability. Theorem 4 then gives a relationship among T, α, β and γ providing a sufficient condition for contractive stability.

Lemma 4. System (50) is uniformly finite-time stable with respect to $(\alpha, \beta, \Gamma, ||\cdot||)$ for any given α and β with

$$0 < \alpha < \beta\sqrt{\frac{k_1}{k_2}} \leq \beta \leq r, \tag{56}$$

if there exist $r > 0$ and a continuously differentiable function $V(t,x)$ with

$$\begin{aligned} &\dot{V}(t,x) \leq 0, \\ &k_1||x||^2 \leq V(t,x) \leq k_2||x||^2, \end{aligned} \tag{57}$$

for all $x \in \overline{B(r)}$, $t \in \Gamma$. Here, $0 < k_1 \leq k_2$ and the norm used is the Euclidean norm.

Proof: The result follows directly from condition (52) of Lemma 3.

□

In the next theorem, we introduce a condition on (50) and a relationship among T, α, β and γ guaranteeing finite-time *contractive* stability.

Theorem 4. System (50) is contractively stable with respect to $(\alpha, \beta, \gamma, \Gamma, ||\cdot||)$ for any triple α, β, γ with

$$\alpha\sqrt{\frac{k_2}{k_1} \cdot \exp(-\frac{k_3}{k_2}T)} \leq \gamma < \alpha < \sqrt{\frac{k_1}{k_2}}\beta < \beta \leq r \tag{58}$$

if there exist $r > 0$ and a continuously differentiable function $V(t,x)$ satisfying the conditions

$$k_1||x||^2 \leq V(t,x) \leq k_2||x||^2, \tag{59}$$

$$k_3||x||^2 \leq -\dot{V}(t,x), \tag{60}$$

for all $x \in \overline{B(r)}$, $t \in \Gamma$. Here, $k_i > 0$, $i = 1, 2, 3$, we employ the Euclidean norm, and the time interval length T is such that

$$T > \frac{k_2}{k_3} \cdot \ln\frac{k_2}{k_1}. \tag{61}$$

Proof: Condition (60) implies that

$$\dot{V}(t,x) \le 0, \quad \forall \ x \in \overline{B(\beta)}, \quad t \in \Gamma.$$

Hence, it is implied by Lemma 4 that (50) is uniformly finite-time stable with respect to $(\alpha, \beta, \Gamma, || \cdot ||)$ for any α, β satisfying condition (58). Next, we prove quasi-contractive stability of the system. From conditions (59) and (60), we have

$$\dot{V}(t,x) \le -\frac{k_3}{k_2} V(t,x), \quad \forall \, x \in \overline{B(r)}, \quad t \in \Gamma.$$

Hence,

$$V(t, \phi(t; t_0, x_0)) \le V(t_0, x_0) \exp(-\frac{k_3}{k_2}(t - t_0)), \quad \forall \, x_0 \in \overline{B(r)}, \quad t \in \Gamma.$$

Then it follows from (59) that

$$||\phi(t; t_0, x_0)||^2 \le \frac{k_2}{k_1} ||x_0||^2 \exp(-\frac{k_3}{k_2}(t - t_0)), \quad \forall \, x_0 \in \overline{B(r)}, \quad t \in \Gamma.$$

Thus, there exists a $t_1 \in \Gamma$ such that $||\phi(t; t_0, x_0)|| < \gamma, \forall \, t \in [t_1, t_0 + T)$ when conditions (58) and (61) hold. According to Definition 3, system (50) is hence quasi-contractively stable with respect to $(\alpha, \gamma, \Gamma, || \cdot ||)$ for any α, γ satisfying (58).

□

VI.2. Application to deployment

In the following discussion, we consider the deployment of the subsatellite of the tethered satellite system. For simplicity, let $\dot{\ell} > 0$ throughout deployment. By Eq. (41), we have

$$\dot{\ell} > 0 \iff -\frac{\Omega r_0}{2}(1 - \frac{r_0^3}{(\hat{r}_m^*(\ell))^3}) \sin \theta^* > 0.$$

From the discussion above and Eq. (42), the condition on θ^* for $\dot{\ell} > 0$ is that either $\theta_2(\ell_f) < \theta^* < \pi$, or $-\frac{\pi}{2} < \theta^* < 0$ (see Figure 3), where $\theta_2(\ell_f)$ solves

$$\cos \theta_2 = -\frac{\ell_f}{2r_0}, \quad 0 < \theta_2 < \pi, \tag{62}$$

and ℓ_f is the desired post-deployment tether length.

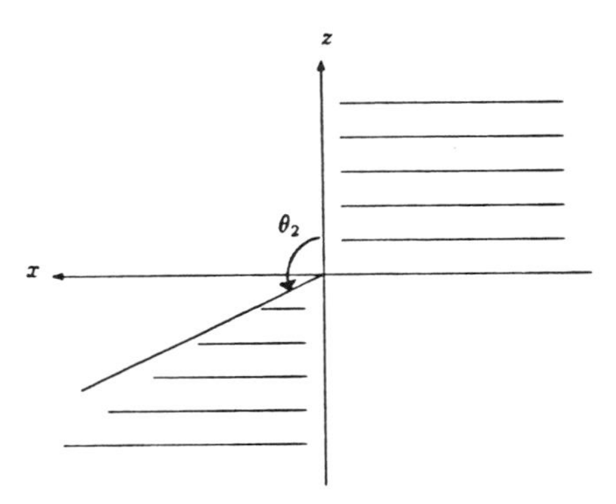

Figure 3. Deployment Regions for θ^* with $\phi^* = 0$

Two strategies for deployment are considered here. The first consists of the constant in-plane angle control law for deployment, and the second involves following the constant in-plane angle control law followed by a stabilizing station-keeping control once the desired in-plane angle is close enough to 0 radians or π radians.

Strategy 1: Constant Angle Control Only

We now consider application of the constant in-plane angle control law discussed above to subsatellite deployment. In the following, ℓ_f denotes the desired final tether length and ℓ_i denotes the initial tether length, which accounts for support by a boom.

From (41), we have

$$\begin{aligned}\frac{\dot{\ell}}{\ell} &= -\frac{\Omega r_0 \sin\theta^*}{2\ell}(1 - \frac{r_0}{(\hat{r}_m^*(\ell))^3}) \\ &= -\frac{\Omega \sin\theta^*}{2} \cdot \frac{r_0}{\ell} \cdot \frac{(\hat{r}_m^*(\ell))^6 - r_0^6}{(\hat{r}_m^*(\ell))^3[r_0^3 + (\hat{r}_m^*(\ell))^3]} \\ &\geq -\Omega \sin\theta^* \cos\theta^* > 0 \end{aligned} \tag{63}$$

for any $\theta^* \in S_d$ and $\ell_i \leq \ell \leq \frac{1}{22} r_0$, where (in radian measure)

$$S_d := \{\theta^* | -0.68 \leq \theta^* < 0, \text{ or } 2.5 \leq \theta^* < \pi\}.$$

Hence, $\frac{\dot{\ell}}{\ell}$ is bounded below for all $\ell_i \leq \ell \leq \ell_f$ and $\theta^* \in S_d$, and similarly for $\dot{\ell}$. Thus, for any $\ell_f > \ell_i$, ℓ will increase past ℓ_f at some $T > 0$. Theorem 5 below asserts that the system will be finite-time contractively stable during deployment over the interval $[t_0, t_0 + T)$, near the equilibrium point $(\theta^*, 0, 0, 0)$ with $\theta^* \in S_d$.

Theorem 5. Suppose $\Omega < 1$, $\ell_f \leq \frac{1}{22} r_0$, and $\Gamma := [t_0, t_0 + T)$. There is an $r > 0$ such that system (33)-(36) is finite-time contractively stable with respect to $(\alpha, \beta, \gamma, \Gamma, ||\cdot||)$ at the equilibrium point $(\theta^*, 0, 0, 0)$ for any α, β, γ and T satisfying (58) and (61), if either of the following two conditions on the desired in-plane angle θ^* holds:

(i) $-0.68 \leq \theta^* < 0$,

(ii) $2.5 \leq \theta^* < \pi$.

Proof: Denote $m := \Omega \sin 2\theta^*$. It is clear from (63) that

$$-\frac{2\dot{\ell}}{\ell} \leq m < 0, \quad \forall \ t \in \Gamma,$$

if either (i) or (ii) holds and $\frac{1}{22} r_0 \leq \ell_f \leq \ell$. Invoking the finite-time stability criteria given in Section VI.1, the stability of the TSS during constant angle deployment can be proved as follows.

Using a general construction [18] for a class of systems of the form (48)-(49), we prove the finite-time contractive stability of (33)-(36) during deployment by employing the Liapunov-like function

$$V(t, \tilde{\theta}, \omega_\theta, \phi, \omega_\phi) = (\frac{2\dot{\ell}}{\ell} + \frac{n_1(t)}{m})\tilde{\theta}^2 + 2\tilde{\theta}\omega_\theta - \frac{1}{m}\omega_\theta^2 + (\frac{2\dot{\ell}}{\ell} + \frac{n_2(t)}{m})\phi^2 + 2\phi\omega_\phi - \frac{1}{m}\omega_\phi^2, \tag{64}$$

where

$$n_1(t) := 2\Omega\frac{\dot{\ell}}{\ell}\cot\theta^* + \frac{3\Omega^2 r_0^5}{(\hat{r}_m^*(\ell))^5}\sin^2\theta^*, \tag{65}$$

$$n_2(t) := -\Omega^2 + 2\Omega\frac{\dot{\ell}}{\ell}\cot\theta^*. \tag{66}$$

Then corresponding to the original system (33)-(36), we have

$$\dot{V}(t, \tilde{\theta}, \omega_\theta, \phi, \omega_\phi) = n_3(t)\tilde{\theta}^2 + n_4(t)\phi^2 + 2(1 + \frac{2\dot{\ell}}{m\ell}) \cdot (\omega_\theta^2 + \omega_\phi^2) + 2(\tilde{\theta} - \frac{\omega_\theta}{m})f_1(t) + 2(\phi - \frac{\omega_\phi}{m})f_2(t), \tag{67}$$

where

$$f_1(t) = -\frac{2\dot{\ell}}{\ell}\Omega + 2\tan\phi(\omega_\theta + \Omega)\omega_\phi - \frac{\Omega^2 r_0 \sin(\theta^* + \tilde{\theta})}{\ell \cos\phi}(1 - \frac{r_0^3}{(\hat{r}_m^*(\ell))^3}) - n_1(t)\tilde{\theta}, \quad (68)$$

$$f_2(t) = -\frac{1}{2}\sin(2\phi)(\omega_\theta + \Omega)^2 - \frac{\Omega^2 r_0}{\ell}\cos(\theta^* + \tilde{\theta})\sin\phi(1 - \frac{r_0^3}{(\hat{r}_m^*(\ell))^3}) - n_2(t)\phi, \quad (69)$$

and

$$n_3(t) = 2n_1(t) + \frac{d}{dt}\{\frac{2\dot{\ell}}{\ell}\} + \frac{1}{m}\cdot\frac{dn_1(t)}{dt}, \quad (70)$$

$$n_4(t) = 2n_2(t) + \frac{d}{dt}\{\frac{2\dot{\ell}}{\ell}\} + \frac{1}{m}\cdot\frac{dn_2(t)}{dt}. \quad (71)$$

First, consider the case in which θ^* satisfies condition (i). After some calculations using (64)-(66), we find that there exist $k_{1,1}$, $k_{1,2} > 0$ (given in the Appendix) such that

$$k_{1,1}||x||^2 \leq V(t, \tilde{\theta}, \omega_\theta, \phi, \omega_\phi) \leq k_{1,2}||x||^2, \ \forall\, t \in \Gamma, \quad (72)$$

where $x = (\tilde{\theta}, \omega_\theta, \phi, \omega_\phi)^T$ and the norm indicated is the Euclidean norm. Moreover, by choosing $k_{1,3} := 0.132\Omega^2$ and

$$r = \sup_{t\in\Gamma}\{\ ||x||\ :\ |f_1| \leq \frac{mk_{1,3}||x||}{2(m-1)} \quad \text{and} \quad |f_2| \leq \frac{mk_{1,3}||x||}{2(m-1)}\}, \quad (73)$$

we have

$$-\dot{V}(t, \tilde{\theta}, \omega_\theta, \phi, \omega_\phi) \geq k_{1,3}||x||^2, \ \forall\, t \in \Gamma, \quad x \in \overline{B(r)}. \quad (74)$$

Thus, conditions (59)-(60) are satisfied and the conclusion follows from Theorem 4.

Similarly, for the case in which θ^* satisfies condition (ii), we have

$$k_{2,1}||x||^2 \leq V(t, \tilde{\theta}, \omega_\theta, \phi, \omega_\phi) \leq k_{2,2}||x||^2, \ \forall\, t \in \Gamma, \quad (75)$$

where $k_{2,1}$, $k_{2,2} > 0$ are also specified in the Appendix. By choosing $k_{2,3} := 0.0442\Omega^2$ and

$$r = \sup_{t\in\Gamma}\{\ ||x||\ :\ |f_1| \leq \frac{mk_{2,3}||x||}{2(m-1)} \quad \text{and} \quad |f_2| \leq \frac{mk_{2,3}||x||}{2(m-1)}\}, \quad (76)$$

we guarantee that

$$-\dot{V}(t, \tilde{\theta}, \omega_\theta, \phi, \omega_\phi) \geq k_{2,3}||x||^2, \ \forall\, t \in \Gamma, \quad x \in \overline{B(r)}. \quad (77)$$

The conclusion again follows from Theorem 4.

□

The finite-time contractive stable regions of the desired in-plane angle θ^* for constant in-plane angle deployment are given in Theorem 5. In addition, a relationship between the time-interval T, the bound of initial disturbances and the final contracted region is set up in Theorem 4. Furthermore, the simulation results given in Section VII.2 show that the criteria given in Theorem 5 are not vacuous.

Strategy 2: Station-Keeping Control Included

A tension control law has been designed in Section III to regulate the tether length, while ensuring the out-of-plane angle $\phi = 0$ and the in-plane angle $\theta = 0$ or $(\theta = \pi)$. Combining the result of Theorem 5 with the station-keeping control strategies of Theorems 1 and 2, a switching control law for deployment is constructed as follows:

Step 1. Apply the constant angle control law (41) for the first step subsatellite deployment, in which the desired in-plane angle θ^* satisfies the conditions of Theorem 5 and is close to 0 (or π).

Step 2. Apply the tension control law given in Theorem 1 (or Theorem 2) once the tether length is sufficiently near the desired length ℓ_f.

Theorem 5 implies that the initial disturbance in the state of the TSS can be attenuated. Specifically, with the desired in-plane angle sufficiently near 0 or π, the system state can be steered to the domain of attraction of the station-keeping control mode in Step 1. Hence, the tether length will be regulated to the desired length upon switching to the station-keeping stabilization control when the tether length is sufficiently near the desired value. Simulation results of a typical system given in Section VII.2 demonstrate the asymptotic stability of the TSS using this algorithm.

VII. Simulation Results

Many simulation examples for tethered satellite systems in the station-keeping mode have been presented in [13]. In this section, we present simulation results only for deployment and retrieval.

A TSS with following characteristics is considered :

- Orbital radius $r_0 = 6598$ km,
- Subsatellite mass $m = 170$ kg,
- Orbital angular velocity $\Omega = 0.0011781$ rad/sec.

In the following discussion, $\tilde{\theta} = \theta - \theta^*$ denotes the differential of the in-plane angle, ℓ_f denotes the desired final tether length, $\tilde{\ell} = \ell - \ell_f$ denotes the differential of the tether length and F_ℓ denotes the applied tension control force.

VII.1. Retrieval

As discussed in Section V, the set of candidate in-plane angles for constant angle retrieval is given by

$$S_r := \{\theta | 0 < \theta < \frac{\pi}{2} \quad \text{or} \quad -\pi < \theta < \theta_1(\ell_i)\},$$

where $\theta_1(\ell_i)$ is defined in (43). Let the initial state of the system be $\phi = 0.01$, $\tilde{\theta} = -0.01$, and $\omega_\theta = \omega_\phi = 0$. The initial tether length ℓ_i is assumed to be 10 km. It is observed from Figures 4 and 5 that the equilibrium point $(\theta^*, 0, 0, 0)$ is unstable during retrieval with a desired in-plane angle of $\theta^* = -3.0$ and $\theta^* = -1.6$, respectively. As mentioned in [13], since the tether is not in reality rigid, the applied tension control force cannot be positive (to rule out compression). However, Figure 5(d) shows that a positive tension control force F_ℓ occurs during some time interval. Thus, if a constant angle control law is applied during retrieval, then not only will the system be unstable, but tether compression may also occur.

It is also found that when the desired in-plane angle satisfies $\theta^* \in S_1 := \{-2.1 < \theta^* < \theta_1(\ell_i)\}$ for constant angle retrieval, the applied tension control force F_ℓ can assume positive values during some time-intervals, i.e., compression may occur. The system response for $\theta^* = -2.1$ and constant angle retrieval is depicted in Figure 6, where F_ℓ is found (see Figure 6(d)) to at times be very close to 0 but is never positive.

Similar simulation results are found for the region $0 < \theta^* < \frac{\pi}{2}$ for constant angle retrieval. The equilibrium point $(\theta^*, 0, 0, 0)$ is found to be unstable during retrieval and compression of the tether may occur in case $1.0 < \theta^* < \frac{\pi}{2}$. The system responses are not shown.

VII.2. Deployment

According to Theorem 5, the set of candidate θ^* for stable deployment is

$$S_d = \{\theta | -0.68 \leq \theta < 0, \quad \text{or} \quad 2.5 \leq \theta < \pi\}.$$

Let the initial disturbance of the system be $\phi = 0.01$, $\tilde{\theta} = -0.01$, and $\omega_\theta = \omega_\phi = 0$. The initial tether length is assumed to be $\ell_i = 10$ m, which is provided by a boom. First, the system response during deployment (applying constant in-plane angle control only) are depicted in Figures 7 and 8, with $\theta^* = -0.68$, and $\theta^* = 2.5$, respectively. It is observed from the system responses that, for instance, the differential of the in-plane angle $\tilde{\theta}$ and the out-of-plane angle ϕ decay during deployment.

The switching control strategy, which involves both constant angle control and station-keeping control, is applied to deploy a subsatellite from the satellite with the desired final tether length $\ell_f = 10$ km. The first example concerns deploying the subsatellite upward (i.e., away from the Earth) by applying constant angle control with $\theta^* = -0.015$ for the first 260,500 seconds, and applying the station-keeping control thereafter. The applied tension control force for station-keeping is governed by

$$F_\ell = -m(U + h_1\tilde{\ell} + h_2\dot{\ell}), \tag{78}$$

where $U = 0.041019$, $h_1 = 3.1\Omega^2$ and $h_2 = 0.0034$. The responses of the system during constant angle deployment are shown in Figure 9. At time $t = 260,500$ seconds, we have

- out-of-plane angle $\phi = -7.01636 \times 10^{-6}$, and $\dot{\phi} = 1.70633 \times 10^{-8}$ rad/sec
- in-plane angle $\theta = -0.0150051$,and $\dot{\theta} = 5.61812 \times 10^{-10}$ rad/sec
- actual tether length $\ell = 9.97617$ km and $\dot{\ell} = 2.63603 \times 10^{-4}$ km/sec.

With these values, the applied tension control law is switched to the station-keeping control and governed by Eq. (78). The system responses governed by (78) are depicted in Figure 10.

Another example for deploying the subsatellite downward (i.e., toward the Earth) is implemented by applying constant angle control for the first 235,300 seconds with $\theta^* = 3.125$, then switched to the station-keeping control governed by Eq. (78). At time $t = 235,300$ seconds, we have

- out-of-plane angle $\phi = -2.01378 \times 10^{-6}$, and $\dot{\phi} = -2.35517 \times 10^{-8}$ rad/sec
- in-plane angle $\theta = 3.1250$, and $\dot{\theta} = 8.96566 \times 10^{-9}$ rad/sec
- actual tether length $\ell = 9.88531$ km and $\dot{\ell} = 2.90670 \times 10^{-4}$ km/sec.

The system responses are shown in Figures 11 and 12.

Acknowledgment

The authors thank Professors P.S. Krishnaprasad and J.H. Maddocks for helpful discussions. This research was supported in part by the Air Force Office of Scientific Research under URI Grant AFOSR-87-0073, and by the NSF under Grants ECS-86-57561 and CDR-85-00108.

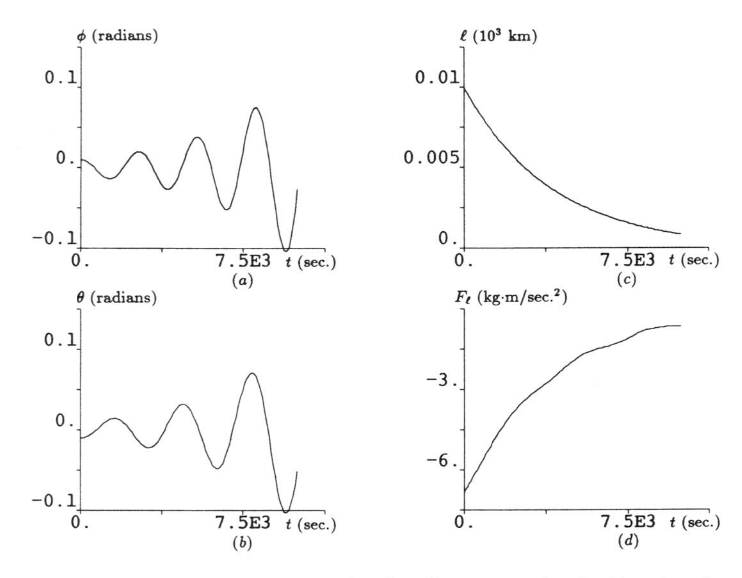

Figure 4. Simulation Results for Constant Angle Retrieval with $\theta^* = -3.0$ radians

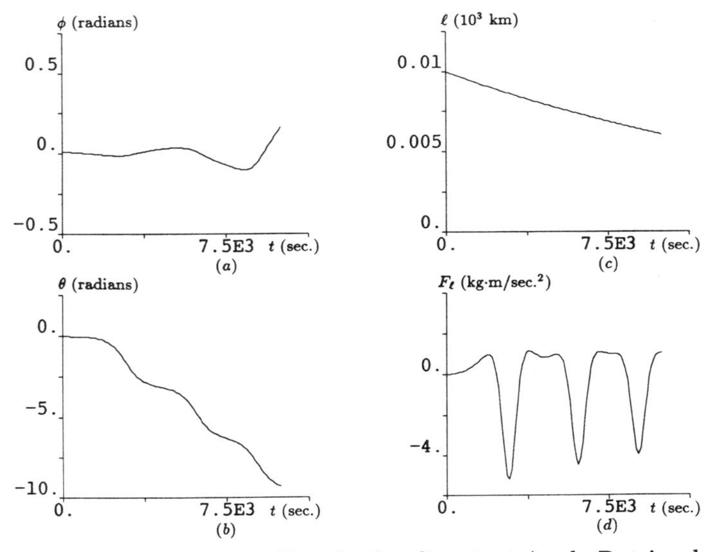

Figure 5. Simulation Results for Constant Angle Retrieval with $\theta^* = -1.6$ radians

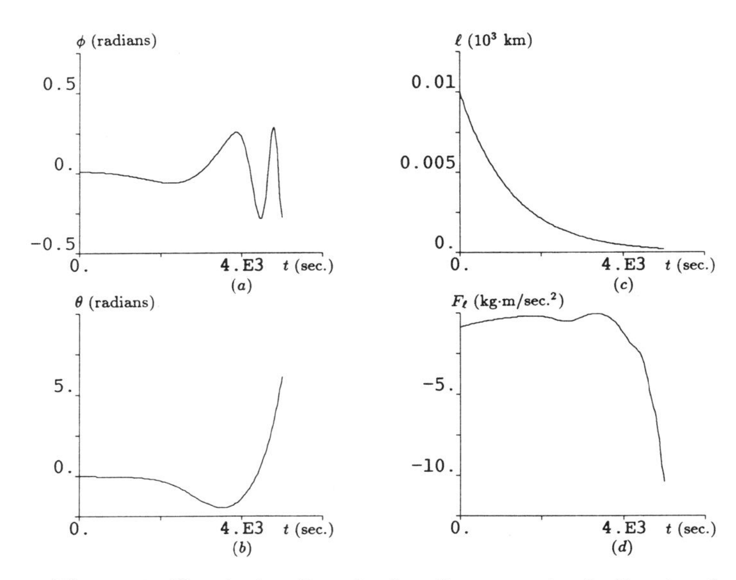

Figure 6. Simulation Results for Constant Angle Retrieval with $\theta^* = -2.1$ radians

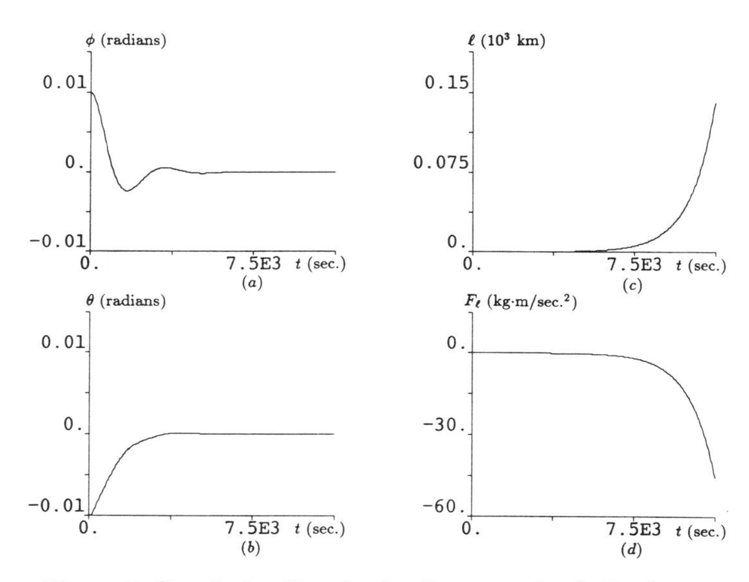

Figure 7. Simulation Results for Constant Angle Deployment with $\theta^* = -0.68$ radians

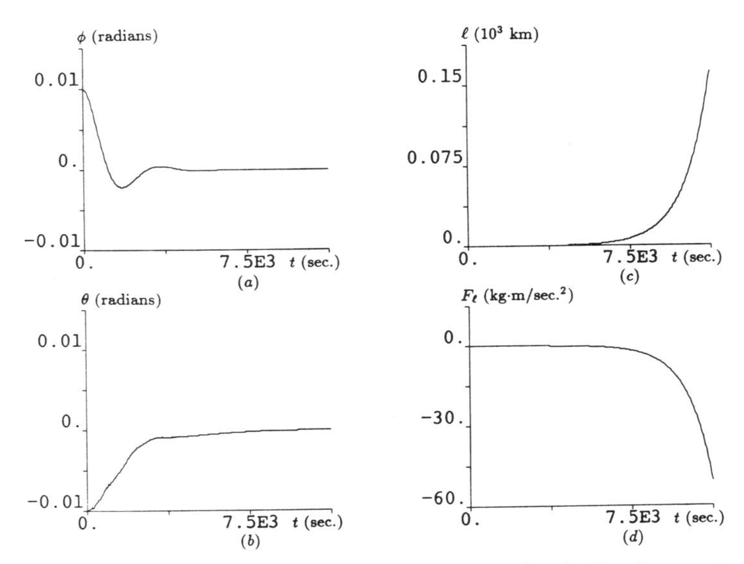

Figure 8. Simulation Results for Constant Angle Deployment with $\theta^* = 2.5$ radians

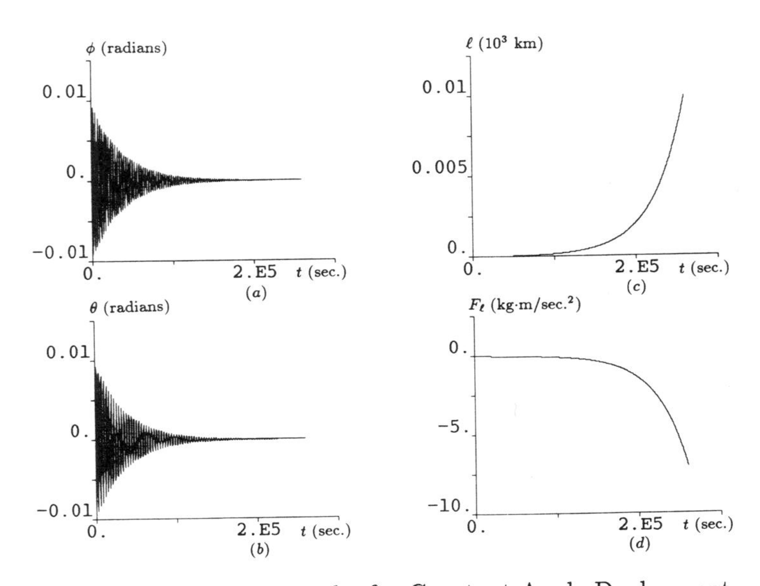

Figure 9. Simulation Results for Constant Angle Deployment with $\theta^* = -0.015$ radians

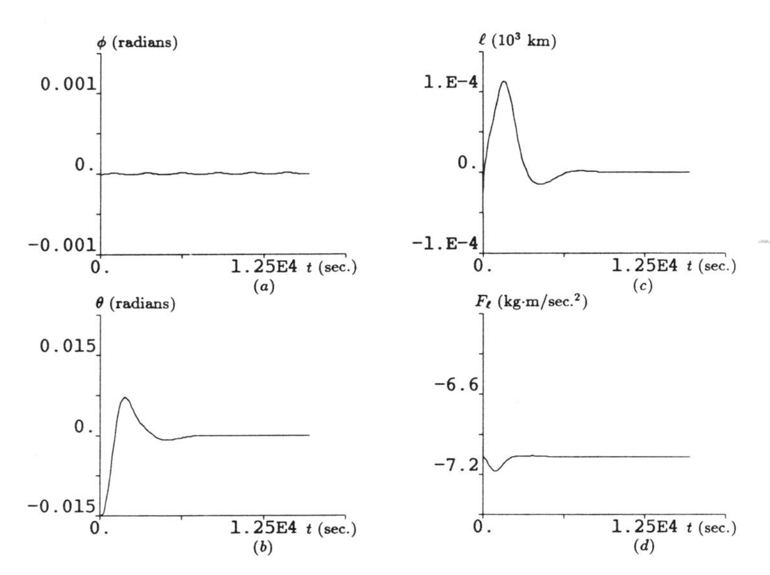

Figure 10. Simulation Results for Station-Keeping with $\theta^* = 0$ radians

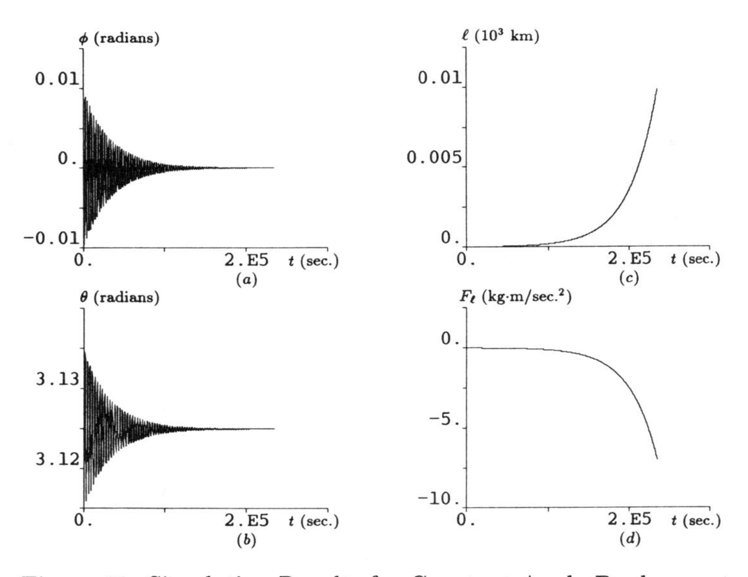

Figure 11. Simulation Results for Constant Angle Deployment with $\theta^* = 3.125$ radians

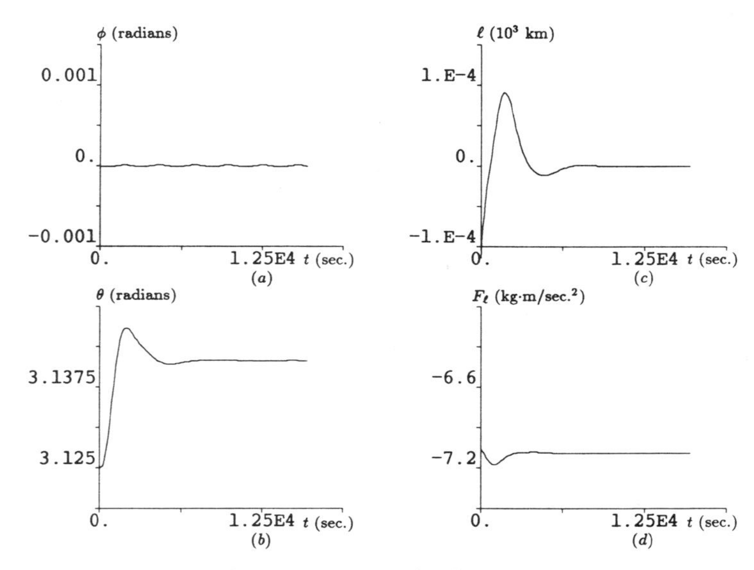

Figure 12. Simulation Results for Station-Keeping with $\theta^* = \pi$ radians

Appendix

The values of $k_{i,j}$, $i = 1, 2$ and $j = 1, 2$ are as given below.

$$k_{1,1} = \frac{0.295575407 \csc^2 2\theta^*}{k_{1,2}},$$

$$k_{1,2} = \frac{l_1 + l_2 + \sqrt{(l_1 + l_3)^2 + 4}}{2},$$

$$k_{2,1} = \frac{0.498328311 \csc^2 2\theta^*}{k_{2,2}},$$

$$k_{2,2} = \frac{l_4 + l_5 + \sqrt{(l_4 + l_6)^2 + 4}}{2},$$

where

$$l_1 = -3\Omega \cos\theta^* \sin\theta^* - 0.5\Omega \sin\theta^*,$$

$$l_2 = -\frac{\Omega^2(3\cos^2\theta^* + 0.5\cos\theta^*) + 1}{\Omega \sin 2\theta^*},$$

$$l_3 = -\frac{\Omega^2(3\cos^2\theta^* + 0.5\cos\theta^*) - 1}{\Omega \sin 2\theta^*},$$

$$l_4 = -3.68961\Omega \cos\theta^* \sin\theta^*,$$

$$l_5 = -\frac{3.6896061\Omega^2 \cos^2\theta^* + 1}{\Omega \sin 2\theta^*},$$

$$l_6 = -\frac{3.6896061\Omega^2 \cos^2\theta^* - 1}{\Omega \sin 2\theta^*}.$$

References

[1] C.C. Rupp and J.H. Laue, "Shuttle/tethered satellite system," *The Journal of the Astronautical Sciences*, Vol. 26, pp. 1-17, 1978.

[2] D.A. Arnold, "The behavior of long tethers in space," *The Journal of the Astronautical Sciences*, Vol. 35, pp. 3-18, 1987.

[3] J.B. Eades, Jr. and H. Wolf, *Tethered Body Problems and Relative Motion Orbit Determination*, Final Report, Contract NAS5-21453, Analytical Mechanics Associates, Inc., Seabrook, MD, 1972.

[4] T.R. Kane and A.K. Banerjee, "Tether deployment dynamics," *The Journal of the Astronautical Sciences*, Vol. 30, pp. 347-365, 1982.

[5] C.C. Rupp, *A Tether Tension Control Law for Tethered Subsatellites Deployed Along Local Vertical*, NASA TM X-64963, 1975.

[6] A.K. Misra and V.J. Modi, "Deployment and retrieval of shuttle supported tethered satellites," *J. Guidance*, Vol. 5, pp. 278-285, 1980.

[7] P.M. Bainum, S. Woodard and J. Juang, "The development of optimal control laws for orbiting tethered platform systems," AAS 85-360, pp. 41-55, 1985.

[8] W.P. Baker, J.A. Dunkin, Z.J. Galaboff, K.D. Johnston, R.R. Kissel, M.H. Rheinfurth and P.L. Siebel, *Tethered Subsatellite Study*, NASA TM X-73314, 1976.

[9] E.H. Abed and D.-C. Liaw, "On the stabilization of tethered satellite systems," *Abstracts of the Second Conf. on Non-Linear Vibrations, Stability, and Dynamics of Structures and Mechanisms*, VPI, Blacksburg, VA, June 1988.

[10] R.W. Brockett, *Finite Dimensional Linear Systems*, John Wiley, New York, 1970.

[11] D.-C. Liaw and E.H. Abed, "Stability analysis and control of tethered satellites," *Model Determination for Large Space Systems Workshop*, Rept. JPL D-5574, Jet Propulsion Laboratory, Pasadena, CA, pp. 308-330, 1988.

[12] NASA, Mission Planning and Analysis Division, *Space Station Mission Requirements Data Base*, Rept. JSC-32072, NASA Space Station Program Office, Reston, VA, May 1988.

[13] D.-C. Liaw and E.H. Abed, "Nonlinear stabilization of tethered satellites," in *Proc. 27th IEEE Conf. Decision and Control*, Austin, TX, pp. 1738-1745, 1988.

[14] J.L. Kaplan, "Converse theorems for finite-time stability and practical stability," *IEEE Trans. Circuit Theory*, Vol. CT-20, pp. 66-67, 1973.

[15] E.H. Abed and J.-H. Fu, "Local feedback stabilization and bifurcation control, I. Hopf bifurcation," *Systems and Control Letters*, Vol. 7, pp. 11-17, 1986.

[16] D.-C. Liaw and E.H. Abed, *Stabilization of Tethered Satellites During Station-Keeping*, Rept. TR 88-72, Systems Research Center, Univ. of Maryland, College Park, 1988. (Submitted for publication.)

[17] W. Hahn, *Stability of Motion*, Springer-Verlag, New York, 1967.

[18] D.-C. Liaw and E.H. Abed, "Bounds for finite-time stability," in preparation.

Contemporary Mathematics
Volume 97, 1989

Quantum Control Theory I

E. B. Lin

ABSTRACT. This paper outlines some results recently obtained concerning quantum mechanical control systems. The relationship between classical and quantum control systems are discussed in the framework of the Lagrangian and Hamiltonian formalisms. The problem of constructing variational principles for a given second -order differential equation is considered. This leads to a generalization of the Helmholtz conditions for the inverse problem in classical mechanics.

1. INTRODUCTION. Recently there has been an enormous advance in our understanding of the microworld. However, the new insights of system control theory have not been brought to bear in full force on the more complicated problems of mechanics. In fact, a natural extension of the concepts of modern physics to other field is provided by the framework of system theory[4,5,10]. Quantum physics and control are two rich and well-developed theories. It is necessary to combine them effectively, in particular, to establish the relation between physics and control theory in the specific area of quantum phenomena.

Motivated by [15] we formulate a more generalized quantum mechanical control system. The approach is to via Lagrangian and Hamiltonian systems, which are basically the equations of motion for dynamical problems in terms of a set of local coordinates. Under some assumptions, we have a correspondence between Lagrangian and Hamiltonian systems.

We then show how classical Hamiltonian control systems may be quantized to quantum mechanical control systems. This raised the question which classical control systems actually can be formulated as Hamiltonian control systems and so may be quantized. It is, therefore, of considerable interest to characterize explicitly those systems which arise from a problem in the calculus of variations and which may subsequently submit to variational methods. This is the so-called inverse problem in the calculus of variations. The availability of a variational principle can provide an important tool for analyzing a given system.

0271-4132/89 $1.00 + $.25 per page

2. CONTROL SYSTEMS. Consider a Hamiltonian system as any system with inputs and outputs defined on a symplectic manifold (M, ω) where ω is the symplectic form, and having the form

$$(1) \qquad \dot{x} = X_{H_o}(x) + \sum_{j=1}^{m} u_j X_{H_j}(x), \qquad x \in M$$
$$y_i = H_i(x), \qquad 1 \leqslant i \leqslant m$$

where X_{H_i} the Hamiltonian vector fields on M with Hamiltonian function H_i determined from

$$\omega(X_{H_i}, Z) = dH_i(Z)$$

for Z an arbitrary vector field on M. In fact, a nonlinear control system (2) may be realized as the Hamiltonian system (1).

$$(2) \qquad \dot{x} = f(x) + \sum_{i=1}^{m} u_i g_i(x), \qquad x \in M$$
$$y_i = h_i(x), \qquad 1 \leqslant i \leqslant k$$

where f, g_i are vector fields on manifold M and h_i are functions on M, with k not necessarily equal m.

The inverse problem in Newtonian mechanics concerns the system of equations with external forces, or inputs u_i,

$$(3) \qquad R_i(y, \dot{y}, \ddot{y}) = -u_i, \qquad 1 \leqslant i \leqslant n, \qquad y \in R^n$$

where $\partial R_i / \partial \ddot{y}_j$ is a nonsingular matrix in some domain in $R^n \times R^n \times R^n$, and asks under what conditions on R_i there exists a function $L(y,\dot{y})$ such that on this domain

$$(4) \qquad \frac{d}{dt}\frac{\partial L}{\partial \dot{y}_i} - \frac{\partial L}{\partial y_i} = R_i(y,\dot{y},\ddot{y}), \qquad 1 \leqslant i \leqslant n$$

where L is the Lagrangian function.

So, (3) may be written as a set of Euler-Lagrangian equations with external forces

$$(5) \qquad \frac{d}{dt}\frac{\partial L}{\partial \dot{y}_i} - \frac{\partial L}{\partial y_i} = -u_i, \qquad 1 \leqslant i \leqslant n$$

for some Lagrangian function $L(y,\dot{y})$. Since $\partial R_i / \partial \ddot{y}_j$ and hence $\partial^2 L / \partial \dot{y}_i \dot{y}_j$ is nonsingular, using the Legendre transformation we may write these equations in the form

$$(6)\qquad \dot{q}_i = \frac{\partial H}{\partial p_i} \qquad\qquad 1 \leqslant i \leqslant n$$

$$-\dot{p}_i = \frac{\partial H}{\partial q_i} + u_i, \qquad y_i = q_i \qquad 1 \leqslant i \leqslant n,$$

where H is the corresponding Hamiltonian. Setting $H_i = q_i$ these equations may now be written as

$$(7)\qquad \dot{q}_i = \frac{\partial H}{\partial p_i} + \sum_j u_j \frac{\partial H_j}{\partial p_i} \qquad\qquad 1 \leqslant i \leqslant n$$

$$-\dot{p}_i = \frac{\partial H}{\partial q_i} + \sum_j u_j \frac{\partial H_j}{\partial q_i}, \quad y_i = q_i, \qquad 1 \leqslant i \leqslant n.$$

Given the class of all input-output maps arising from general nonlinear control systems (2) with k = m, the Hamiltonian realization problem is to determine that subclass which also arises from general Hamiltonian control systems (1). In fact, the conditions enabling the Newtonian system of equations to be expressed as Lagrangian equations, also solves the Hamiltonian realization problem for a special class of systems. Finding necessary and sufficient conditions for a nonlinear input-output map to be realized by a Hamiltonian system, was solved by Jakubczyk[9], while already in Crouch & Irving [3] conditions were found for a finite Volterra series to be realizable by a (nilpotent) Hamiltonian system.

More precisely, we give the following definitions.

DEFINITION. A <u>Lagrangian control system</u> consists of :

(i) a smooth manifold M;

(ii) a fiber bundle E over T(M);

(iii) a system in local coordinates takes the form

$$(8)\qquad \frac{d}{dt}\frac{\partial L(x,\dot{x},u)}{\partial \dot{x}_i} - \frac{\partial L(x,\dot{x},u)}{\partial x_i} = 0,$$

$$y(t) = h(x,\dot{x});$$

where L: E $\longrightarrow$ R is called the Lagrangian and h: T(M)$\longrightarrow$Y maps the state space into the output manifold Y.

The Lagrangian is in terms of a set of local coordinates: position, linear momentum and input coordinates. In contrast to these coordinates, if we consider generalized coordinates and generalized momenta we have the following Hamiltonian approach.

DEFINITION. A <u>Hamiltonian control system</u> consists of :

(i) a symplectic manifold M;

(ii) a fiber bundle E over M;

(iii) a smooth function H: E $\longrightarrow$ R;

(iv) a set of input state evolution equations in local coordinates takes the form

$$\dot{x} = \frac{\partial H(x,p,u)}{\partial p} \quad , \quad \dot{p} = \frac{-\partial H(x,p,u)}{\partial x} . \tag{9}$$

The cotangent bundle of any manifold M admits a natural symplectic structure and we can construct Hamiltonian systems on the cotangent bundle. On the other hand, Lagrangian control systems are constructed on the tangent bundle. In fact, we can identify the tangent bundle of any manifold with its cotangent bundle. By introducing local coordinates on a cotangent bundle of a smooth manifold M, $x_1 \dots x_n, p_1 \dots p_n$ with natural symplectic structure $\Sigma\, dx_i dp_i$, we have the following theorem. [10]

THEOREM 1 Let M be a smooth manifold. If the fiber derivative of the Lagrangian L, DL: TM $\longrightarrow T^*M$ is a diffeomorphism then the Lagrangian control system can be converted into a Hamiltonian control system and vice versa. Moreover, any of the above Lagrangian or Hamiltonian control systems on a cotangent bundle of a smooth manifold M with natural symplectic structure is quantizable.

REMARK. In theorem 1 we assume DL:TM $\longrightarrow T^*M$ and DH:$T^*M \longrightarrow$ TM are diffeomorphisms. It is interesting to know the results when they are not diffeomorphisms, which usually happens in some control systems[2].

3. VARIATIONAL PRINCIPLES. We outline some methods to construct Lagrangians [1,13,14,15] for linear and quasi-linear control systems.

Let us be more specific. Consider the following control systems:

Second-order linear control systems,

$$\frac{d^2x}{dt^2} = A\frac{dx}{dt} + Bx + Cu \tag{10}$$

where $x \in R^n$ is the state vector, $u \in R^d$ is the control vector and A, B and C are real constant matrices of appropriate order;

First-order linear control systems,

$$\frac{dx}{dt} = Ax + Bu \tag{11}$$

where $x \in R^n$ is the state vector, $u \in R^d$ is the control vector and A and B are real matrices of appropriate order.

We quote the following two theorems for constructing Lagrangian for control systems (10) and (11). [10, 13, 14,15]

THEOREM 2 A Lagrangian for the analytic representation of system (10) is given by

$$L = K(t,\dot{x}) + D_i(t,x)\dot{x}_i + E(t,x),$$

where the functions K, D_i and E are a solution of the linear system of partial differential equations

$$\frac{\partial^2 K}{\partial \dot{x}_i \, \partial \dot{x}_j} = T_{ij} \quad ,$$

$$\frac{\partial D_i}{\partial x_j} - \frac{\partial D_j}{\partial x_i} = \frac{1}{2}\left(\frac{\partial S_i}{\partial x_j} - \frac{\partial S_j}{\partial x_i}\right) + \left(\frac{\partial^2 K}{\partial x_i \partial \dot{x}_j} - \frac{\partial^2 K}{\partial \dot{x}_i \partial x_j}\right)$$

$$= U_{ij}(t,x)$$

and

$$\frac{\partial E}{\partial x_i} = \frac{\partial D_i}{\partial t} - S_i - \frac{\partial K}{\partial x_i} + \frac{\partial^2 K}{\partial \dot{x}_i \partial t} + \left[\frac{\partial^2 K}{\partial x_i \partial \dot{x}_j} + \frac{1}{2}\left(\frac{\partial S_i}{\partial \dot{x}_j} - \frac{\partial S_j}{\partial \dot{x}_i}\right)\right]x_j$$

$$= V_i(t,x),$$

and K, D, E, S and T are as follows:

$$K(t,\dot{x}) = \dot{x}_i \int_0^1 dr' \; \{ [\int_0^1 drT_{ij}(t)] \; \dot{x}_j\}(t, r\dot{x}),$$

$$D_i = [\int_0^1 drU_{ij}(t,rx)r]x_j$$

and

$$E = [\int_0^1 drV_i(t,rx) \;]x_i .$$

S and T are obtained from the following conventional Lagrange equations:

$$\frac{d}{dt}\frac{\partial L}{\partial \dot{x}} - \frac{\partial L}{\partial x} = T(t) \left(\frac{d^2x}{dt^2} - A\frac{dx}{dt} - Bx - Cu \right)$$

$$= T(t)\frac{d^2x}{dt^2} + S(t,u,x,\dot{x}) \tag{12}$$

and the r.h.s. of equation (12) is self-adjoint, i.e.

$$T_{ij} = T_{ji}$$

$$\frac{\partial S_i}{\partial \dot{x}_j} + \frac{\partial S_j}{\partial \dot{x}_i} = 2\left(\frac{\partial}{\partial t} + \dot{x}_k\frac{\partial}{\partial x_k}\right)T_{ij},$$

$$\frac{\partial S_i}{\partial x_j} - \frac{\partial S_j}{\partial x_i} = \frac{1}{2}\left(\frac{\partial}{\partial t} + \dot{x}_k\frac{\partial}{\partial x_k}\right)\left(\frac{\partial S_i}{\partial \dot{x}_j} - \frac{\partial S_j}{\partial \dot{x}_i}\right) \; ; \; (i,j,k=1,2,\ldots,n).$$

THEOREM 3 A necessary and sufficient condition for system (11) to admit an analytic representation in terms of the conventional Lagrange equations

$$\frac{d}{dt}\frac{\partial L}{\partial \dot{x}} - \frac{\partial L}{\partial x} = T(t)\left(\frac{dx}{dt} - Ax - Bu\right)$$
$$= f(t,x,\dot{x})$$

is that the r.h.s. of the above equation is self-adjoint, i.e.

$$\frac{\partial f_i}{\partial \dot{x}_j} = -\frac{\partial f_j}{\partial \dot{x}_i},$$
$$\frac{\partial f_i}{\partial x_j} = \frac{\partial f_j}{\partial x_i} - \frac{d}{dt}\frac{\partial f_j}{\partial x_i}; \qquad (i,j = 1,2,\ldots,n).$$

The Lagrangian for the analytic representation is given by

$$L(t,x,\dot{x}) = -x_i \int_0^1 dr f_i(t,rx,r\dot{x}).$$

REMARK. If the equation is not self-adjoint, one can multiply the equation by a suitable factor to make equation self-adjoint in some cases.

Quasi-linear case: [1]

Let us consider a single second-order differential operator T in the independent variables $x^1, \ldots, x^n$ and a dependent variable u which is linear in the second derivatives of u, i.e., T is of the form

$$(13) \qquad T = T(x^k, u, u_k, u_{kd})$$
$$= A^{ij}(x^k, u, u_k)u_{ij} + B(x^k, u, u_k),$$

where A^{ij} and B are smooth functions of 2n+1 arguments and that the A^{ij} are the entries of a symmetric matrix.

The problem is to construct a Lagrangian

$$(14) \qquad L = L(x^k, u, u_k)$$

such that the Euler-Lagrangian equation

$$(15) \qquad E(L)(s) = 0$$

derived from the fundamental integral

$$(16) \qquad I(s) = \int L\left(x^k, s, \frac{\partial s}{\partial x^k}\right) dx^1 dx^2 \ldots dx^n$$

is equivalent to the following given equation

$$(17) \qquad T\left(x^k, s, \frac{\partial s}{\partial x^k}, \frac{\partial^2 s}{\partial x^k \partial x^d}\right) = 0,$$

i.e., every solution of (15) is a solution of (17) and conversely.

In (15), $E(L)$ denotes the Euler-Lagrange operator

$$(18) \qquad E(L) = -\frac{\partial L}{\partial u} + D_i \frac{\partial L}{\partial u_i},$$

where D_i denotes total differentiation with respect to x^i, i.e.,

$$D_i = \frac{\partial}{\partial x^i} + u_i \frac{\partial}{\partial u} + u_{ij} \frac{\partial}{\partial u_i} .$$

A direct and simple calculation shows that if T is an Euler-Lagrange operator derived from a Lagrangian of the form (14) , then T satisfies the integrability conditions

$$(19) \qquad \frac{\partial T}{u_i} = D_j \frac{\partial T}{\partial u_{ij}} .$$

If T is a linear operator, then one may check that this condition requires that T be formally self-adjoint.

Should T satisfy (19), then T is the Euler-Lagrange operator derived from the Lagrangian

$$L = -\int_0^1 uT(x^i, tu, tu_i, tu_{ij})dt.$$

One may choose a factor(integrating factor) for T satisfying (19). We define a variational integrating factor as a nonzero function

$$f = f(x^k, u, u_k)$$

such that fT is an Euler-Lagrange operator, i.e.,

$$(20) \qquad fT = E(L).$$

The fundamental equations which relate f and T when (20) holds are given by the following lemma.

LEMMA 4 Let $f = e^g$. Then f is a variational integrating factor for the differential operator (13) if and only if the equations

$$(21) \qquad \frac{\partial g}{\partial u_k} A^{ij} - \frac{\partial g}{\partial u_i} A^{kj} = \frac{\partial A^{kj}}{\partial u_i} - \frac{\partial A^{ij}}{\partial u_k}$$

and

$$(22) \qquad \left(\frac{\partial g}{\partial x^i} + \frac{\partial g}{\partial u} u_i\right) A^{ij} - \frac{\partial g}{\partial u_j} B = \frac{\partial B}{\partial u_j} - \frac{\partial A^{ij}}{\partial x^i} - \frac{\partial A^{ij}}{\partial u} u_i$$

are satisfied.

Moreover, we have the following uniqueness theorem.

THEOREM 5 Let $U \subset R^{n+1}$ be an open connected set and let $f(x^k, u, u_k)$

be a variational integrating factor for the differential operator

$$T = A^{ij}u_{ij} + B$$

on U. If $\mathrm{rank}(A^{ij}) \geqslant 2$ on $J^1(U)$, then f is unique up to a multiplicative function $h(x^k)$ which satisfies

$$A^{ij}\frac{\partial h}{\partial x^j} = 0. \tag{23}$$

4. QUANTIZATION. The states of a quantum mechanical system are represented by unit vectors (the particle wave function) in a Hilbert space, and the observables corresponding to self-adjoint operators on the Hilbert space. Specifying the form of the Hamiltonian determined by the physical nature of the system, we have the corresponding Hamiltonian operator associated to an observable. Accordingly, we have Schrodinger's representation, which via the particle dynamics, shows how the particle wave function evolves in space and time under a specific set of circumstances. As pointed out in[the Hamiltonian of a control system can be decomposed into $H = H_o + uH_1$, where H_o is the Hamiltonian of the system without control and H_1 describes the Hamiltonian of the interaction of the controls with guided system. We therefore have the corresponding Hamiltonian in quantum mechanical system as $\overline{H} = \overline{H}_o + u\overline{H}_1$, where

$$\overline{H} \bullet s = \frac{\hbar}{i}(X_H s) + (H - \Sigma\, p_i\frac{\partial H}{\partial p_i})s .$$

More precisely, in terms of local coordinates $q_1,\ldots,q_n,p_1,\ldots,p_n$ on M with canonical symplectic structure $\Sigma\, dp_i dq_i$. Assign to q_j the operator $\overline{q}_j = q_j$ (Multiplication by q_j) and to p_j the operator $\overline{p}_j = -i\hbar\frac{\partial}{\partial q_j}$.

5. EXAMPLES. There are many interesting examples in [10,15]. Here we illustrate two examples. Example 1 shows that every differential equation may be imbedded in a Hamiltonian system by adjoining additional ('adjoint') variables. Example 2 shows how to quantize a controlled harmonic oscillator in one dimension.

(1) Consider the oscillator with cubic damping

$$\ddot{x} + x + u\dot{x}^3 = 0$$

First write the above equation as a first-order system:

$$\dot{x}_1 = x_2 ,$$

$$\dot{x}_2 = -x_1 - ux_2^3 .$$

Now, use the additional 'adjoint' variables y_1, y_2 to define the Hamiltonian

$$H(x_1, x_2, y_1, y_2) = y_1x_2 + y_2(-x_1 - ux_2^3).$$

Then, with $\dot{x}_1 = \partial H/\partial y_1$, $\dot{x}_2 = \partial H/\partial y_2$ one again obtains the original first-order system and, in view of $\dot{y}_1 = -\partial H/\partial x_1$, $\dot{y}_2 = -\partial H/\partial x_2$, the adjoint differential equations

$$\dot{y}_1 = y_2,$$

$$\dot{y}_2 = -y_1 + 3uy_2x_2^2 .$$

In this manner, the original second-order differential equation which was not of Hamiltonian type has been embedded in a fourth-order Hamiltonian system.

(2) Consider a controlled harmonic oscillator in one dimension,

$$\ddot{x} + \omega^2 x = u(t).$$

This equation is self-adjoint. From Theorem 2, we obtain

$$L = \frac{\dot{x}^2}{2} - \frac{\omega^2}{2} x^2 + u(t)x.$$

Thus $p = L_{\dot{x}} = \dot{x}$, and by Legendre transformation

$$H = p\dot{x} - L = \frac{p^2}{2} + \frac{\omega^2}{2} x^2 - u(t)x.$$

Substituting $p \longrightarrow -i\hbar\, \partial/\partial x$ and $x \longrightarrow x$ we have the Schrödinger equation

$$i\hbar \frac{\partial}{\partial t} \psi(t) = \left(-\frac{\hbar^2}{2} \frac{\partial^2}{\partial x^2} + \frac{\omega^2 x^2}{2} - u(t)x\right) \psi(t).$$

It is recognized that this equation corresponds to the well-defined quantum mechanical problem of a particle in an oscillator well, subject to a uniform gravitational field whose overall strength and direction $u(t)$ is a controllable function of time.

6. ADDITIONAL REMARKS. There have been diverse investigations on quantum mechanical systems[6,8,16]. It stimulates the theoretical study of quantum systems from the point of view of system theory.

As not much has been known yet about the quantum mechanical control systems. Some major objectives such as the adaptation of the notions of controllability, observability, identification, realization, and feedback to the quantum domain are under investigations[11,12].

BIBLIOGRAPHY

1. I. M. Anderson and T. E. Duchamp, Variational principles for second-order quasi-linear scalar equations, J. Diff. Equat. 51(1984), 1-47.
2. R. W. Brockett, "Control Theory and Analytical Mechanics" vol VII,Lie Groups: History Frontiers and Applications, pp. 1-48. Math. Sc. Press.(1977)
3. P. E. Crouch and M. Irving, On Finite Volterra Series which admit Hamiltonian Realizations, Math. Systems Theory 17(1984), 293-318.
4. A. G. Butkovskiy and Yu. I. Samoilenko, Control of quantum systems, Automat. Remote Cont. 40(4),(1979), 485-502.
5. A. G. Butkovskiy and Yu. I. Samoilenko, Control of quantum systems II, Automat. Remote Cont. 40(5),(1979), 629-645.
6. A. G. Butkovskiy and Ye. I. Pustil'nykova, The method of seeking finite control for quantum mechanical processes, CISM Courses and Lectures 294 (1987) 347-359.
7. J. W. Clark and T. J. Tarn, Quantum nondemolition filtering, CISM Courses and Lectures 294, (1987) 331-346.
8. G. M. Huang, T. J. Tarn and J. W. Clark, On the controllability of quantum-mechanical systems, J. Math. Phys. 24(1983), 2608-2618.
9. B. Jakubczyk, Hamiltonian realizations of nonlinear systems, Proc. MTNS 85, Stockholm(1985).
10. E. B. Lin, Quantum mechanical control systems, Math. Comput. Modelling, (1988).
11. E. B. Lin, Controllability of quantum mechanical control systems, Modeling and Simulation,19(1988), 879-882.
12. E. B. Lin, Quantum Control Theory II, in preparation.
13. R. M. Santilli, Necessary and sufficient conditions for the existence of a Lagrangian in field theory, Annals of Phys. 103, (1977), 354-408.
14. R. M. Santilli,"Foundations of Theoretical Mechanics I" , Springer, New York(1978).
15. T. J. Tarn, G. Huang and J. W. Clark, Modelling of quantum mechanical control systems, Math. Modelling 1, (1980), 109-121.
16. A. J. Van Der Schaft, Hamiltonian and quantum mechanical control systems, CISM Courses and Lectures 294, (1987), 277-296.

Department of Mathematics
University of Toledo
Toledo, Ohio 43606

Contemporary Mathematics
Volume **97**, 1989

Cartan-Hannay-Berry Phases and Symmetry

J. Marsden, R. Montgomery**, and T. Ratiu****

Dedicated to Roger Brockett on the occasion of his 50th birthday

Abstract

We give a systematic treatment of the treatment of the classical Hannay-Berry phases for mechanical systems in terms of the holonomy of naturally constructed connections on bundles associated to the system. We make the costructions using symmetry and reduction and, for moving systems, we use the Cartan connection. These ideas are woven with the idea of Montgomery [1988] on the averaging of connections to produce the Hannay-Berry connection.

§1 *Introduction*

In this paper we give some of the results of the work of Montgomery [1988], and of Marsden, Montgomery, and Ratiu [1989] on the use of symmetry and reduction in the theory of Hannay-Berry phases for mechanical systems. We have in mind both classical and quantum systems, but this paper will only be concerned with classical mechanical systems. We intend this paper to be a short version of the cited works which should help interested readers get the ideas quickly. For the most part proofs will be omitted. Readers well versed in connection theory will be able to supply proofs themselves; for details, the longer version can be consulted.

The work is motivated by that of Hannay and Berry on phases for parametrized families of integrable systems. There are, however, many systems to which the Hannay-Berry construction

---

*Research partially supported by NSF grant DMS 8702502, DOE Contract DE-AT03-88ER-12097 and MSI at Cornell University.

**Research partially supported by an NSF postdoctoral fellowship and NSF grant DMS 8702502.

***Research partially supported by NSF grant DMS 8701318-01.

AMS Subject Classification 58F, 70H.

does not literally apply, for example, the motion of a bead on a slowly rotating hoop. (Hannay and Berry get around this by regarding the problem as the limit of a sequence of families of integrable systems, namely the limit of infinite potential constraining forces, but we are able to treat it in a simple, direct way). This paper shows how to put these and other systems consistently into the framework of connections on bundles and reduction of Hamiltonian systems with symmetry. What we do is take the parameter space M to be a space of motions of an ambient space ($\mathbb{R}^2$ for the hoop) and the phase space P to be the cotangent bundle of the constrained configuration space. Physically, the key observation, due to Jeeva Anandan, is that when comparing two different constrained systems one must compare points with the same momentum *relative to a fixed inertial frame*.

Mathematically, there are two key important observations in the present paper. The first is that connections can be averaged; in Montgomery [1988] the context was that of a fixed connection relative to a varying group action. In this paper, we average a connection varying with respect to a fixed (M independent) group action. The second key observation is that the correct connection to average is one due to Cartan [1923]. The ***Cartan connection*** encodes the ficticious forces—the centrifugal, Euler, and Coriolis forces—due to an accelerating inertial frame, namely the one attached to the moving constrained system. Putting these two observations together we find that the correct connection encoding the phase shift of Hannay and Berry for constrained systems is obtained by averaging the Cartan connection. This connection we call the ***Cartan-Hannay-Berry connection.***

§2 *Moving Systems*

Consider a Riemannian manifold $\mathcal{S}$, a submanifold Q, a space M of embeddings of Q into $\mathcal{S}$, and let $m_t \in M$ be a given curve. If a particle in Q is following a curve $q(t)$, and if Q moves by superposing the motion m_t, then the path of the particle in $\mathcal{S}$ is given by $m_t(q(t))$. Thus its velocity in $\mathcal{S}$ is given by

$$T_{q(t)}m_t \cdot \dot{q}(t) + \mathcal{Z}_t(m_t(q(t))) \tag{1}$$

where $\mathcal{Z}_t(m_t(q)) = \frac{d}{dt} m_t(q)$. Consider the Lagrangian on TQ of the form

$$L_{m_t}(q, v) = \frac{1}{2} \| T_{q(t)}m_t \cdot v + \mathcal{Z}_t(m_t(q))^T \|^2 - V(q) - U(m_t(q)) \tag{2}$$

where V is a given potential on Q, U is a given potential on $\mathcal{S}$, and ^{T} denotes projection onto the tangent space to the (moving) image of Q. Taking the Legendre transform of (2), we get a Hamiltonian with momentum

$$p = (T_{q(t)}m_t \cdot v + \mathcal{Z}_t(m_t(q))^T)^\flat \tag{3}$$

where $^\flat$ denotes the index lowering operation determined by the metric on $\mathcal{S}$. Physically, if $\mathcal{S}$ is $\mathbb{R}^3$, then p is the inertial momentum (see the hoop example in §3). This extra term $\mathcal{Z}_t(m_t(q))^T$

leads to a connection called the ***Cartan connection*** on the bundle $Q \times M \to M$, with horizontal lift defined to be $\mathcal{Z}(m) \mapsto (Tm^{-1}\mathcal{Z}(m)^T, \mathcal{Z}(m))$. (See for example, Marsden and Hughes [1983] for an account of some aspects of Cartan's contributions.) The Hamiltonian picks up a cross term and so takes the form

$$H_{\mathcal{Z}_t}(q, p) = \frac{1}{2} \|p\|^2 - \mathcal{P}(\mathcal{Z}_t) + \frac{1}{2} \| \mathcal{Z}_t^T \|^2 + V(q) + U(m_t(q)) \tag{4}$$

where the cross term is $\mathcal{P}(\mathcal{Z}_t)(q, p) = \langle p, \mathcal{Z}_t^T(q)\rangle$. The Hamiltonian vector field of this cross term $X_{\mathcal{P}(\mathcal{Z}_t)}$ represents the noninertial forces and also has the natural interpretation as a horizontal lift of the vector field $\mathcal{Z}_t$ relative to a certain connection on the bundle $T^*Q \times M \to M$, which we also call the ***Cartan connection.***

Let G be a Lie group which acts on T^*Q in a Hamiltonian fashion and leaves H_0 (defined by setting $\mathcal{Z} = 0$) in (4)) invariant. In our examples G is either $\mathbb{R}$ acting by the flow of H_0 (the hoop), or a subgroup of the isometry group of Q which leaves V and U invariant, and acts on T^*Q by cotangent lift (the pendulum). In any case, we assume G has an invariant measure relative to which we can average.

Assuming the "averaging principle" (cf. Arnold [1978], for example) we replace $H_{\mathcal{Z}_t}$ by its G-average,

$$\langle H_{\mathcal{Z}_t}\rangle (q, p) = \frac{1}{2} \|p\|^2 - \langle \mathcal{P}(\mathcal{Z}_t)\rangle + \frac{1}{2}\langle \| \mathcal{Z}_t^T \|^2\rangle + V(q) + \langle U(m_t(q))\rangle. \tag{5}$$

In (5) we shall assume the term $\frac{1}{2}\langle \| \mathcal{Z}_t^T \|^2\rangle$ is small and discard it. Thus, consider

$$\mathcal{H}(q, p, t) = \frac{1}{2} \|p\|^2 - \langle \mathcal{P}(\mathcal{Z}_t)\rangle + V(q) + \langle U(m_t(q))\rangle = \mathcal{H}_0(q, p) - \langle \mathcal{P}(\mathcal{Z}_t)\rangle + \langle U(m_t(q))\rangle \tag{6}$$

where $\mathcal{H}_0 = \frac{1}{2} \|p\|^2 + V(q)$. We shall consider the dynamics on $T^*Q \times M$ given by the vector field

$$(X_{\mathcal{H}}, \mathcal{Z}_t) = (X_{\mathcal{H}_0} - X_{\langle \mathcal{P}(\mathcal{Z}_t)\rangle} + X_{\langle U \circ m_t\rangle}, \mathcal{Z}_t) . \tag{7}$$

The vector field representing the extra terms in this representation due to the superposed motion of the system, namely

$$\mathrm{hor}(\mathcal{Z}_t) = (- X_{\langle \mathcal{P}(\mathcal{Z}_t)\rangle} , \mathcal{Z}_t) \tag{8}$$

has a natural interpretation as the horizontal lift of $\mathcal{Z}_t$ relative to a connection on $T^*Q \times M$, which is obtained by averaging the Cartan connection and so we call it the ***Cartan-Hannay-Berry*** (CHB) ***connection.*** The holonomy of this connection is the ***Hannay-Berry phase*** of a slowly moving constrained system.

§3 *The ball in the hoop and the Foucault pendulum*

We now give two examples of the formalism. The procedures used in the ball in the hoop example are due to J. Anandan.

Example 1 Ball in the Hoop Consider a hoop (not necessarily circular) in which a bead slides without friction, as in Figure 1 below. As the bead is sliding, the hoop is rotated in its plane through an angle $\theta(t)$ with angular velocity $\omega(t) = \dot{\theta}(t)\mathbf{k}$. Let s denote the arc length along the hoop, measured from a reference point on the hoop and let $\mathbf{q}(s)$ be the vector from the origin to the corresponding point on the hoop; thus the shape of the hoop is determined by this function $\mathbf{q}(s)$. The unit tangent vector is $\mathbf{q}'(s)$ and the position of the reference point $\mathbf{q}(s(t))$ relative to an inertial frame in space is $R_{\theta(t)}\mathbf{q}(s(t))$, where R_θ is the rotation in the plane of the hoop through an angle θ.

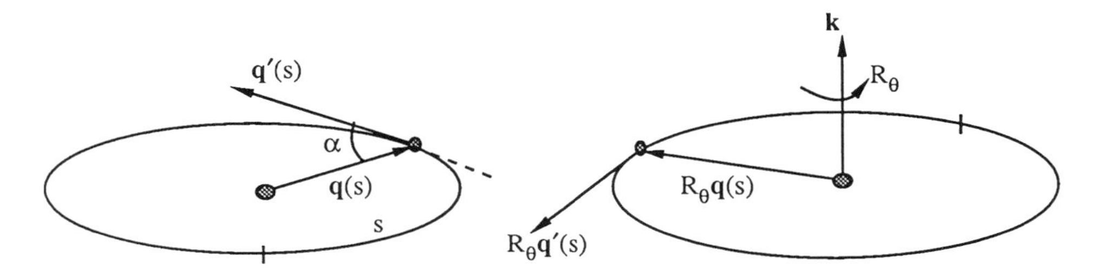

Figure 1 The ball in the hoop

The configuration space is diffeomorphic to the circle $Q = S^1$ with length L the length of the hoop. The Lagrangian $L(s, \dot{s}, t)$ is simply the kinetic energy of the particle; i.e., since

$$\frac{d}{dt} R_{\theta(t)}\, \mathbf{q}(s(t)) = R_{\theta(t)}\mathbf{q}'(s(t))\, \dot{s}(t) + R_{\theta(t)}[\omega(t) \times \mathbf{q}(s(t))] \,,$$

we set

$$L(s, \dot{s}, t) \;=\; \frac{1}{2}m \,\|\, \mathbf{q}'(s)\, \dot{s} + \omega \times \mathbf{q}(s) \,\|^2. \tag{1}$$

The Legendre transformation yields

$$p = m\dot{s} + \mathbf{q}'(s) \cdot [\omega(t) \times \mathbf{q}(s)] = \textit{inertial momentum}$$

and

$$H = \frac{1}{2m} p^2 - \mathcal{P} + \frac{1}{2} \| \, (\omega \times q)^T \, \|^2 \,.$$

Here $' = \frac{d}{ds}$, ^{T} denotes projection onto the hoop's tangent, and

$$\mathcal{P} = p\mathbf{q}'(s) \cdot (\omega(t) \times \mathbf{q}(s)) = p \, \| \mathbf{q} \| \, \sin \alpha(s) \tag{2}$$

(see Figure 1). $\mathcal{P} ds$ is the instantaneous slippage of the bead due to the hoop's rotation.

For ω small relative to p it is a good approximation to replace H by

$$\mathcal{H} = \frac{1}{2m} p^2 - \langle \mathcal{P} \rangle . \tag{3}$$

$$\langle \mathcal{P} \rangle = \frac{1}{L} \oint \mathcal{P}(s) ds = \frac{2A}{L} p\omega \tag{4}$$

where A is the hoop's area and L its length. (This is seen by Stokes' theorem: $\mathbf{d}(r \sin \alpha \, ds) = r \cos \alpha \, d\alpha \wedge ds = 2dA$.) Now

$$X_{\mathcal{H}} = \frac{1}{m} p \frac{\partial}{\partial q} - \omega \frac{2A}{L} \frac{\partial}{\partial q} \tag{5}$$

is the (approximate) dynamic vector field. The second term in (5) gives the Hannay-Berry angle. After one full revolution in the space of motions ($\int \omega dt = 2\pi$) the integral of this second term is

$$-\frac{4\pi A}{L} = \text{Hannay-Berry angle for hoop.} \tag{6}$$

Equation (6) gives the holonomy of the CHB connection, and expresses the displacement of a bead of inertial momentum $\mathcal{P}$ after one full revolution of the hoop, compared to where it would have been after this same amount of time with the same inertial momentum, but *without rotating* the hoop.

Example 2 The Foucault Pendulum The Foucault pendulum is a spherical pendulum at co-latitude α on the surface of the Earth. Denote by $\mathbf{q}$ the position of the pendulum on the sphere of radius ℓ, the length of the pendulum arm (see Figure 2).

Let $\mathbf{r}_0$ denote the vector from the center of the Earth to the point of suspension of the pendulum. Due to the rotation of the Earth the point $\mathbf{q}$ moves to the point $R_t(\mathbf{r}_0 + \mathbf{q})$, where R_t is the rotation about the Oz-axis. Let $\omega = \frac{T}{2\pi}$ denote the angular velocity of the Earth's rotation. The potential energy of the pendulum is $V(\mathbf{q}) = mgl \, \mathbf{q} \cdot \hat{\mathbf{r}}_0$, where $\hat{\mathbf{r}}_0 = \frac{\mathbf{r}_0}{\| \mathbf{r}_0 U \|}$. The velocity of the point $\mathbf{q}$ on the sphere during Earth's rotation is therefore

$$\dot{\mathbf{q}} + R_t[\omega \times (\hat{\mathbf{r}}_0 + \mathbf{q})] \tag{7}$$

where we identify ω with the vector $\omega \mathbf{k}$.

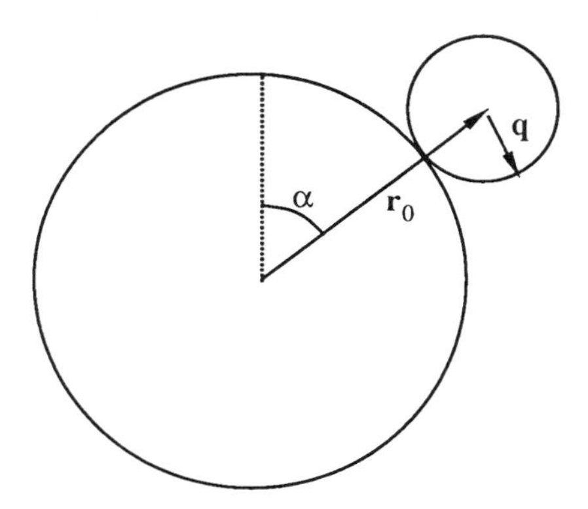

Figure 2

The Lagrangian is

$$L = \frac{1}{2}m \| \omega \times (\hat{\mathbf{r}}_0 + \mathbf{q}) + \dot{\mathbf{q}} \|^2 - V(\mathbf{q}) \tag{8}$$

so that the Legendre transformation gives

$$\mathbf{p} = m(\dot{\mathbf{q}} + \omega \times (\hat{\mathbf{r}}_0 + \mathbf{q}))^T = m[\dot{\mathbf{q}} + (\omega \times (\hat{\mathbf{r}}_0 + \mathbf{q}))^T] \tag{9}$$

by identifying T^*S^2 with TS^2 via the standard metric; here $(\omega \times (\hat{\mathbf{r}}_0 + \mathbf{q}))^T$ denotes the tangential component of $\omega \times (\hat{\mathbf{r}}_0 + \mathbf{q})$. Then

$$\dot{\mathbf{q}} = \frac{1}{m}\mathbf{p} - [\omega \times (\hat{\mathbf{r}}_0 + \mathbf{q})]^T \tag{10}$$

and

$$\dot{\mathbf{q}} + \omega \times (\hat{\mathbf{r}}_0 + \mathbf{q}) = \frac{1}{m}\mathbf{p} + [\omega \times (\hat{\mathbf{r}}_0 + \mathbf{q})]^\perp , \tag{11}$$

where $(\omega \times (\hat{\mathbf{r}}_0 + \mathbf{q}))^\perp$ denotes the normal component of $\omega \times (\hat{\mathbf{r}}_0 + \mathbf{q})$ to the sphere. Therefore the Hamiltonian is given by

$$\begin{aligned} H = \mathbf{p}\dot{\mathbf{q}} - L &= \frac{1}{m}\| \mathbf{p} \|^2 - \frac{1}{m}\mathbf{p} \cdot (\omega \times (\hat{\mathbf{r}}_0 + \mathbf{q}))^T \\ &\quad - \left(\frac{1}{2m}\| \mathbf{p} \|^2 + \frac{m}{2}\| (\omega \times (\hat{\mathbf{r}}_0 + \mathbf{q}))^\perp \|^2 - V(q)\right) \\ &= \frac{1}{2m}\| \mathbf{p} \|^2 + V(\mathbf{q}) - \mathcal{P} - \frac{m}{2}\| (\omega \times (\hat{\mathbf{r}}_0 + \mathbf{q}))^\perp \|^2 \end{aligned} \tag{12}$$

where

$$\mathcal{P} = \frac{1}{m}\mathbf{p} \cdot (\omega \times (\mathbf{r}_0 + \mathbf{q}))^T = \frac{1}{m}\mathbf{p} \cdot (\omega \times (\mathbf{r}_0 + \mathbf{q})) \tag{13}$$

since $\mathbf{p}$ is tangential.

Letting M be the space S^1 of embeddings of the sphere of radius ℓ in $\mathbb{R}^3$ obtained by rotating a fixed initial sphere tangent to the Earth at co-latitude α, as in Figure 2, we recognize that $\mathcal{P}$ is the Hamiltonian defining the induced Cartan connection on $T^*S^2 \times S^1$ (see §**4**).

Since $(\omega \times \mathbf{q})^\perp = 0$, we have $[\omega \times (\mathbf{r}_0 \times \mathbf{q})]^\perp = (\omega \times \mathbf{r}_0)^\perp$ = constant. In the equations of motion one can drop this constant. Had we considered a Foucault pendulum on an ellipsoid, this term would not be constant, but it would be of order ω^2, which is the general case. According to the averaging principle, one ignores this term.

Let $S^1_{r_0}$ be another circle, acting on the phase space of the pendulum by rotation about the $\mathbf{r}_0$ axis. As in the ball in the hoop, and consistent with the theory sketched in §**4**, the horizontal lift of the induced Hannay-Berry connection is given by the Hamiltonian vector field of $\mathcal{P}$, i.e., by $\langle p \cdot [\omega \times (\mathbf{r}_0 \times \mathbf{q})]\rangle$. But if $\mathbf{v}$ is any constant vector,

$$\langle \mathbf{p} \cdot \mathbf{v}\rangle = \langle \mathbf{p} \cdot \hat{\mathbf{r}}_0\rangle(\mathbf{v} \cdot \hat{\mathbf{r}}_0) \tag{14}$$

since the $S^1_{r_0}$-action over which we average has $\mathbf{r}_0$ as its axis of rotation. Setting $\mathbf{v} = \omega \times \mathbf{r}_0$, this implies that

$$\langle \mathbf{p} \cdot (\omega \times \mathbf{r}_0)\rangle = 0 ,$$

and hence

$$\langle \mathbf{p} \cdot [\omega \times (\mathbf{r}_0 + \mathbf{q})]\rangle = \langle \mathbf{p} \cdot (\omega \times \mathbf{q})\rangle .$$

Let $I = \mathbf{p} \cdot (\mathbf{r}_0 \times \mathbf{q})$ be the momentum map of the $S^1_{r_0}$-action. By (14)

$$\langle \mathbf{p} \cdot (\omega \times \mathbf{q})\rangle = \langle \omega \cdot (\mathbf{q} \times \mathbf{p})\rangle = [(\mathbf{q} \times \mathbf{p}) \cdot \mathbf{r}_0]\,\langle \omega \cdot \hat{\mathbf{r}}_0\rangle = I\,\omega \cos\alpha$$

since $\omega \cdot \hat{\mathbf{r}}_0$ is constant. Thus,

$$\langle \mathcal{P}\rangle = I\,\omega\cos\alpha \tag{15}$$

and so the horizontal lift of ω is given by $(-X_{\langle\mathcal{P}\rangle}, \omega) = \left(-\,\omega\cos\alpha\,\frac{\partial}{\partial\theta}, \omega\right)$, where $\frac{\partial}{\partial\theta}$ is the infinitesimal generator of the $S^1_{r_0}$-action, corresponding to I. Therefore the horizontal lift of the curve $\varphi(t) = \omega t$ in M is determined by the differential equation

$$\dot{\theta} = -\,\omega\cos\alpha$$

so that if this curve is a loop parametrized on $\left[0, \frac{T}{2\pi}\right]$ we get

$$\theta(T) - \theta(0) = -\int_0^T \omega \cos\alpha \, dt = -\omega T \cos\alpha = -2\pi\cos\alpha \tag{16}$$

which is the deviation of the plane of oscillation in the laboratory frame (i.e. a frame fixed on the Earth) of the Foucault pendulum during 24 hours. Equation (16) is essentially Foucault's classic formula.

For example, if we are at the equator, where $\alpha = \frac{\pi}{2}$, there is no deviation. If we are at the North Pole, where $\alpha = 0$, the plane rotates in the opposite direction to that of the Earth's rotation, performing a full circular motion. These results are in an inertial frame. The usual Foucault result is for a lab frame attached to the Earth.The two frames are related by $\mathbf{q}_{\text{inertial}} = R_t(\mathbf{r}_0 + \mathbf{q}_{\text{lab}})$. In 24 hours this means we must add 2π to our $\Delta\theta$ to get the angle shift in the lab frame.

For a nice elementary discussion of Foucault's pendulum also see Berry [1988].

§4 *Construction and Properties of the Hannay-Berry Connection*

In this section we summarize the properties and construction of a general Hannay-Berry connection and specialize this to the case of moving systems. The general theory is a generalization of that developed for trivial bundles with symplectic fiber by Golin, Knauf, and Marmi [1989] and Montgomery [1988]. The freedom we allow of choosing an arbitrary background connection is essential in the construction of the CHB-connection discussed below. These background connections need not be principal connections but they are Ehresmann connections.

§4A *Ehresmann Connections and Holonomy*

If $\pi : E \to M$ is a locally trivial fiber bundle, an ***Ehresmann connection*** is a smooth subbundle H of TE such that $H \oplus V = TE$, where $V = \ker T\pi$ is called the ***vertical subbundle*** and H is the ***horizontal subbundle*** of the connection. Alternatively, an Ehresmann connection is given by a one-form $\Gamma \in \Omega^1(E, V)$ on E with values in V and which is the identity on V, or by a horizontal lift, i.e., a map hor sending smooth vector fields on M to smooth sections of H.

Let $m(t)$, $t \in [0, 1]$ be a smooth path in M. A ***horizontal lift*** of $m(t)$ is a smooth path $p(t)$ in E such that $\pi(p(t)) = m(t)$ and the tangent vector $\dot{p}(t)$ to $p(t)$ is horizontal for every $t \in [0, 1]$. If $\pi : F \to M$ is a locally trivial fiber bundle, then given a smooth path $m(t)$, in M with $m_0 = m(0)$, $0 \le t \le 1$ and $p \in \pi^{-1}(m_0)$, there is a unique horizontal lift $p(t)$ of $m(t)$ satisfying $p(0) = p$. The map sending p to $p(1)$ defines, by uniqueness of horizontal lift, a bijection from $\pi^{-1}(m_0)$ to $\pi^{-1}(m_1)$. By smooth dependence of solutions of ordinary differential equations on initial conditions, it follows that this map is a diffeomorphism; it is called the ***parallel transport operator***.

Now let $m(t), t \in [0, 1]$ be closed path in M satisfying $m_1 = m_0$. The diffeomorphism of $\pi^{-1}(m_0)$ onto itself given by parallel transport along $m(t)$ is called the ***holonomy of the path*** $m(t)$. It is easy to see that parallel transport sends juxtaposition of loops based on m_0 into the composition of diffeomorphisms of $\pi^{-1}(m_0)$. Thus the holonomy operation is a group homomorphism of $\mathcal{L}(m_0)$, the loop group at m_0, to the diffeomorphism group of the fiber $\pi^{-1}(m_0)$; its image $\mathcal{H}(m_0)$ is called the ***holonomy group at*** m_0. It is straightforward to see that if M is connected, all holonomy groups are conjugate: $\mathcal{H}(m_0)$ and $\mathcal{H}(m_1)$ are conjugate by the parallel transport along any path connecting m_0 to m_1. Thus if M is connected, we speak of $\mathcal{H}$, the ***holonomy group of the connection***.

If $\pi : E \to M$ is a principal bundle with abelian structure group and Ω is the curvature of a principal connection Γ, the holonomy is given by the group element

$$\text{holonomy} = \exp\left(-\iint d\Gamma\right) = \exp\left(-\iint \Omega\right),$$

where the integral is taken over any two-dimensional submanifold in M whose boundary is $m(t)$, assuming one exists.

§4B *The Hannay-Berry Connection*

We summarize below the main properties and the construction procedure of the Hannay-Berry connection. The proofs will be omitted since they are straightforward generalizations of those in Montgomery [1988] once the following theorem is proved.

4.1 Theorem Averaging of Connections *Let* $\pi : E \to M$ *be a fiber bundle and let* $\Gamma \in \Omega^1(E, V)$ *be an Ehresmann connection. Suppose a compact Lie group* G *acts on* E *by bundle transformations, not necessarily covering the identity. Then the average* $\langle\Gamma\rangle$ *of* Γ (see equation (3)) *is also an Ehresmann connection. Moreover the* G*-action commutes with the action of parallel translation with respect to* $\langle\Gamma\rangle$.

The proof is a consequence of the following two facts: **1** The average of a collection of points sitting inside a convex set is again an element of that set. **2** The set of connections is a convex (in fact affine) subset of the set of (2,1) tensor field on E. (For example, these two facts are used when one averages a Riemannian metric to obtain a new metric for which G acts by isometries.)

Let $\pi : E \to M$ be a Poisson fiber bundle, i.e., π is a surjective submersion, all fibers are Poisson manifolds, and the transition functions are Poisson maps. Let G be a Lie group. A ***family of Hamiltonian G-actions*** on E is a smooth (left) G-action on E such that each fiber $\pi^{-1}(m)$ is invariant under the action and the action restricted to each fiber is Hamiltonian, i.e. it is Poisson and it admits a fiberwise momentum map $\mathbf{I} : E \to \mathfrak{g}^*$. This means that for each $\xi \in \mathfrak{g}$, we have

$$\xi_E(p) = X_{I^\xi}(p) \tag{1}$$

where I^ξ denotes the real valued function defined by $I^\xi(p) = \mathbf{I}(p) \cdot \xi$, for $p \in E$ and $\xi \in \mathfrak{g}$, $\xi_E(p) = \frac{d}{d\varepsilon}\Big|_{t=0} (\exp \varepsilon\xi \cdot p)$ is the infinitesimal generator of the action defined by ξ and $X_{I^\xi}(p)$ is the Hamiltonian vector field on the fiber through p defined by the function $I^\xi : E \to \mathbb{R}$ *restricted* to this fiber. Since the action on each fiber is Hamiltonian, the symplectic leaves of the fiber are G-invariant. Also, note that the Casimirs are G-invariant.

An Ehresmann connection on $\pi : E \to M$ is called ***Poisson*** if its horizontal lift hor_0 is a Poisson bracket derivation, i.e.

$$(\mathrm{hor}_0 Z)[\{f, h\}] = \{(\mathrm{hor}_0 Z)[f], h\} + \{f, (\mathrm{hor}_0 Z)[h]\} \tag{2}$$

for all $f, h : E \to \mathbb{R}$ and $Z \in \mathfrak{X}(M)$. Equivalently, this says that

$$\mathbf{D}_0\{f, h\} \cdot X = \{\mathbf{D}_0 f \cdot X, h\} + \{f, \mathbf{D}_0 h \cdot X\} \tag{2'}$$

for all $X \in \mathfrak{X}(E)$, where $\mathbf{D}_0$ is the covariant differentiation defined by hor_0.

For example, let

$$\mathrm{hor}_0 Z = X_{-\mathcal{P}\cdot Z} + Z \tag{2''}$$

be the Cartan connection on $E = T^*Q \times M$ defined in §**2**. Then

$$\mathbf{D}_0 f \cdot Z = \{f, -\mathcal{P} \cdot Z\}.$$

(The bracket is defined by restricting f to $T^*Q \times \{m\}$.) It follows from the Jacobi identity that the Cartan connection is Poisson.

4.2 Definition *The Hannay-Berry* (HB) *connection induced by a Poisson-Ehresmann connection* hor_0 *is the Ehresmann connection on* $\pi : E \to M$ *obtained by averaging* hor_0. *(See Equation* (3).) *We will let* $\mathbf{D}$ *denote its covariant derivative,* hor *its horizontal lift, and* $\gamma \in \Omega^1(E, \ker T\pi)$ *its connection one-form.*

If hor_0 is the Cartan connection, then we will call the resulting average the Cartan-Hannay-Berry (CHB) connection. Its horizontal lift was given in §**2**:

$$\mathrm{hor} \cdot Z = X_{-\langle \mathcal{P} \cdot Z\rangle} + Z.$$

If λ is any tensor field defined along (as opposed to on) a G-invariant submanifold of E, its ***average*** $\langle\lambda\rangle$ is the smooth G-*i*nvariant tensor field of the same type defined along the same submanifold by

$$\langle\lambda\rangle = \frac{1}{|G|}\int_G (\Phi_g^* \lambda)\, dg\,, \quad \text{where} \tag{3}$$

$\Phi : G \times E \to E$ is the G-action on E. Note that $\langle\lambda\rangle$ is a G-invariant tensor field.

The following two propositions summarize the important properties of the HB connection.

4.3 Proposition *Suppose* G *is compact and connected. Then the* HB *connection satisfies the following properties:*

i *It is Poisson.*

ii *If* $v \in T_mM$ *then its* HB *horizontal lift is given by*

$$\text{hor } v = \langle \text{hor}_0\, v\rangle.$$

iii $\text{hor } Z = \text{hor}_0 Z + X_{K\cdot Z}$, *for a smooth function* $K\cdot Z : E \to \mathbb{R}$ *and* $Z \in \mathfrak{X}(M)$.

iv *The connection one-form of the* HB *connection is given by*

$$\Gamma(v) = \Gamma_0(v) - X_{K\cdot T\pi(v)}(p), \quad \text{for } v \in T_p E\,. \tag{4}$$

v $\langle K\cdot Z\rangle$ *is a fiberwise Casimir function.*

vi $\mathbf{D}\langle\lambda\rangle = \langle \mathbf{D}\lambda\rangle = \langle \mathbf{d}\lambda\rangle \circ \mathbf{P}_{\text{hor}}$, *for any* $\lambda \in \Omega^k(E)$, $k = 0, 1, \ldots,$ *where* $\mathbf{P}_{\text{hor}}$ *is the horizontal projection relative to* hor.

vii $\mathbf{D}I^\xi \cdot \text{hor } Z = \langle \mathbf{D}_0 I^\xi \cdot \text{hor } Z\rangle$ *is a Casimir for each* $\xi \in \mathfrak{g}$, $Z \in \mathfrak{X}(M)$. *Moreover,* $\mathbf{DI} = \langle \mathbf{D}_0\, \mathbf{I}\rangle$ *and hence* $\mathbf{DI} = 0$ *iff* $\langle \mathbf{D}_0\, \mathbf{I}\rangle = 0$.

Remark Property **vi** holds for any averaged connection. The rest of the properties are consequences of the following general principle:

If E *has structure group* $\mathcal{G}$, *and both* Γ_0 *and* G *preserve this structure, then* $\langle\Gamma_0\rangle$ *also preserves this structure.* In the HB case, $\mathcal{G}$ is the group of Hamiltonian automorphisms of the fiber.

4.4 Proposition *Let* hor *be the horizontal lift of an Ehresmann connection on* $\pi : E \to M$ *satisfying*

a $\mathbf{DI} = 0$, *where* $\mathbf{D}$ *is the covariant differentiation given by* hor ; *this says that parallel translation relative to* hor *preserves the level sets of* $\mathbf{I}$;

b $\text{hor } Z = \text{hor}_0 Z + X_{K\cdot Z}$ *for some smooth function* $K\cdot Z : E \to \mathbb{R}$, *where* $Z \mapsto K\cdot Z$ *is linear;*

c $\langle K \cdot v\rangle$ *is a Casimir function on* $\pi^{-1}(m)$ *where* $v \in T_mM$; *replacing* $K \cdot v$ *by* $K \cdot v - \langle K \cdot v\rangle$, *we can assume that this Casimir is zero.*

Then **i** *such a connection is unique;*

ii *such a connection exists if and only if the 'adiabatic condition'*

$$\langle \mathbf{D}_0 \mathbf{I}\rangle = 0 \tag{A}$$

holds, in which case the connection equals the Hannay-Berry connection.

Remarks 1 According to Proposition **4.3iii** the covariant derivative of a function f with respect to the HB-connection is

$$\mathbf{D}f \cdot u = \mathbf{D}_0 f \cdot u + \{f, K \cdot u\}. \tag{B}$$

2 We call condition (A) the 'adiabatic condition' because in the context of a family of completely integrable systems this equality is the content of the classical adiabatic theorem.

3 If G is semisimple and $\mathbf{I}$ is equivariant the 'adiabatic condition' (A) automatically holds. Consequently, the HB connection satisfies properties **a, b,** and **c** of Proposition **4.4.**

Indeed, properties **vi** and **vii** of Proposition **4.3** we have $\mathbf{DI} = \langle \mathbf{DI}\rangle = \mathbf{D}\langle \mathbf{I}\rangle$. By equivariance $\langle \mathbf{I}\rangle^{\xi} = \mathbf{I}^{\langle\xi\rangle}$, for every $\xi \in \mathfrak{g}$. Now $\langle\xi\rangle$ is an Ad-invariant vector. Since G is semi-simple, the adjoint representation is irreducible and so $\langle\xi\rangle = 0$; consequently $\langle \mathbf{I}\rangle = 0$, and thus $\mathbf{DI} = 0$.

4 Suppose, for each $v \in T_mM$ there is a function $\tilde{K} \cdot v : \pi^{-1}(m) \to \mathbb{R}$ satisfying

$$\mathbf{d}I^{\xi} \cdot \mathrm{hor}_0 v + \{I^{\xi}, \tilde{K} \cdot v\} = 0 \tag{2}$$

for all $\xi \in \mathfrak{g}$. Set $K \cdot v = \tilde{K} \cdot v - \langle \tilde{K} \cdot v\rangle$ and $\mathrm{hor}\, v = \mathrm{hor}_0 v + X_{K\cdot v}$. Assume $\langle \mathbf{D}_0 \mathbf{I}\rangle = 0$, so hor defines the HB connection. This is proved by verifying the conditions in Proposition **4.4.**

We now specialize to the case of the Cartan connection, ending with the main theorem of this section. Assume that a Lie group G acts on T^*Q on the left with equivariant momentum map $\mathbf{I}: T^*Q \to \mathfrak{g}^*$. Then G defines a family of Hamiltonian actions on $T^*Q \times M$ by letting G act trivially on M. Its parametrized momentum map is simply $\mathbf{I}$ thought of as a function of two variables, independent of M. Notice that the adiabatic condition

$$\langle \mathbf{D}_0 \mathbf{I}\rangle = 0$$

holds for the Cartan connection. This is a consequence of the following lemma, applied to the functions $f = \mathcal{P} \cdot Z$.

4.5 Lemma *Let* $\pi : E \to M$ *be a Poisson fiber bundle endowed with a family of Hamiltonian* G-*actions with equivariant parametrized momentum map* $\mathbf{I}: E \to \mathfrak{g}^*$. *Then for any* $f: E \to \mathbb{R}$ *we have* $\langle\{\mathbf{I}^\xi, f\}\rangle = 0$ *for all* $\xi \in \mathfrak{g}$.

Proof Apply the Fubini theorem and use the fact that $\left\langle \frac{\partial f}{\partial \theta} \right\rangle = 0$ when $G = S^1$. ■

4.6 Main Theorem *The* CHB *connection is a Poisson connection for* $T^*Q \times M \to M$. *Its horizontal lift is given by*

$$\text{hor } Z = (-X_{\langle \mathcal{P} \cdot Z \rangle}, Z)$$

and satisfies

$$\mathbf{DI} = 0;$$

together with the rest of properties **a, b, c** of Proposition **4.4**, and **i - vii** of Proposition **4.3**. *Its holonomy defines the* ***Hannay angles*** *of a slowly moving constrained mechanical system.*

This follows from the adiabatic condition and Propositions **4.3** and **4.4**.

Recall that the holonomy of a closed loop relative to an Ehresmann connection is the diffeomorphism of the fiber given by parallel translation. In the case of the Hannay-Berry connection induced by a Cartan connection, the fiber is T^*Q. Thus if $c(t)$ is a closed loop of embeddings of Q in $\mathcal{S}$ the differential equations for the horizontal lift of $c(t)$ in T^*Q are Hamilton's equations for the Hamiltonian $-\left\langle \mathcal{P} \cdot \frac{dc}{dt} \right\rangle$.

§5 *Concluding Remarks*

An expanded version of the present paper is available. In addition to the theory for moving systems described here, we also discuss the following.

§5A *Symplectic Fiber Bundles*

Suppose the bundle $E \to M$ of **§4** is a symplectic fiber bundle, that is, its fibers are symplectic, so that each fiber has a symplectic form ω_f and these fit together smoothly. It is a basic fact that Poisson connections hor_0 (with Hamiltonian holonomy) are in one-to-one correspondence with closed extensions ω of the fiber 2-form ω_f to all of E. The correspondence is

$$\text{Hor}_0 = \text{Vert}^\omega ,$$

where $^{\omega}$ denotes the ω-orthogonal complement. (This fact is due to Gotay, *et al.* [1983] and was presented to us in this more digestible fashion by V. Guillemin.)

This means that our averaged connection can be obtained by averaging the 2-form:

$$\mathrm{Hor} = \mathrm{Vert}^{\langle\omega\rangle}.$$

This approach has many simplifying virtues, and is the one taken by Golin, *et al.* [1988]. For example, if $\omega = -\mathbf{d}\theta$, it immediately leads to

$$I^{\xi} = \langle \mathbf{i}_{\xi_E}\theta \rangle$$

for the momentum map. One checks directly that this is the standard formula $\oint p\,dq$ for actions in the case where the group is abelian. Also, these automatically satisfy the adiabatic condition $\langle \mathbf{D}_0\mathbf{I}\rangle = 0$.

§5B *Normal Forms*

Suppose in **§2**, that the external potential U is independent of M, and that we are averaging over the flow of H_0. Then

$$[X_{H_0}, \mathrm{hor}\, Z_t] = 0,$$

and our approximate dynamics is defined by the vector field

$$(X_{\mathcal{H}}, Z_t) = X_{H_0} + \mathrm{hor}\, Z_t.$$

It follows that in this case the HB connection $\mathrm{hor}\, Z_t$, can be viewed as the second term in a normal form for the true dynamics (2.4).

To see this, recall that the normal form for a (time-independent) vector field $X = X_0 + \varepsilon X_1 + O(\varepsilon^2)$, ε a formal parameter, is obtained by changing variables using the flow of X_0. In the new coordinates $X = X_0 + \varepsilon Y_1$ where $[X_0, Y_1] = 0$ and Y_1 is the average of X_1 with respect to the flow of X_0. See Churchill, Kummer, Rod [1983], or Cushman [1988] for more regarding this point of view on normal forms.

§5C *Reconstruction of Reduced Motions*

We begin as in **§2** with a simple mechanical system $H = K + V$, together with a group G acting in such a way as to preserve K and V. Here K is kinetic energy on T^*Q. We have the momentum map $\mathbf{I} : T^*Q \to \mathfrak{g}^*$, and the reduced spaces $(T^*Q)_\mu = \mathbf{I}^{-1}(\mu)/G_\mu$ for $\mu \in \mathfrak{g}^*$; see Marsden and Weinstein [1974].

Motions with momentum μ push down to solutions to the reduced Hamiltonian system on $(T^*Q)_\mu$. (One example of this is the planar Kepler problem. There $\mathbf{I}$ is angular momentum, and $(T^*Q)_\mu$ is coordinatized by the radius and the radial momentum variables.) The basic problem is to reconstruct the original dynamics from the reduced dynamics. Consider the bundle

$$\mathbf{I}^{-1}(\mu) \to (T^*Q)_\mu \tag{R}$$

with fiber G_μ representing the "angles" we ignore in reducing. The idea is then to ***horizontally*** lift the solution curve on the reduced space, to a curve $\bar{c}(t)$ in $\mathbf{I}^{-1}(\mu)$. Writing

$$c(t) = g(t) \cdot \bar{c}(t),$$

we get a first order differential equation for the desired angles $g(t) \in G_\mu$. If G_μ is abelian (as it must be generically) this equation is solvable in quadratures, and the reconstruction is complete.

The new ingredient needed is a connection with which to obtain the horizontal curve $\bar{c}$. We construct this connection out of the natural connection on $Q \to Q/G_\mu$. This is the connection for which

$$\mathrm{Hor}_q = (T_q G_\mu \cdot q)^\perp ,$$

where $\perp$ is taken relative to the kinetic energy metric. (For this to be an honest connection, we must assume that the G action is free.)

§5D *Tower of Bundles*

The HB connection and the 'reconstruction connection' just described can be put together to define a connection on each post of the tower of bundles

$$\mathbf{I}^{-1}(\mu) \to \mathbf{I}^{-1}(\mu)/G_\mu \to M. \tag{T}$$

The HB connection is a connection for $\mathbf{I}^{-1}(\mu) \to M$, and the reconstruction connection gives a fiber-wise connection on

$$\mathbf{I}^{-1}(\mu)_m \to \mathbf{I}^{-1}(\mu)_m/G_\mu$$

for each $m \in M$. The two connections can be put together thus defining connections on either $\mathbf{I}^{-1}(\mu) \to \mathbf{I}^{-1}(\mu)/G_\mu$, or, on the bundle of sympelctic reduced spaces $\mathbf{I}^{-1}(\mu)/G_\mu \to M$. It is hoped that these connections can help give fairly complete descriptions of the dynamics in the averaging approximation. This situation appears to be appropriate, for example, for the dynamics of slowly moving coupled rigid bodies (e.g., coupled rigid bodies in orbit about the earth). The first bundle in (T) is the bundle corresponding to internal versus overall rotational modes and the second to the parametrized motion of the system. See also **§5E**.

§5E *An Optimal Control Problem*

Go back to the setting of **§5C** but now view it as the set-up for a control problem. The desired result will be a holonomy $g \in G$. The control parameter space will be Q/G, based at q_0, whose holonomy is g. The optimal control problem is to find the shortest such loop.

One of the motivating problems is the following. Let Q be the configuration space for a sytstem of coupled planar rigid bodies relative to an inertial frame. $G = \mathbf{SO}(2)$ represents rigid rotations. Then Q/G is the set of ***shapes***, which can be parameterized by the hinge angles, which we assume that we can control.

The holonomy g resulting from a given shape change is exactly the rigid rotation which the system undergoes after a given manipulation of the hinge joints. the connection with respect to which we calculate the holonomy is the constraint: "angular momentum equals zero". This is the same connection as the one alluded to in **§5C**, i.e. the connection defined by kinetic energy on Q.

The optimal control question has been answered by Montgomery [1989] in the sense that it has been reduced to an o.d.e. Somewhat surprisingly this is the same o.d.e. which governs the motion of a "charged particle in an electromagentic field" (on the space Q/G).

References

V. Arnold [1978] *Mathematical Methods of Classical Mechanics*, Graduate Texts in Mathematics, **60**, Springer Verlag.

M. Berry [1984] Quantal phase factors accompanying adiabatic changes, *Proc. Roy. Soc. London* **A 392**, 45-57.

M. Berry [1985] Classical adiabatic angles and quantal adiabatic phase, *J. Phys. A: Math. Gen.* **18**, 15-27.

M. Berry [1988] The geometric phase. *Scientific American*, December, 1988.

M. Berry and J. Hannay [1988] Classical non-adiabatic angles, *J. Phys. A: Math. Gen* **21**, 325-333.

E. Cartan [1923] Sur les varietes a connexion affine et theorie de relativite generalizée, *Ann. Ecole Norm. Sup.* **40**, 325-412, **41**, 1-25.

R. Churchill, M. Kummer, D. Rod [1983] On averaging, reduction, and symmetry in Hamiltonian system, *J. Diff. Eq.* **49**, 359-414.

R. Cushman [1988] A survey of normalization techniques applied to perturbed Keplerian system, *Dept. of Math., Univ. of Utrecht, preprint* **496**.

S. Golin, A. Knauf, S. Marmi [1988] The Hannay angles: geometry, adiabaticity, and an example, preprint.

M. Gotay, R. Lashof, J. Sniatycki, A. Weinstein [1983] Closed forms on symplectic fiber bundles, *Comm Math. Helv.* **58**, 617-621.

J. Hannay [1985] Angle variable holonomy in adiabatic excursion of an integrable Hamiltonian, *J. Phys. A: Math. Gen* **18**, 221-230.

J.E. Marsden, and T. J.R. Hughes [1983] *Mathematical Foundations of Elasticity*, Prenctice Hall.

J.E. Marsden, R. Montgomery, T. Ratiu [1989] Reduction, Symmetry, and Berry's Phase in Mechanics, preprint.

J.E. Marsden and A. Weinstein [1974] Reduction of symplectic manifolds with symmetry. *Rep. Math. Phys.* **5**, 121-130

R. Montgomery [1988] The connection whose holonomy is the classical adiabatic angles of Hannay and Berry and its generalization to the non-integrable case, *Comm. Math. Phys.* **120**, 269-294

R. Montgomery [1989] Shortest loops with a fixed holonomy, *MSRI Berkeley, preprint.*

Jerrold E. Marsden
Department of Mathematics
University of California
Berkeley, CA 94720

Richard Montgomery
MSRI
1000 Centennial Drive
Berkeley, CA 94720

Tudor Ratiu
Department of Mathematics and MSRI
University of California
Santa Cruz, CA 94305

MSRI
1000 Centennial Drive
Berkeley, CA 94720

Contemporary Mathematics
Volume **97**, 1989

Block Diagonalization and the Energy-Momentum Method

J. E. Marsden*, J. C. Simo**, D. Lewis*** and T. A. Posbergh**

Dedicated to Roger Brockett on the occasion of his 50th birthday

Abstract

We prove a geometric generalization of a block diagonalization theorem first found by the authors for rotating elastic rods. The result here is given in the general context of simple mechanical systems with a symmetry group acting by isometries on a configuration manifold. The result provides a choice of variables for linearized dynamics at a relative equilibrium which block diagonalizes the second variation of an augmented energy - these variables effectively separate the rotational and internal vibrational modes. The second variation of the effective Hamiltonian is block diagonal, separating the modes completely, while the symplectic form has an off diagonal term which represents the dynamic interaction between these modes. Otherwise, the symplectic form is in a type of normal form. The result sets the stage for the development of useful criteria for bifurcation as well as the stability criteria found here. In addition, the techniques should apply to other systems as well, such as rotating fluid masses.

* Research partially supported by AFOSR/DARPA contract F49620-87-C0118 and MSI at Cornell University.

** Research partially supported by AFOSR contract 2-DJA-544 and 2-DJA-771.

*** Research partially supported by MSI at Cornell University.

AMS Subject Classification 58F, 70H

§1 *Introduction*

In this paper we present a block diagonalization theorem which is designed for the analysis of stability and bifurcation of rotating systems, or more generally, of relative equilibria. The context of the discussion is the energy-momentum method for mechanical systems with symmetry. Simo, Posbergh and Marsden [1989] and Lewis and Simo [1989] discovered crucial special cases of the block diagonalization theorem for uniformly rotating systems, including general nonlinear elasticity and geometrically exact rods. Our purpose is to abstract these examples and prove a general geometric theorem. We expect these general results will be important for rotating gravitational fluid masses as well.

For rotating systems the result says that a splitting of coordinates can be *explicitly* found on a linearized level which represent the rotational and internal vibrational modes. In these coordinates, the second variation of an augmented Hamiltonian is block diagonal. Of course coordinates can always be found in principle to do this, but we are able to do it explicitly enough to give useful stability and, we believe, bifurcation criteria. On the other hand, the symplectic form does *not* block diagonalize, indicating that the rotational and internal modes are in fact dynamically coupled. However, for purposes of the stability calculation, block diagonalization of the augmented energy is what is important. The off diagonal terms in the symplectic form (sometimes called *Coriolis coupling terms*) are, however, sufficiently simple that they should be useful for studying the dynamic interaction of the rotational and internal vibrational modes.

For rotating pseudo-rigid bodies, Lewis and Simo [1989] noticed that the computation of the definiteness of the second variation is considerably simplified by our result - in this case the simplification saves considerable computation time. In their case, the symbolic and numerical manipulation needed would normally require testing a full 14×14 matrix for definiteness; block diagonalization techniques, however, reduces this to testing a 6×6 matrix for nonisotropic bodies or a 3×3 matrix for the isotropic case.

According to Jellinek and Li [1989], "the general problem of separation and characterization of the overall rotation in any (not necessarily rigid or near rigid) N-body system is among the few still unsolved problems of traditional classical mechanics." Jellinek and Li are able to achieve results for the N-body problem by elimination of the coupling from the expression for the energy in an instantaneous fashion. We have been able to achieve a similar result for the general case of rotating structures, be they N-body systems, coupled rigid bodies, or elastic or fluid structures. This flexibility is achieved through a general geometric approach.

Acknowledgements We thank Tony Bloch, P.S. Krishnaprasad, Richard Montgomery, George Patrick, and Tudor Ratiu for helpful comments.

§2 *The Energy-Momentum Method*

We begin our work in the context of standard mechanical systems with symmetry before any reductions have taken place. In other words, we begin with a *symplectic* manifold (P, Ω) rather than a *Poisson* manifold. In fact, shortly we shall specialize to the case of $P = T^*Q$ and a Hamiltonian of the form kinetic plus potential.

Let G be a Lie group acting symplectically on P with an equivariant momentum mapping

$$\mathbf{J} : P \to \mathfrak{g}^* \tag{1}$$

(see Abraham and Marsden [1978], Marsden [1981] or Marsden, Weinstein, Ratiu, Schmid, and Spencer [1983] for the standard definitions and results used here).

Let $H : P \to \mathbb{R}$ be a given G-invariant Hamiltonian. A point $z_e \in P$ is called a ***relative equilibrium*** if there is a $\xi \in \mathfrak{g}$, the Lie algebra of G, such that the curve

$$z(t) := \exp(t\xi)\, z_e \tag{2}$$

is the dynamical orbit of the Hamiltonian vector field X_H of H, with initial condition $z(0) = z_e$. Let $\mu_e = \mathbf{J}(z_e)$, the momentum at equilibrium. The energy-momentum method relies on the following result.

2.1 Relative Equilibrium Theorem *A point z_e is a relative equilibrium iff there is a $\xi \in \mathfrak{g}$ such that z_e is a critical point of $H_\xi : P \to \mathbb{R}$, where*

$$H_\xi(z) = H(z) - \langle \mathbf{J}(z) - \mu_e, \xi \rangle . \tag{3}$$

In (3), the Lie algebra element $\xi \in \mathfrak{g}$ may be regarded as a Lagrange multiplier. Since $\mathbf{J}$ is conserved by the flow of X_H, the set $\mathbf{J} - \mu_e = 0$ is preserved, so one may regard it as a (non-holonomic) constraint set. It also follows that $\xi \in \mathfrak{g}_{\mu_e}$, the isotropy algebra of μ_e (with respect to the coadjoint action). Thus,

$$\delta H_\xi(z_e) = 0 \tag{4}$$

may be regarded as a (constrained) *variational principle for relative equilibria.*

The relative equilibrium theorem is readily verified. Of course it has a long history, going back to Lagrange and Poincaré for rotating systems. Like many basic results, it has been rediscovered in a number of contexts by various authors. Early references in our context are Arnold [1966], Smale [1970] and Marsden and Weinstein [1974]. As we shall state below, the relative equilibrium theorem sometimes specializes to the ***principle of symmetric criticality*** (Palais [1979]).

The energy-momentum test for formal stability of a relative equilibrium z_e proceeds as follows (see Holm *et al.* [1985] for the meaning of formal stability and related references).

Energy-Momentum Method

1 Choose $\xi \in \mathfrak{g}$ such that $\delta H_\xi(z_e) = 0$

2 Choose a linear subspace $\mathcal{S} \subset T_{z_e}P$ such that

i $\mathcal{S} \subset \ker T\mathbf{J}(z_e)$ and

ii $\mathcal{S}$ complements $T_{z_e}(G_{\mu_e} \cdot z_e)$ in $\ker T\mathbf{J}(z_e)$, where $G_{\mu_e} \subset G$ is the isotropy subgroup of μ_e.

3 Test $\delta^2 H_\xi(z_e)$ for definiteness as a bilinear form on $\mathcal{S}$.

The energy-momentum method "covers" the energy-Casimir method (Holm *et al.* [1985]) in the sense that if the latter applies and gives formal stability, so does the former. One difficulty with the energy-Casimir method is that on the reduced space P/G there may not be enough Casimirs to make the method effective; in particular, it may not be possible to obtain the analogue $\delta(H + C)(z_e) = 0$ of (4). This difficulty is genuine for the case of geometrically exact rods, for instance. See Simo, Posbergh and Marsden [1989] for further details.

The fact that $\delta^2 H_\xi(z_e)$ drops to the reduced space follows from the next lemma.

2.2 Gauge Invariance Lemma

$$\delta^2 H_\xi(z_e)(\eta_P(z_e), \delta z) = 0 \tag{5}$$

for all $\delta z \in \ker T\mathbf{J}(z_e)$ *and* $\eta \in \mathfrak{g}$, *where* η_P *denotes the infinitesimal generator of the group action on* P.

This follows readily from invariance of H and equivariance of $\mathbf{J}$. One can view (5) as a block diagonalization result on the unconstrained tangent space $T_{z_e}P$, but it does not yield block diagonalization *within* the constrained subspace $\mathcal{S}$ in the energy-momentum method. It is the latter that we are concerned with here.

One can identify any choice of $\mathcal{S}$ with the tangent space to the ***reduced space***

$$P_{\mu_e} = \mathbf{J}^{-1}(\mu_e)/G_{\mu_e}$$

at $[z_e]$ (assuming, as we shall, that μ_e is a regular and generic value; c.f. Weinstein [1984]), but it is easier to do our analysis directly on $T_{z_e}T^*Q$ rather than on the quotient space. This is the usual situation found in constrained optimization problems although we shall see that keeping in mind the geometry of the quotient space will play a useful role.

§3 *Simple Mechanical Systems*

Let Q be a configuration manifold and $P = T^*Q$ the associated phase space with its canonical symplectic structure. In the finite dimensional case, we denote cotangent coordinates on T^*Q by (q^i, p_i). (When we use coordinates, we assume Q is finite dimensional, although the results are *not* restricted to this case.) Coordinates on the velocity phase space TQ are similarly denoted $(q^i, \dot{q}^i)$.

Let g denote a Riemannian metric on Q; in coordinates, we write the components of g as g_{ij} as usual, and we write g^{ij} for the inverse tensor. Let $K : TQ \to \mathbb{R}$ denote the corresponding kinetic energy,

$$K(q, \dot{q}) = \frac{1}{2} g_{ij}(q)\, \dot{q}^i \dot{q}^j , \tag{1}$$

and let $V : Q \to \mathbb{R}$ be a given potential.

Assume G acts on Q (by a left action) and hence on T^*Q by the cotangent lift, so the equivariant momentum map is given by

$$\langle \mathbf{J}, \xi\rangle(\alpha_q) = \langle \alpha_q, \xi_Q(q)\rangle. \tag{2}$$

In coordinates, we define the ***action coefficients*** $A^i_a(q)$ by writing

$$[\xi_Q(q)]^i = A^i_a(q)\, \xi^a \tag{3}$$

where a, b, c, ... denote coordinate indices for the Lie algebra $\mathfrak{g}$. Thus (2) becomes

$$J_a(q, p) = p_i\, A^i_a(q). \tag{4}$$

We assume that G acts on Q by isometries and that the potential V is G-invariant. For elasticity, for instance, this is the requirement of material frame indifference. Note that (3) of §**2** reads

$$H_\xi(q, p) = \frac{1}{2} g^{ij} p_i\, p_j + V(q) - p_i\, A^i_a(q)\, \xi^a + \langle \mu_e, \xi\rangle\, . \tag{5}$$

Define the ***moment of inertia tensor*** $\mathbb{I}$ for the system ***locked at*** $q \in Q$ by

$$\mathbb{I}_{ab}(q) = g_{ij}(q)\, A^i_a(q)\, A^j_b(q) \tag{6}$$

(alternatively, in terms of the q-dependent inner product $\langle \xi, \eta \rangle := \langle \xi_Q(q), \eta_Q(q) \rangle$ on $\mathfrak{g}$ we have $\langle \xi, \eta \rangle = \mathbb{I}_{ab}(q)\xi^a\eta^b$), define the ***augmented potential*** V_ξ by

$$V_\xi(q) = V(q) - \tfrac{1}{2}\mathbb{I}_{ab}(q)\, \xi^a\xi^b. \tag{7}$$

We note that the function V_ξ is not the same as the ***amended*** potential in the sense of Smale [1970]. This is the function $V_\mu(q) = V(q) + \frac{1}{2}\mathbb{I}^{ab}(q)\, \mu_a\mu_b$ which also plays an important role in this story, but will be the subject of other investigations.

One can readily verify the following (see Abraham and Marsden [1978] and Palais [1979]) by writing out the conditions $\delta H_\xi = 0$ in **2.1**. A more elegant argument is, however, given below.

3.1 Principle of Symmetric Criticality *A point* $z_e = (q_e, p_e) = (q^i, p_i)$ *is a relative equilibrium if and only if there is a* $\xi \in \mathfrak{g}_{\mu_e}$ *such that*

i $p_i = g_{ij} A^j_a \xi^a$ (*i.e.,* p_e *is the Legendre transform of* $\xi_Q(q_e)$) (8a)

and

ii q_e *is a critical point of* V_ξ. (8b)

This is useful for carrying out the computations that follow. We also observe that V_ξ is G_ξ-invariant, and so induces a function on Q/G_ξ. Define the one-form A^ξ on Q by $A^\xi = A^\xi_i dq^i$, where

$$A^\xi_i(q) = g_{ij}(q)\, A^j_a(q)\, \xi^a \tag{9}$$

or abstractly, $A^\xi(q) = [\xi_Q(q)]^\flat$, where $^\flat$ denotes the index lowering operation with respect to the metric g_{ij}. In other words, $A^\xi(q)$ is the Legendre transform of $\xi_Q(q)$. We remark that A may be viewed as a G-connection for the bundle $Q \to Q/G$ and that this connection plays an important role in Berry's phase; cf. Marsden, Montgomery and Ratiu [1989]. Now notice that at equilibrium, (8a) says

$$p_e = A^\xi(q_e)\,. \tag{10}$$

Also note that

$$H_\xi(q, p) = K_\xi(q, p) + V_\xi(q) + \langle \mu_e, \xi \rangle , \tag{11}$$

where $K_\xi(q, p) = \frac{1}{2} \| p - A^\xi(q) \|^2$, and V_ξ is given by (7). By (10), K_ξ has a critical point at z_e. *Thus, (8b) is a consequence of the relative equilibrium theorem and (11).*

In the energy-momentum method we shall use a special choice of $\mathcal{S}$, namely

$$\mathcal{S} = \left\{ v_{z_e} \in T_{z_e}T^*Q \mid T\pi_Q \cdot v_{z_e} \text{ is g-orthogonal to } T(G_{\mu_e} \cdot q_e) \text{ and } v_{z_e} \in \ker[T\mathbf{J}(z_e)] \right\}, \tag{12a}$$

where $\pi_Q : T^*Q \to Q$ is the canonical projection. Letting coordinates on $T(T^*Q)$ be denoted

$$(q^i, p_i, \delta q^i, \delta p_i),$$

(12) reads, with the help of (8a),

$$\mathcal{S} = \Big\{ (q^i, p_i, \delta q^i, \delta p_i) \mid g_{ij}(\delta q)^i A^j_a \chi^a = 0 \text{ for all } \chi \in \mathfrak{g}_{\mu_e} \text{ and}$$
$$(\delta p)_i A^i_a + g_{ij} A^j_b \xi^b \frac{\partial A^i_a}{\partial q^k} (\delta q)^k = 0 \Big\} \tag{12b}$$

§4 *Rigid Variations*

One version of the cotangent bundle reduction theorem (see Abraham and Marsden [1978] and Kummer [1981], Montgomery [1986] and references therein) states that the reduced space $(T^*Q)_{\mu_e}$ is a symplectic bundle over $T^*(Q/G)$ with fiber the coadjoint orbit through μ_e. Thus there is an isomorphism

$$T_{[z_e]}(T^*Q)_{\mu_e} \cong \mathfrak{g}/\mathfrak{g}_{\mu_e} \times T_{[z_e]}(T^*(Q/G)) \cong \mathfrak{g}/\mathfrak{g}_{\mu_e} \times (\mathcal{V}_{INT} \times \mathcal{V}^*_{INT})$$

where $\mathcal{V}_{INT}$ is a model space for Q/G. For $G = \mathbf{SO}(3)$, $\mathcal{V}_{INT}$ *models the configuration space for the internal modes*, while $\mathfrak{g}/\mathfrak{g}_e \cong T_{\mu_e}\mathcal{O}_{\mu_e}$ *models the phase space for rigid modes.* Our goal is to realize this decomposition explicitly, in such a way that $\delta^2 H_\xi(z_e)$ block diagonalizes. The bundle $(T^*Q)_\mu \to T^*(Q/G)$ with fiber $\mathcal{O}_\mu$ also has a natural connection (Montgomery [1986]). Unfortunately, our decomposition is not simply the horizontal-vertical split for this connection. We shall need a construction which is somewhat more sophisticated, but is similar in spirit.

We will define two subspaces $\mathcal{S}_{\mathrm{RIG}}$ and $\mathcal{S}_{\mathrm{INT}}$ of $\mathcal{S}$ and further subspaces $\mathcal{W}_{\mathrm{INT}}$ (isomorphic with $\mathcal{V}_{\mathrm{INT}}$) and $\mathcal{W}^*_{\mathrm{INT}}$ (isomorphic with $\mathcal{V}^*_{\mathrm{INT}}$) of $\mathcal{S}_{\mathrm{INT}}$ such that

$$\mathcal{S} = \mathcal{S}_{\mathrm{RIG}} \oplus \mathcal{S}_{\mathrm{INT}} = \mathcal{S}_{\mathrm{RIG}} \oplus (\mathcal{W}_{\mathrm{INT}} \oplus \mathcal{W}^*_{\mathrm{INT}}), \tag{1}$$

relative to which $\delta^2 H_\xi(z_e)$ will be block diagonal. As above, the first component $\mathcal{S}_{\mathrm{RIG}} \cong \mathfrak{g}/\mathfrak{g}_{\mu_e}$ of $\mathcal{S}$ is isomorphic to the tangent space to the coadjoint orbit through μ_e. As we shall see, this component also carries the coadjoint orbit symplectic structure.

The first component $\mathcal{S}_{\mathrm{RIG}}$ will be defined in terms of rigid variations. This will be done by going back to Q temporarily, defining rigid variations there, and then using the Legendre transformation to transfer the information over to the cotangent bundle. To carry this out, let

$$\mathfrak{g}_Q = \{\eta_Q(q) \in TQ \mid \eta \in \mathfrak{g} \text{ and } q \in Q\} \tag{2}$$

and let $T\mathfrak{g}_Q \subset T(TQ)$ be its tangent bundle.

4.1 Definition *Let* $V_{\mathrm{RIG}} = s(T\mathfrak{g}_Q)$ *where* $s : T^2Q \to T^2Q$ *is the canonical involution. Alternatively,* V_{RIG} *consists of double tangents of curves denoted by* $\Delta\dot{q}$ *(identified with velocity variations of superposed rigid body motions in the case of* $\mathbf{SO}(3)$).

$$\Delta\dot{q} = \frac{d}{d\varepsilon}\bigg|_{\varepsilon=0} \frac{d}{dt}\bigg|_{t=0} \exp(\varepsilon\eta(t))\, q(t),$$

where $\eta(t)$ *is a curve in* $\mathfrak{g}$ *with* $\eta(0) = \eta$ *and* $q(t)$ *is a curve in* Q. (The canonical involution in effect swaps the order of differentiation.)

In coordinates, if we write elements of V_{RIG} as

$$(q^i, \dot{q}^i, \Delta q^i, \Delta\dot{q}^i), \tag{3a}$$

then we find that

$$\Delta q^i = A^i_a \eta^a \quad \text{and} \quad \Delta\dot{q}^i = \frac{\partial A^i_a}{\partial q^k}\, \dot{q}^k \eta^a + A^i_a \zeta^a. \tag{3b}$$

An intrinsic way of writing the split (3b) and hence the definition of V_{RIG} is the following:

$$V_{\mathrm{RIG}} = T_{v_e}(G \cdot v_e) \oplus \mathrm{vert}_{v_e}(\mathfrak{g}_Q) \tag{3c}$$

where $v_e = (q_e, \dot{q}_e)$ is the relative equilibrium in TQ. Here, $\mathbb{F}L(v_e) = z_e$, where $\mathbb{F}L : TQ \to T^*Q$ is the Legendre transform given by

$$p_i = g_{ij}\,\dot{q}^j \tag{4}$$

and $\mathrm{vert}_{v_e}(\mathfrak{g}_Q)$ denotes the vertical lift of the bundle $\mathfrak{g}_Q$ at the point v_e.

Next, let $T\mathbb{F}L : T(TQ) \to T(T^*Q)$ be the tangent map to the Legendre transformation (4), and set

$$\mathcal{S}_{RIG} = T\mathbb{F}L \cdot V_{RIG} \cap \mathcal{S}, \tag{5}$$

where $\mathcal{S}$ is defined by (12) of §3. If we let $\mathfrak{g}_{\mu_e}^{\perp}$ denote the (q-dependent) orthogonal complement of $\mathfrak{g}_{\mu_e}$ in the metric $\mathbb{I}_{ab}$, then one finds that $\mathcal{S}_{RIG}$ is parametrized by elements $\eta \in \mathfrak{g}_{\mu_e}^{\perp}$ as follows: we write elements of $\mathcal{S}_{RIG}$ as

$$(q^i, p_j, \Delta q^i, \Delta p_j), \tag{6a}$$

where

$$\Delta q^i = A_a^i\, \eta^a, \tag{6b}$$

and

$$\Delta p_i = -\frac{\partial A_a^k}{\partial q^i}\eta^a\, p_k + g_{ij}\, A_a^j\, \zeta^a, \tag{6c}$$

where $\eta \in \mathfrak{g}_{\mu_e}^{\perp}$ and $\zeta \in \mathfrak{g}$; the condition that (6a) belongs to $\ker(T_{z_e}\mathbf{J})$ is equivalent to the relation

$$\zeta^a = \mathbb{I}^{ab}(\mathrm{ad}_\eta^*\, \mu_e)_b \tag{7}$$

i.e., $\zeta^\flat = \mathrm{ad}_\eta^* \mu_e$, so ζ is determined by η. One checks that $\zeta \in \mathfrak{g}_{\mu_e}^{\perp}$ as well.

§5 *The Internal Vibration Space*

Now we define a complement to $\mathcal{S}_{RIG}$ in $\mathcal{S}$. We will do this by a constructive procedure that can be effectively carried out in examples. To define this complement, to be denoted $\mathcal{S}_{INT}$, we first describe $\mathcal{V}_{INT}$.

Recall that the amended potential V_ξ is given by

$$V_\xi = V + L_\xi , \tag{1a}$$

where

$$L_\xi(q) = -\tfrac{1}{2}\langle \xi_Q(q), \xi_Q(q)\rangle . \tag{1b}$$

For mechanical systems undergoing stationary rotations about ξ, i.e., $G = \mathbf{SO}(3)$ and G_{μ_e} = rotations about the axis μ_e, which is parallel to ξ, we note that L_ξ gives the potential of the centrifugal force. Now define $\mathcal{V}_{INT}$ as the subspace on which V_ξ or, equivalently, L_ξ ***looks objective*** in the sense of nonlinear elasticity (cf. Marsden and Hughes [1983]). More precisely:

5.1 Definition

$$\mathcal{V}_{INT} = \Big\{ \delta q \in T_{q_e}Q \mid \big\langle \delta q, (\pounds_{\eta_Q} \mathbf{d}L_\xi)(q)\big\rangle = 0 \quad \text{for all} \quad \eta \in \mathfrak{g}_{\mu_e}^{\perp} \text{ and } \langle \delta q, \chi_Q(q_e)\rangle = 0 \text{ for all } \chi \in \mathfrak{g}_{\mu_e} \Big\}, \tag{2}$$

where the first pairing is the natural pairing between vectors and one forms while the second is the metric inner product, and $\pounds$ *denotes the Lie derivative.*

Since V_ξ has a critical point at q_e (by the principal of symmetric criticality) and V is G-invariant, we find that

$$\big\langle \delta q, (\pounds_{\eta_Q} \mathbf{d}L_\xi)(q)\big\rangle = \delta^2 V_\xi(q_e)(\delta q, \eta_Q(q)) \tag{3}$$

and so we see that the geometric condition (2) is exactly what is needed to block diagonalize $\delta^2 V_\xi(q_e)$ within $\mathcal{S}$. In coordinates, the first condition on δq^i defining $\mathcal{V}_{INT}$ is the geometric condition

$$\delta q^i \eta^a \xi^b \xi^c \frac{\partial}{\partial q^i}\Big[A^k_a \frac{\partial}{\partial q^k}(A^\ell_b A^m_c g_{\ell m})\Big] = 0; \tag{2'}$$

the second condition is just the defining condition on $\mathcal{S}$. Notice that the space $\mathcal{V}_{INT}$ is a model space for the quotient Q/G. Now we are ready to define $\mathcal{S}_{INT}$.

5.2 Definition $\mathcal{S}_{INT} = \{\delta z \in T_{z_e}T^*Q \mid \delta q \in \mathcal{V}_{INT}$ *and* $\delta z \in \ker[T\mathbf{J}(z_e)]\} \subset \mathcal{S}$. (4)

Assuming that the quadratic form in η and $\overline{\eta}$ given by

$$A^i_d \eta^d \overline{\eta}^a \xi^b \xi^c \frac{\partial}{\partial q^i}\left[A^k_a \frac{\partial}{\partial q^k}(A^\ell_b A^m_c g_{\ell m})\right] \tag{2''}$$

is nondegenerate, we get the following result.

5.3 Proposition $\mathcal{S} = \mathcal{S}_{RIG} \oplus \mathcal{S}_{INT}$.

In fact, the condition of nondegeneracy of the form (2″) implies that $\mathcal{S}_{RIG} \cap \mathcal{S}_{INT} = \{0\}$, the spanning follows from the dimension count $\dim \mathcal{S}_{RIG} = \dim(\mathfrak{g}/\mathfrak{g}_\mu)$ and the fact that $\mathcal{S}_{INT}$ is determined by $\dim(\mathfrak{g}/\mathfrak{g}_\mu)$ equations.

As we shall see, the condition of nondegeneracy of (2″) is the same as the condition that the second variation of the Hamiltonian H_ξ restricted to $\mathcal{S}_{RIG}$ is nondegenerate. In fact, for stability, we want to assume that this is positive definite.

Now we want to insert the space $\mathcal{V}_{INT}$ into the space $\mathcal{S}_{INT}$. To do so, we shall use the condition $T\mathbf{J}(z_e) \cdot \delta z = 0$. This condition gives us a way of determining δp in terms of δq in such a way that the corresponding δz lies in the space $\ker T\mathbf{J}(z_e)$. This condition in coordinates is as follows:

$$\delta p_i A^i_a(q) + p_i \frac{\partial A^i_a}{\partial q^k} \delta q^k = 0. \tag{5a}$$

Now, let $\mathcal{W}^*_{INT}$ be defined as the set of vertical variations $(\delta q, \delta p) = (0, \delta p)$ which annihilate the infinitesimal action; i.e., $\delta p_i A^i_a(q) = 0$. Clearly, $\mathcal{W}^*_{INT} \subset \mathcal{S}_{INT}$. Next, let

$$\mathcal{T} = \{\delta p_i \in T^*_{q_e}Q \mid g^{ij} dp_i \pi_j = 0 \text{ for all } \pi_i \text{ such that } \pi_i A^i_a(q) = 0\}$$

$$= \{\delta p_i \in T^*_{q_e}Q \mid \delta p_i = g_{ij} A^i_b \chi^b \text{ for some } \chi \in \mathfrak{g}\}.$$

Now given $\delta q^i \in \mathcal{V}_{INT}$, we can *uniquely solve* (5a) for $\delta p_i \in \mathcal{T}$. In fact, we find that

$$\chi^b = -\mathbb{I}^{ba} p_i \frac{\partial A^i_a}{\partial q^k} \delta q^k. \tag{5b}$$

Using (5b), construct the corresponding pair $(\delta q^i, \delta p_i)$ (using the metric and vertical lift makes this intrinsic). This defines the space $\mathcal{W}_{INT}$. By construction, $\mathcal{W}_{INT} \subset \mathcal{S}_{INT}$ and $\mathcal{W}_{INT}$ is isomorphic to $\mathcal{V}_{INT}$. Thus, we have acheived the split

$$\mathcal{S}_{INT} = \mathcal{W}_{INT} \oplus \mathcal{W}^*_{INT}. \tag{6}$$

Remark This way of injecting $\mathcal{V}_{\rm INT}$ into $\ker T\mathbf{J}(z_e)$ is closely related to the map

$$\alpha_{\mu_e} : Q \to \mathbf{J}^{-1}(\mu_e);\ q \mapsto (\mathbb{I}(q)^{-1}\mu_e)_Q(q)^{\flat}$$

which occurs in the cotangent bundle reduction theorem (Smale [1970], Abraham and Marsden [1978], Kummer [1981]).

We remark here that even if G is abelian (for instance, $G = S^1$ in the case of planar coupled rigid bodies), the decompositions are not trivial; while $\mathcal{S}_{\rm RIG} = \{0\}$ in this case, $\mathcal{S}_{\rm INT} = \mathcal{W}_{\rm INT} \oplus \mathcal{W}^*_{\rm INT}$ is still not a trivial decomposition.

Next, we give a characterization of $\mathcal{V}_{\rm INT}$ in terms of superposed motions.

5.4 Proposition *Let $q_\varepsilon \in Q$ be a curve tangent to δq at q_e, let $\eta \in \mathfrak{g}^\perp_{\mu_e}$ and let $\eta_\varepsilon = \mathrm{Ad}_{\exp(\varepsilon\xi)}\eta$. Then $\mathcal{V}_{\rm INT}$ is characterized by those δq orthogonal to $T_{q_e}(G_{\mu_e} \cdot q_e)$ and satisfying*

$$\frac{d}{d\varepsilon}\left\langle \xi_Q(q_\varepsilon), (\eta_\varepsilon)_Q(q_\varepsilon)\right\rangle\Big|_{\varepsilon=0} = 0 \tag{7}$$

or, equivalently,

$$\frac{d}{d\varepsilon}\left\langle \xi_Q(\exp(\varepsilon\xi)q_\varepsilon)\,,\ \eta_Q(\exp(\varepsilon\xi)q_\varepsilon)\right\rangle\Big|_{\varepsilon=0} = 0\,. \tag{8}$$

This is verified by a direct coordinate calculation. We can lift this expression to get an alternative characterization of $\mathcal{S}_{\rm INT}$. We consider the momentum map $\mathbf{J}$ restricted to $\mathfrak{g}^\perp_{\mu_e}$ and regarded as a function on TQ. In other words, for $\zeta \in \mathfrak{g}^\perp_{\mu_e}$, set

$$J(\zeta)(\delta q) = \langle \zeta_Q(q), \delta q\rangle = g_{ij}\, A^i_a\, \zeta^a (\delta q)^j. \tag{9}$$

In what follows, we shall assume the following condition is satisfied, which is automatically true for $\mathbf{SO}(3)$: $\zeta \in \mathfrak{g}^\perp_{\mu_e} \Rightarrow [\xi, \zeta] \in \mathfrak{g}^\perp_{\mu_e}$. (This property is only needed for a few alternative formulas and will be investigated for general groups in a future publication.) Now consider the condition

$$\frac{d}{dt} J(\zeta)(\delta q)\Big|_{t=0} = 0, \tag{10}$$

where ζ is to evolve as $\dot\zeta = [\xi, \zeta]$ which is consistent with (8a) and $\zeta \in \mathfrak{g}^\perp_{\mu_e}$; here ξ is the Lie algebra element giving the relative equilibrium. Equation (10) defines a condition on T(TQ). We shall regard it as a condition on $T_{z_e}(T^*Q)$ via the Legendre transform. For simplicity we still write the resulting condition as $\dot{\mathbf{J}} = 0$.

5.5 Proposition

$$\mathcal{S}_{\mathrm{INT}} = \{\dot{\mathbf{J}}(z_e) = 0\} \cap \mathcal{S}. \tag{11}$$

Remark The split (6) appears to be not the same as, but related to the complement to the vertical space relative to a natural connection on the coadjoint orbit bundle $(T^*Q)_\mu \to T^*(Q/G)$. In this regard we note that the metric naturally induced on Q/G is Wilson's G-matrix (see Wilson, Decius and Cross [1955]). Our decomposition appears to be finer than the one proposed by Guichardet [1984] and discussed by Iwai [1988]. Notice that we have connections on all levels of this *tower of bundles*

$$T^*Q \supset \mathbf{J}^{-1}(\mu) \to (T^*Q)_\mu \to T^*(Q/G)$$

where $\mathbf{J}^{-1}(\mu) \to (T^*Q)_\mu$ is regarded as a G_μ bundle and $(T^*Q)_\mu \to T^*(Q/G)$ is regarded as an $\mathcal{O}_\mu$ bundle, where $\mathcal{O}_\mu$ is the coadjoint orbit through μ.

The Guichardet-Iwai results appear to be largely concerned with the bundle $\mathbf{J}^{-1}(\mu) \to (T^*Q)_\mu$; the fact that the reduced space $(T^*Q)_\mu$ still has the factor $\mathcal{O}_\mu$ seems to be the reason the connection on the G_μ bundle $\mathbf{J}^{-1}(\mu) \to (T^*Q)_\mu$ is not sufficient to completely isolate the vibrational modes from the rotational ones. We believe that the $\mathcal{O}_\mu$ bundle fills this gap.

§ 6 *Block Diagonalization*

Now $H_\xi = K_\xi + V_\xi + \langle \mu_e, \xi \rangle$ and we have arranged for V_ξ to be block diagonal. As far as K_ξ is concerned, we compute in coordinates that

$$K_\xi = \frac{1}{2} g^{ij}(p_i - g_{ik}A^k_a\xi^a)(p_j - g_{jm}A^m_b\xi^b). \tag{1a}$$

Thus, since $p_i = g_{ik} A^k_a \xi^a$ at equilibrium, we get

$$\delta^2 K_\xi(z_e) \cdot (\delta z, \delta \overline{z}) = g^{ij}\, \delta p_i\, \delta \overline{p}_j\,. \tag{1b}$$

Regarding the block diagonalization of $\delta^2 K_\xi$ on $\mathcal{S}_{\mathrm{RIG}} \oplus \mathcal{S}_{\mathrm{INT}}$, we shall use some further interesting identities.

The block diagonalization results for $\delta^2 H_\xi$ follow from two basic formulas:

6.1 Proposition *Let* $\Delta z \in \mathcal{S}_{RIG}$ *and* $\delta z \in T_{z_e}P$. *Then*

$$\delta^2 H_\xi(z_e)(\Delta z, \delta z) = \frac{d}{dt}\langle \zeta_Q(q), \delta q\rangle - \langle [\xi, \eta], \delta \mathbf{J}(z_e) \cdot \delta z\rangle , \tag{2a}$$

where Δz *has associated* η *and* ζ *as in* (3b) *and* (7) *of* §**4**.

6.2 Proposition *Let* δz_1 *and* $\delta z_2 \in \mathcal{S}_{INT}$; *then*

$$\delta^2 H_\xi(z_e)(\delta z_1, \delta z_2) = \delta^2 K_\xi(z_e) \cdot (\delta z_1, \delta z_2) + \delta^2 V_\xi(q_e)(\delta q_1, \delta q_2) \tag{2b}$$

Proposition **6.1**, which is proved by direct calculation, shows that $\delta^2 H_\xi(z_e)$ *block diagonalizes* on $\mathcal{S}_{RIG} \oplus \mathcal{S}_{INT}$, i.e., if $\Delta z \in \mathcal{S}_{RIG}$ and $\delta z \in \mathcal{S}_{INT}$, then

$$\delta^2 H_\xi(z_e)(\Delta z, \delta z) = 0 . \tag{3}$$

Proposition **6.2** then follows from our earlier calculations. It also follows that if $\Delta z \in \mathcal{S}_{RIG}$ and $\Delta \bar{z} \in \mathcal{S}_{RIG}$, then

$$\delta^2 H_\xi(z_e)(\Delta z, \Delta \bar{z}) = \frac{d}{dt}\langle \zeta_Q(q), \bar{\zeta}_Q(q)\rangle \tag{4}$$

which is a generalization of the rigid body second variation formula for motion on the coadjoint orbit $\mathcal{O}_{\mu_e}$ with the metric $\mathbb{I}_{ab}$. (Recall that ζ are determined by η and μ_e by equation (7) of §**4**.) We summarize:

6.3 Theorem *The relative equilibrium* z_e *is formally stable* (*with* $\delta^2 H_\xi(z_e)$ *on* $\mathcal{S}$ *positive definite*) *iff*

i $\frac{d}{dt}\langle \zeta_Q(q), \bar{\zeta}_Q(q)\rangle$ *is positive definite on* $\mathcal{S}_{RIG}$

and **ii** $\delta^2 H_\xi(q_e)$ *is positive definite on* $\mathcal{W}_{INT}$.

A ***sufficient*** *condition for* **ii** *is*

ii* $\delta^2 V_\xi(q_e)$ *is positive definite on* $\mathcal{V}_{INT}$.

We note that condition **i** is sufficient for the nondegeneracy condition (2″) of §**5**.

We note that $\delta^2 V_\xi(q_e)$ separates (in coordinates on $\mathcal{V}_{INT}$) into $\delta^2 V(q_e)$ plus a term which is quadratic in ξ. Thus, **ii** is equivalent to a condition of the form $\| \xi \| \leq \sqrt{\lambda_{min}}$, where $\| \ \|$ is a suitable norm and λ_{min} is the minimum (non-zero) eigenvalue of $\delta^2 V(q_e)$; one has to take care here because V itself need not have a critical point at q_e, so $\delta^2 V(q_e)$ does *not* make intrinsic sense. To see how this works in examples, see Simo, Posbergh and Marsden [1989] and Lewis and Simo [1989].

Interestingly, the effective potential V_μ which was defined earlier, is such that *the conditions in the preceeding theorem become sharp*; that is, they are not only sufficient for stability, they are necessary as well. This point is important for bifurcation analysis, and will be the subject of future investigations.

As far as the symplectic form Ω is concerned, we have

6.4 Theorem *Let* $\Delta z \in \mathcal{S}_{\mathrm{RIG}}$ *and* $\delta z \in T_{z_e}P$. *Then*

$$\Omega(z_e)(\Delta z, \delta z) = -\langle \eta, \delta J(z_e) \cdot \delta z \rangle + \langle \zeta_Q(q_e), \delta q \rangle. \tag{5}$$

Notice that in an appropriate sense, $\delta^2 H_\xi(z_e)$ on $\mathcal{S}_{\mathrm{RIG}} \times T_{z_e}P$ is the time derivative of the symplectic form Ω!

From (5) and (7) of §4 one finds that on $\mathcal{S}_{\mathrm{RIG}} \times \mathcal{S}_{\mathrm{RIG}}$, Ω gives the coadjoint orbit symplectic form

$$\Omega(z_e)(\Delta z, \Delta \bar{z}) = -\langle \mu_e, [\eta, \bar{\eta}] \rangle, \tag{6}$$

while on $\mathcal{S}_{\mathrm{RIG}} \times \mathcal{S}_{\mathrm{INT}}$ we have the cross terms

$$\Omega(z_e)(\Delta z, \delta z) = \langle \zeta_Q(q_e), \delta q \rangle, \tag{7}$$

which depend on the δq components alone. In summary, we have

$$\begin{array}{cccc} & \mathfrak{g}/\mathfrak{g}_{\mu_e} & \mathcal{W}_{\mathrm{INT}} & \mathcal{W}^*_{\mathrm{INT}} \\ \delta^2 H_\xi(z_e) = & \left[\begin{array}{c} \left[\begin{array}{c}\text{Generalized}\\ \text{Rigid Body}\\ \text{Second Variation}\end{array}\right] \\ \begin{array}{c}0\\0\end{array} \end{array}\right. & \begin{array}{c} 0 \\ \left[\begin{array}{cc} \delta^2 H_\xi(z_e) & 0 \\ 0 & \delta^2 K_\xi(z_e) \end{array}\right. \end{array} & \left.\begin{array}{c} 0 \\ \end{array}\right]\left.\begin{array}{c} \\ \end{array}\right] \end{array}$$

for the second variation of the augmented energy. In this matrix, note that $\delta^2 H_\xi(z_e)$ is given by the expression (2b) and so positive definiteness of $\delta^2 V_\xi(q_e)$ implies postive definiteness of $\delta^2 H_\xi(z_e)$ and hence, assuming the rigid body second variation is positive (the rotation is about the shortest axis), positive definiteness of the entire matrix. As far as the symplectic form is concerned, we have the form

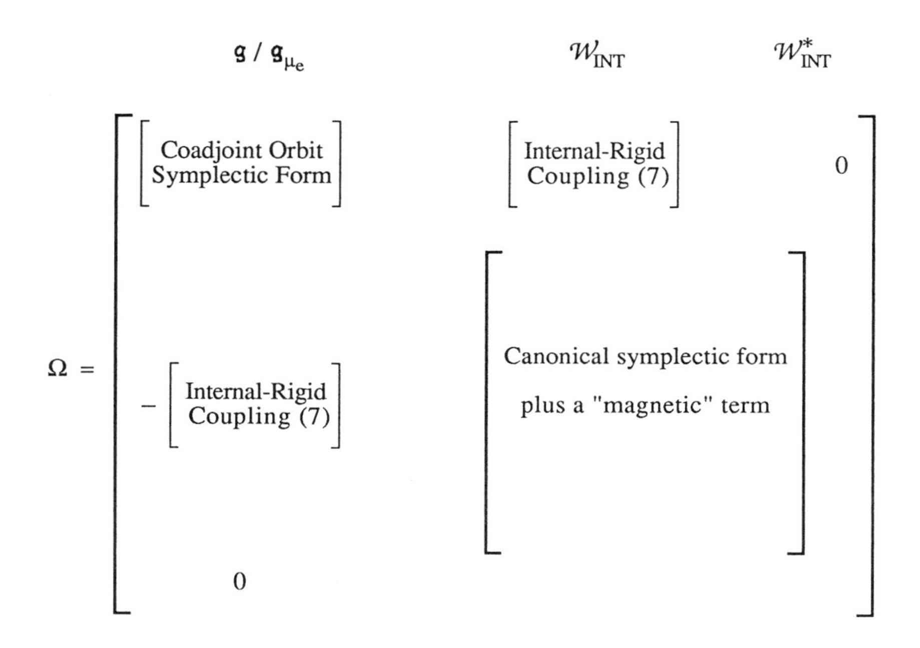

For information on the "magnetic" term, and its interpretation as a curvature, we refer the reader to Kummer [1981]. Also, we presume that the coupling terms can be interpreted in terms of the curvature of the connection on the coadjoint bundle $T^*Q \to T^*(Q/G)$ that we used to define the splitting of $\mathcal{S}_{INT}$; see Montgomery [1986] and Lewis, Marsden, Montgomery and Ratiu [1986].

References

R. Abraham and J. Marsden [1978] *Foundations of Mechanics*. Second Edition, Addison-Wesley Publishing Co., Reading, Mass.

V.I. Arnol'd [1966] Sur la géometrie differentielle des groupes de Lie de dimenson infinie et ses applications a L'hydrodynamique des fluids parfaits. *Ann. Inst. Fourier. Grenoble* **16**, 319-361.

C. Eckart [1935] Some studies concerning rotating axes and polyatomic molecules. *Phys. Rev.***47**, 552-558.

A. Guichardet [1984] On rotation and vibration motions of molecules, *Ann. Inst. H. Poincaré* **40**, 329-342.

D.D. Holm, J.E. Marsden, T. Ratiu and A. Weinstein [1985] Nonlinear stability of fluid and plasma equilibria. *Physics Reports* **123**, 1-116.

T. Iwai [1988] A geometric setting for classical molecular dynamics. *Ann. Inst. H. Poincaré* (to appear).

J. Jellinek and D.H. Li [1989] Separation of the energy of overall rotation in any N-body system. *Phys. Rev. Lett.* **62**, 241-244.

M. Kummer [1981] On the construction of the reduced phase space of a Hamiltonian system with symmetry. *Indiana Univ. Math. J.* **30**, 281-291.

D. Lewis and J.C. Simo [1989] Nonlinear stability of rotating pseudo-rigid bodies (preprint).

D.Lewis, J.E. Marsden, R. Montgomery and T. Ratiu [1986] The Hamiltonian structure for dynamic free boundary problems. *Physica* **18D**, 391-404.

J.E. Marsden [1981] *Lectures on geometric methods in mathematical physics. SIAM, CBMS Conf. Series,* **37**.

J.E. Marsden and T.J.R. Hughes [1983] *Mathematical Foundations of Elasticity,* Prentice Hall.

J.E. Marsden, T. Ratiu, R. Schmid, R.G. Spencer and A. Weinstein [1983] Hamiltonian systems with symmetry, coadjoint orbits and plasma physics. *Proc. IUTAM-ISIMM Symposium on "Modern Developments in Analytical Mechanics"* Torino, June 7-11, 1982, *Atti della Academia della Scienze di Torino* **117**, 289-340.

J.E. Marsden, R. Montgomery and T. Ratiu [1989] Reduction, symmetry, and Berry's phase in mechanics (preprint); also, see their article in these proceedings.

J.E. Marsden and A. Weinstein [1974] Reduction of symplectic manifolds with symmetry. *Rep. Math. Phys.* **5**, 121-130.

J.E. Marsden, A. Weinstein, T. Ratiu, R. Schmid, and R.G. Spencer [1983] Hamiltonian systems with symmetry, coadjoint orbits and plasma physics. *Proc. IUTAM-IS1MM Symposium on "Modern Developments in Analytical Mechanics,"* Torino, June 7-11, 1982, *Atti della Academia della Scienze di Torino* **117**, 289-340.

R. Montgomery [1986] *The bundle picture in mechanics*. Thesis, U.C. Berkeley.

R.S. Palais [1979] The principle of symmetric criticality, *Comm. Math. Phys.* **69**, 19-30.

J.C. Simo, T.A. Posbergh and J.E. Marsden [1989] Nonlinear stability of elasticity and geometrically exact rods by the energy-momentum method (preprint).

S. Smale [1970] Topology and Mechanics. *Inv. Math.* **10**, 305-331, **11**, 45-64.

A. Weinstein [1984] Stability of Poisson-Hamilton equilibria. *Cont. Math. AMS* **28**, 3-14.

E.B. Wilson, J.C. Decius and P.C. Cross [1955] *Molecular vibrations*. McGraw Hill (reprinted by Dover).

Jerrold E. Marsden
Department of Mathematics
University of California
Berkeley, California 94720

Juan C. Simo, Thomas A. Posbergh
Division of Applied Mechanics
Stanford University
Stanford, CA 94305.

Debra Lewis
Theoretical & Applied Mech.
Cornell University
Ithaca New York 14853

Contemporary Mathematics
Volume **97**, 1989

The Dynamics of Two Coupled Rigid Bodies in Three Space

George W. Patrick

Abstract

The dynamics of two rigid bodies coupled by an ideal spherical joint is studied. Attention is restricted to the case of two identical bodies with two equal moments of inertia. The system admits the symmetry group $S^1 \times S^1 \times SO(3)$; all relative equilibria are explicitly computed. Minor modifications of the energy-Casimir method are used to obtain stability criteria.

1 Notation

Let G be a Lie group with Lie algebra $\mathfrak{g}$. Write left multiplication by $g \in G$ as L_g. If Φ is a left action of G on a manifold M, the infinitesimal generator of Φ corresponding to ξ is

$$\operatorname{ign}(\Phi, \xi)(m) = \left.\frac{d}{dt}\right|_{t=0} \Phi(\exp(\xi t), m) = T\Phi_m \xi, \quad m \in M \ .$$

In what follows, differential calculus on Lie groups will be necessary, and it will be convenient to take advantage of the trivializations $TG \cong \mathfrak{g} \times G$ and $T^*G \cong \mathfrak{g}^* \times G$ by left translation. Suppose a basis $\{\xi_i, i = 1, \ldots, n\}$ of $\mathfrak{g}$ is specified. Define the maps

$$v_g \in T_gG \mapsto {v_g}^{\cup} \in R^n \ : \qquad v_g = TL_g \sum_{i=1}^{n} ({v_g}^{\cup})^i \xi_i \ ,$$

$$(g, a) \in G \times R^n \mapsto (g, a)^{\cap} \in T_gG \ : \qquad (g, a)^{\cap} = TL_g \sum_{i=1}^{n} a^i \xi_i \ .$$

Mathematics subject classification (1985 revision) 70E99 (58f05, 58f10).

In the second map, if the first coordinate g is omitted, it is to be assumed the identity of G. Thus, $\mathfrak{g}$ is identified with R^n by the linear isomorphism $\xi \mapsto \xi^{\cup}$ and its inverse $a \mapsto a^{\cap}$. The cotangent bundle T^*G is similarly trivialized with the dual basis. Another general notation: for a function $f : G \to R$ define

$$\nabla f(g) = (\boldsymbol{d}f(g))^{\cup} = \left.\frac{d}{dt}\right|_{t=0} \begin{pmatrix} f\left(g\exp(t\xi_1)\right) \\ \vdots \\ f\left(g\exp(t\xi_n)\right) \end{pmatrix} ,$$

so that if $v_g \in T_gG$,

$$\langle \boldsymbol{d}f(g), v_g\rangle = \nabla f(g) \cdot v_g^{\cup} .$$

For example, consider the Lie group $SO(3)$; if $a \in R^3$, define the maps

$$a^{\wedge} = \begin{pmatrix} 0 & -a^3 & a^2 \\ a^3 & 0 & -a^1 \\ -a^2 & a^1 & 0 \end{pmatrix} , \qquad \begin{pmatrix} 0 & -a^3 & a^2 \\ a^3 & 0 & -a^1 \\ -a^2 & a^1 & 0 \end{pmatrix}^{\vee} = a ,$$

and note the useful identities

$$\begin{aligned} (a^{\wedge}b^{\wedge} - b^{\wedge}a^{\wedge})^{\vee} &= a^{\wedge}b = -b^{\wedge}a = a \times b , \\ Ba^{\wedge}B^{-1} &= (Ba)^{\wedge}, \quad B \in SO(3) . \end{aligned}$$

The Lie algebra of $SO(3)$ is $\mathfrak{so}(3)$, the pairs $(\mathrm{Id}, \dot{A})$, where $\dot{A}$ is antisymmetric, with Lie bracket $[(\mathrm{Id}, \dot{A}_1), (\mathrm{Id}, \dot{A}_2)] = (\mathrm{Id}, \dot{A}_1\dot{A}_2 - \dot{A}_2\dot{A}_1)$. Thus, if $\{\boldsymbol{i}, \boldsymbol{j}, \boldsymbol{k}\}$ is the usual basis of R^3; $\{(\mathrm{Id}, \boldsymbol{i}^{\wedge}), (\mathrm{Id}, \boldsymbol{j}^{\wedge}), (\mathrm{Id}, \boldsymbol{k}^{\wedge})\}$ is a basis for $\mathfrak{so}(3)$, and one easily computes

$$\begin{aligned} (A, \dot{A})^{\cup} &= (A^{-1}\dot{A})^{\vee} \\ (A, \Omega)^{\cap} &= (A, A\Omega^{\wedge}) \end{aligned}$$

Yet another example is R^3 with the usual basis $\{\boldsymbol{i}, \boldsymbol{j}, \boldsymbol{k}\}$, and then ∇ is the usual gradient of functions. A product of Lie groups with specified bases has a basis with the obvious ordering; the maps $^{\cup}$ and $^{\cap}$ above will refer to this basis without mention.

2 Kinematics

This section recalls the description of the free motion of two rigid bodies coupled by an ideal spherical joint derived in [4]. Fix an inertial frame, and let ρ_1 and ρ_2 be two distributions on R^3 of order zero and compact support representing the densities of the two bodies in some reference configuration; without loss of generality, assume that the spherical joint in this configuration is located at the origin. Take as configuration space the manifold $SO(3)^2 \times R^3$: a point $X \in R^3$ on body $i \in \{1,2\}$ in the reference configuration has position $A_i X + w$ in the configuration (A_1, A_2, w).

Let m_i be the mass and S_i be the center of mass of body i in the reference configuration:

$$m_i = \int \rho_i(X)\, dX \,,$$
$$S_i = \frac{1}{m_i} \int X \rho_i(X)\, dX \,,$$

and let $m = m_1 + m_2$ be the total mass of the system. Also, let I_i be the coefficient of inertia matrix of body i in the reference configuration, but with respect to the center of mass, instead of the origin:

$$I_i = \int (X - S_i)(X - S_i)^t \rho_i(X)\, dX \,.$$

The tangent vector $(A_i, w, \dot{A}_i, \dot{w})$ describes a state with kinetic energy

$$\begin{aligned} KE &= \frac{1}{2}\mathrm{trace}\left(\dot{A}_1 I_1 \dot{A}_1^t\right) + \frac{1}{2}\mathrm{trace}\left(\dot{A}_2 I_2 \dot{A}_2^t\right) + \frac{m_1}{2}|\dot{A}_1 S_1|^2 + \frac{m_2}{2}|\dot{A}_2 S_2|^2 \\ &\quad + \left(m_1 \dot{A}_1 S_1 + m_2 \dot{A}_2 S_2\right) \cdot \dot{w} + \frac{m}{2}|\dot{w}|^2 \,. \end{aligned}$$

This function is also the total energy E and the lagrangian L, since the kinetic energy is manifestly quadratic in the velocities and the potential energy is zero.

The group R^3 acts on configuration space by addition to the variable w, and this action lifts to an action on velocity phase space (also addition to w) under which the total energy E is invariant. Standard symplectic reduction applies, and in this case offers no difficulty. Equivalently, one may assume that the total linear momentum is zero, and then use the fact that the center of mass of the system is stationary to eliminate w and $\dot{w}$. The result is the mechanical system with configuration space $Q = SO(3)^2$ and lagrangian

$$\begin{aligned} L &= \frac{1}{2}\text{trace}\left(\dot A_1 I_1 \dot A_1^t\right) + \frac{1}{2}\text{trace}\left(\dot A_2 I_2 \dot A_2^t\right) + \frac{m_1}{2}|\dot A_1 S_1|^2 + \frac{m_2}{2}|\dot A_2 S_2|^2 \\ &\quad - \frac{1}{2m}|m_1 \dot A_1 S_1 + m_2 \dot A_2 S_2|^2 \\ &= \frac{1}{2}\text{trace}\left(\dot A_1 I_1 \dot A_1^t\right) + \frac{1}{2}\text{trace}\left(\dot A_2 I_2 \dot A_2^t\right) + \frac{m_1 m_2}{2m}|\dot A_1 S_1 - \dot A_2 S_2|^2 . \end{aligned}$$

Consider the action of $SO(3)$ on the configuration space by diagonal left translation: if $B \in SO(3)$ then

$$B \cdot (A_1, A_2) = (BA_1, BA_2) .$$

This action lifts to an action on momentum phase space and also to an action on velocity phase space under which the lagrangian L is invariant. For these lifted actions, one has the quotient maps

$$\begin{aligned} \tau : TQ \to (R^3)^2 \times SO(3) : \quad & (A_i, \dot A_i) \mapsto \left((A_i{}^{-1}\dot A_i)^\vee, A_1{}^{-1}A_2\right) , \\ \acute\tau : T^*Q \to (R^3)^2 \times SO(3) : \quad & (\mu_g, \nu_h) \mapsto (\mu_g{}^\cup, \nu_h{}^\cup, g^{-1}h) . \end{aligned}$$

The Legendre transformation FL is equivariant and hence induces a map $\bar{FL} : (R^3)^2 \times SO(3) \to (R^3)^2 \times SO(3)$; FL is hyperregular ([1]) if and only if $\bar{FL}$ is a diffeomorphism. In this case, the hamiltonian H is defined, and there is the following diagram of functions and Poisson maps

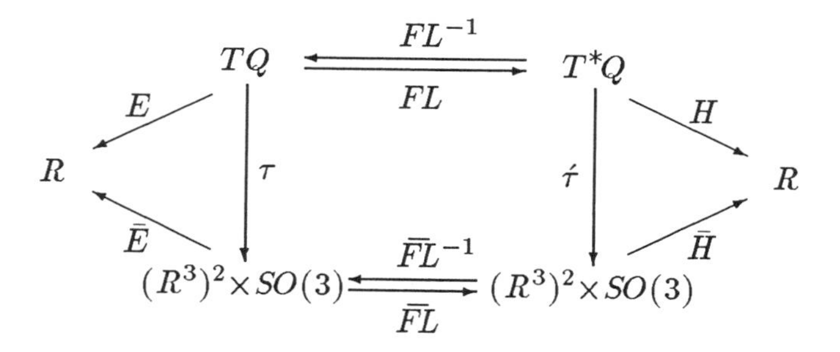

Let $J_i = \text{trace}(I_i)\text{Id} - I_i$ be the moment of inertia matrices, and $J(A)$ be the $SO(3)$ dependent matrix

$$J(A) = \begin{pmatrix} J_1 - \dfrac{m_1 m_2}{m}(S_1{}^\wedge)^2 & \dfrac{m_1 m_2}{m} S_1{}^\wedge A S_2{}^\wedge \\ \dfrac{m_1 m_2}{m} S_2{}^\wedge A^t S_1{}^\wedge & J_2 - \dfrac{m_1 m_2}{m}(S_2{}^\wedge)^2 \end{pmatrix} .$$

Then the map $\bar{FL}$ and the function $\bar{E}$ are given by

$$\bar{FL}(\Omega_1, \Omega_2, A) = \left(J(A) \begin{pmatrix} \Omega_1 \\ \Omega_2 \end{pmatrix}, A \right),$$

$$\bar{E} = \frac{1}{2} \begin{pmatrix} \Omega_1{}^t & \Omega_2{}^t \end{pmatrix} J(A) \begin{pmatrix} \Omega_1 \\ \Omega_2 \end{pmatrix},$$

so FL is hyperregular if and only if $J(A)$ is invertible. Assuming this, the function $\bar{H}$ is

$$\bar{H}(\pi_1, \pi_2, A) = \frac{1}{2} \begin{pmatrix} \pi_1{}^t & \pi_2{}^t \end{pmatrix} J(A)^{-1} \begin{pmatrix} \pi_1 \\ \pi_2 \end{pmatrix}.$$

The induced Poisson bracket on the quotient T^*Q may be written

$$\{f, h\}(p) = \nabla f(p)^t \, \Pi(\pi_1, \pi_2, A) \, \nabla h(p),$$

where Π is the 9×9 matrix

$$\Pi(\pi_1, \pi_2, A) = \begin{pmatrix} \pi_1{}^\wedge & 0 & A \\ 0 & \pi_2{}^\wedge & -\mathrm{Id} \\ -A^t & \mathrm{Id} & 0 \end{pmatrix}.$$

Finally, a Casimir for this Poisson bracket is the function $|\pi_1 + A\pi_2|^2$. By direct examination of the rank of Π, the symplectic leaves are the constant level sets of this single Casimir. By an obvious linear transformation, a nonzero level set is diffeomorphic to $R^3 \times S^2 \times SO(3)$, while the zero level is diffeomorphic to $R^3 \times SO(3)$.

Even after the reduction by spatial rotations, the nine dimensional, twenty parameter Poisson system of two coupled rigid bodies is complex. The remainder of this article deals with the following simpler special case:

- $I_1 = I_2 = I$, I is diagonal, and $I^{11} = I^{22}$.
- $S_1 = S_2 = S = |S|\boldsymbol{k}$.

The resulting system admits material symmetries: the lagrangian is invariant under the lift of the action of $S^1 \times S^1$ on configuration space by

$$(\theta_1, \theta_2) \cdot (A_1, A_2) = (A_1 \exp(-\theta_1 \boldsymbol{k}^\wedge), A_2 \exp(-\theta_2 \boldsymbol{k}^\wedge)).$$

This action commutes with the action of diagonal left translation. Thus, there are induced actions Ψ and $\acute{\Psi}$ of $S^1 \times S^1$ on the reduced phase spaces such that the quotient maps τ and $\acute{\tau}$ are equivariant:

$$\begin{aligned}
\Psi : \quad & (\theta_1, \theta_2) \cdot (\Omega_i, A) = (\exp(\theta_i \boldsymbol{k}^\wedge)\Omega_i, \exp(\theta_1 \boldsymbol{k}^\wedge) A \exp(-\theta_2 \boldsymbol{k}^\wedge)) , \\
\acute{\Psi} : \quad & (\theta_1, \theta_2) \cdot (\pi_i, A) = (\exp(\theta_i \boldsymbol{k}^\wedge)\pi_i, \exp(\theta_1 \boldsymbol{k}^\wedge) A \exp(-\theta_2 \boldsymbol{k}^\wedge)) .
\end{aligned}$$

Before proceeding with any computations, it is convenient to collect the parameters in the lagrangian. Temporarily write $\epsilon = |S|^2 m_1 m_2 / m$; then with the simplifying assumptions of this section

$$J(A) = \frac{1}{I^{11} + I^{33} + \epsilon} \begin{pmatrix} 1 & 0 & 0 & -\beta A^{22} & \beta A^{21} & 0 \\ 0 & 1 & 0 & \beta A^{12} & -\beta A^{11} & 0 \\ 0 & 0 & \alpha & 0 & 0 & 0 \\ -\beta A^{22} & \beta A^{12} & 0 & 1 & 0 & 0 \\ \beta A^{21} & -\beta A^{11} & 0 & 0 & 1 & 0 \\ 0 & 0 & 0 & 0 & 0 & \alpha \end{pmatrix} ,$$

where

$$\begin{aligned}
(1) \qquad \alpha &= \frac{2I^{11}}{I^{11} + I^{33} + \epsilon} , \\
\beta &= \frac{\epsilon}{I^{11} + I^{33} + \epsilon} .
\end{aligned}$$

The nonzero multiplier of $J(A)$ merely reparametrizes evolution curves, and so may be discarded for the purpose of analyzing the dynamics.

The equations (1) map the set $I^{11} \geq 0$, $I^{33} \geq 0$, $\epsilon \geq 0$ onto the triangular set

$$PS = \{(\alpha, \beta); \alpha \geq 0,\ \beta \geq 0,\ \alpha + 2\beta \leq 2\} ,$$

and in addition to these constraints on α and β, the determinant of the matrix $J(A)$ is

$$\det(J(A)) = \alpha^2 (1 - \beta^2)(1 - \beta^2 (A^{33})^2) ,$$

so FL is a diffeomorphism only for $\alpha \neq 0$ and $\beta \neq 1$.

Henceforth, then,

$$J(A) = \begin{pmatrix} 1 & 0 & 0 & -\beta A^{22} & \beta A^{21} & 0 \\ 0 & 1 & 0 & \beta A^{12} & -\beta A^{11} & 0 \\ 0 & 0 & \alpha & 0 & 0 & 0 \\ -\beta A^{22} & \beta A^{12} & 0 & 1 & 0 & 0 \\ \beta A^{21} & -\beta A^{11} & 0 & 0 & 1 & 0 \\ 0 & 0 & 0 & 0 & 0 & \alpha \end{pmatrix},$$

and

$$\alpha > 0, \quad \beta \geq 0, \quad \alpha + 2\beta \leq 2 .$$

Some intuition on the meaning of α and β may be obtained by examining the boundary of the set PS:

- $\alpha = 0$, $\beta = 1$. Then $I^{11} = I^{33} = 0$; two point masses coupled by massless rods.
- $\alpha = 0$, $\beta \neq 1$. Then $I^{11} = 0$ while $I^{33} \neq 0$; two ideal rods.
- $\alpha + 2\beta = 2$, $\alpha \neq 0$. Then $I^{33} = 0$ while $I^{11} \neq 0$; two ideal disks coupled with massless rods through the centers of the disks.
- $\beta = 0$. Then $S = 0$; the spherical joint lies at the center of mass, so no torque is exerted by the joint, and the bodies are uncoupled.

With the simplification of axial symmetry, the matrices involved are sparse enough that manual computations may be feasible, but instead, use has been made of the MAPLE symbolic manipulator ([3]). What follows is a description of how the computations are done and the results. To avoid indices upon indices in large expressions, write

$$\pi_1 = \begin{pmatrix} x_1 \\ x_2 \\ x_3 \end{pmatrix}, \quad \pi_2 = \begin{pmatrix} y_1 \\ y_2 \\ y_3 \end{pmatrix}, \quad \Omega_1 = \begin{pmatrix} v_1 \\ v_2 \\ v_3 \end{pmatrix}, \quad \Omega_2 = \begin{pmatrix} w_1 \\ w_2 \\ w_3 \end{pmatrix}.$$

3 Relative Equilibria

Although it would be sensible to compute the Poisson reduced phase spaces $(R^3)^2\times SO(3)/S^1\times S^1$, such a direct approach is difficult due to the lack of a simple realization of the quotient. Even besides the issue of global topology, the presence of points with various isotropy indicates that the quotient is not a manifold. The $S^1\times S^1$ symmetry simplifies computations on $(R^3)^2\times SO(3)$, however: one has only to calculate the various values for one point on each $S^1\times S^1$ orbit; the values at other points are available by $S^1\times S^1$ translation. Now the subset $M = (R^3)^2\times\{\exp(\theta \boldsymbol{j}^\wedge); \theta \in [0,\pi]\}$ of $(R^3)^2\times SO(3)$ intersects each orbit at least once; this is just the assertion that the map $R^3 \to SO(3)$ by the Euler angles is onto. Consequently, objects computed below are computed only at points of M; the value at (Ω_i,θ) or (π_i,θ) is to be regarded as the value at $(\Omega_i, \exp(\theta\boldsymbol{j}^\wedge))$ or $(\pi_i, \exp(\theta\boldsymbol{j}^\wedge))$, respectively.

First, the computation of the hamiltonian vector field $\mathrm{hvf}(\bar{H})$: the first 6 components of $\nabla\bar{H}$ are, of course,

$$\begin{pmatrix} \nabla_1\bar{H} \\ \nabla_2\bar{H} \end{pmatrix} = J(\exp(\theta\boldsymbol{j}^\wedge))^{-1}\begin{pmatrix} \pi_1 \\ \pi_2 \end{pmatrix},$$

and the last three components are computed as follows: if $a \in R^3$,

$$\begin{aligned} \nabla_3\bar{H}\cdot a &= \frac{1}{2}\frac{d}{dt}\bigg|_{t=0}\begin{pmatrix} {\pi_1}^t & {\pi_2}^t \end{pmatrix} J\left(\exp(\theta\boldsymbol{j}^\wedge)\exp(ta^\wedge)\right)^{-1}\begin{pmatrix} \pi_1 \\ \pi_2 \end{pmatrix} \\ &= -\frac{1}{2}\begin{pmatrix} {\pi_1}^t & {\pi_2}^t \end{pmatrix} J\left(\exp(\theta\boldsymbol{j}^\wedge)\right)^{-1} \\ &\qquad \left[\frac{d}{dt}\bigg|_{t=0} J\left(\exp(\theta\boldsymbol{j}^\wedge)\exp(ta^\wedge)\right)\right] J\left(\exp(\theta\boldsymbol{j}^\wedge)\right)^{-1}\begin{pmatrix} \pi_1 \\ \pi_2 \end{pmatrix}. \end{aligned}$$

The inner derivative is computed by noticing that $J(A)$ is a constant diagonal matrix plus a matrix which is linear in A; thus, the derivative is found by replacing the diagonal elements by 0 and the matrix A by $\exp(\theta\boldsymbol{j}^\wedge)a^\wedge$. Letting $a = \boldsymbol{i},\boldsymbol{j},\boldsymbol{k}$ consecutively then computes $\nabla_3\bar{H}$; the hamiltonian vector field is found by $\mathrm{hvf}(\bar{H}) = \Pi\,\nabla\bar{H}$. This (Ω_i,θ) dependent element of R^9 is shown in figure (1). The vector field $\mathrm{hvf}(\bar{E})$ is computed by noting that $\mathrm{hvf}(\bar{E}) = T\bar{FL}^{-1}\circ\mathrm{hvf}(\bar{H})\circ\bar{FL}$; for $z_1, z_2, z_3 \in R^3$,

$$
\begin{pmatrix}
\dfrac{\begin{aligned}&-\alpha\beta^2\sin\theta\cos\theta\,x_1x_2-\alpha\beta\sin\theta\,x_1y_2\\ &\quad-(1-\beta^2)(\alpha+\beta^2\cos^2\theta-1)x_2x_3-\alpha\beta^3\sin\theta\cos\theta\,x_2y_1\\ &\quad-\alpha\beta(1-\beta^2)\cos\theta\,x_3y_2-\alpha\beta^2\sin\theta\,y_1y_2\end{aligned}}{\alpha(1-\beta^2)(1-\beta^2\cos^2\theta)} \\[2ex]
\dfrac{\begin{aligned}&(\beta^2+\alpha-1)(1-\beta^2\cos^2\theta)^2x_1x_3-\alpha\beta^2(1-\beta^2)\sin\theta\cos\theta\,x_2^2\\ &\quad-\alpha\beta(1-\beta^2)\sin\theta(1+\beta^2\cos^2\theta)x_2y_2+\alpha\beta(1-\beta^2\cos^2\theta)^2x_3y_1\\ &\quad-\alpha\beta^2(1-\beta^2)\sin\theta\cos\theta\,y_2^2\end{aligned}}{\alpha(1-\beta^2)(1-\beta^2\cos^2\theta)^2} \\[2ex]
0 \\[2ex]
\dfrac{\begin{aligned}&\alpha\beta^2\sin\theta\,x_1x_2+\alpha\beta^3\sin\theta\cos\theta\,x_1y_2+\alpha\beta\sin\theta\,x_2y_1\\ &\quad-\alpha\beta(1-\beta^2)\cos\theta\,x_2y_3+\alpha\beta^2\sin\theta\cos\theta\,y_1y_2\\ &\quad-(1-\beta^2)(\alpha+\beta^2\cos^2\theta-1)y_2y_3\end{aligned}}{\alpha(1-\beta^2)(1-\beta^2\cos^2\theta)} \\[2ex]
\dfrac{\begin{aligned}&\alpha\beta(1-\beta^2\cos^2\theta)^2x_1y_3+\alpha\beta^2(1-\beta^2)\sin\theta\cos\theta\,x_2^2\\ &\quad+\alpha\beta(1-\beta^2)\sin\theta(1+\beta^2\cos^2\theta)x_2y_2+\alpha\beta^2(1-\beta^2)\sin\theta\cos\theta\,y_2^2\\ &\quad+(1-\beta^2\cos^2\theta)^2(\beta^2+\alpha-1)y_1y_3\end{aligned}}{\alpha(1-\beta^2)(1-\beta^2\cos^2\theta)^2} \\[2ex]
0 \\[2ex]
\dfrac{\alpha(1-\beta\cos\theta)y_1+(1-\beta^2)\sin\theta\,x_3+\alpha(\beta-\cos\theta)x_1}{\alpha(1-\beta^2)} \\[2ex]
\dfrac{y_2-x_2}{1+\beta\cos\theta} \\[2ex]
\dfrac{(1-\beta^2)y_3-\alpha\beta\sin\theta\,y_1-(1-\beta^2)\cos\theta\,x_3-\alpha\sin\theta\,x_1}{\alpha(1-\beta^2)}
\end{pmatrix}
$$

Figure 1: The Vector Field $\mathrm{hvf}(\bar{H})^{\cup}$.

$$
\begin{aligned}
&\left(T\bar{FL}\left(((\pi_1,\pi_2,A),(z_1,z_2,z_3))^{\cup}\right)\right)^{\cap}\\
&= \left(\left.\frac{d}{dt}\right|_{t=0}\bar{FL}^{-1}\left(\pi_1+tz_1,\pi_2+tz_2,A\exp(tz_3{}^{\wedge})\right)\right)^{\cap}\\
&= \left(\left.\frac{d}{dt}\right|_{t=0}\left(J\left(A\exp(tz_3{}^{\wedge})\right)^{-1}\begin{pmatrix}\pi_1+tz_1\\ \pi_2+tz_2\end{pmatrix},A\exp(tz_3{}^{\wedge})\right)\right)^{\cap}\\
&= \Bigg(J(A)^{-1}\begin{pmatrix}z_1\\ z_2\end{pmatrix}\\
&\qquad -J(A)^{-1}\left[\left.\frac{d}{dt}\right|_{t=0}J\left(A\exp(tz_3{}^{\wedge})\right)\right]J(A)^{-1}\begin{pmatrix}\pi_1\\ \pi_2\end{pmatrix},z_3\Bigg).
\end{aligned}
$$

Thus, $\operatorname{hvf}(\bar{E})$ is found from this formula with the replacements

$$
A=\exp(\theta\boldsymbol{j}^{\wedge}),\quad \begin{pmatrix}z_1\\ z_2\\ z_3\end{pmatrix}=\operatorname{hvf}(\bar{H}),\quad \begin{pmatrix}\pi_1\\ \pi_2\end{pmatrix}=J\left(\exp(\theta\boldsymbol{j}^{\wedge})\right)\begin{pmatrix}\Omega_1\\ \Omega_2\end{pmatrix},
$$

and the result is displayed in figure (2).

This representation of $\operatorname{hvf}(\bar{E})$ may be used to find the $S^1\times S^1\times SO(3)$ relative equilibria; that is, the points $m\in M$ such that the vector field $\operatorname{hvf}(\bar{E})$ is tangent to the vector space spanned by the $S^1\times S^1$ infinitesimal generators. Two vector fields (written here as row vectors) spanning this space are easily determined:

$$
\begin{aligned}
&\operatorname{ign}(\Psi,(1,0))(\Omega_i,\theta)\\
&= \left[\left.\frac{d}{dt}\right|_{t=0}(\exp(t\boldsymbol{k}^{\wedge})\Omega_1,\Omega_2,\exp(t\boldsymbol{k}^{\wedge})\exp(\theta\boldsymbol{j}^{\wedge}))\right]^{\cup}\\
&= \left[(\Omega_1,\Omega_2,\exp(\theta\boldsymbol{j}^{\wedge})),(\boldsymbol{k}\times\Omega_1,0,\boldsymbol{k}^{\wedge}\exp(\theta\boldsymbol{j}^{\wedge}))\right]^{\cup}\\
&= (\boldsymbol{k}\times\Omega_1,0,(\exp(-\theta\boldsymbol{j}^{\wedge})\boldsymbol{k}^{\wedge}\exp(\theta\boldsymbol{j}^{\wedge}))^{\vee})\\
&= (\boldsymbol{k}\times\Omega_1,0,\exp(-\theta\boldsymbol{j}^{\wedge})\boldsymbol{k})\\
&= \begin{pmatrix}-v_2 & v_1 & 0 & 0 & 0 & 0 & -\sin\theta & 0 & \cos\theta\end{pmatrix},
\end{aligned}
\tag{2}
$$

and similarly

$$
\begin{aligned}
&\operatorname{ign}(\Psi,(0,1))(\Omega_i,\theta)\\
&= \left[\left.\frac{d}{dt}\right|_{t=0}(\Omega_2,\exp(t\boldsymbol{k}^{\wedge})\Omega_2,\exp(\theta\boldsymbol{j}^{\wedge})\exp(-t\boldsymbol{k}^{\wedge}))\right]^{\cup}\\
&= \begin{pmatrix}0 & 0 & 0 & -w_2 & w_1 & 0 & 0 & 0 & -1\end{pmatrix}.
\end{aligned}
\tag{3}
$$

$$
\begin{pmatrix}
\dfrac{-(\alpha+\beta^2-1)v_2v_3-\alpha\beta w_2w_3}{1-\beta^2} \\[2ex]
\dfrac{\begin{array}{l}\beta^2\sin\theta\cos\theta\, v_1^2+(\alpha+\beta^2\cos^2\theta-1)v_1v_3+\beta^2\sin\theta\cos\theta\, v_2^2 \\ \quad -\beta\sin\theta\, w_1^2+\alpha\beta\cos\theta\, w_1w_3-\beta\sin\theta\, w_2^2\end{array}}{1-\beta^2\cos^2\theta} \\[2ex]
0 \\[2ex]
\dfrac{-\alpha\beta v_2v_3-(\alpha+\beta^2-1)w_2w_3}{1-\beta^2} \\[2ex]
\dfrac{\begin{array}{l}\beta\sin\theta\, v_1^2+\alpha\beta\cos\theta\, v_1v_3+\beta\sin\theta\, v_2^2-\beta^2\sin\theta\cos\theta\, w_1^2 \\ \quad +(\alpha+\beta^2\cos^2\theta-1)w_1w_3-\beta^2\sin\theta\cos\theta\, w_2^2\end{array}}{1-\beta^2\cos^2\theta} \\[2ex]
0 \\[2ex]
-v_1\cos\theta+v_3\sin\theta+w_1 \\[2ex]
-v_2+w_2 \\[2ex]
-v_1\sin\theta-v_3\cos\theta+w_3
\end{pmatrix}
$$

Figure 2: The Vector Field $\mathrm{hvf}(\bar{E})^{\cup}$

If $v_1 = v_2 = w_1 = w_2 = 0$ and $\theta = 0$ or $\theta = \pi$, then $\mathrm{hvf}(\bar{E})$ is tangent to the one dimensional space spanned by (2) and (3), so these are relative equilibria; otherwise (2) and (3) are independent. In the latter case, the relative equilibria can be realized as the zero level of some functions on M by finding a 7×9 matrix R with independent row vectors which annihilate (2) and (3), and then setting $R \cdot \mathrm{hvf}(\bar{E}) = 0$. Implementing this gives all the relative equilibria on M; they are, in four classes,

(4a) $\theta = 0,\ \Omega_1 = \Omega_2$;

(4b) $\theta = \pi,\ \Omega_1 = \exp(\pi \boldsymbol{j}^\wedge)\Omega_2$;

(4c) $v_1 = v_2 = w_1 = w_2 = 0$;

(4d) $\theta \neq 0,\ \theta \neq \pi,\ v_2 = 0,\ w_2 = 0,$
$$\cos\theta\, v_1^2 - \alpha \sin\theta\, v_1 v_3 - (1 + \beta\cos\theta) v_1 w_1 + \beta w_1^2 = 0,$$
$$\beta v_1^2 - (1 + \beta\cos\theta) v_1 w_1 + \alpha \sin\theta\, w_1 w_3 + \cos\theta\, w_1^2 = 0\ .$$

One way to avoid unnecessary consideration of separate relative equilibria which are $S^1\times S^1$ equivalent is to work with a set which intersects no $S^1\times S^1$ orbit more than once. Some relative equilibria in the first three classes may be discarded for this purpose; the $\theta \neq 0$ and $\theta \neq \pi$ subset of M has this property already, so no adjustments are required for relative equilibria in the fourth class. Note also that equations (4d) are linear in v_3 and w_3, and may be solved for these variables. The resulting relative equilibria, in six disjoint classes, are

(5a) $\theta = 0,\ v_1 = v_2 = w_1 = w_2 = 0,\ v_3 = t_1,\ w_3 = t_2$;

(5b) $\theta = \pi,\ v_1 = v_2 = w_1 = w_2 = 0,\ v_3 = t_1,\ w_3 = t_2$;

(5c) $\theta = 0,\ v_1 = w_1 = t_1,\ v_2 = w_2 = 0, v_3 = w_3 = t_2,\ 0 < t_1$;

(5d) $\theta = \pi,\ v_1 = w_1 = t_1,\ v_2 = w_2 = 0, v_3 = -w_3 = t_2,\ 0 < t_1$;

(5e) $\theta = t_3,\ v_1 = v_2 = w_1 = w_2 = 0,\ v_3 = t_1,\ w_3 = t_2,\ 0 < t_3 < \pi$;

(5f) $\theta = t_3,\ v_1 = t_1,\ v_2 = 0,\ v_3 = -\dfrac{(t_1\cos t_3 - t_2)(\beta t_2 - t_1)}{\alpha t_1 \sin t_3},$
$$w_1 = t_2,\ w_2 = 0, w_3 = \frac{(t_2\cos t_3 - t_1)(\beta t_1 - t_2)}{\alpha t_2 \sin t_3},$$
$$t_1 \neq 0,\ t_2 \neq 0,\ 0 < t_3 < \pi\ .$$

Interpreting these equilibria is not difficult; note that

$$\tau(\mathrm{Id}, \exp(\theta \boldsymbol{j}^\wedge), \Omega_1{}^\wedge, \exp(\theta \boldsymbol{j}^\wedge)\Omega_2{}^\wedge) = (\Omega_1, \Omega_2, \exp(\theta \boldsymbol{j}^\wedge)) ,$$

and that any lagrangian evolution curve is second order. Then one wishes to find scalars σ_1 and σ_2, and a vector $\sigma_T \in R^3$ such that

$$\begin{aligned} c(t) &= (\exp(t\sigma_T{}^\wedge)\exp(t\sigma_1 \boldsymbol{k}^\wedge), \exp(t\sigma_T{}^\wedge)\exp(\theta \boldsymbol{j}^\wedge)\exp(t\sigma_2 \boldsymbol{k}^\wedge), \\ &\qquad \exp(t\sigma_T{}^\wedge)\Omega_1{}^\wedge \exp(t\sigma_1 \boldsymbol{k}^\wedge), \exp(t\sigma_T{}^\wedge)\exp(\theta \boldsymbol{j}^\wedge)\Omega_2{}^\wedge \exp(t\sigma_2 \boldsymbol{k}^\wedge)) \\ &= \frac{d}{dt}(\exp(t\sigma_T{}^\wedge)\exp(t\sigma_1 \boldsymbol{k}^\wedge), \exp(t\sigma_T{}^\wedge)\exp(\theta \boldsymbol{j}^\wedge)\exp(t\sigma_2 \boldsymbol{k}^\wedge)) . \end{aligned}$$

Performing the differentiation at $t = 0$ gives two equations for σ_T:

$$\begin{aligned} \sigma_T &= \Omega_1 - \sigma_1 \boldsymbol{k} , \\ \sigma_T &= \exp(\theta \boldsymbol{j}^\wedge)(\Omega_2 - \sigma_1 \boldsymbol{k}) . \end{aligned} \tag{6}$$

Since for all relative equilibria, $\Omega_1 \cdot \boldsymbol{j} = \Omega_2 \cdot \boldsymbol{j} = 0$, one can conclude that $\sigma_T \cdot \boldsymbol{j} = 0$; this observation, together with the fact that the bodies lie along $\boldsymbol{k}$ and $\exp(\theta \boldsymbol{j}^\wedge)\boldsymbol{k}$ in this configuration, shows the following fact: *in a state of motion which is an $SO(3)\times S^1 \times S^1$ relative equilibria, the axis of total rotation and the axes of the bodies are coplanar.* Furthermore, the $S^1 \times S^1$ equivariance of the quotient map τ implies

$$\begin{aligned} \tau \circ c(t) &= \tau(\mathrm{Id} \exp(t\sigma_1 \boldsymbol{k}^\wedge), \exp(\theta \boldsymbol{j}^\wedge)\exp(t\sigma_2 \boldsymbol{k}^\wedge), \Omega_1{}^\wedge \exp(t\sigma_2 \boldsymbol{k}^\wedge), \\ &\qquad \exp(\theta \boldsymbol{j}^\wedge)\Omega_2{}^\wedge \exp(t\sigma_2 \boldsymbol{k}^\wedge)) \\ &= (-t\sigma_1, -t\sigma_2)\cdot(\Omega_1, \Omega_2, \exp(\theta \boldsymbol{j}^\wedge)) . \end{aligned}$$

Differentiation at $t = 0$ gives

$$-\sigma_1 \,\mathrm{ign}\,(\Psi, (1,0))(\Omega_i, \theta) - \sigma_2 \,\mathrm{ign}\,(\Psi, (0,1))(\Omega_i, \theta) = \mathrm{hvf}(\bar{E})(\Omega_i, \theta) ,$$

which may be used to solve for σ_1 and σ_2 in each of the classes (5a)–(5f):

$$\begin{aligned} &(5a): \quad \sigma_1 - \sigma_2 = v_3 - w_3 \\ &(5b): \quad \sigma_1 + \sigma_2 = v_3 + w_3 \\ &(5c) \text{ and } (5d): \quad \sigma_1 = -\frac{[\mathrm{hvf}(\bar{E})]_2}{v_1}, \ \sigma_2 = -\frac{[\mathrm{hvf}(\bar{E})]_5}{w_1} \\ &(5e) \text{ and } (5f): \quad \sigma_1 = \frac{[\mathrm{hvf}(\bar{E})]_7}{\sin\theta}, \ \sigma_2 = \frac{[\mathrm{hvf}(\bar{E})]_7 \cos\theta}{\sin\theta} + [\mathrm{hvf}(\bar{E})]_9 . \end{aligned}$$

Once σ_1 and σ_2 are known, σ_T can be found from equations (6). In correspondence with the list (5a)–(5f), the result is

$$\sigma_1 = t_1,\ \sigma_2 = t_2,\ \sigma_T = 0\ ; \tag{7a}$$

$$\sigma_1 = t_1,\ \sigma_2 = t_2,\ \sigma_T = 0\ ; \tag{7b}$$

$$\sigma_1 = \sigma_2 = -\frac{(\alpha+\beta-1)t_2}{1-\beta},\ \sigma_T = t_1\boldsymbol{i} + \frac{\alpha t_2}{1-\beta}\boldsymbol{k}\ ; \tag{7c}$$

$$\sigma_1 = -\sigma_2 = \frac{(\alpha-\beta-1)t_2}{1-\beta},\ \sigma_T = -t_1\boldsymbol{i} + \frac{\alpha t_2}{1-\beta}\boldsymbol{k}\ ; \tag{7d}$$

$$\sigma_1 = t_1,\ \sigma_2 = t_2,\ \sigma_T = 0 \tag{7e}$$

$$\begin{aligned} \sigma_1 &= -\frac{(t_1\cos t_3 - t_2)(\beta t_2 - (1-\alpha)t_1)}{\alpha t_1 \sin t_3}, \\ \sigma_2 &= -\frac{(t_2\cos t_3 - t_1)(\beta t_1 - (1-\alpha)t_2)}{\alpha t_2 \sin t_3}, \\ \sigma_T &= t_1\boldsymbol{i} + \frac{t_1\cos t_3 - t_2}{\sin t_3}\boldsymbol{k}\ . \end{aligned} \tag{7f}$$

For rough intuition, then, the relative equilibria are as follows:

- Classes (5a),(5b) and (5e). The angle that the bodies make with the joint is arbitrary, they are both spinning, and there is no overall rotation.
- Class (5c). The bodies are aligned on top of one another, with identical spin, and the entire ensemble is rotating about some axis which is tilted with respect to the body axes.
- Class (5d). The bodies are aligned opposed to one another, with opposite spin, so that they rotate together as a unit, and the entire ensemble is rotating about some axis which is tilted with respect to the body axes.
- Class (5f). The joint angle is arbitrary, the bodies are spinning, and there is some overall rotation depending on the spin and the joint angle.

4 Stability

Fixing a relative equilibria, there is obtained an equilibria e on one of the Poisson reduced phase spaces $(R^3)^2 \times SO(3)/S^1 \times S^1$. Call the relative equilibria stable if the reduced hamiltonian restricted to the symplectic leaf through e is definite at e; absence of this condition does not, of course, imply dynamic instability of e. This section presents necessary and sufficient conditions for the stability of the relative equilibria (5a)–(5f). The methods used here are minor modifications of the energy-Casimir method ([6],[7]). As an illustration, a sketch of the argument for the class (5f) is provided.

Consider, then, the class (5f). To begin with, there is required a chart containing the relative equilibria; fortunately, one is readily available which contains the entire class. As noted in §3, the subset

$$\tilde{M} = (R^3)^2 \times \{\exp(\theta \mathbf{j}^\wedge); \theta \in (0, \pi)\}$$

of M intersects each $S^1 \times S^1$ orbit only once. As $\tilde{M}$ is a submanifold projecting to an open subset, it can be used as a chart on $(R^3)^2 \times SO(3)/S^1 \times S^1$; the quotient map is found by following an orbit to its unique intersection with N. An advantage of this scheme: the projection of a given $S^1 \times S^1$ invariant function is simply it's restriction to N. Applying this to the Poisson reduced phase space obtained from TQ gives the hamiltonian

$$\tilde{E} = \frac{1}{2}|v|^2 + \frac{1}{2}|w|^2 - \beta(v_1 w_1 + v_2 w_2) ,$$

and three Casimirs: $c_1 = \alpha v_3$, $c_2 = \alpha w_3$, and

$$\begin{aligned} c_3 \;=\; & (1 - \beta \cos\theta)({v_1}^2 + {w_1}^2) + \alpha^2({v_3}^2 + {w_3}^2) \\ & + 2(\beta^2 \cos\theta - 2\beta + \cos\theta) v_1 w_1 + 2\alpha^2 v_3 w_3 \cos\theta \\ & + 2\alpha(v_1 w_3 - w_1 v_3 + \beta v_1 v_3 - \beta w_1 w_3) \sin\theta . \end{aligned}$$

Fix values of the parameters t_1, t_2, t_3 in (5f) corresponding to an equilibria $e \in N$. For convenience, set $z = t_1/t_2$ and $t = \cos t_3$, so $z \neq 0$ and $t \in (-1, 1)$. Already one needs to distinguish two cases: as noted in §2, the symplectic leaf changes dimension at the zero level of c_3. The Casimir c_3 evaluated on the relative equilibria (5f) is

$$\frac{{t_2}^4(\beta z^2 - 2z + \beta)^2(z^2 - 2zt + 1)}{z^2 t} ,$$

which is nonzero if and only if $\beta z^2 - 2z + \beta \neq 0$. Assume this—the zero case is similar but easier. The energy-Casimir method requires constants κ_j, $j \in \{1,2,3\}$ such that

$$\tilde{E}_c = \tilde{E} + \sum \kappa_j c_j$$

satisfies $\boldsymbol{d}\tilde{E}_c(e) = 0$. This is accomplished by setting $\kappa_1 = \sigma_1$, $\kappa_2 = \sigma_2$ (from (7f)), and

$$\kappa_3 = \frac{t_1 t_2}{2(\beta {t_1}^2 - 2t_1 t_2 + \beta {t_2}^2)} .$$

The four (linearly independent) row vectors of

$$\begin{pmatrix} 0 & 0 & 0 & 0 & 1 & 0 & 0 \\ 0 & 0 & 0 & 0 & 0 & 1 & 0 \\ 0 & 0 & 0 & 0 & 0 & 0 & 1 \\ 0 & 0 & \beta t_1 - t_2 & t_1 - \beta t_1 & 0 & 0 & 0 \end{pmatrix} \tag{8}$$

are annihilated by $\nabla c_i(e)$, and hence span the tangent space V to the symplectic leaf through e. As $\tilde{E}_c$ and $\tilde{E}$ differ by a constant on this leaf, and $\boldsymbol{d}\tilde{E}_c(e) = 0$, the second variation of $\tilde{E}|V$ may be computed by evaluating second variation of $\tilde{E}$ on the basis (8). The resulting 4×4 block diagonal matrix consists of two 2×2 blocks, the upper of which is (up to multiplication by a scalar)

$$T_1 = \begin{pmatrix} \beta z^2 + (\beta^2 t^2 - 2\beta t - 1)z + \beta & -\beta^2 t z^2 + (1 + \beta^2 t^2)z - \beta^2 t \\ -\beta^2 t z^2 + (1 + \beta^2 t^2)z - \beta^2 t & \beta z^2 + (\beta^2 t^2 - 2\beta t - 1)z + \beta \end{pmatrix}$$

and the lower of which is (up to multiplication by a scalar with the same sign)

$$T_2 = \begin{pmatrix} \dfrac{(z-\beta)(\beta z - 1)(z_2 - 2zt + 1)}{(1-\beta^2)\sqrt{1-t^2}} & (1-z^2)(1-\beta t) \\ (1-z^2)(1-\beta t) & \begin{array}{l} [-2z^2(1-\beta^2)t + \beta z^4 \\ \quad - (1+3\beta^2)z^3 + 6\beta z^2 \\ \quad - (1+3\beta^2)z + \beta]\sqrt{1-t^2} \end{array} \end{pmatrix} .$$

Note that

$$\det T_1 = \beta(1-\beta^2t^2)(z^2-2zt+1)(\beta z^2-2z+\beta)\,,$$

and so the conditions

$$\beta \neq 0, \quad \beta z^2 - 2z + \beta > 0 \tag{9}$$

are necessary for stability. They are also sufficient: view $[T_1]^{11}$ as a quadratic polynomial in z. Write

$$[T_1]^{11} = \beta z^2 - 2z + \beta + (1-\beta t)^2 z\,, \tag{10}$$

so $[T_1]^{11} > 0$ if $z > 0$. Also, the sum of the roots of (10) is $2-(1-\beta t)^2 > 0$, while the product of the roots is 1, so both roots are positive or nonreal, hence $[T_1]^{11} > 0$ for $z < 0$ as well. As already noted, $\det T_1 > 0$, so T_1 is positive definite. Similarly, writting the product of the first two factors of $[T_2]^{11}$ as

$$\beta z^2 - 2z + \beta + (1+\beta^2)z$$

shows $[T_2]^{11} > 0$. Furthermore,

$$\begin{aligned}(1-\beta^2)\det T_2 \;=\; & [-(1-\beta^2)(\beta z^2-2z+\beta)^2 z^2]t^2 \\ & + [-2\beta z(\beta z^2-2z+\beta)(z^2-2\beta z-1)^2]t \\ & + [(\beta z^2-2z+\beta)(\beta z^6-4\beta^2z^5+\beta(3\beta^2+4)z^4 \\ & - (3\beta^2+1)z^3+\beta(3\beta^2+4)z^2-4\beta^2z+\beta)]\,,\end{aligned}$$

and since $\det T_2$ is a concave-down quadratic polynomial in t, the two evaluations

$$\begin{aligned}\det T_2|_{t=1} &= \beta(z-1)^2(z^2-2\beta z+1)^2 \geq 0\,, \\ \det T_2|_{t=-1} &= \beta(z+1)^2(z^2-2\beta z+1)^2 \geq 0\,,\end{aligned}$$

show that $\det T_2$ is positive for $t \in (0,1)$. Thus, T_2 is positive definite, and conditions (9) are sufficient for stability.

The second of conditions (9) admits an interpretation in terms of the geometry of the bodies and the axis of total rotation along σ_T. If θ_i are the angles from σ_T to body i, elementary geometry and (7f) show $z = \sin\theta_1/\sin\theta_2$. Then (9) become $\beta = 0$ or

$$\frac{\sin\theta_1}{\sin\theta_2} < \frac{\beta}{1+\sqrt{1-\beta^2}} \quad\text{or}\quad \frac{\sin\theta_1}{\sin\theta_2} > \frac{1+\sqrt{1-\beta^2}}{\beta}\,, \tag{11}$$

inequalities which bound the bodies away from the overlapped configuration characterized by $\theta_1 = \theta_2$.

The remaining classes of relative equilibria are similar, except for (5a) and (5b), since these reside at points of the unreduced phase space with nontrivial isotropy, and hence at points where the reduced phase space is singular ([2]). This difficulty may be overcome by reducing by the subgroup of material symmetries $S^1 \times S^1$, instead of the full symmetry group, and by adding certain conserved quantities to the analogue of $\tilde{E}$ (see the exposition of the stability of the Lagrange top in [6]). The relative equilibria and their stability conditions are summarized in the following

Theorem. *The base integral curves of all relative equilibria are the $S^1 \times S^1 \times SO(3)$ translates of the curves in configuration space $SO(3) \times SO(3)$*

$$t \to (\exp(t\sigma_T{}^\wedge)\exp(t\sigma_1 \boldsymbol{k}^\wedge), \exp(t\sigma_T{}^\wedge)\exp(t_3\boldsymbol{j}^\wedge)\exp(t\sigma_2\boldsymbol{k}^\wedge)) ,$$

where the total rotation σ_T and spins σ_i are given by (7a)–(7f) in conjunction with (5a)–(5f). In this state of motion, the axis of total rotation and the body axes are coplanar. A relative equilibria is stable if either of the following hold:

- *The total rotation is zero, the total angular momentum is nonzero, and no body has zero spin.*
- *The bodies are bounded away from the overlap configuration by (11) and the joint is not at the bodies centers of mass (i.e. $\beta \neq 0$).*

5 Further Remarks

Typically, in mechanical systems where symmetry is absent, equilibria are isolated and of finite number. For example, in the 7 dimensional Poisson space $\tilde{M}$, the classes (5e) and (5f) are 3 dimensional, while the generic symplectic leaf is 4 dimensional, so transversal intersection at isolated points is expected. Thus, there is the following question: *how many equilibria are on each symplectic reduced phase space?* The definition of this question is not affected by singularities in the reduced phase space if one agrees to count, on constant levels of the momentum map, inequivalent relative equilibria on the original phase space (or on any other well defined partial reduction).

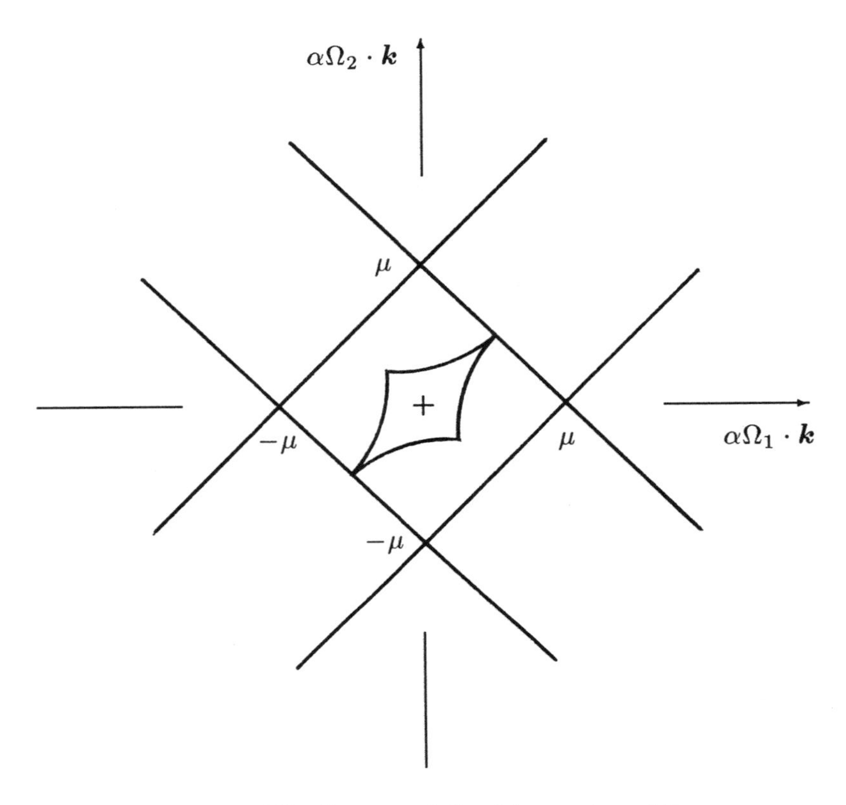

Figure 3: The Image of the Momentum Map.

Refer to figure (3). This diagram represents a plane whose points are values of the momentum map

$$(\Omega_1, \Omega_2, A) \to (\alpha\Omega_1 \cdot \boldsymbol{k}, \alpha\Omega_2 \cdot \boldsymbol{k}) \tag{12}$$

arising from the $S^1 \times S^1$ material symmetries on a fixed symplectic leaf

$$\bar{F}L^{-1}\left\{(\pi_i, A); |\pi_1 + A\pi_2|^2 = \mu\right\}, \ \mu \neq 0$$

in the reduced phase space $(R^3)^2 \times SO(3)$. The straight lines of slope ± 1 through $(\pm\mu, 0)$ are values of the momentum map (12) whose level sets contain points with nontrivial isotropy; the affine structure is predicted by

[5], §27. The following remarks are based mostly on exact calculations, but with some support of numerical data.

- The diagram is merely scaled as α and μ change.
- Connected regions of the diagram correspond to level sets with the same equilibria count. This count is finite, except at the points $(\pm\mu, 0)$ and $(0, \pm\mu)$, on whose level sets reside infinitely many equilibria of class (5e) in which one of the bodies has zero spin.
- Let E_β be the region bounded by the curved lines and containing the origin. The set E_0 exactly fills the square containing the origin, and E_β monotonely decreases toward the line segment from $(-\mu/2, -\mu/2)$ to $(\mu/2, \mu/2)$ as β increases towards 1.
- The equilibria may be naturally organized so as to have moments in regions with boundary the affine elements of the diagram, *except for the equilibria of class (5f) that are not stable in the sense of §4*. It is these equilibria that give rise to E_β.

Finally, I record here that the characteristic polynomials for the linearized vector fields about (5a)–(5e) have been computed. For a given set of equilibria, it is possible to superimpose plots of those that are spectrally stable upon figure (3), and thereby obtain numerical evidence for the existence of hamiltonian-Hopf bifurcations: the spectral stability changes when the equilibria count does not, as the values of (12) or of β are varied.

Acknowledgements

The author thanks Jerrold Marsden and Robert Grossman for their suggestions during innumerable discussions on this work. Thanks to Ian Emmons, who read the manuscript and caught many errors.

References

[1] R. Abraham & J. E. Marsden, *Foundations of Mechanics,* Second Edition, Addison-Wesley, Reading, 1978.

[2] J. Arms, J. Marsden & V. Moncrief, Bifurcations of momentum mappings, *Comm. Math. Phys.* **78**, 1981, 455–478.

[3] B. W. Char, K. O. Geddes, G. H. Gonnet, M. B. Monagan, & S. M. Watt, *First Leaves: a Tutorial Introduction to Maple,* Second Edition, WATCOM Publications Ltd., Waterloo, Ontario, Canada, 1988.

[4] R. Grossman, K. S. Krishnaprasad, & J. E. Marsden, The dynamics of two coupled rigid bodies, *Dynamical Systems Approaches to Nonlinear Problems in Systems and Circuits*, SIAM, 1988, 373–378.

[5] V. Guillemin & S. Sternberg, *Symplectic techniques in physics,* Cambridge University Press, Cambridge, 1984.

[6] D. Holm, J. E. Marsden, T. Ratiu, & A. Weinstein, Stability of rigid body motion using the energy-Casimir method, *Cont. Math., AMS* **28**, 1984, 15–23.

[7] P. S. Krishnaprasad & J. E. Marsden, Hamiltonian structures and stability of rigid bodies with flexible attachments, *Arch. Rat. Mech. Anal.* **98**, 1987, 71–93.

DEPARTMENT OF MATHEMATICS
UNIVERSITY OF CALIFORNIA AT BERKELEY
BERKELEY, CALIFORNIA 94720
U.S.A.

Contemporary Mathematics
Volume 97, 1989

NONSMOOTH OPTIMIZATION ALGORITHMS FOR THE DESIGN OF CONTROLLED FLEXIBLE STRUCTURES[1]

E. Polak

Department of Electrical Engineering and
Computer Sciences
University of California
Berkeley, Ca. 94720

ABSTRACT

First we show that both open-loop and closed-loop optimal control problems can be expressed in the form of nonsmooth optimization problems. Then we present the basics of a class of nonsmooth optimization algorithms which solve constrained optimization problems involving the maxima of differentiable functions. The described algorithms are shown to be natural extensions of the method of centers.

1. INTRODUCTION

It is reasonably obvious that the solution of constrained open-loop optimal control problems requires the use of appropriate optimization algorithms. It is less obvious that the design of compensators for closed-loop systems can be cast as an optimization problem which can be treated effectively by optimization algorithms. In this paper, we make a strong case for the use of *nonsmooth* optimization algorithms in the solution of constrained open-loop optimal control problems, as well as of linear feedback system design problems, and we present an introduction to a set of appropriate nonsmooth optimization algorithms.

Feedback is used to achieve various desirable properties in a control system, such as stability, disturbance attenuation, and low sensitivity to changes in the plant. Since these properties depend on the shape of various closed-loop system responses, all control system design techniques are at least partially based on response shaping. In the sixties and seventies, the most popular control system design methods were based on weighted least-squares unconstrained minimization in the form of linear-quadratic regulator (LQR) theory (see e.g. [Kwa.1]). The main drawback of the LQR approach is that it tends to result in poor stability robustness and output-disturbance rejection (see e.g., [Doy.1]). Furthermore, the least-squares approach does not permit imposition of hard bounds on system responses. A more recent approach for shaping a *single* frequency response is to minimize not a weighted quadratic

[1] The research reported herein was sponsored in part by the National Science Foundation under grant ECS-8121149, the Air Force Office of Scientific Research grant AFOSR-83-0361, the State of California MICRO Program and the General Electric Co.

0271-4132/89 $1.00 + $.25 per page

norm, as in the LQR approach, but a weighted sup-norm (H^∞-norm) (for a survey see [Fra.1]). This is usually done in conjunction with a compensator parametrization which makes all transfer functions affine in the design parameter, and hence reduces the response shaping to a *convex, unconstrained optimization problem in* H^∞.

However, most closed-loop system design problems require shaping of *several* frequency and time domain responses, some of which may be subject to *hard* constraints. For example, while minimizing the norm of the sensitivity matrix over the bandwidth of the feedback system, the norm of the transfer matrix from the command input to the plant input has to be upper-bounded. Otherwise the command input can drive the plant input outside the linearization region, which may lead to performance deterioration and even instability. The requirement of simultaneous shaping several frequency responses can be dealt with in various ways. For example, in [Doy.2], loop transformations and weighting functions are used to transform the multiloop shaping problem into a problem of unconstrained minimization of the norm of an affine matrix function in H^∞. This approach can be quite conservative. Furthermore, when there are hard bounds on the norms of some of transfer function matrices, the weighting approach cannot be used, because it is not known how to transform a constrained H^∞ minimization problem into an unconstrained one by using weights.

A second approach, first presented in [Kwa.2], is based on the fact that many essential design objectives can be formulated as bounds on the weighted sensitivity matrix $S(j\omega)$ and/or on the weighted complementary sensitivity matrix $T(j\omega) = I - S(j\omega)$. It is concluded in [Kwa.2] that a balance between conflicting design objectives can be achieved by minimizing a performance criterion of the form $\sup_{\omega} [\ |V(j\omega)S(j\omega)|^2 + |W(j\omega)T(j\omega)|^2\ ]$, where $V(\cdot)$ and $W(\cdot)$ are weighting functions selected by the designer. This approach can also be conservative.

From a designer's point of view, both LQR and the above approaches to complex design problem solution suffer from the drawback that they use design weights which are very difficult to select. This drawback can be further accentuated by the fact that the solution of a weight-dependent, unconstrained optimization problem, can be very sensitive to the weights, which implies that whenever a constrained problem is somehow converted to an unconstrained one by means of weights, a large amount of time may have to be devoted to weight selection.

In [Pol.7], the reader will find a formulation of finite dimensional, linear, time invariant feedback-system design, subject to various hard constraints, as convex, nonsmooth optimization problems. In [Pol.7], affine compensator parametrizations are used, as in the H^∞ approach. In this paper we use a direct compensator parametrization which enables us to select the degree of the compensator, because affine parametrization would result in an infinite dimensional compensator. As a trade-off, we give up problem convexity. We show the mathematical unity of both open- and closed-loop optimal control problems and we present a sequence of progressively more complex algorithms for their solution.

For further reading on nonsmooth optimization and optimal control algorithms, we refer the reader to [Gon.1, Kiw.1, Kle.1, May.1, May.2, Pir.1, Pir.2, Pol.3, Pol.4, Pol.5, Pol.6, Pol.10].

2. FORMULATION OF OPTIMAL DESIGN PROBLEMS

We propose to consider both open-loop optimal control and closed-loop optimal control of flexible structures. By *open-loop optimal control* we mean the computation of optimal open-loop controls which take a structure from an initial state to a desired state subject to various constraints on the control and state, while by *closed-loop optimal control* we mean the computation of optimal, finitely parametrized, finite-dimensional closed-loop compensators, subject to constraints on various time- and frequency-domain constraints.

2.1. CANONICAL FORMS

Our first task is to show that these problems can be cast in the form of the two canonical problems, below. Note that the two problems differ only in the space on which they are defined. For the design of finite-dimensional closed-loop compensators, which is a problem with a finite dimensional design vector, we adopt the canonical form

$$\min\{\ \psi^0(x)\ \mid\ \psi^j(x) \le 0,\ j \in \mathbf{m},\ \ x \in \mathbf{X}\ \}, \tag{2.1a}$$

where $\mathbf{m} \triangleq \{\ 1,2,\ldots,m\ \}$, $\mathbf{X} \subset \mathbb{R}^n$ is a set with a very tractable description, e.g., $\mathbf{X} = \mathbb{R}^n$ or

$$\mathbf{X} \triangleq \{ x \in \mathbb{R}^n \mid |x^k| \leq b^k,\ k = 1,2,3,\ldots,n \}, \tag{2.1b}$$

and, with $\mathbf{M} \triangleq \{ 0,1,2,\ldots,m \}$,

$$\psi^j(x) = \max_{y_j \in \mathbf{Y}_j} \phi^j(x,y_j),\ \forall\ j \in \mathbf{M}, \tag{2.1c}$$

where $\phi^j:\mathbb{R}^n \times \mathbb{R} \to \mathbb{R}$. We will assume that the functions $\phi^j(\cdot,\cdot)$ and their gradients $\nabla_x\phi^j(\cdot,\cdot)$ are Lipschitz continuous on bounded sets. In addition, we will assume that the intervals $\mathbf{Y}_j = [a_j,b_j] \subset \mathbb{R}$ are compact. We note that when $\mathbf{Y}_j$ contains only one point, i.e., $a_j = b_j$, the function $\psi^j(x) = \phi^j(x,a_j) \triangleq f^j(x)$ is differentiable (otherwise it need not be); thus we see that the formulation (2.1a-c) allows that some of the $\psi^j(x)$ are ordinary differentiable functions.

Similarly, for open-loop optimal control, we adopt the canonical form

$$\min\{ \psi^0(u,T) \mid \psi^j(u,T) \leq 0,\ j \in \mathbf{m},\ (u,T) \in \mathbf{U} \times \mathbf{T} \}, \tag{2.2a}$$

where $\mathbf{U} \triangleq \{ u \in L_2^p[0,1] \mid u(t) \in U\ \forall\ t \in [0,1] \}$, $U \subset \mathbb{R}^p$ either is compact or else equal to $\mathbb{R}^p$, $\mathbf{T} \triangleq [T_0,T_f]$, and for $j \in \mathbf{M}$,

$$\psi^j(u,T) = \max_{t \in \mathbf{Y}_j} \phi^j(u,T,t), \tag{2.2b}$$

with $\phi^j:L_2^p[0,1] \times \mathbb{R}_+ \times \mathbb{R} \to \mathbb{R}$. We will assume that the functions $\phi^j(\cdot,\cdot,\cdot)$ and their gradients[2] $\nabla_u\phi^j(\cdot,\cdot,\cdot)$ are Lipschitz continuous on bounded sets. In addition, we will assume that the intervals $\mathbf{Y}_j \subset \mathbb{R}$ are compact.

2.2. TRANSCRIPTION OF OPEN-LOOP OPTIMAL CONTROL PROBLEMS INTO CANONICAL FORM

Since the transcription of open-loop optimal problems into the form (2.2a) is simpler than the transcription of closed-loop optimal control problems into the form (2.1a), we will do it first. Although in practice one usually computes with second order dynamics, the explanation becomes simpler if we adopt as our model for the dynamics the first order differential equation

$$\dot{z}(t) = TAz(t) + Th(z(t),u(t)),\ \ t \in [0,1],\ \ z(0) = z_0\ ,\ \ T > 0,\ \ u \in \mathbf{U}, \tag{2.3}$$

---

[2] By gradients we mean the kernels of linear functionals which play an analogous role to gradients of functions defined on $\mathbb{R}^n$, i.e., the gradient of $f:L_2^p[0,1] \to \mathbb{R}$ is defined by the property that $\lim_{t\to 0} [f(u + t\delta u) - f(u) - t\langle\nabla f(u),\delta u\rangle_2]/t = 0$, where $\langle\cdot,\cdot\rangle_2$ denotes the $L_2^p[0,1]$ scalar product.

where the *state vector* $z(t)$ is an element of a Hilbert space, H, so that (2.3) can, in fact, be a *partial* differential equation, and the control $u(t) \in \mathbb{R}^p$ is finite dimensional. We will assume that A is an infinitesimal generator of a C_0 semigroup[3] and that the operator $h(\cdot,\cdot)$ is bounded and continuously differentiable. The parameter T is a time-scaling parameter which enables us to convert free time problems into fixed time form, as well as to avoid some well known pathological behavior of the discretizations that are needed in solving optimal control problems. When (2.3) represents an ODE, $A = 0$ holds. We note that (2.3) can be used to represent a broad class of dynamical systems described either by ODEs or PDEs, including PDEs derived using Lagrangian dynamics.

Obviously, we must assume that (2.3) has a weak solution, which we will denote by $z^{u,T}(t)$. In addition, we will assume that the differential, $\delta z^{u,T}(t;\delta u,\delta T)$, of this solution, with respect to the control u and the time scaling parameter T, is given by the weak solution of linearized equation

$$\delta\dot{z}(t) = T[A + \frac{\partial h(z^{u,T}(t),u(t))}{\partial z}]\delta z(t) + T\frac{\partial h(z^{u,T}(t),u(t))}{\partial u}\delta u(t) + [A + h(z^{u,T}(t),u(t))]\delta T,$$

$$t \in [0,1], \quad \delta z(0) = 0 \, . \tag{2.4}$$

Now consider the optimal control problem with control, end point and state space constraints:

$$\min_{(u,T)}\{\, g^0(z^{u,T}(1)) \mid g^j(z^{u,T}(1)) \le 0,\ j = 1,2,\ldots,m_1;\ \ g^j(z^{u,T}(t)) \le 0,\ \forall\ t \in [0,1],$$

$$j = m_1 + 1, m_1 + 2,\ldots,m;\ \ u(t) \in U \ \forall\ t \in [0,1], T \in [T_o,T_f] \,\}, \tag{2.5}$$

where $T_o > 0$ is assumed to be very small and $T_f < \infty$. We assume that the control $u(\cdot)$ is an $L_2^p[0,1]$ function, that the set $U \subset \mathbb{R}^p$ is compact and that all the functions $g^j : H \to \mathbb{R}$ are continuously differentiable. Next, we define $\mathbf{Y}_j = \{\ 1\ \}$ for $j = 1,2,\ldots,m_1$ and $\mathbf{Y}_j = [0,1]$ for $j = m_1 + 1,\ldots,m_1 + m_2$ (so that $m = m_1 + m_2$), and we define $\phi^j : L_2^p[0,1] \times \mathbb{R}_+ \times \mathbb{R} \to \mathbb{R}$, $j = 0,1,\ldots,m$, by $\phi^j(u,T,t) = g^j(z^{u,T}(t))$ for $j = 1,2,\ldots,m_1$, and by $\phi^j(u,T,t) = g^j(z^{u,T}(t))$ for $j = m_1 + 1,\ldots,m$, and, finally, if for $0,1,2,\ldots,m$, we define the functions $\psi^j : L_2^p[0,1] \times \mathbb{R}_+ \to \mathbb{R}$ by $\psi^j(u,T) = \max_{t \in Y_j} \phi^j(u,T,t)$, then we see that problem (2.5) assumes the form (2.2a).

[3] For a discussion of semigroup theory see [Bal.1] or [Paz.1].

2.3. TRANSCRIPTION OF FEEDBACK-SYSTEM DESIGN INTO CANONICAL FORM

Next we turn to the more arduous task of transcribing closed-loop optimal control problems into the form (2.1a). Consider the n_i–input - n_o–output feedback system S, shown in Fig. 1. We assume that the plant is described by a linear, time-invariant differential equation in a Hilbert space E:

$$\dot{z}_p(t) = A_p z_p(t) + B_p e_2(t), \tag{2.6a}$$

$$y_2(t) = C_p z_p(t) + D_p e_2(t), \tag{2.6b}$$

where $z_p(t) \in E$, $e_2(t) \in \mathbb{R}^{n_i}$, $y_2(t) \in \mathbb{R}^{n_o}$, for $t \geq 0$. We will assume that the operators $B_p:\mathbb{R}^{n_i} \to E$, $C_p:E \to \mathbb{R}^{n_o}$ and $D_p:\mathbb{R}^{n_i} \to \mathbb{R}^{n_o}$ are bounded, and that A_p may be an unbounded operator from E to E, with domain dense in E, which generates a C_0 semigroup, $\{e^{A_p t}\}_{t \geq 0}$. We will denote the *spectrum* of A_p by $\sigma(A_p)$, and we will denote the *resolvent set* of A_p by $\rho(A_p)$. Referring to [Paz.1], we find that there exist $M \in (1,\infty)$ and $\gamma \in \mathbb{R}$ such that

$$\| e^{A_p t} \| \leq M e^{\gamma t}, \quad \forall\ t \geq 0\ ; \tag{2.7}$$

furthermore $\rho(A_p)$ contains the half-plane, $\{s \in \mathbb{C} \mid \mathrm{Re}\, s > \gamma\}$. We will denote the domain and the range of A_p by $D(A_p)$ and $R(A_p)$, respectively.

We define the *transfer function* of the plant, $G_p(s)$, by

$$G_p(s) \triangleq C_p(sI - A_p)^{-1} B_p + D_p, \ \forall\ s \in \rho(A_p)\ . \tag{2.8}$$

It follows from [Kat.1, Theorem III 6.7], that $G_p(s)$ is analytic on $\rho(A_p)$. In addition, it is shown in [Jac.1] that for $s \in \{s \in \mathbb{C} \mid \mathrm{Re}\, s > \gamma\}$, $G_p(s)$ is equal to the Laplace transform of $\{C_p e^{A_p t} B_p + D_p \delta(t)\}_{t \geq 0}$ and $\lim_{\substack{|s| \to \infty \\ \mathrm{Re}\, s > \gamma}} G_p(s) \to D_p$.

We will assume that the compensator is *finite dimensional, linear, and time-invariant*, with state equations

$$\dot{z}_c(t) = A_c z_c(t) + B_c e_1(t), \tag{2.9a}$$

$$y_1(t) = C_c z_c(t) + D_c e_1(t)\ , \tag{2.9b}$$

where $z_c(t) \in \mathbb{R}^{n_c}$, $e_1(t) \in \mathbb{R}^{n_o}$, $y_1(t) \in \mathbb{R}^{n_i}$ and A_c, B_c, C_c and D_c are matrices of appropriate dimension. We will assume that a number of the elements of the matrices A_c, B_c, C_c and D_c are to be determined by optimization. We group all these elements into a single *design vector* $x \in \mathbb{R}^n$ and will assume that all the compensator matrices are continuously differentiable in this vector. From now on we will show the dependence of the matrices A_c, B_c, C_c and D_c on x explicitly. Similarly, we will refer to the corresponding closed-loop system as $S(x)$.

We note that the compensator transfer function is given by

$$G_c(x,s) = C_c(x)(sI_{n_c} - A_c(x))^{-1}B_c(x) + D_c(x) \ . \tag{2.10}$$

To ensure well-posedness of the feedback system, we assume that $det(I_{n_i} + D_c(x)D_p) \neq 0$. Finally, we define the Hilbert space $H = E \times \mathbb{R}^{n_c}$, on which the inner product is defined as follows: for $u = (z_p,z_c)$, $v = (z'_p,z'_c)$ in H

$$\langle u,v \rangle_H = \langle z_p,z'_p \rangle_E + \langle z_c,z'_c \rangle_{\mathbb{R}^{n_c}} \ . \tag{2.11}$$

Since $e_1 = u - y_2 - d_o$ and $e_2 = y_1 + d_i$, where d_o is the plant output disturbance and d_i is the plant input disturbance, the state equations for the feedback system are given by

$$\begin{bmatrix} \dot{z}_p \\ \dot{z}_c \end{bmatrix} = A(x)\begin{bmatrix} z_p \\ z_c \end{bmatrix} + B(x)\begin{bmatrix} u \\ d_o \\ d_i \end{bmatrix} , \tag{2.12a}$$

$$\begin{bmatrix} e_1 \\ e_2 \\ y \end{bmatrix} = C(x)\begin{bmatrix} z_p \\ z_c \end{bmatrix} + D(x)\begin{bmatrix} u \\ d_o \\ d_i \end{bmatrix} , \tag{2.12b}$$

where

$$A(x) = \begin{bmatrix} A_p - B_pD_c(x)(I_{n_o} + D_pD_c(x))^{-1}C_p & B_p(I_{n_i}+D_c(x)D_p)^{-1}C_c(x) \\ -B_c(x)(I_{n_o} + D_pD_c(x))^{-1}C_p & A_c(x)-B_c(x)(I_{n_o} + D_pD_c(x))^{-1}D_pC_c(x) \end{bmatrix} \tag{2.12c}$$

$$B(x) = \begin{bmatrix} B_pD_c(x)(I_{n_o}+D_pD_c(x))^{-1} & -B_pD_c(x)(I_{n_o}+D_pD_c(x))^{-1} & B_p(I_{n_i} + D_c(x)D_p)^{-1} \\ B_c(x)(I_{n_o} + D_pD_c(x))^{-1} & -B_c(x)(I_{n_o} + D_pD_c(x))^{-1} & -B_c(x)(I_{n_o} + D_pD_c(x))^{-1}D_p \end{bmatrix} , \tag{2.12d}$$

$$C(x) = \begin{bmatrix} -(I_{n_o} + D_pD_c(x))^{-1}C_p & -(I_{n_o} + D_pD_c(x))^{-1}D_pC_c(x) \\ -D_c(x)(I_{n_o} + D_pD_c(x))^{-1}C_p & (I_{n_i} + D_c(x)D_p)^{-1}C_c(x) \\ (I_{n_o} + D_pD_c(x))^{-1}C_p & (I_{n_o} + D_pD_c(x))^{-1}D_pC_c(x) \end{bmatrix} , \tag{2.12e}$$

$$D(x) = \begin{bmatrix} (I_{n_o} + D_pD_c(x))^{-1} & -(I_{n_o} + D_pD_c(x))^{-1} & -(I_{n_o} + D_pD_c(x))^{-1}D_p \\ D_c(x)(I_{n_o} + D_pD_c(x))^{-1} & -D_c(x)(I_{n_o} + D_pD_c(x))^{-1} & (I_{n_i} + D_c(x)D_p)^{-1} \\ I_{n_o} - (I_{n_o} + D_pD_c(x))^{-1} & -(I_{n_o} + D_pD_c(x))^{-1} & (I_{n_o} + D_pD_c(x))^{-1}D_p \end{bmatrix} . \quad \textbf{(2.12f)}$$

The domain of A is given by $D(A) = D(A_p) \times \mathbb{R}^{n_c} \subset H$. It follows from [Paz.1, p. 76], that because, with the exception of A_p, all the operators in the matrix A are bounded, and because A_p generates a C_0-semigroup, the operator A also generates a C_0–semigroup, $\{e^{At}\}_{t \geq 0}$.

A. Frequency-Domain Performance Specifications

First, since the feedback system (2.12a,b) has 3 inputs and 3 outputs, we write its transfer function $G(x,s) = C(x)(sI - A(x))^{-1}B(x) + D(x)$, which is defined for all $s \in \rho(A(x))$, in block form, as follows:

$$G(x,s) = \begin{bmatrix} G_{11}(x,s) & G_{12}(x,s) & G_{13}(x,s) \\ G_{21}(x,s) & G_{22}(x,s) & G_{23}(x,s) \\ G_{31}(x,s) & G_{32}(x,s) & G_{33}(x,s) \end{bmatrix} . \quad (2.13)$$

We will use a "hat" to denote the Laplace transforms of various functions: e.g., $\hat{u}(s)$ is the Laplace transform of $u(t)$.

(i) Stability Constraint. Our first and most important performance requirement is closed-loop system stability. Let $z = [z_p, z_c] \in H$. Then we recall (see [Paz.1]) that the *mild solution* of (2.12a) is given by

$$z(t) = e^{A_pt}z_0 + \int_0^t e^{A_p(t-\tau)}Bu(\tau)d\tau . \quad (2.14a)$$

We therefore define the *exponential stability* of the feedback system $S(P,K)$ in terms of the semigroup $\{e^{At}\}_{t \geq 0}$, as follows.

Definition 2.1: For any $\alpha \geq 0$, the feedback system $S(x)$ is α–stable if there exist $M \in (0,\infty)$ and $\alpha_0 > \alpha$ such that

$$\|e^{A(x)t}\|_H \leq Me^{-\alpha_0 t} \; \forall \; t \geq 0 . \quad (2.14b)$$

■

The class of plants that we can deal with are characterized by the following definition:

Definition 2.2: [Jac.1] Given an $\alpha \geq 0$, the pair (A_p, B_p) $((A_p, C_p))$ is *α–stabilizable* (*α–detectable*) if there exists a bounded linear operator $K: E \to \mathbb{R}^{n_i}$ $(F: \mathbb{R}^{n_o} \to E)$ such that $A_p - B_p K$ $(A_p - FC_p)$ is the infinitesimal generator of an α–stable C_0–semigroup. ■

It was shown in [Jac.1] that a plant is α–stabilizable and α–detectable if and only if there exists a finite dimensional compensator with $D_c = 0$ such that the feedback system is α–stable.

For any $\alpha \geq 0$, we define the *stability region* $D_{-\alpha} \triangleq \{s \in \mathbb{C} \mid \mathrm{Re}(s) < -\alpha\}$. Let $U_{-\alpha} = \{s \in \mathbb{C} \mid \mathrm{Re}(s) \geq -\alpha\}$, $\partial U_{-\alpha} = \{s \in \mathbb{C} \mid \mathrm{Re}(s) = -\alpha\}$ and let $U^o_{-\alpha} = \{s \in \mathbb{C} \mid \mathrm{Re}(s) > -\alpha\}$.

Proposition 2.1: [Jac.1] For any $\alpha \geq 0$, the plant is α–stabilizable and α–detectable if and only if there exists a decomposition of $E = E_- + E_+$, with E_+ *finite-dimensional*, which induces a decomposition of the plant (2.6a,b), of the form

$$\frac{d}{dt}\begin{bmatrix} z_{p-}(t) \\ z_{p+}(t) \end{bmatrix} = \begin{bmatrix} A_{p-} & 0 \\ 0 & A_{p+} \end{bmatrix}\begin{bmatrix} z_{p-}(t) \\ z_{p+}(t) \end{bmatrix} + \begin{bmatrix} B_{p-} \\ B_{p+} \end{bmatrix} u(t) , \tag{2.15a}$$

$$y(t) = [C_{p-} \quad C_{p+}]\begin{bmatrix} z_{p-}(t) \\ z_{p+}(t) \end{bmatrix} + Du(t), \tag{2.15b}$$

such that $\sigma(A_{p+}) \subset U_{-\alpha}$, (A_{p+}, B_{p+}) is controllable, (A_{p+}, C_{p+}) is observable, and A_{p-} is the infinitesimal generator of an α–stable C_0–semigroup on E_-. ■

In view of the above, we will restrict ourselves to feedback systems in which the plant is α–stabilizable and α–detectable for some $\alpha > 0$.

The relationship between α–stability of the feedback system and α–stabilizability of the plant is established in the following result:

Proposition 2.2: [Jac.1] Suppose that the plant is α–stabilizable and α–detectable for some $\alpha > 0$. Then, for any $\alpha \geq 0$, the feedback system is α–stable if and only if $U_{-\alpha}$ is contained in $\rho(A)$. ■

We are finally on the way to defining a computational stability criterion which can be expressed in the form of an inequality of the type appearing in problem (2.1a). First we define the *characteristic function* $\chi: \mathbb{C} \to \mathbb{C}$, of the feedback system $S(x)$, by

$$\chi(x,s) \triangleq \det(sI_{n_+} - A_{p+})\det(sI_{n_c} - A_c(x))\det(I_{n_i} + G_c(x,s)G_p(s)) \ , \tag{2.16}$$

where A_{p+} is defined as in (2.15a) and n_+ is the dimension of A_{p+}. Next, for any function f: $\mathbb{C} \to \mathbb{C}$, we define $Z(f(s)) \triangleq \{s \in \mathbb{C} \mid f(s) = 0\}$ to be its set of zeros. in [Har.1] we find the following pair of crucial results:

Proposition 2.3: [Har.1] The system $S(x)$ is stable if and only if $Z(\chi(x,s)) \subset D_{-\alpha}$. ■

Theorem 2.1: [Har.1] Let n_+ and n_c be the dimensions of the matrices A_{p+} in (2.15b) and $A_c(x)$ in (2.9a), respectively. Then $Z(\chi(x,s)) \subset D_{-\alpha}$ if and only if there exists an integer $N_n > 0$, and polynomials $d_0(s)$ and $n_0(s)$, of degree $N_d = N_n + n_s$ and N_n, respectively, with $n_s = n_c + n_+$, such that

$$(i) \ \ Z(d_0(s)) \subset D_{-\alpha} \ , \quad Z(n_0(s)) \subset D_{-\alpha}, \tag{2.17a}$$

$$(ii) \ \ \mathrm{Re}\left[\frac{\chi(x,s)n_0(s)}{d_0(s)}\right] > 0 \quad \forall \ s \in \partial U_{-\alpha} \ . \tag{2.17b}$$

■

In practice the test (2.17a,b) can be used only as a sufficient condition of stability, because one is forced to choose in advance the degree N_d of the polynomial $d_0(s)$. Furthermore, the polynomials $d_0(s)$ and $n_0(s)$ must be parametrized in such a way that satisfaction of (2.17a) can be ensured by satisfying a simple set of inequalities. We note that when $a,b \in \mathbb{R}$, $Z[(s + \alpha) + a)] \subset D_{-\alpha}$ if and only if $a > 0$, and $Z[(s + \alpha)^2 + a(s + \alpha) + b] \subset D_{-\alpha}$ if and only if $a > 0$, $b > 0$. Hence, assuming that the degree of $d_0(s)$ is odd, we set

$$d_0(s,y_d) \triangleq ((s+\alpha) + a_0)\prod_{i=1}^{m}((s + \alpha)^2 + a_i(s + \alpha) + b_i), \tag{2.18}$$

where $y_d \triangleq [a_0,a_1,a_2, \cdots ,a_m,b_1,b_2, \cdots ,b_m]^T \in \mathbb{R}^{2m+1}$ and $N_d = 2m+1$. When the degree of $d_0(s)$ is even, the first factor in (2.18) is omitted. The polynomial $n_0(s)$, which is of degree $N_n = N_d - n_s$ can be parametrized similarly, with corresponding parameter vector y_n.

It now follows from Theorem 2.1 that the requirement of closed-loop stability reduces to solving the following set of inequality constraints:

$$\varepsilon - y_d^i \leq 0, \qquad \text{for } i = 1,2, \cdots ,N_d, \tag{2.19a}$$

$$\varepsilon - y_n^i \leq 0, \qquad \text{for } i = 1,2, \cdots ,N_n, \tag{2.19b}$$

$$\varepsilon - \mathrm{Re}\left(\frac{\chi(x,-\alpha + j\omega)n_0(y_n,-\alpha + j\omega)}{d_0(y_d,-\alpha + j\omega)}\right) \le 0, \quad \forall\ \omega \in [0,\infty)\ , \tag{2.19c}$$

where y_d^i is the i–th element of y_d, y_n^i is the i–th element of y_n and $\varepsilon > 0$ is small. If we define $\bar{x} = (x, y_n, y_d)$, and

$$\psi^1(\bar{x}) \triangleq \sup_{\omega \in [0,\infty)} \left[\varepsilon - \mathrm{Re}\left(\frac{\chi(x,-\alpha + j\omega)n_0(y_n,-\alpha + j\omega)}{d_0(y_d,-\alpha + j\omega)}\right) \right], \tag{2.19d}$$

we see that (2.19c) is of the form

$$\psi^1(\bar{x}) \le 0\ . \tag{2.19e}$$

The evaluation of $\chi(x,-\alpha + j\omega)$ requires the evaluation of the closed-loop system frequency response. For an elementary treatment of how this computation can be carried out see the Appendix and [Wuu.1].

Now that we have established a set of inequalities ensuring stability, dealing with the remaining performance requirement is relatively easy.

(ii) Command Tracking and Output Disturbance Rejection. Suppose that the desired bandwidth for the feedback system is $[0,\omega_c]$. Both good tracking of the input u and good rejection of the output disturbance d_o, over this frequency interval, as well as reasonable behavior outside, can be achieved by making small $\|G_{11}(x,j\omega)\|$, the norm of the transfer function from the command input $\hat{u}$ to the tracking error $\hat{e}_1$. Therefore we define the performance function $\psi^2 : \mathbb{R}^n \to \mathbb{R}$ by

$$\psi^2(x) \triangleq \sup_{\omega \in [0,\omega_f]} \{\ \bar{\sigma}\,[G_{11}(x,j\omega)\,] - b_{fo}(\omega)\ \}, \tag{2.20a}$$

where $b_{fo}(\cdot)$ is a piecewise continuous bound function and $\bar{\sigma}\,[A]$ denotes the largest singular value of A, and ω_f is the highest frequency over which quantitative design is required. We can now express the command tracking and output disturbance rejection requirement as an inequality:

$$\psi^2(x) \le 0\ . \tag{2.20b}$$

It remains to choose the bound function $b_{fo}(\cdot)$. Since by an extension of Bode's Integral Theorem to multivariable systems [Boy.1], it follows that for every frequency interval of nonzero measure over which the feedback system attenuates output disturbances, there must exist an interval of nonzero length over which the system amplifies output disturbances, we must let $b_{fo}(\cdot)$ exceed 1 over some frequency interval outside the system bandwidth. Therefore a simple choice for the bound function would be to set

$$0 < b_{fo}(\omega) = b_1 << 1, \quad \text{if } \omega \leq \omega_c$$
$$b_{fo}(\omega) = b_2 > 1 \quad \text{if } \omega > \omega_c \ . \tag{2.21}$$

(iii) Input Disturbance Rejection. To obtain good input disturbance rejection over the feedback system bandwidth $[0,\omega_c]$, we must keep small $\| G_{33}(x,j\omega) \|$, the norm of the transfer function from the input disturbance $\hat{d}_i(s)$ to the output $\hat{y}(s)$. Therefore we define the performance function $\psi^3 : \mathbb{R}^n \to \mathbb{R}$ by

$$\psi^3(x) \triangleq \sup_{\omega \in [0,\omega_f]} \{ \bar{\sigma} [G_{33}(x,j\omega)] - b_{fi}(\omega) \}, \tag{2.22a}$$

where $b_{fi}(\cdot)$ is a piecewise continuous bound function. Hence the input disturbance rejection requirement reduces to the inequality:

$$\psi^3(x) \leq 0 \ . \tag{2.22b}$$

(iv) Plant-Input Saturation Avoidance. Since a large plant input, e_2, or state can drive the plant out of the operating region for which the linear model is valid, it is important to keep the plant input and state small, for otherwise deterioration of performance and instability may occur. Hence, to limit saturation effects produced by command inputs or output disturbances, we define the performance function $\psi^4 : \mathbb{R}^n \to \mathbb{R}$ by

$$\psi^4(x) \triangleq \sup_{\omega \in [0,\omega_f]} \{ \bar{\sigma}[G_{21}(x,j\omega)] - b_s \} \tag{2.23a}$$

where $b_s > 0$ is a suitable bound for the plant input power spectrum amplitude. The plant-input saturation avoidance requirement can now be formulated as

$$\psi^4(x) \leq 0 \ . \tag{2.23b}$$

(v) Stability Robustness to Plant Uncertainty. Plant models always have some uncertainty in them. Since closed loop stability is a fundamental requirement, the design process must take into account not only the *nominal* model, which was used to set up (2.19a-c), but also model uncertainty. The following result (see [Doy.2], [Che.1]) gives a characterization of plant uncertainty and a corresponding condition for stability robustness.

Theorem 2.2: Consider the feedback system in Fig. 1, and assume that the compensator has been chosen so that this system is 0-stable for the *nominal* plant transfer function $G_p(s)$. Let $l:\mathbb{R}_+ \to \mathbb{R}_+$ be a continuous, strictly positive "tolerance function" such that for some $k \in \mathbb{N}, \omega_0 \in \mathbb{R}_+$, $l(\omega) > 1/\omega^k$ for all $\omega > \omega_0$. Let the set

$$\mathbf{G}_l \triangleq \{ \tilde{G}_p \in \mathbb{R}(s)^{n_o \times n_i} \mid \tilde{G}_p(j\omega) = G_p(j\omega) + \delta G_p(j\omega),$$

$$\bar{\sigma}[\delta G_p(j\omega)] < l(\omega),\ \omega \in \mathbb{R}_+\ ,\ N_{\tilde{G}_p} = N_{G_p} \} \tag{2.24a}$$

where $N_{\tilde{G}_p}$, N_{G_p} denote the number of unstable poles of $\tilde{G}_p$ and G_p, respectively[4] . Then, the feedback system in Fig. 1 is 0-stable for all $\tilde{G}_p \in \mathbf{G}_l$, if and only if

$$\bar{\sigma}[G_{21}(x,j\omega)] \leq 1/l(\omega) \quad \forall\ \omega \geq 0\ . \tag{2.24b}$$

■

Hence, if we define the function $\psi^5:\mathbb{R}^n \to \mathbb{R}$ by

$$\psi^5(x) \triangleq \sup_{\omega \in [0,\infty)} \{ \bar{\sigma}[G_{21}(x,j\omega)] - 1/l(\omega) \} \tag{2.25a}$$

then for all x such that

$$\psi^5(x) \leq 0 \tag{2.25b}$$

holds, the compensator $K(x)$ will stabilize not only the plant G_p, but also any plant $\tilde{G}_p \in \mathbf{G}_l$.

B. Time Domain Performance Specifications

Frequency domain performance specifications are inadequate when "hard" time-domain bounds need to be satisfied at various points in the feedback loop. For example, it has been traditional to impose bounds, in terms of rise time, overshoot, and settling time specifications on feedback system zero-state step responses. In general, the satisfaction of such time-response specifications cannot be insured by shaping transfer functions in frequency domain.

[4] A larger set of plant perturbations is considered in [Che.1].

We will denote the zero-state responses of the system (2.12a,b) by $e_1(t,u,d_o,d_i)$, $e_2(t,u,d_o,d_i)$, and $y(t,u,d_o,d_i)$, respectively. We will continue to denote components of a vector by superscripts.

(i) Time Domain Responses. Suppose that we are required to ensure that the step response, as measured in the output $y^1(t)$, when $u(t) = u_s(t) \triangleq (1,0,0....,0)$, $d_o(t) \equiv 0$, $d_i(t) \equiv 0$, is contained in a window defined by upper and a lower, piecewise continuous bound functions, $\overline{b}(t)$, and $\underline{b}(t)$, respectively. Let $\psi^6 : \mathbb{R}^n \to \mathbb{R}$ be defined by

$$\psi^6(x) \triangleq \max_{t \in [0,t_f]} \{ y^1(t,u_s,0,0) - \overline{b}(t) \}, \tag{2.26a}$$

and let $\psi^7 : \mathbb{R}^n \to \mathbb{R}$ be defined by

$$\psi^7(x) \triangleq \max_{t \in [0,t_f]} \{ \underline{b}(t) - y^1(t,u_s,0,0) \} \tag{2.26b}$$

Then the step response specification reduces to the pair of inequalities:

$$\psi^6(x) \le 0 , \qquad \psi^7(x) \le 0 . \tag{2.26c}$$

More generally, we may require that the plant output track, within a tolerance, a given command input, say $u(t) = u_d(t)$. Let $b_d(\cdot)$ be a piecewise continuous tolerance function, and let $\psi^8 : \mathbb{R}^n \to \mathbb{R}$ be defined by

$$\psi^8(x) \triangleq \max_{t \in [0,t_f]} \{ \| y(t,u_d,0,0) - u_d(t) \|^2 - b_d(t) \}. \tag{2.27a}$$

Then the tracking requirement becomes

$$\psi^8(x) \le 0 . \tag{2.27b}$$

(ii) Output Disturbance Rejection. There may be a need to limit the effect of "persistent" output disturbances on the plant output (see [Vid.2], [Dah.1]). Frequency-domain disturbance specifications may be inadequate to meet this need, and one must deal with this problem in the time-domain by limiting the induced sup-norm of the operator that takes plant output disturbances into plant outputs. Clearly, the zero-state response of the feedback system in Fig. 1 to an output disturbance (assuming that all other inputs are zero) has the form

$$y(t,0,d_o,0) = \int_0^t K_{d_o}(x,t-\tau)d_o(\tau)d\tau \ . \tag{2.28a}$$

Let $\psi^9 : \mathbb{R}^n \rightarrow \mathbb{R}$ be defined by

$$\psi^9(x) \triangleq \sup_{t \in [0,\infty)} \int_0^t \| K_{d_o}(x,\tau) \| \, d\tau - b_o , \tag{2.28b}$$

where $b_o > 1$. Then, the requirement that

$$\psi^9(x) \leq 0 , \tag{2.28c}$$

ensures that no output disturbance $d_o(\cdot)$ with norm less than 1 will produce a feedback system output of norm larger than b_o.

(iii) Plant Saturation Avoidance. The frequency-domain saturation avoidance inequality (2.23b) does not limit the time-domain amplitude of the plant input when the command input has significant spectral content outside the closed-loop system bandwidth. We can ensure that the plant input does not exceed required bounds by limiting the size of the induced sup-norm of the operator that takes the command input into the plant input. This results in an inequality similar to (2.28c).

C. Formulation of Optimal Design Problem

First of all, it is not clear that a set of specifications, such as those stated above, is consistent. Hence it may be desirable to solve first a "phase I" problem of the form:

$$\nu = \min_{x \in \mathbf{X}} \max_{j \in \mathbf{m}} \psi^j(x) \ . \tag{2.29}$$

If the value ν turns out to be negative, all the specifications can be satisfied. If ν turns out to be positive, some compromise must be reached either by relaxing the bounds in the definitions of the $\psi^j(\cdot)$, or by increasing the compensator dimension until satisfaction is obtained. Once this has been done, then one of the performance functions can be designated as the cost function, while the others become constraint functions in a design problem of the form (2.1a). Alternatively, weights can be introduced into the minimax problem (2.29) as a way of obtaining a compromise design.

3. ALGORITHMS

Firstly, for an extensive treatment of nonsmooth optimization algorithms we refer the reader to [Pol.3]. In this paper we will content ourselves with an introduction to the subject. The easiest way to explain algorithms for solving (2.1a) and (2.2a) is to proceed in stages which take us from a *conceptual* algorithm for solving finite dimensional problems of the form (2.1a), to an *implementable* algorithm for solving finite dimensional problems of the form (2.1a), to an *implementable* algorithm which solves optimal control problems, with control and state space constraints, and either ODE or PDE dynamics, of the form (2.2a). We will not treat separately the problem (2.29) since an algorithm for it is obtained from one for solving (2.1a), by setting $\psi^0(x) \equiv -\infty$, or by observing that it is equivalent to the problem below, defined on $\mathbb{R}^{n+1}$, in which we denote vectors by $\bar{x} = (x, x^{n+1}) = (x^1, \cdots, x^{n+1})$:

$$\min\{ x^{n+1} \mid \psi^j(x) - x^{n+1} \leq 0,\ j \in \mathbf{m},\ x \in \mathbf{X} \}. \tag{3.1}$$

Let the *steering parameter*[5] $\gamma \geq 1$ be given and let

$$\psi(x) \triangleq \max_{j \in \mathbf{m}} \psi^j(x), \tag{3.2a}$$

$$\psi_+(x) \triangleq \max\{ 0, \psi(x) \}, \tag{3.2b}$$

and, for any $z \in \mathbb{R}^n$, let the parametrized function $F_z(x)$ be defined by

$$F_z(x) \triangleq \max\{ \psi^0(x) - \psi^0(z) - \gamma\psi_+(z), \psi^j(x) - \psi_+(z),\ j \in \mathbf{m} \}. \tag{3.2c}$$

Our first observation is that for any $z \in \mathbb{R}^n$, $F_z(z) = 0$. Our second observation is that if x^* is a local minimizer for (2.1a), then it follows from the fact that (i) $\psi(x) > 0$ when x is infeasible for (2.1a), and (ii) $\psi^0(x) > \psi^0(x^*)$ when x is feasible, but not optimal for (2.1a), that x^* must also be a local minimizer for the problem

$$\min_{x \in \mathbf{X}} F_{x^*}(x)\ . \tag{3.2d}$$

Hence, as we shall now show, the function $F_z(\cdot)$ provides a very useful means for obtaining a first order

[5] An examination of (3.2c) shows that the value of γ and, in fact, the term $\psi_+(z)$ has no effect at *feasible points*. We shall see later that their inclusion enables us to construct a phase I - phase II algorithm which does not require a feasible starting point.

optimality condition for the problem (2.1a)[6].

Now, suppose that given $z \in \mathbb{R}^n$, we approximate each function $\phi^j(x,y_j)$, $j = 0,1,2,\ldots,m$, around z by the following *first order convex* approximation:

$$\hat{\phi}_z^j(x,y_j) \triangleq \phi^j(z,y_j) + \langle \nabla_x \phi^j(z,y_j),(x-z)\rangle + \tfrac{1}{2}\|x-z\|^2 . \tag{3.3a}$$

Then $\psi^j(x)$ is approximated around z by the *first order convex* approximation

$$\hat{\psi}_z^j(x) \triangleq \max_{y_j \in \mathbf{Y}_j} \hat{\phi}_z^j(x,y_j), \; j = 0,1,2,\ldots,m, \tag{3.3b}$$

and, in turn, $F_z(x)$ is approximated around z, by the *first order convex* approximation

$$\hat{F}_z(x) \triangleq \max\{ \hat{\psi}_z^0(x) - \psi^0(z) - \gamma\psi_+(z), \hat{\psi}_z^j(x) - \psi_+(z), j \in \mathbf{m} \}. \tag{3.3c}$$

Referring to [Pol.3] we find the following first order optimality condition for (2.1a), where we find it convenient to replace x in (3.3c) by $x^* + h$:

Theorem 3.1 :[Pol.3] If x^* is a local minimizer for (2.1a), then

$$\theta(x^*) \triangleq \min_{h \in \mathbf{X} - \{x^*\}} \hat{F}_{x^*}(x^* + h) = 0 . \tag{3.4a}$$

■

It is shown in [Pol.3] that $\theta(\cdot)$ is continuous; it follows by inspection that $\theta(x) \le 0$ for all $x \in \mathbb{R}^n$. Furthermore, it is shown in [Pol.3] that at any $x \in \mathbb{R}^n$ such that $\theta(x) < 0$, the vector

$$\eta(x) \triangleq \arg \min_{h \in \mathbf{X} - \{x\}} \hat{F}_x(x + h), \tag{3.4b}$$

has the following property: if $\psi(x) > 0$, then $\eta(x)$ is a descent direction for $\psi(\cdot)$; if $\psi(x) \le 0$, then $\eta(x)$ is a descent direction for $\psi^0(\cdot)$ along which the constraints will not be violated for some distance. The search direction function $\eta(\cdot)$ can be shown to be continuous (see [Pol.3]).

Phase I - phase II methods of feasible directions, such as the ones described in [Pol.3], as well as the optimal control algorithm that we will present, can be seen as progressively more complex imple-

---

[6] In [Cla.1] the reader will find optimality conditions for (2.1a) in the more familiar form involving generalized gradients and multipliers, emanating from the fact that if x^* is optimal for (3.2d), then $dF_{x^*}(x^*; x - x^*) \ge 0$ must hold for all $x \in \mathbf{X}$. It is shown in [Pol.3] that the conditions given in this paper, which were derived specifically for use in algorithm construction, are equivalent to the ones in [Cla.1].

mentations of the following *conceptual* method, which we have derived by extension, from the Huard method of centers [Hua.1], for solving (2.1a).

Algorithm 3.1 (Phase I - Phase II Method of Centers) :

Parameters : $\gamma \geq 1$.

Data : $x_0 \in \mathbb{R}^n$,

Step 0 : Set $i = 0$.

Step 1 : Compute $x_{i+1} = \arg \min_{x \in \mathbf{X}} F_{x_i}(x)$.

Step 2 : Set $i = i + 1$ and go to Step 1. ■

Since $F_{x_i}(x_i) = 0$, $F_{x_i}(x_{i+1}) < 0$. Hence if $\psi(x_{i_0}) \leq 0$ for some i_0, then $\psi(x_i) \leq 0$ for all $i \geq i_0$. When all the functions are convex, it can be shown that the value of γ controls the speed with which the above algorithm approaches the *feasible set*, $\{ x \mid \psi(x) \leq 0 \}$: the larger γ, the faster the algorithm drives the iterates x_i into the feasible set.

Theorem 3.2 : Suppose that for every $x \in \mathbb{R}^n$ which is not a local minimizer of (2.1a),

$$M(x) \triangleq \min_{x' \in \mathbf{X}} F_x(x') < 0, \tag{3.4c}$$

and that either $\mathbf{X}$ is compact or that the level sets of $F_x(\cdot)$ are compact. If $\{ x_i \}_{i=0}^{\infty}$ is an infinite sequence constructed by Algorithm 3.1, then every accumulation point x^* of $\{ x_i \}_{i=0}^{\infty}$ is a local minimizer for (2.1a).

Proof : First, referring to [Ber.1], we conclude that because of the compactness assumption, $M(\cdot)$ is continuous. Next, suppose that $\{ x_i \}_{i=0}^{\infty}$ is an infinite sequence constructed by Algorithm 3.1 which has an accumulation point x^* such that $M(x^*) < 0$, i.e., x^* is not a local minimizer for (2.1a). Then there exists an infinite subset $K \subset \mathbb{N}$ such that the subsequence $\{ x_i \}_{i \in K}$ converges to x^*, and hence, by continuity of $M(\cdot)$, there exists an i_o such that $M(x_i) \leq M(x^*)/2 < 0$ for all $i \geq i_o$, $i \in K$. Now there are two possibilities.

First suppose that $\psi(x_i) > 0$ for all $i \in \mathbb{N}$. Then, because $\psi(x_{i+1}) - \psi(x_i) \leq M(x_i) \leq 0$ holds for all i, the sequence $\{ \psi(x_i) \}_{i=0}^{\infty}$ is monotone decreasing, and hence, since $\psi(\cdot)$ is continuous and since $\{ x_i \}_{i=0}^{\infty}$ has an accumulation point, the sequence $\{ \psi(x_i) \}_{i=0}^{\infty}$ must converge. However, this contradicts the fact that $\psi(x_{i+1}) - \psi(x_i) \leq M(x_i) \leq M(x^*)/2 < 0$ for all $i \geq i_o$, $i \in K$.

Next, suppose that there exists an i_1 such that $\psi(x_{i_1}) \leq 0$. Then, for all $i \geq \max\{i_o, i_1\}$, we must have $\psi(x_{i+1}) \leq M(x_i) \leq 0$ and $\psi^0(x_{i+1}) - \psi^0(x_i) \leq M(x_i) < 0$, so that the sequence $\{ \psi^0(x_i) \}_{i=0}^{\infty}$ is monotone decreasing. Since $\{ x_i \}_{i=0}^{\infty}$ has an accumulation point, the sequence $\{ \psi^0(x_i) \}_{i=0}^{\infty}$ must converge. However, this contradicts the fact that $\psi^0(x_{i+1}) - \psi^0(x_i) \leq M(x_i) \leq M(x^*)/2 < 0$ for all $i \geq i_o$, $i \in K$. ■

It should be obvious that the unconstrained minimax problems, in Step 1 of the Method of Centers 3.1, are hardly any easier to solve than the original problem (2.1a). However, a very efficient algorithm can be obtained by replacing the computation in Step 1 of this method by the approximation indicated in (3.4a,b) and supplementing it by an Armijo type step size rule [Arm.1], as follows[7]:

Algorithm 3.2 (Phase I - Phase II Method of Feasible Directions) :

Parameters : $\gamma > 1$, α, $\beta \in (0,1)$.

Data : $x_0 \in \mathbb{R}^n$.

Step 0 : Set $i = 0$.

Step 1 : Compute the the *optimality function* value $\theta_i = \theta(x_i)$, and the corresponding *search direction* $\eta_i = \eta(x_i)$.

Step 2 : Compute the *step size* λ_i:

$$\lambda_i = \max\{ \beta^k \mid k \in \mathbb{N},\ F_{x_i}(x_i + \beta^k \eta_i) \leq \beta^k \alpha \theta_i \} . \tag{3.5a}$$

Step 3 : Set $x_{i+1} = x_i + \lambda_i \eta_i$, set $i = i + 1$ and go to Step 1. ■

The Armijo step size rule is well known to be efficient and is used in many algorithms. The sensitivity to the value of γ and the convergence properties of the Phase I - Phase II Method of Feasible

[7] The phase I - phase II algorithms reported in [Pol.3] use a different step size rule when $\psi(x_i) > 0$ and when $\psi(x_i) < 0$. The simplified algorithm in this paper (see [Pol.9]) is only slightly less efficient, because it evaluates the cost function at infeasible points.

Directions 3.2 are quite similar to those of the Phase I - Phase II Method of Centers 3.1 (for a proof see [Pol.9]):

Theorem 3.3 : Suppose that for every $x \in \mathbf{X}$ such that $\psi(x) \geq 0$, $\theta(x) < 0$. If $\{ x_i \}_{i=0}^{\infty}$ is an infinite sequence constructed by Algorithm 3.2, then every accumulation point x^* of $\{ x_i \}_{i=0}^{\infty}$ satisfies the first order optimality condition, for (2.1a), $\psi(x^*) \leq 0$, $\theta(x^*) = 0$.

Proof : First, we recall that it can be shown that both $\theta(\cdot)$ and $\eta(\cdot)$ are continuous. Next, suppose that $\{ x_i \}_{i=0}^{\infty}$ is an infinite sequence constructed by Algorithm 3.2 which has an accumulation poin x^* such that $\theta(x^*) < 0$[8]. Then there exists an infinite subset $K \subset \mathbb{N}$ such that the subsequence $\{ x_i \}_{i \in K}$ converges to x^*, and hence, by continuity of $\theta(\cdot)$, there exists an i_o such that $\theta(x_i) \leq \theta(x^*)/2 < 0$ for all $i \geq i_o$, $i \in K$. Next, since $dF_{x^*}(x^*;\eta(x^*)) \leq \theta(x^*)$, it follows that there exists a $k^* < \infty$ such that

$$F_{x^*}(x^* + \beta^{k^*}\eta(x^*)) - \beta^{k^*}\alpha\theta(x^*) < 0 \,. \tag{3.5b}$$

Since $\theta(\cdot)$, $\eta(\cdot)$, and $F_z(x)$ are all continuous, it follows from (3.5b) that there exists a finite $i_1 \geq i_o$, such that for all $i \in K$, $i \geq i_1$, $\lambda_i \geq \beta^{k^*}$ and hence for all $i \in K$, $i \geq i_1$,

$$F_{x_i}(x_{i+1}) \leq \beta^{k^*}\alpha\theta(x^*)/2 < 0 \,. \tag{3.5c}$$

As in the proof of Theorem 3.1, there are now two possibilities. First suppose that $\psi(x_i) > 0$ for all $i \in \mathbb{N}$. Then, because $\psi(x_{i+1}) - \psi(x_i) \leq F_{x_i}(x_{i+1}) \leq 0$ holds for all i, the sequence $\{ \psi(x_i) \}_{i=0}^{\infty}$ is monotone decreasing, and hence, since $\psi(\cdot)$ is continuous and since $\{ x_i \}_{i=0}^{\infty}$ has an accumulation point, the sequence $\{ \psi(x_i) \}_{i=0}^{\infty}$ must converge. However, this contradicts the fact that $\psi(x_{i+1}) - \psi(x_i) \leq F_{x_i}(x_{i+1}) \leq \beta^{k^*}\alpha\theta(x^*)/2 < 0$ for all $i \geq i_1$, $i \in K$.

Next, suppose that there exists an i_2 such that $\psi(x_{i_2}) \leq 0$. Then, for all $i \geq \max\{i_1, i_2\}$, we must have $\psi(x_{i+1}) \leq F_{x_i}(x_{i+1}) \leq 0$ and $\psi^0(x_{i+1}) - \psi^0(x_i) \leq F_{x_i}(x_{i+1}) \leq 0$, so that the sequence $\{ \psi^0(x_i) \}_{i=0}^{\infty}$ is monotone decreasing. Since $\{ x_i \}_{i=0}^{\infty}$ has an accumulation point, the sequence $\{ \psi^0(x_i) \}_{i=0}^{\infty}$ must converge. However, this contradicts the fact that $\psi^0(x_{i+1}) - \psi^0(x_i) \leq \beta^{k^*}\alpha\theta(x^*)/2 < 0$ for all $i \geq i_2$, $i \in K$. ■

---

[8] The case where $\psi(x^*) > 0$ is eliminated by our assumption.

When the functions $\psi^j(\cdot)$ in (2.1a) are all differentiable, Algorithm 3.2 can be used directly. However, when the functions $\psi^j(\cdot)$ are max functions, then neither these functions nor the optimality function $\theta(\cdot)$ can be evaluated exactly on a digital computer in finite time. Hence, for such problems Algorithm 3.2 must be viewed as a *conceptual algorithm*. To construct an implementable version, we make use of the theory developed in [Kle.1], which allows us to discretize the intervals $\mathbf{Y}_j$ adaptively, as follows. For $j = 1,2,...,m$, let $l_j \triangleq (b_j - a_j)$ be the length of the interval $\mathbf{Y}_j$. Next, for any positive integer q, we define the corresponding discretized versions of the functions used by Algorithm 3.2:

$$\mathbf{Y}_{jq} \triangleq \{\, a_j \,,\, a_j + \frac{l_j}{q} \,,\, a_j + \frac{2l_j}{q} \,,\, \dots \,,\, b_j \,\}, \tag{3.6a}$$

$$\psi_q^j(x) \triangleq \max_{y \in \mathbf{Y}_{jq}} \phi^j(x,y), \tag{3.6b}$$

$$F_{q\,z}(x) \triangleq \max\{\, \psi_q^0(x) - \psi_q^0(z) - \gamma\psi_{q+}(z), \psi_q^j(x) - \psi_{q+}(z),\ j \in \mathbf{m} \,\}, \tag{3.6c}$$

$$\hat{\psi}_{q\,z}^j(x) \triangleq \max_{y \in \mathbf{Y}_{jq}} \hat{\phi}_z^j(x,y), \tag{3.6d}$$

$$\hat{F}_{q\,z}(x) \triangleq \max\{\, \hat{\psi}_{q\,z}^0(x) - \psi_q^0(z) - \gamma\psi_{q+}(z), \hat{\psi}_{q\,z}^j(x) - \psi_{q+}(z),\ j \in \mathbf{m} \,\}, \tag{3.6e}$$

$$\theta_q(z) \triangleq \min_{x \in \mathbf{X}} \hat{F}_{q\,z}(x), \tag{3.6f}$$

$$\eta_q(z) \triangleq \arg \min_{h \in \mathbf{X} - \{z\}} \hat{F}_{q\,z}(z + h)\,. \tag{3.6g}$$

The following easy to prove result assures that the implementable version of Algorithm 3.2, to be stated shortly, satisfies the requirements of the theory in [Kle.1].

Proposition 3.1 : [Bak.1] There exists a $K < \infty$ such that for all $z,\, x \in \mathbf{X}$,

$$|\psi^j(x) - \psi_q^j(x)| \le \frac{K}{q},\quad j = 0,1,2,...,m, \tag{3.7a}$$

$$|F_z(x) - F_{q\,z}(x)| \le \frac{K}{q}, \tag{3.7b}$$

$$|\hat{\psi}_z^j(x) - \hat{\psi}_{q\,z}^j(x)| \le \frac{K}{q},\quad j = 0,1,2,...,m, \tag{3.7c}$$

$$|\hat{F}_z(x) - \hat{F}_{q\,z}(x)| \le \frac{K}{q}, \tag{3.7d}$$

$$|\theta(x) - \theta_q(x)| \leq \frac{K}{q}, \tag{3.7e}$$

$$\|\eta(x) - \eta_q(x)\|^2 \leq \frac{K}{q}. \tag{3.7f}$$

We can now state an implementable version of Algorithm 3.2 which increases the discretization of the intervals $\mathbf{Y}_j$ whenever the reduction per iteration in constraint violation, or cost, as appropriate, drops below a preassigned level, which we will call ε.

Algorithm 3.3 (Implementable Phase I - Phase II Method of Feasible Directions) :

Parameters : $q \in \mathbb{N}$, $\varepsilon > 0$, $\gamma \geq 1$, α, $\beta \in (0,1)$.

Data : $x_0 \in \mathbb{R}^n$.

Step 0 : Set $i = 0$.

Step 1 : Compute the the *optimality function* value $\theta_i = \theta_q(x_i)$, and the corresponding *search direction* $\eta_i = \eta_q(x_i)$.

Step 2 : Compute the *step size* λ_i:

$$k_i = \arg\max\{ \beta^k \mid k \in \mathbb{N},\ F_{q\,x_i}(x_i + \beta^k\eta_i) \leq \beta^k\alpha\theta_i \}. \tag{3.8a}$$

If

$$F_{q\,x_i}(x_i + \beta^{k_i}\eta_i) > -\varepsilon, \tag{3.8b}$$

replace q by $2q$, ε by $\varepsilon/2$ and go to Step 1. Else set $\lambda_i = \beta^{k_i}$.

Step 3 : Set $x_{i+1} = x_i + \lambda_i\eta_i$, set $i = i + 1$ and go to Step 1. ■

When $\mathbf{X} = \mathbb{R}^n$, the search direction, η_i, in the above algorithm (as well as in Algorithm 3.2) can be computed quite efficiently using the algorithms in [Hoh.1, Hig.1]; when $\mathbf{X}$ is polyhedral, Polyak's constrained Newton algorithm [Pol.10] can be used (see [Pol.5]), after (3.5a) has been converted to dual form. Since the discretization rule that we have described satisfies the assumptions in [Kle.1], we obtain the following result.

Theorem 3.4 : Suppose that for every $x \in \mathbf{X}$ such that $\psi(x) \geq 0$, $\theta(x) < 0$. If Algorithm 3.3 jams up, cycling between Step 1 and Step 2, at a point x_k, then x_k satisfies the first order condition $\psi(x_k) \le 0$ and $\theta(x_k) = 0$. If Algorithm 3.3 constructs an infinite sequence $\{ x_i \}_{i=0}^{\infty}$, then every accumulation point x^* of $\{ x_i \}_{i=0}^{\infty}$ satisfies the first order optimality condition, for (2.1a), $\psi(x^*) \le 0$, $\theta(x^*) = 0$.

Proof: According to the theory in [Kle.1], because (3.7a) holds, we only need to show that for every $\hat{x} \in \mathbb{R}^n$ such that $\theta(\hat{x}) < 0$, there exist a $\hat{\rho} > 0$, a $\hat{\delta} > 0$ and a $\hat{q} \in \mathbb{N}_+$ such that for all $x \in \mathbf{B}(\hat{x},\hat{\rho}) \triangleq \{ x \in \mathbf{X} \mid \|x - \hat{x}\| \le \hat{\rho} \}$, if $x_i \in \mathbf{B}(\hat{x},\hat{\rho})$, $q \ge \hat{q}$, and k_i is constructed according to (3.8a), then

$$F_{q\,x_i}(x_i + \beta^{k_i}\eta(x_i)) \le -\hat{\delta} \,. \tag{3.9a}$$

Now, referring to the proof of Theorem 3.3, we see that for Algorithm 3.2 there exists a $\hat{\rho} > 0$, a $\hat{\delta} > 0$ and a $\hat{k} < \infty$ such that for all $x_i \in \mathbf{B}(\hat{x},\hat{\rho})$,

$$F_{x_i}(x_i + \beta^{\hat{k}}\eta(x_i)) - \beta^{\hat{k}}\alpha\theta(x_i) \le -2\hat{\delta} \,. \tag{3.9b}$$

Now, it follows from (3.7a-f) that there exists a $\hat{q} < \infty$, such that if $q \ge \hat{q}$, $x_i \in \mathbf{B}(\hat{x},\hat{\rho})$ and k_i is computed according to (3.8a), as appropriate, then $k_i \le \hat{k}$ must hold. The desired result now follows from the continuity of $\theta(\cdot)$ and (3.7a-f). ■

We will obtain an algorithm for solving optimal control problems of the form (2.2a) by formal extension of Algorithm 3.3. First, to obtain a first order optimality condition for problem (2.2a), we need to obtain an analogue of the expressions (3.2a-c), (3.3a-c) and (3.4a-b). Clearly, analogues of (3.2a-c) are obtained by replacing x in (3.2a-c) by the pair (u,T). Next, the analogue of (3.3a) is seen to be given by

$$\hat{\phi}^j_{u',T'}(u,T,t) \triangleq g^j(z^{u',T'}(t)) + \langle \nabla g^j(z^{u',T'}(t), \delta z^{u',T'}(t;u(t) - u'(t),T - T')\rangle + \tfrac{1}{2}\int_0^1 \|u(t) - u'(t)\|^2 dt + \tfrac{1}{2}|T - T'|^2, \; j = 0,1,2,\ldots,m, \tag{3.10}$$

Next, the analogues of (3.3b -c) are seen to be

$$\hat{\psi}^j_{u',T'}(u,T) \triangleq \max_{t \in \mathbf{Y}_j} \hat{\phi}^j_{u',T'}(u,T,t), \; j = 0,1,2,\dots,m, \tag{3.11a}$$

$$\hat{F}_{u',T'}(u,T) \triangleq \max\{ \hat{\psi}^0_{u',T'}(u,T) - \psi^0(u',T') - \gamma\psi_+(u',T'), \hat{\psi}^j_{u',T'}(u,T) - \psi_+(u',T'), j \in \mathbf{m} \}. \tag{3.11b}$$

The analogues of (3.4a-b), defining the optimality function $\theta(\cdot)$ and search direction function $\eta(\cdot,\cdot)$ are:

$$\theta(u',T') \triangleq \min_{(u,T) \in \mathbf{U} \times \mathbf{T}} \hat{F}_{u',T'}(u,T), \tag{3.12a}$$

$$\eta(u',T') \triangleq \arg \min_{(u,T) \in \mathbf{U} \times \mathbf{T} - \{u',T'\}} \hat{F}_{u',T'}(u' + u, T' + T) . \tag{3.12b}$$

Theorem 3.1 assumes the following form for problem (2.2a):

Theorem 3.5 : If (u^*,T^*) is a local minimizer for (2.2a), then

$$\theta(u^*,T^*) = 0 . \tag{3.13}$$

■

Since $\phi^j(u',T',t) \triangleq g^j(z^{u',T'}(t))$, our first observation is that expressions such as

$$d\phi^j((u',T',t);(u(t) - u'(t),T - T')) = \langle \nabla g^j(z^{u',T'}(t)), \delta z^{u',T'}(t;u(t) - u'(t),T - T') \rangle \tag{3.14}$$

can be computed using adjoints. Our second observation is that the numerical solution of ordinary or partial differential equations requires discretization of at least one variable and hence that we cannot utilize the analogue of Algorithm 3.3 without addressing this source of difficulty. To ensure that our final implementable algorithm has the desired convergence properties, we must use discretizations in the solution of the ODEs or PDEs which guarantee that the relations (3.7a - d) are satisfied, with x replaced by u,T. Again we receive guidance from the theory in [Kle.1], where the discretizations are worked out for ODEs. Hence we will only consider the case of PDEs here. To make matters concrete, we may assume that H is the space of r–times differentiable functions, $z(s)$, from $[0,1]$ into $\mathbb{R}^p$.

First we introduce a set of orthogonal spline functions $\{ \zeta^i_{q_s}(\cdot) \}_{i=0}^{2^{q_s}} \subset H$, for "spatial" discretization[9], write $z^{u,T}(t,s)$ in the form

[9] For many dynamical systems, a system of second order PDEs, coupled with ODEs, is a more "natural" description than (3.10a). In that case all calculations are carried out with the original dynamics. As a result, since the weak form of a solution is used, it is often possible to use splines that are only $r/2$–times differentiable, which results in considerable computational simplification. Also, Newmark's method is then used for temporal discretization. See [Str.1] for details.

$$z^{u,T}(t,s) = \sum_{i=0}^{2^{q_s}} \zeta^i_{q_s}(s)\omega^i_{q_s}(t,u,T) \,. \tag{3.15a}$$

and compute the projection Πz_0 of z_0 onto the subspace of H spanned by the splines. Let $\omega_{q_s} = (\omega^0_{q_s},\ldots,\omega^{2^{q_s}}_{q_s})$, and let Z_{q_s} be a matrix with columns $\zeta^i_{q_s}$, $i = 0,1,\ldots,2^{q_s}$. Then (3.15a) can be written in the shorter form

$$z^{u,T}(t,s) = Z_{q_s}(s)\omega_{q_s}(t,u,T) \,. \tag{3.15b}$$

On the subspace spanned by the splines, our dynamics have the form

$$Z_{q_s}(s)\dot{\omega}_{q_s}(t,u,T) = T[AZ_{q_s}(s)\omega_{q_s}(t,u,T) + h(Z_{q_s}(s)\omega_{q_s}(t,u,T),u(t))], \tag{3.15c}$$

$$Z_{q_s}(s)\omega_{q_s}(0,u,T) = \Pi z_o(s) \,,\ \ \forall\ s \in [0,1] \,. \tag{3.15d}$$

Next we use the orthogonality of the splines to set up the differential equations for the functions $\omega^i_{q_s}(t,u,T)$:

$$\dot{\omega}^i_{q_s}(t,u,T) = \int_0^1 \langle\, \zeta^i_{q_s}(s), T[AZ_{q_s}(s)\omega_{q_s}(t,u,T) + h(Z_{q_s}(s)\omega_{q_s}(t,u,T),u(t))]\rangle\, ds,\ \ i = 0,1,2,3,\ldots,2^{q_s}, \tag{3.15e}$$

$$\omega^i_{q_s}(0,u,T) = \int_0^1 \langle\, \zeta^i_{q_s}(s), \Pi z_o(s)\rangle\, ds \,,\ \ i = 0,1,2,3,\ldots,2^{q_s} \,. \tag{3.15f}$$

Then (3.15e,f) can be written as a first order vector differential equation in which the function $F(\cdot,\cdot)$ is defined by (3.15c):

$$\dot{\omega}(t) = TF(\omega(t),u(t),T) \,,\ \ \forall\ t \in [0,1],\ \ \omega(0) = \omega_0 \,. \tag{3.16a}$$

Finally, we discretize the normalized time interval $[0,1]$ into 2^{q_t} equal intervals, set $\Delta_{q_t} \triangleq 1/2^{q_t}$, and replace (3.16a) by the difference equation resulting from the use of the Euler method of integration:

$$\omega((k+1)\Delta_{q_t}) = \omega(k\Delta_{q_t}) + TF(\omega(k\Delta_{q_t}),u(k\Delta_{q_t}),T) \,,\ \ \forall\ k = 0,1,2,\ldots,2^{q_t},\ \ \omega(0) = \omega_0 \,. \tag{3.16b}$$

We are now ready to relate this construction to the quantities defined in (3.6a-g). First, let

$$p_{q_t}(t) \triangleq \begin{cases} 1 \text{ for } t \in [0,\Delta_{q_t}] \\ 0 \text{ for } t > \Delta_{q_t} \end{cases} , \tag{3.17a}$$

let $\mathbf{U}_{q_t} \subset \mathbf{U}$ be the set of controls which are constant over our time grid, i.e., if $u(t) \in \mathbf{U}_{q_t}$, then for a sequence of vectors $\{ u_k \}_{k=0}^{2^{q_t}-1} \subset U$,

$$u(t) = \sum_{k=0}^{2^{q_t}-1} u_k p_{q_t}(t - k\Delta_{q_t}) , \tag{3.17b}$$

and, finally, let $\omega_{q_t}(k\Delta_{q_t},u,T)$ denote the solution of (3.16b) corresponding to a control in $\mathbf{U}_{q_t}$. If we let $\mathbf{Y}_{jq_t} \triangleq \{ 1 \}$ for $j = 0,1,2,\ldots,m_1$, and $\mathbf{Y}_{jq_t} = \mathbf{Y} \triangleq \{ 0,\frac{1}{\Delta_{q_t}},\frac{2}{\Delta_{q_t}},\ldots,1 \}$, for $j = m_1 + 1,\ldots m$, then, for any $u \in \mathbf{U}_{q_t}$, we can define

$$\psi^j_{q_s,q_t}(u) \triangleq \max_{t \in \mathbf{Y}_{jq_t}} g^{j-m_1}(Z_{q_s}\omega_{q_t}(t,u,T)),\ j = 0,1,2,\ldots,m . \tag{3.18}$$

Next, the sensitivities of the difference equation (3.16b) to perturbations in the control and scale factor are given by the solution $\delta\omega^{u,T}_{q_t}(t,u,T)$, of the linearized difference equation:

$$\begin{aligned}\delta\omega((k+1)\Delta_{q_t}) = \delta\omega(k\Delta_{q_t}) &+ T\frac{\partial F(\omega(k\Delta_{q_t}),u(k\Delta_{q_t}),T)}{\partial\omega}\delta\omega(k\Delta_{q_t}) \\ &+ T\frac{\partial F(\omega(k\Delta_{q_t}),u(k\Delta_{q_t}),T)}{\partial u}\delta u(k\Delta_{q_t}) \\ &+ F(\omega(k\Delta_{q_t}),u(k\Delta_{q_t}),T)\delta T,\ \forall\ k = 0,1,2,\ldots,2^{q_t},\ \ \delta\omega(0) = 0 .\end{aligned} \tag{3.19}$$

Hence, given any (u',T'), $(u,T) \in \mathbf{U}_{q_t} \times \mathbf{T}$, we define

$$\begin{aligned}\hat{\phi}^j_{q_s,q_t\,(u',T')}(u,T,t) \triangleq\ & g^j(Z_{q_s}\omega^{u',T}_{q_t}(t)) + \langle\nabla g^j(Z_{q_s}\omega^{u',T}_{q_t}(t)),Z_{q_s}\delta\omega^{u',T}_{q_t}(t;u(t) - u'(t),T - T')\rangle \\ &+ ½\int_0^1 \|u(t) - u'(t)\|^2 dt + ½|T - T'|^2, t \in \mathbf{Y}_{q_t}\ \ j = 0,1,2,\ldots,m,\end{aligned} \tag{3.20}$$

In turn, these definitions lead to the following ones: For any $(u,T) \in \mathbf{U}_{q_t} \times \mathbf{T}$,

$$\hat{\psi}^j_{q_s,q_t\,(u',T')}(u,T) \triangleq \max_{t \in Y_{jq_t}} \hat{\phi}_{j\;q_s,q_t\,(u',T')}(u,T,t), \tag{3.21a}$$

$$\hat{F}_{q_s,q_t\,(u',T')}(u,T) \triangleq \max\{\, \hat{\psi}^0_{q_s,q_t\,(u',T')}(u,T) - \psi^0_{q_s,q_t}(u',T') - \gamma\psi_{q_s,q_t\,+}(u',T'),$$

$$\hat{\psi}^j_{q_s,q_t\,(u',T')}(u,T) - \psi_{q_s,q_t\,+}(u',T'),\ j \in \mathbf{m}\,\}, \tag{3.21b}$$

$$\theta_{q_s,q_t}(u',T') \triangleq \min_{(u,T) \in \mathbf{U}_{q_t} \times \mathbf{T}} \hat{F}_{q_s,q_t\,(u',T')}(u,T), \tag{3.21c}$$

$$\eta_{q_s,q_t}(u',T') \triangleq \arg \min_{(u,T) \in \mathbf{U}_{q_t} \times \mathbf{T} - \{u',T'\}} \hat{F}_{q_s,q_t\,(u',T')}(u' + u, T' + T)\,. \tag{3.21d}$$

It takes some work to show that the following result is true (see [Bak.1]):

Proposition 3.2 : There exists a $K < \infty$ such that for any positive integer q, if $\max\{\, q_s, q_t \,\} > q$, then for all $(u',T'),\ (u,T) \in \mathbf{U}_{q_t} \times \mathbf{T}$,

$$|\psi^j(u,T) - \psi^j_{q_s,q_t}(u,T)| \le \frac{K}{q},\quad j = 0,1,2,\ldots,m, \tag{3.22a}$$

$$|\hat{\psi}^j_{(u',T')}(u,T) - \hat{\psi}^j_{q_s,q_t\,(u',T')}(u,T)| \le \frac{K}{q},\quad j = 0,1,2,\ldots,m, \tag{3.22b}$$

$$|\theta(u,T) - \theta_{q_s,q_t}(u,T)| \le \frac{K}{q}, \tag{3.22c}$$

$$\|\eta(u,T) - \eta_{q_s,q_t}(u,T)\|^2 \le \frac{K}{q}\,. \tag{3.22d}$$

■

With these developments out of the way, we can now state our implementable optimal control algorithm. A close examination will show that the algorithm below constructs a finite dimensional problem, in which the design vector is the sequence of vector coefficients $\{\, u_k \,\}_{k=0}^{2^{q_t}-1}$ which defines a control in $\mathbf{U}_{q_t}$, to be solved by Algorithm 3.2 until the discretization test requires that the discretization be refined.

Algorithm 3.4 (Implementable Phase I - Phase II Optimal Control Algorithm) :

Parameters : $q_s,\ q_t \in \mathbb{N},\ \varepsilon > 0,\ \gamma > 1,\ \alpha,\ \beta \in (0,1)$.

Data : A vector coefficient sequence $u_0 = (u_0^0, \ldots, u^{2^{q_t}-1}) \in \mathbb{R}^{p2^{q_t}}$, defining the control $u_0(t)$ via (3.17b), and a scaling parameter T_0.

Step 0 : Set $i = 0$.

Step 1 : Compute the the *optimality function* value $\theta_i = \theta_{q_s,q_t}(u_i,T_i)$, and the corresponding *search direction* $\eta_i = \eta_{q_s,q_t}(u_i,T_i)$.[10]

Step 2 : Compute the *step size* λ_i:

$$k_i = \arg\max\{ \beta^k \mid k \in \mathbb{N},\ F_{q_s,q_t}((u_i,T_i) + \beta^k\eta_i) \le \beta^k\alpha\theta_i \}\ . \tag{3.23a}$$

If

$$F_{q_s,q_t}((u_i,T_i) + \beta^{k_i}\eta_i) > -\varepsilon, \tag{3.23b}$$

replace q_s, q_t by $2q_s$, $2q_t$, respectively, replace ε by $\varepsilon/2$ and go to Step 1. Else set $\lambda_i = \beta^{k_i}$.

Step 3 : Set $(u_{i+1},T_{i+1}) = (u_i,T_i) + \lambda_i\eta_i$, set $i = i + 1$ and go to Step 1. ■

The convergence properties of the above optimal control algorithm are quite analogous to those of Algorithm 3.3 and depend on Proposition 3.2:

Theorem 3.6 : Suppose that for every $(u,T) \in \mathbf{U} \times \mathbb{R}_+$ such that $\psi(u,T) \ge 0$, $\theta(u,T) < 0$. If Algorithm 3.4 jams up, cycling between Step 1 and Step 2, at a point u_k,T_k, then u_k,T_k satisfies the first order condition $\psi(u_k,T_k) \le 0$ and $\theta(u_k,T_k) = 0$. If Algorithm 3.4 constructs an infinite sequence $\{ (u_i,T_i) \}_{i=0}^{\infty}$, then every accumulation point (u^*,T^*) of $\{ (u_i,T_i) \}_{i=0}^{\infty}$ satisfies the first order optimality condition, for (2.1a), $\psi(u^*,T^*) \le 0$, $\theta(u^*,T^*) = 0$. ■

We recall that optimal control problems, such as (3.11a) do not necessarily have solutions in **U**. Similarly, the sequence of controls $u_i(t)$ constructed by Algorithm 3.4 need not have accumulation points in **U**. This difficulty can be resolved by showing that the conclusions of Theorem 3.5 are valid in the space of relaxed controls (for a proof of this fact see [Bak.1]). Alternatively, one may resort to arguments involving infimizing sequences, as in [Pol.6].

---

[10] See [Bak.1] for an efficient procedure, based on Polyak's constrained Newton method [Pol.4], for computing both θ_i and η_i.

4. CONCLUSION

We have shown that nonsmooth optimization algorithms can be used for solving both open-loop and closed-loop complex optimal control problems involving both open-loop and closed-loop systems. By comparison with other methods in the literature, the design procedure that we have presented for closed-loop systems has the advantage that it can deal with time- and frequency-domain specifications simultaneously, including L^1 -type specifications. Furthermore, it makes possible design by selection and tuning of bounds on responses, which is a much more direct process than the use of weights common to such methods as linear quadratic regulator theory. Of particular significance to the design of finite dimensional controllers for flexible structures is the fact that our procedure does not require modal truncation of partial differential equation models and that it therefore avoids destabilizing "spill-over" effects which plague many other approaches.

5. APPENDIX: EVALUATION OF THE CHARACTERISTIC FUNCTION

The design of a feedback system by means of a nonsmooth optimization algorithm, such as Algorithm 3.3, requires a large number of evaluations of the characteristic function $\chi(x,-\alpha + j\omega)$ and of its partial derivatives with respect to x^i for many values of ω. Hence it is important to perform these operations as efficiently as possible. In the discussion below, we follow the presentation in [Pol.2, Har.1].

Referring to (2.16), we see that the evaluation of $\chi(x,s)$, involves the evaluation of the determinants $\det(sI_{n_c} - A_c(x))$ and $\det(I_{n_i} + G_c(x,s)G_p(s))$. The simplest situation occurs when the matrix $A_c(x)$ is diagonalizable, i.e., when there exists a matrix of eigenvectors $V(x)$ such that $\Lambda(x) = V(x)^{-1}A_c(x)V(x)$, where $\Lambda(x) = diag(\lambda_1(x),\ldots,\lambda_{n_c}(x))$, with the $\lambda_j(x)$ the eigenvalues of the matrix $A_c(x)$. In this case, considerable computational savings result from the use of the two formulae

$$\det[sI_{n_c} - A_c(x)] = \det[sI_{n_c} - \Lambda(x)] = \prod_{j=1}^{n_c}[s - \lambda_j(x)], \tag{5.1a}$$

$$G_c(x,s) = C_c(x)V(x)[sI_{n_c} - \Lambda_c(x)]^{-1}V^{-1}(x)B_c(x) + D_c(x)\ . \tag{5.1b}$$

When diagonalization cannot be used, one can simplify the computation of the required determinants by first reducing $A_c(x)$ to upper Hessenberg form $H_c(x)$ by means of an orthogonal similarity transformation: $H_c(x) = U(x)^T A_c(x) U(x)$, where $U(x)$ is a Hermitian matrix. This results in

$$\det[sI_{n_c} - A_c(x)] = \det[sI_{n_c} - H(x)], \tag{5.2a}$$

$$G_c(x,s) = C_c(x)U(x)(sI_{n_c} - H(x))^{-1}U(x)^T B_c(x) + D_c(x) . \tag{5.2b}$$

Next we need to deal with the evaluation of the plant matrix transfer function $G_p(-\alpha + j\omega)$ for frequencies ω. Since we do not wish to expose ourselves to spillover effects resulting from modal truncation, we propose to evaluate this matrix transfer function by solving two-point boundary value problems which are most conveniently produced by Laplace transformation of the original partial differential equations describing the plant, and thus bypassing a transcription into the form (2.6a,b). We shall illustrate this process by an example.

The planar bending motion of a flexible beam of unit length, which is fixed at one end and carries a particle with mass M attached to the other end is described by (see [Har.1]), can be described by a partial differential equation of the form,

$$m\frac{\partial^2 w(t,x)}{\partial t^2} + cI\frac{\partial^5 w(t,x)}{\partial x^4 \partial t} + EI\frac{\partial^4 w(t,x)}{\partial x^4} = \sum_{j=1}^{n_i} f^j(t)\zeta^j(x,x^j), \quad t \geq 0, \quad 0 \leq x \leq 1 , \tag{5.3a}$$

with boundary conditions

$$w(t,0) = 0 \ , \ \frac{\partial w}{\partial x}(t,0) = 0, \tag{5.3b}$$

$$J\frac{\partial^3 w}{\partial x \partial t^2}(t,1) + cI\frac{\partial^3 w}{\partial x^2 \partial t}(t,1) + EI\frac{\partial^2 w}{\partial x^2}(t,1) = 0 \ , \quad M\frac{\partial^2 w}{\partial t^2}(t,1) - cI\frac{\partial^4 w}{\partial x^3 \partial t}(t,1) - EI\frac{\partial^3 w}{\partial x^3}(t,1) = 0, \tag{5.3c}$$

where x is the distance along the undeformed-beam centroidal line, $w(t,x)$ is the vibration along the cross section principal axis (y–axis), $f^j(t)$ is a control force, $\zeta^j(x,x^j)$ is the influence function of the j–th actuator which is located at x^j, m is the distributed mass per unit length of the beam, c is the material viscous damping coefficient, E is Young's modulus, M is the end mass, I is the beam sectional moment of inertia with respect to y-axis, EI is the beam flexural stiffness in the direction of y-axis, J is the inertia of the end mass in the direction of y-axis, and n_i is the number of inputs.

The output sensors can be assumed to satisfy

$$y^i(t) = \int_0^1 \kappa^i(\nu, z^i) w(t,\nu) d\nu \ , \quad t \geq 0 \ , \quad 1 \leq i \leq n_o, \tag{5.4}$$

where n_o is the number of the sensors, and $\kappa^i(\nu, z^i)$ is the distribution function of the i-th sensor and z_i is the location of the i-th sensor.

It can be shown that the plant described by (5.3a-c), (5.4) can be transcribed into the form (2.6a,b) with the associated hypotheses satisfied [Gib.1]. In fact, the corresponding operator A_p generates an analytic semigroup [Hua.2].

Taking the Laplace transforms of the partial differential equations (5.3a) - (5.3c) and (5.4) with respect to time, we obtain, for each value of $s = -\alpha + j\omega$, the two-point boundary value problem involving an *ordinary differential equation*:

$$(cIs + EI)\frac{d^4 W(x,s)}{dx^4} + ms^2 W(x,s) = \sum_{j=1}^{n_i} F^j(s)\zeta^j(x,x^j), \quad 0 \leq x \leq 1 \ , \tag{5.5a}$$

with boundary conditions

$$W(s,0) = 0 \ , \quad \frac{dW}{dx}(s,0) = 0 \ , \tag{5.5b}$$

$$(cIs + EI)\frac{d^2 W}{dx^2}(s,1) + Js^2 \frac{dW}{dx}(s,1) = 0 \ , \qquad (cIs + EI)\frac{d^3 W}{dx^3}(s,1) - Ms^2 W(s,1) = 0 \ . \tag{5.5c}$$

The Laplace transforms of the outputs are given by

$$Y^i(s) = \int_0^1 \kappa^i(\nu, z^i) W(s,\nu) d\nu \ , \quad 1 \leq i \leq n_o, \tag{5.6}$$

where $W(x,s)$, $F^j(s)$ and $Y^i(s)$ are the Laplace transforms of $w(t,x)$, $f^j(t)$ and $y^i(t)$, respectively. Hence the (k,l)-th element of the matrix $G_p(s)$ can be obtained by setting $F^l(s) = 1$ and $F^j(s) = 0$ for all other j, then solving (5.5a) - (5.5d), and evaluating (5.6) for $i = k$. The boundary value problem (5.5a) - (5.5d) can be solved by means of shooting methods (see [Kel.1], [Pol.8]).

Next, we turn to the computation of the partial derivatives of $\chi(x,s)$. This requires the calculation of the partial derivatives of $\det[s - A_c(x)]$ and $\det[I_{n_i} + G_c(x)G_p(s)]$. When the eigenvalues $\lambda_j(x)$ of $A_c(x)$ are distinct, they are differentiable [Kat.1] and their partial derivatives are given by

$$\frac{\partial \lambda_j(x)}{\partial x^i} = \langle u_j, \frac{\partial A_c(x)}{\partial x^i} v_j \rangle / \langle u_j, v_j \rangle\,, \tag{5.7a}$$

where v_j and u_j are the right and left eigenvectors, respectively, of $A_c(x)$, corresponding to the eigenvalue $\lambda_j(x)$. In this case, the partial derivatives of $\det[sI_{n_c} - A_c(x)]$ can be computed making use of the following formula [Pol.2]:

$$\frac{\partial \det[sI_{n_c} - A_c(x)]}{\partial x^i} = \sum_{j=1}^{n_c} \{- \frac{\partial \lambda_j(x)}{\partial x^i} \prod_{\substack{k=1 \\ k \neq j}}^{n_c} [s - \lambda_k(p_c)]\} = \det[sI_{n_c} - A_c(x)] \sum_{j=1}^{n_c} - \frac{\partial \lambda_j(x)}{\partial x^i} \frac{1}{s - \lambda_j(x)}\,. \tag{5.7b}$$

When the eigenvalues of $A_c(x)$ are not distinct, the computation of its partial derivative requires a more general formula which can be found in [Pol.2]. The computation of the partial derivatives of $\det[I_{n_i} + G_c(x,s)G_p(s)]$ can also be carried out by making use of a formula analogous to (5.7b), provided that the matrix $\left[I_{n_i} + G_c(x,s)G_p(s)\right]$ has distinct eigenvalues. When the eigenvalues of $(I_{n_i} + G_c(x,s)G_p(s))$ are not distinct, the computation of its partial derivative becomes considerably more difficult. Fortunately, this is not very likely to be the case in practice.

6. REFERENCES

[Arm.1] Armijo, L., "Minimization of Functions Having Continuous Partial Derivatives", *Pacific J. Math,* Vol. 16, pp. 1-3, 1966

[Bak.1] T. Baker,"Algorithms for Optimal Control of Systems Described by partial and Ordinary Differential Equations", Ph.D. Thesis, *University of California, Berkeley,* 1988.

[Bal.1] A. V. Balakrishnan, *Applied Functional Analysis,* Springer-Verlag, 1981.

[Ber.1] Berge, C., *Topological Spaces,* Macmillan, New York, N.Y., 1963. Wiley-Interscience, New York, N.Y., 1983.

[Boy.1] S. Boyd and C.A. Desoer, "Subharmonic Functions and Performance Bounds on Linear Time-Invariant Feedback Systems", University of California, Berkeley, E.R.L. Memorandum, M84/51, June 11, 1984.

[Che.1] M.J. Chen and C.A. Desoer, "Necessary and Sufficient Condition for Robust Stability of Linear Distributed Feedback Systems", *International Journal of Control* Vol. 35, No. 2, pp 255-267, 1982.

[Cla.1] F. H. Clarke, Optimization and Nonsmooth Analysis, Wiley-Interscience, New York, 1983.

[Dah.1] M.A. Dahleh and J.B. Pearson, "l^1 -Optimal Feedback Controllers for MIMO Discrete-Time Systems",
IEEE Transactions on Automatic Control, Vol. AC-32, pp 314-322, 1987.

[Doy.1] J. C. Doyle, "Guaranteed Margins for LQG Regulators", *IEEE Transactions on Automatic Control,* Vol. AC-23, pp. 756-757, 1971.

[Doy.2] J.C. Doyle and G. Stein, "Multivariable Feedback Design: Concepts for a Classical/Modern Synthesis", *IEEE Transactions on Automatic Control,* Vol. AC-26, pp. 4-16, 1981.

[Fra.1] B. A. Francis and A J. C. Doyle, "Linear Control Theory with an H^∞ Optimality Criterion" *SIAM Journal on Control and Optimization,* Vol. 25, No.4, pp.815-844, 1987.

[Gib.1] J. S. Gibson, "An Analysis of Optimal Modal Regulation: Convergence and Stability," *SIAM J. Control and Optimization*, Vol. 19, No. 5, pp. 686-707, Sept. 1981.

[Gon.1] C. Gonzaga, E. Polak and R. Trahan, "An Improved Algorithm for Optimization Problems with Functional Inequality Constraints", *IEEE Trans. on Automatic Control,* Vol. AC-25, No. 1, pp. 49-54 1979.

[Jac.1] C. A. Jacobson and C. N. Nett, "Linear State-Space Systems in Infinite-Dimensional Space: The role and Characterization of Joint Stabilizability/Detectability," *IEEE Trans. Automat. Contr.*, Vol. 33, No. 6, pp. 541-549, June 1988.

[Har.1] Y-P. Harn and E. Polak, "On the Design of Finite Dimensional Controllers for Infinite Dimensional Feedback-Systems via Semi-Infinite Optimization", *IEEE Trans. Automat. Contr,* in press.

[Hua.1] P. Huard, Programmation Mathematic Convex, *Rev. Fr. Inform. Rech. Operation.*, Vol. 7, pp. 43-59, 1968.

[Hua.2] F. Huang, "On the Mathematical Model for Linear Elastic Systems with Analytic Damping", *SIAM J. Control and Optimization*, Vol. 26,

[Hig.1] J. E. Higgins and E. Polak, "Minimizing Pseudo-Convex Functions on Convex Compact Sets", *University of California, Berkeley, Electronics Research Laboratory* Memo No. UCB/ERL M88/22, March 1988. To appear in *Journal of Optimization Theory and Applications,*

[Hoh.1] B. von Hohenbalken, "Simplicial Decomposition in Nonlinear Programming Algorithms", *Mathematical Programming,* Vol. 13, pp. 49-68, 1977

[Kiw.1] K. C. Kiwiel, *Methods of Descent for Nondifferentiable Optimization,* Springer-Verlag, Berlin-Heidelberg-New York-Tokyo, 1985.

[Kat.1] T. Kato, *Perturbation Theory for Linear Operators*, Springer-Verlag, 1983.

[Kel.1] H. B. Keller, *Numerical Methods for Two Point Boundary Value Problems,* Blaisdell, New York, 1968.

[Kle.1] R. Klessig and E. Polak, "An Adaptive Algorithm for Unconstrained Optimization with Applications to Optimal Control", *SIAM J. Control,* Vol. 11, No. 1, pp. 80-94, 1973.

[Kwa.1] H. Kwakernaak and R. Sivan, *Linear Optimal Control Systems*, Wiley-Interscience, New York, 1972

[Kwa.2] H. Kwakernaak, "Minimax Frequency Domain Performance and Robustness Optimization of Linear Feedback Systems", *IEEE Transactions on Automatic Control,* Vol. AC-30, No. 10, pp. 994-1004, 1985.

[May.1] D. Q. Mayne and E. Polak "An Exact Penalty Function Algorithm for Optimal Control Problems with Control and Terminal Inequality Constraints, Part 1", *Journal of Optimization Theory and Applications*, Vol. 32 No. 2, pp. 211-246, 1980.

[May.2] D. Q. Mayne and E. Polak "An Exact Penalty Function Algorithm for Optimal Control Problems with Control and Terminal Inequality Constraints, Part 2", *Journal of Optimization Theory and Applications*, Vol. 32 No. 3, pp. 345-363, 1980.

[Nyq.1] H. Nyquist, "Regeneration Theory", *Bell Syst. Tech. J.* vol.2, pp.126-147, Jan. 1932.

[Paz.1] A. Pazy, *Semigroups of Linear Operators and Applications to Partial Differential Equations*, Springer-Verlag, 1983.

[Pir.1] O. Pironneau and E. polak, "On the Rate of Convergence of Certain Methods of Centers", *Mathematical Programming,* Vol. 2, No. 2, pp. 230-258, 1972.

[Pir.2] O. Pironneau and E. Polak, " A Dual Method for Optimal Control Problems with Initial and Final Boundary Constraints", *SIAM J. Control,* Vol. 11, No. 3, pp. 534-549, 1973.

[Pol.1] E. Polak, "A Modified Nyquist Stability Test for Use in Computer Aided Design", *IEEE Trans. Automat. Contr*, Vol. AC-29, No.1, pp.91-93, 1984.

[Pol.2] E. Polak and T. L. Wuu, "On the Design of Stabilizing Compensators Via Semi-Infinite Optimization", *IEEE Trans. Automat. Contr.*, (to appear Feb. 1989).

[Pol.3] E. Polak, "On the Mathematical Foundations of Nondifferentiable Optimization in Engineering Design", *SIAM Review*, pp. 21-91, March 1987.

[Pol.4] E. Polak, R. Trahan and D. Q. Mayne, "Combined Phase I - Rhase II Methods of Feasible Directions", *Mathematical Programming* No. 1, pp. 32-61, 1979.

[Pol.5] E. Polak, D. Q. Mayne and J. Higgins, "A superlinearly Convergent Algorithm for Min-Max Problems", *University of California, Berkeley, Electronics Research Laboratory* Memo No. M86/103, Nov. 15, 1986.

[Pol.6] E. Polak and Y. Y. Wardi, "A Study of Minimizing Sequences", *SIAM J. Control and Optimization*, Vol. 22, No. 4, pp. 599-609, 1984.

[Pol.7] E. Polak and S. E. Salcudean, "On The Design of Linear Multivariable Feedback Systems via Constrained Nondifferentiable Optimization in H^∞ Spaces", *IEEE Trans on Automatic Control*, in press.

[Pol.8] E. Polak, *Computational Methods in Optimization: A Unified Approach*, Academic Press, 1971.

[Pol.9] E. Polak and L. He, "A Unified Phase I-Phase II Method of Feasible Directions for Semi-Infinite Optimization", *University of California, Berkeley, Electronics Research Laboratory* Memo No. UCB/ERL M89/7, Feb. 1989.

[Pol.10] B. T. Polyak, *Introduction to Optimization*, (in Russian) Nauka, Moscow, 1983

[Str.1] G. Strang and George F., *An Analysis of the Finite Element Method*, Prentice Hall, Englewood Cliffs, N. J., 1973

[Vid.1] M. Vidyasagar, "Optimal Rejection of Persistent Bounded Disturbances", *IEEE Transactions on Automatic Control*, Vol. AC-31, pp. 527-534, 1986.

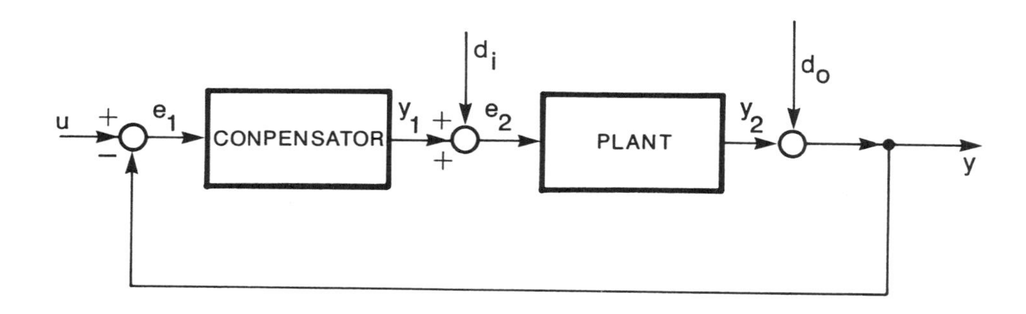

Fig. 1 Configuration of Feedback System S(x).

Contemporary Mathematics
Volume 97, 1989

Stability Analysis of a Rigid Body with Attached Geometrically Nonlinear Rod by the Energy–Momentum Method *

T. A. POSBERGH †, J. C. SIMO† &
J. E. MARSDEN ‡

January 30, 1989

Abstract

This paper applies the energy-momentum method to the problem of nonlinear stability of relative equilibria of a rigid body with attached flexible appendage in a uniformly rotating state. The appendage is modeled as a geometrically exact rod which allows for finite bending, shearing and twist in three dimensions. Application of the energy-momentum method to this example depends crucially on a special choice of variables in terms of which the second variation block diagonalizes into blocks associated with rigid body modes and internal vibration modes respectively. The analysis yields a nonlinear stability result which states that relative equilibria are nonlinearly stable provided that; (i) the angular velocity is bounded above by the square root of the minimum eigenvalue of an associated linear operator and, (ii) the whole assemblage is rotating about the minimum axis of inertia.

§1. Introduction

This paper discusses the application of the energy-momentum method to the case of a rotating rigid body with an attached, flexible appendage. The model for the appendage we have chosen is referred to a a geometrically exact rod model and

---

* Paper presented by T. A. Posbergh. 1985 Subject Classification 70K, 58F.

† Division of Applied Mechanics, Stanford University, Stanford CA 94305. Research supported by AFOSR contract numbers 2-DJA-544 and 2-DJA-771 with Stanford University.

‡ Department of Mathematics, University of California, Berkeley, CA 94720 and Cornell University, Ithaca, NY 14853-7901. Research partially supported by DOE contract DE-AT03-88ER-12097 and MSI at Cornell University.

0271-4132/89 $1.00 + $.25 per page

is discussed in detail in SIMO [1985] and SIMO, MARSDEN, & KRISHNAPRASAD [1988]. Because the formulation satisfies exactly all the invariance requirements under superposed rigid body motions, exactly captures without simplification all the dynamic effects, and places no restrictions on the degree of allowable deformations, the rod model is said to be *geometrically exact*. Use of this class of models avoids the potential for unphysical results which may appear in more *ad-hoc* linearized models; see SIMO & VU-QUOC [1988b].

For our stability analysis we use the *energy-momentum method*, introduced in SIMO, POSBERGH & MARSDEN [1989]. This method constitutes a systematic application of the relative equilibrium theorem (see ARNOLD [1978] or ABRAHAM & MARSDEN [1978]), and represents an extension of the *energy-Casimir method* introduced by ARNOLD [1966] and further developed in HOLM, MARSDEN, RATIU & WEINSTEIN [1985]. The energy-Casimir method was applied to rigid bodies with flexible attachments by KRISHNAPRASAD & MARSDEN [1987] using rod models accounting for extension and shear, but precluding bending deformation. The energy-momentum method was applied to this example in POSBERGH & SIMO [1988].

In contrast to the energy-Casimir method, for a Hamiltonian system with symmetry we work directly in the material representation as opposed to the convective (or reduced) representation. Thus, instead of using Casimirs, one employs directly the momentum map as the conserved quantity The success of the method relies crucially on the choice of a particular set of variables, introduced in SIMO, POSBERGH & MARSDEN [1989], which block diagonalizes the second variation $\delta^2 H_\xi$. Conceptually, this choice of variables enforces automatically conservation of the momentum constraint along with gauge symmetries, and separates the overall infinitesimal rigid body modes from the internal vibration modes (including shear and torsion) of the rod. Further geometric aspects underlying this parameterization are examined in the paper of MARSDEN, SIMO, LEWIS & POSBERGH [1989] in this proceedings.

§2. The Energy-Momentum Method

In this section we give a brief outline of the energy-momentum method; for further details see SIMO, POSBERGH & MARSDEN [1989].

§2A General Formulation and Relative Equilibria

We consider a mechanical system with configuration manifold Q and phase space $P = T^*Q$, where T^*Q is the cotangent bundle. The Hamiltonian $H: P \to \mathbf{R}$ corresponds to the total energy of the system. Let $\boldsymbol{X}_H: P \to TP$ denote the Hamiltonian vector field associated with H; *i.e.*,

$$dH(z) \cdot \delta z = \Omega(\boldsymbol{X}_H(z), \delta z), \qquad \text{for all } z \in P \text{ and } \delta z \in T_z P, \tag{2.1}$$

where Ω is the canonical symplectic two-form on P. Hamilton's equations are then formulated abstractly as $\dot{z} = \boldsymbol{X}_H(z)$.

In addition we have a symmetry group G which acts on P by canonical transformations, along with the corresponding Lie algebra $\mathcal{G}$. The action of the group G on Q will be denoted $\Psi: G \times Q \to Q$. Associated with this action is the corresponding infinitesimal generator

$$\xi_Q(q) := \frac{d}{dt}\bigg|_{t=0} \Psi(\exp(t\xi), q), \qquad q \in Q. \tag{2.2}$$

The action Ψ on Q induces, by cotangent lift, a symplectic action on P.

The momentum map for the action of G on P is denoted by $\boldsymbol{J}: P \to \mathcal{G}^*$. We recall that associated with this G-action, for any $\xi \in \mathcal{G}$ one has a Hamiltonian vector field $X_{J(\xi)}: P \to TP$ with Hamiltonian function $J(\xi): P \to \mathbf{R}$ defined in terms of the momentum map by the relation

$$J(\xi)(z) = \langle \boldsymbol{J}(z), \xi \rangle \qquad \xi \in \mathcal{G}, \tag{2.3}$$

where $\langle \cdot, \cdot \rangle$ denotes the pairing between $\mathcal{G}$ and $\mathcal{G}^*$. The function $J(\xi)$ is then given by the standard formula:

$$J(\xi)(\alpha_q) = \langle \alpha_q, \xi_Q(q) \rangle, \qquad \alpha_q \in P, \tag{2.4}$$

which as a special case reproduces the usual linear and angular momentum. We denote by q an element in configuration space Q, and by p an element in T_q^*Q, the cotangent space for a particular configuration. Thus $z := (q, p) \in P$.

Following terminology due to Poincaré, a point $z_e \in P$ is a *relative equilibrium* if the trajectory of Hamilton's equations through z_e is given by

$$z(t) = \exp[t\xi] \cdot z_e, \qquad \text{for some} \qquad \xi \in \mathcal{G}, \tag{2.5}$$

a condition which states that the dynamic orbit through z_e equals the group orbit through z_e. A basic result exploited below is that the relative equilibria of a mechanical system with Hamiltonian H and momentum map $\boldsymbol{J}$ for the symplectic action of a Lie group G on the phase space P are the critical points of the *energy-momentum functional* $H_\xi: P \to \mathbf{R}$ defined as

$$H_\xi := H - \langle \boldsymbol{J} - \mu_e, \xi \rangle \tag{2.6}$$

where $\mu_e = \boldsymbol{J}(z_e)$ is the value of the momentum map at the sought relative equilibrium. In mechanical terms this result, known as the relative equilibrium theorem, provides a variational characterization of the relative equilibrium as the stationary point of the energy (Hamiltonian) subject to the side constraint of constant momentum. Within the context of this constrained optimization problem, formal stability of a relative equilibria is then concluded by examining the definiteness of the second variation $\delta^2 H_\xi$ restricted to the subspace defined by the side constraint $\boldsymbol{J}(z_e) - \mu_e = 0$ modulo neutral directions due to group invariance. This subspace is isomorphic to the quotient space

$$\mathcal{S} \cong \ker[T_{z_e} J(z_e)] \Big/ T_{z_e}(G_{\mu_e} \cdot z_e), \tag{2.7}$$

where $G_{\mu_e} \cdot z_e$ denotes the orbit of the (isotropy) subgroup G_{μ_e} of G that leaves μ_e invariant (under the coadjoint action of G on $\mathcal{G}^*$), and $T_{z_e}(G_{\mu_e} \cdot z_e)$ is the

Box 2.1. The Energy-Momentum Method

1. (*First variation*) Construct $H_\xi = H - [J(\xi) - \langle \mu_e, \xi \rangle]$ and find $z_e \in P$ and $\xi \in \mathcal{G}$ such that

$$dH_\xi(z_e) \cdot \delta z = 0, \qquad \text{and} \quad J(z_e) - \mu_e = 0,$$

for all $\delta z \in T_{z_e} P$ (No restrictions placed on δz at this stage).

2. (*Admissible variations for second variation test*) Choose a linear subspace $\mathcal{S} \subset T_{z_e} P$ such that
   - i. $dJ(\xi)(z_e) \cdot \delta z = 0$ for all $\delta z \in \mathcal{S}$.
   - ii. $\mathcal{S}$ complements $T_{z_e}(G_{\mu_e} \cdot z_e)$ in $[\ker dJ(\xi)(z_e)]$; *i.e.*, every variation $\delta z \in T_{z_e} P$ satisfying i. is uniquely written as

$$\delta z = v + \underbrace{\chi_P(z_e)}_{\textit{tangent to orbit}},$$

for some $v \in \mathcal{S}$ and $\chi \in \mathcal{G}_{\mu_e}$ (so that $\chi_P(z_e) \in T_{z_e}(G_{\mu_e} \cdot z_e)$).

3. Test for definiteness of the second variation $\delta^2 H_\xi$ on $\mathcal{S}$; *i.e.*

$$\delta^2 H_\xi(z_e) \cdot (v, v) > 0,$$

for all $v \in \mathcal{S}$. Definiteness implies *formal stability* of $z_e \in P$.

Box 2.1. Procedure for Stability Analysis

tangent space at z_e to this orbit. The criterion for formal stability then takes the following form

$$\boxed{z_e \in P \text{ formally stable} \;\Leftrightarrow\; \delta^2 H(z_e) \cdot (\delta z, \delta z) > 0 \text{ for } \delta z \in \mathcal{S}.} \tag{2.8}$$

Below, in equation (2.27) we will show how to specifically choose $\mathcal{S}$.

In this paper, we apply this method to the stability analysis of relative equilibria of a uniformly rotating rigid body coupled to a rod. For this problem

$G = SO(3)$ and $\mathcal{G} \in so(3) \cong \mathbf{R}^3$. Moreover, we have

$$\left.\begin{aligned} G_{\mu_e} &= \{\exp[t\hat{\xi}] \in SO(3) \mid t \in \mathbf{R}\}, \\ \mathcal{G}_{\mu_e} &= \{\xi \in \mathbf{R} \text{ with } \xi \times \mu_e = 0\}. \end{aligned}\right\} \tag{2.9}$$

That is, G_{μ_e} is the group of rotations about ξ, and $\mathcal{G}_{\mu_e}$ is the line along ξ (equivalently, the one dimensional space of infinitesimal rotations about ξ). Note that $\dim[\mathcal{G}_{\mu_e}(z_e)] = 1$.

To define the constraint subspace $\mathcal{S} \subset T_{z_e}P$, we first will need to enforce the condition **i** in Box 2.1, *i.e.*, $TJ(z_e) \cdot \delta z = 0$, for any $(z_e; \delta z) \in \ker[T_{z_e}J(z_e)]$. This condition places *three* restrictions on the variations in $T_{z_e}P$. The *additional constraint* **ii** in Box 2.1 that variations in $\ker[T_{z_e}J(z_e)]$ be taken modulo the *one dimensional* subspace $\mathcal{G}_{\mu_e}(z_e)$ introduces another restriction and leads to the dimension count

$$\operatorname{codim}[\mathcal{S}] = 4. \tag{2.10}$$

To perform the second variation test in Box 2.1 we introduce a decomposition of the of the constraint subspace $\mathcal{S}$ of the form

$$\mathcal{S} = \mathcal{S}_{RIG} \oplus \mathcal{S}_{INT} \tag{2.11}$$

which results in a block diagonal structure of the second variation of the energy-momentum functional H_ξ restricted to $\mathcal{S}$,

$$\delta^2 H_\xi \Big|_{\mathcal{S}\times\mathcal{S}} = \begin{bmatrix} \begin{bmatrix} 2\times 2 \text{ rigid} \\ \text{body block} \end{bmatrix} & 0 \\ 0 & \begin{bmatrix} \text{Internal vibration} \\ \text{block} \end{bmatrix} \end{bmatrix}. \tag{2.12}$$

We outline below the basic steps involved in the construction of this decomposition following the construction given in SIMO, POSBERGH & MARSDEN [1989]. We direct the reader to this later reference for further details. Abstract and geometric aspects underlying this construction are examined in MARSDEN, SIMO, LEWIS & POSBERGH [1989].

§2B The Block Diagonalization Theorem and the Second Variation Test

Assume a Hamiltonian function H of the form $H = V + K$, where $V: Q \to \mathbf{R}$ is the potential energy, and $K: P \to \mathbf{R}$ is the kinetic energy of the system. We further assume that K defines on Q an inner product denoted by

$$\langle\cdot,\cdot\rangle_g : TQ \times TQ \to \mathbf{R}. \tag{2.13}$$

For instance, for finite dimensional Hamiltonian systems, we have

$$K = \tfrac{1}{2}\langle p^\sharp, p^\sharp\rangle_g = \tfrac{1}{2} p_i g^{ij}(q) p_j, \tag{2.14}$$

where $g(q) = g_{ij}\, dq^i \otimes dq^j$ is a given Riemannian metric on Q.

Step 1. *Reformulation of the energy-momentum functional.* Define a modified potential $V_\xi : Q \to \mathbb{R}$ by the expression

$$\left.\begin{aligned} V_\xi(q) &:= V(q) + L_\xi(q), \\ L_\xi(q) &:= -\tfrac{1}{2}\langle \xi_Q(q), \xi_Q(q)\rangle_g . \end{aligned}\right\} \tag{2.15}$$

It can be shown that the critical points of V_ξ are precisely the relative equilibrium configurations $q_e \in Q$ (see MARSDEN, SIMO, LEWIS & POSBERGH [1989]). Accordingly, if $\delta V_\xi / \delta q$ denotes the functional derivative of V_ξ defined in the standard fashion as

$$dV_\xi(q) \cdot \delta q = \langle \delta q, \frac{\delta V_\xi}{\delta q} \rangle, \qquad \text{for all } q \in Q, \tag{2.16}$$

we have the critical point condition

$$\left. \frac{\delta V_\xi}{\delta q} \right|_{(q_e)} = 0. \tag{2.17}$$

For stationary rotations about $\xi \in \mathcal{G}$, the term L_ξ gives the potential energy associated with the centrifugal force.

Next, define a potential function $K_\xi : P \to \mathbb{R}$ by the expression

$$K_\xi(z) := \tfrac{1}{2} \| p - \mathsf{FL}(\xi_Q(q)) \|^2_{g^{-1}}, \qquad z = (q, p) \in P, \tag{2.18}$$

where $\mathsf{FL}: TQ \to P$ is the Legendre transformation and $\| \cdot \|_{g^{-1}}$ is the norm induced by (2.14) on T_q^*Q. It is evident that K_ξ also has critical points at the relative equilibria $z_e \in P$. Furthermore, we have

$$\boxed{H_\xi = V_\xi + K_\xi + \langle \mu_e, \xi \rangle.} \tag{2.19}$$

Finally, observe that the second variations of V_ξ and K_ξ make *intrinsic sense* at a relative equilibrium $z_e \in P$.

Step 2. *The tangent space of admissible variations for V_ξ.* Recall that G acts on Q by *isometries*, and that V is *left invariant* under the full group G at *any* configuration $q \in Q$ (for $G = SO(3)$ this is the condition of frame indifference). The term L_ξ, on the other hand, is left invariant under the full G *only at a relative equilibrium* $q_e \in Q$ (*i.e.*, the Lie derivative of V_ξ in the direction of any η_Q, evaluated at q_e vanishes). Thus, in general, L_ξ and consequently V_ξ, is invariant only under the action of the isotropy subgroup $G_{\mu_e} \subset G$. Let $\mathcal{G}_{\mu_e} \subset \mathcal{G}$ be the corresponding Lie subalgebra; *i.e.*,

$$\mathcal{G}_{\mu_e} := \{ \zeta \in \mathcal{G} \mid ad^*_{\mu_e}(\zeta) = 0 \}. \tag{2.20}$$

The space of admissible variations $\mathcal{V} \subset T_q Q$ for V_ξ at q, is then the tangent space to the orbit space Q/G_{μ_e} which can be realized as

$$\mathcal{V} := T_qQ \big/ T_q(G_{\mu_e} \cdot z_e) \cong \{ \delta q \in T_qQ \mid \langle \delta q, \zeta_Q(q) \rangle_g = 0, \ \zeta \in \mathcal{G}_{\mu_e} \}. \tag{2.21}$$

Step 3. *Split of $\mathcal{V}$: Block-diagonalization of V_ξ.* Construct a decomposition of $\mathcal{V}$,

$$\mathcal{V} = \mathcal{V}_{RIG} \oplus \mathcal{V}_{INT}, \tag{2.22}$$

into infinitesimal rigid body variations, and 'deformation' variations as follows. Let $\mathcal{G}_{\mu_e}^{\perp} \subset \mathcal{G}$ be the g-dependent orthogonal complement of $\mathcal{G}_{\mu_e}$ in the kinetic energy inner product; *i.e.*,

$$\mathcal{G}_{\mu_e}^{\perp} := \{ \eta \in \mathcal{G} \mid \langle \eta_Q(q), \zeta_Q(q) \rangle_g = 0, \ \zeta \in \mathcal{G}_{\mu_e} \}, \tag{2.23}$$

so that $\mathcal{G} = \mathcal{G}_{\mu_e} \oplus \mathcal{G}_{\mu_e}^{\perp}$. Recall that a superposed rigid body variation is of the form $\eta_Q(q) \in T_qQ$, with $\eta \in \mathcal{G}$. Thus, in view of (2.21) and (2.22) we set

$$\mathcal{V}_{RIG} := \{ \eta_Q(q_e) \in T_{q_e}Q \mid \eta \in \mathcal{G}_{\mu_e}^{\perp} \} \subset \mathcal{V}. \tag{2.24}$$

Note that the requirement $\eta \in \mathcal{G}_{\mu_e}^{\perp}$ furnishes the condition which ensures that indeed $\mathcal{V}_{RIG} \subset \mathcal{V}$.

To construct $\mathcal{V}_{INT}$, recall that V is G-left invariant whereas V_ξ is only G_{μ_e}-left invariant. However, V_ξ *is* infinitesimally $\mathcal{G}$ invariant, but the *body force* $\delta V_\xi/\delta q$ need not be. The quantity capturing the lack of invariance of $\delta V_\xi/\delta q$ under G/G_{μ_e} is

$$\begin{aligned} \mathcal{L}_{\eta_Q(q_e)} \frac{\delta V_\xi}{\delta q}(q_e) &= \mathcal{L}_{\eta_Q(q_e)} \frac{\delta L_\xi}{\delta q}(q_e) \\ &:= \frac{d}{d\epsilon}\bigg|_{\epsilon=0} \Psi_{\exp[-\epsilon\eta]} \left(\frac{\delta L_\xi}{\delta q} \left(\Psi_{\exp[\epsilon\eta]}(q_e) \right) \right) \qquad \text{for all } \eta \in \mathcal{G}_{\mu_e}^{\perp}, \end{aligned} \tag{2.25}$$

where $\mathcal{L}_a b$ denotes the Lie derivative of b in the direction a. We define $\mathcal{V}_{INT} \subset \mathcal{V}$ by the condition

$$\mathcal{V}_{INT} := \{ \delta q \in \mathcal{V} \mid \langle \delta q, \mathcal{L}_{\eta_Q(q_e)} \frac{\delta L_\xi}{\delta q}(q_e) \rangle_g = 0, \text{ for } \eta \in \mathcal{G}_{\mu_e}^{\perp} \}. \tag{2.26}$$

Note that the number of constraints in (2.26) equals $\dim[\mathcal{G}_{\mu_e}^{\perp}] = \dim[\mathcal{V}_{RIG}]$. Furthermore, by construction $\mathcal{V}_{INT} \cap \mathcal{V}_{RIG} = \{0\}$ so that (2.22) holds.

Step 4. *Split of $\mathcal{S}$: Block-diagonalization of H_ξ.* Conditions **i** and **ii** in Box 2.1 lead to the following concrete realization of $\mathcal{S}$ as a (constrained) subspace of $T_{z_e}P$:

$$\mathcal{S} := \{\delta z = (\delta q, \delta p) \in T_{z_e}P \mid T_{z_e}J(z_e)\cdot \delta z = 0, \text{ and } \delta q \in \mathcal{V},\ \zeta \in \mathcal{G}_{\mu_e}\}. \quad (2.27)$$

The split (2.22) then induces a split $\mathcal{S} = \mathcal{S}_{RIG} \oplus \mathcal{S}_{INT}$ via the Legendre transformation as follows. $\mathcal{S}_{RIG}$ will now be identified with the tangent space at z_e of superposed G/G_{μ_e}-motions on motions starting at z_e. (For $G = SO(3)$, these are superposed infinitesimal rigid body motions modulo motions about μ_e).

Let $t \mapsto z(t) = (q(t), p(t)) \in P$ be a motion starting at $z(t)|_{t=0} = z_e$. Consider a superposed G/G_{μ_e}-motion which, by definition, is given by

$$\left.\begin{aligned} q^+(t) &= \Psi_{g(t)}(q(t)), \\ p^+(t) &= \mathsf{FL}\Big(\frac{d}{dt}\Psi_{g(t)}(q(t))\Big), \end{aligned}\right\} \quad (2.28)$$

where $t \mapsto g(t)$ is a motion in G/G_{μ_e}. Here Ψ is the action induced by $g(t)$ on Q. The tangent space at $z_e \in P$ to all superposed G/G_{μ_e}-motions is obtained by linearizing (2.28) at $t = 0$ as follows. Consider the one parameter family of G/G_{μ_e}-motions given, for $t \mapsto \eta(t) \in \mathcal{G}_{\mu_e}^{\perp}$, by

$$\epsilon \mapsto g(\epsilon, t) := \exp[\epsilon\eta(t)] \in G\big/G_{\mu_e}. \quad (2.29)$$

Then define $\Delta z = (\Delta q, \Delta p) \in T_{z_e}P$ by the expressions

$$\Delta q := \frac{d}{d\epsilon}\bigg|_{\epsilon=0} \Psi_{\exp[\epsilon\eta(t)]}(q(t))\bigg|_{t=0} = \eta_Q(q_e), \quad (2.30)$$

so that $\Delta q \in \mathcal{V}_{RIG}$, and

$$\Delta p := \frac{d}{d\epsilon}\bigg|_{\epsilon=0} \mathsf{FL}\left(\frac{d}{dt}\Psi_{\exp[\epsilon\eta(t)]}(q(t))\right). \quad (2.31)$$

It can be shown that $\Delta z = (\Delta q, \Delta p)$ given by formulae (2.30) and (2.31) actually lies in $T_{z_e}P/T_{z_e}(G_{\mu_e}\cdot z_e)$. Since $\mathcal{V}_{RIG} \subset \mathcal{V}$ it follows from (2.27) that the restriction to $\ker[T_{z_e}J(z_e)]$ completes the construction of $\mathcal{S}_{RIG}$; *i.e.*,

$$\boxed{\mathcal{S}_{RIG} := \{\Delta z = (\Delta q, \Delta p) \mid T_{z_e}J(z_e)\cdot \Delta z = 0\}.} \quad (2.32)$$

One can show that $\mathcal{S}_{RIG}$ defined by (2.32) is parametrized solely in terms of elements $\eta \in \mathcal{G}_{\mu_e}^{\perp}$; hence,

$$\dim[\mathcal{S}_{RIG}] = \dim[\mathcal{V}_{RIG}] = \dim[\mathcal{G}_{\mu_e}^{\perp}]. \quad (2.33)$$

Finally, we define $\mathcal{S}_{INT}$ by setting

$$\boxed{\mathcal{S}_{INT} := \{\delta z = (\delta q, \delta p) \in \mathcal{S} \mid \delta q \in \mathcal{V}_{INT}\}.} \quad (2.34)$$

One can easily show that $\mathcal{S}_{RIG} \cap \mathcal{S}_{RIG} = \{0\}$ so that one indeed has

$$\mathcal{S} = \mathcal{S}_{RIG} \oplus \mathcal{S}_{INT}. \tag{2.35}$$

With this construction in hand we have the following basic result

The block-diagonalization theorem. *Let $z_e = (q_e, p_e) \in P$ be a relative equilibrium. Further, let $\mathcal{V}_{RIG}, \mathcal{V}_{INT} \subset \mathcal{V}$ and $\mathcal{S}_{RIG}, \mathcal{S}_{INT} \subset \mathcal{S}$ be constructed as above. Then*

$$\text{i.} \qquad \delta^2 V_\xi(q_e) \cdot (\eta_Q(q_e), \delta q) = 0, \tag{2.36a}$$

$$\text{ii.} \qquad \delta^2 K_\xi(z_e) \cdot (\Delta z, \delta z) = 0, \tag{2.36b}$$

for all $\eta_Q(q_e) \in \mathcal{V}_{RIG}$, $\delta q \in \mathcal{V}_{INT}$ and $\Delta z \in \mathcal{S}_{RIG}$, $\delta z \in \mathcal{S}_{INT}$. Consequently, in view of (2.19) we also have

$$\text{iii.} \qquad \delta^2 H_\xi(z_e) \cdot (\Delta z, \delta z) = 0. \tag{2.36c}$$

■

Remarks

1. The condition $\delta^2 H_\xi(z_e)|_{\mathcal{S}_{RIG}\times\mathcal{S}_{RIG}} > 0$ leads to stability requirements that generalize the classical stability conditions for a rigid body in stationary rotation.

2. From expression (2.18) and the fact that at a relative equilibrium one has

$$p_e = \mathsf{FL}(\xi_Q(q_e)), \tag{2.37}$$

it is easily concluded that

$$\delta^2 K_\xi(z_e)\Big|_{\mathcal{S}_{INT}\times\mathcal{S}_{INT}} > 0. \tag{2.38}$$

Consequently, one has the estimate

$$\delta^2 H_\xi(z_e)\Big|_{\mathcal{S}_{INT}\times\mathcal{S}_{INT}} > \delta^2 V_\xi(q_e)\Big|_{\mathcal{V}_{INT}\times\mathcal{V}_{INT}}. \tag{2.39}$$

Thus, positive definiteness of $\delta^2 V_\xi(q_e)|_{\mathcal{V}_{INT}\times\mathcal{V}_{INT}}$ ensures positive definiteness of $\delta^2 H_\xi(z_e)|_{\mathcal{S}_{INT}\times\mathcal{S}_{INT}}$.

3. The proof of the block-diagonalization result (2.36a) follows from the identity

$$\langle \delta q, \nabla_{\eta_Q(q_e)} \frac{\delta V_\xi}{\delta q}(q_e) \rangle_g = d^2 V_\xi(q_e) \cdot (\eta_Q(q_e), \delta q), \tag{2.40}$$

the defining condition in (2.26), and relation (2.25). ■

With the aid of the block diagonalization theorem the second variation test for formal stability in Box 2.1 takes a remakably simple form: Formal stability of

$z_e = (q_e, p_e)$ is implied by the conditions

$$\boxed{\begin{array}{lll} \text{i.} & \delta^2 H_\xi(z_e)\Big|_{\mathcal{S}_{RIG}\times\mathcal{S}_{RIG}} & > 0, \\ \text{ii.} & \delta^2 V_\xi(q_e)\Big|_{\mathcal{V}_{INT}\times\mathcal{V}_{INT}} & > 0. \end{array}} \tag{2.41}$$

Condition **ii** requires that the lowest eigenvalue of $\delta^2 V_\xi(z_e)$ restricted to $\mathcal{V}_{INT}$ be positive. It can be shown (see SIMO, POSBERGH & MARSDEN [1989]) that this condition is in turn implied by the requirement that $|\xi|$ be less that the lowest natural frequency of the system at the relative equilibrium configuration $q_e \in Q$.

§3. The Rigid Body

In this section we outline the notation and mechanical setup for the rigid body. For a more complete discussion we refer to ABRAHAM & MARSDEN [1978] or ARNOLD [1978]. Most of the concepts and notation will be used again for the geometrically exact rod in the next section. For an application of the energy momentum method to the rigid body alone see SIMO, POSBERGH & MARSDEN [1989].

§3A. Notation for the Rotation Group

The rotation group $SO(3)$ consists of all orthogonal linear transformations of Euclidean three space to itself which have determinant one. Its Lie algebra, denoted $so(3)$ consists of all 3×3 skew matrices, which we identify with $\mathbf{R}^3$ by the isomorphism $\hat{}: \mathbf{R}^3 \to so(3)$ defined by

$$\boldsymbol{\Omega} \mapsto \hat{\boldsymbol{\Omega}} = \begin{bmatrix} 0 & -\Omega^3 & \Omega^2 \\ \Omega^3 & 0 & -\Omega^1 \\ -\Omega^2 & -\Omega^1 & 0 \end{bmatrix}, \tag{3.1}$$

where $\boldsymbol{\Omega} = (\Omega^1, \Omega^2, \Omega^3)$. One checks that for any vector $\boldsymbol{r}$, $\hat{\boldsymbol{\Omega}}\boldsymbol{r} = \boldsymbol{\Omega} \times \boldsymbol{r}$ and, $\hat{\boldsymbol{\Omega}}\hat{\boldsymbol{\Theta}} - \hat{\boldsymbol{\Theta}}\hat{\boldsymbol{\Omega}} = (\boldsymbol{\Omega} \times \boldsymbol{r})\hat{}$. These give the usual identification of the Lie algebra $so(3)$ with $\mathbf{R}^3$ and the Lie algebra bracket with the cross product of vectors.

Given $\boldsymbol{\Lambda} \in SO(3)$, let $\hat{\boldsymbol{v}}_{\boldsymbol{\Lambda}}$ denote an element of the tangent space to $SO(3)$ at $\boldsymbol{\Lambda}$. Since $SO(3)$ is a submanifold of $GL(3)$, the general linear group, we can identify $\hat{\boldsymbol{v}}_{\boldsymbol{\Lambda}}$ with a 3×3 matrix, which we denote with the same letter. Linearizing the defining (submersive) condition $\boldsymbol{\Lambda}\boldsymbol{\Lambda}^T = \mathbf{1}$, gives

$$\boldsymbol{\Lambda}\hat{\boldsymbol{v}}_{\boldsymbol{\Lambda}}^T + \hat{\boldsymbol{v}}_{\boldsymbol{\Lambda}}\boldsymbol{\Lambda}^T = 0, \tag{3.2}$$

which defines $T_{\boldsymbol{\Lambda}}SO(3)$. We can identify $T_{\boldsymbol{\Lambda}}SO(3)$ with $so(3)$ by the following isomorphism: Given $\hat{\theta} \in so(3)$ and $\Lambda \in SO(3)$ we define $(\Lambda, \hat{\theta}) \mapsto \hat{\theta}_\Lambda \in T_\Lambda SO(3)$ through right translations by setting

$$\hat{\theta}_\Lambda := T_e R_\Lambda \cdot \hat{\theta} \cong (\Lambda, \hat{\theta}\Lambda). \tag{3.2b}$$

Thus $\hat{\theta}_\Lambda$ is the right invariant extension of $\hat{\theta}$.

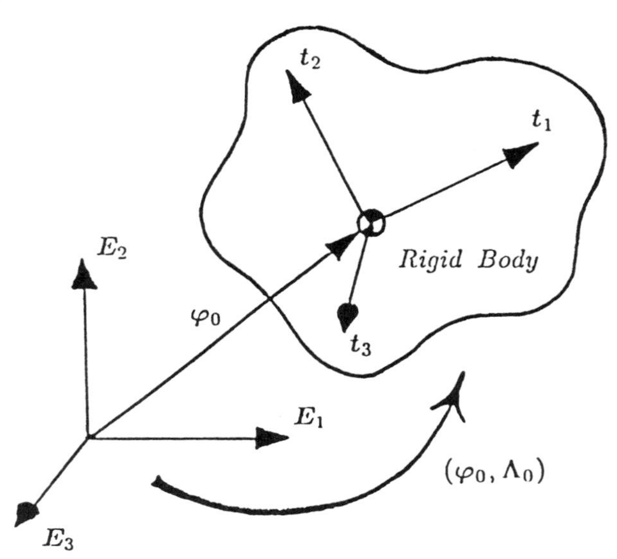

Figure 3.1. Rigid Body

Often, the base point is omitted and with an abuse in notation we write $\hat{\theta}\Lambda$ for $\hat{\theta}_\Lambda$.

The dual of the Lie algebra $so(3)$ is identified with $\mathbf{R}^3$ via the standard dot product:

$$\pi \cdot \theta = \tfrac{1}{2}\mathrm{tr}[\hat{\pi}^T \hat{\theta}]. \tag{3.3}$$

This extends to the *left-invariant* pairing on $T_\Lambda SO(3)$ given by

$$\langle \hat{\pi}_\Lambda, \hat{\theta}_\Lambda \rangle = \tfrac{1}{2}\mathrm{tr}[\hat{\pi}_\Lambda^T \hat{\theta}_\Lambda] = \tfrac{1}{2}\mathrm{tr}[\hat{\pi}^T \hat{\theta}] = \pi \cdot \theta. \tag{3.4}$$

We shall, thereby, write elements of $so(3)^*$ as $\hat{\pi}$ with $\pi \in \mathbf{R}^3$ and elements of $T^*_\Lambda SO(3)$ in spatial representation as

$$\hat{\pi}_\Lambda = (\Lambda, \hat{\pi}\Lambda). \tag{3.5}$$

§3B. *Rigid Body Dynamics*

In general, any configuration of the rigid body is described by a position, $\varphi_0 \in \mathbf{R}$, and an orientation, $\Lambda_0 \in SO(3)$ in ambient space (see Figure 3.1). Thus, we let

$$\mathcal{C}_0 := \{\Phi_0 := (\varphi_0, \Lambda_0) \in \mathbf{R}^3 \times SO(3)\} \tag{3.6}$$

be the configuration manifold for the rigid body. Associated with this configuration manifold we have the collection of tangent spaces $T_{\Phi_0}\mathcal{C}_0$ for $\Phi_0 \in \mathcal{C}_0$ defined as

$$T_{\Phi_0}\mathcal{C}_0 := \left\{ \mathbf{v}_{\Phi_0} := (v_{\varphi_0}, \hat{\omega}_{\Lambda_0}) \in \mathbf{R}^3 \times T_{\Lambda_0}SO(3) \right\}. \tag{3.7}$$

The *phase space* for the rigid body is the cotangent bundle $T^*\mathcal{C}_0 = \left\{ \cup_{\Phi_0 \in \mathcal{C}_0} T^*_{\Phi_0}\mathcal{C}_0 \right\}$ where

$$T^*_{\Phi_0}\mathcal{C}_0 := \left\{ \mathbf{p}_{\Phi_0} := (p_{\varphi_0}, \hat{\pi}_{\Lambda_0}) \in \mathbf{R}^3 \times T^*_{\Lambda_0}SO(3) \right\}. \tag{3.8}$$

The tangent space $T_{\Phi_0}\mathcal{C}_0$ and the cotangent space $T^*_{\Phi_0}\mathcal{C}_0$ are in duality through the pairing given by

$$\left\langle \mathbf{v}_{\Phi_0}, \mathbf{p}_{\Phi_0} \right\rangle = p_{\varphi_0} \cdot v_{\varphi_0} + \tfrac{1}{2}\mathrm{tr}(\hat{\pi}^T_{\Lambda_0}\hat{\omega}_{\Lambda_0}). \tag{3.9}$$

The Hamiltonian is the mapping $H: P \to \mathbf{R}$ corresponding to the kinetic energy of a free rigid body. Thus

$$H = \tfrac{1}{2}\pi_0 \cdot \mathbf{I}_B{}^{-1}\pi_0 + \tfrac{1}{2}M_B^{-1}\|p_0\|^2. \tag{3.10}$$

where π_0 is the spatial angular momentum vector, p_0 is the spatial linear momentum vector, M_B is the mass of the rigid body and $\mathbf{I}_B := \Lambda_0 \mathbf{J}_B \Lambda_0^T$ is the *time dependent* inertia tensor (in spatial coordinates) and $\mathbf{J}_B$ is the *constant inertia dyadic* given by

$$\mathbf{J}_B = \int_{\mathcal{B}} \rho_{ref}(\boldsymbol{X})[\|\boldsymbol{X}\|^2 \mathbf{1} - \boldsymbol{X} \otimes \boldsymbol{X}]\, d^3\boldsymbol{X}. \tag{3.11}$$

Here, $\mathcal{B} \subset \mathbf{R}^3$ is the reference configuration of the rigid body and $\rho_{ref} : \mathcal{B} \to \mathbf{R}$ the reference density.

§4. Geometrically Exact Rod Models

In this section we outline the geometrically exact rod model used in this paper. For a more complete discussion see SIMO [1985]. For application of the energy-momentum method to rods see SIMO, POSBERGH & MARSDEN [1989].

§4A. *Kinematic Idealization. Canonical Phase Space*

We assume that the placement in $\mathbf{R}^3$ at time t of a rod-like body is defined as the set

$$\mathcal{S}_t := \{ x \in \mathbf{R}^3 \mid x = \varphi(S,t) + \sum_{\alpha=1}^{2} \xi^\alpha \boldsymbol{t}_\alpha(S,t);\ \text{where } 0 \le S \le L \text{ and } (\xi^1, \xi^2) \in A \}, \tag{4.1}$$

where $A \subset \mathbf{R}^2$ is a given compact set. The map $\varphi_t: [0, L] \to \mathbf{R}^3$ given by $\varphi_t(S) = \varphi(S, t)$ defines the position at time $t \in \mathbf{R}_+$ of a curve referred to as the *line of centroids*. The vector fields $\boldsymbol{t}_\alpha(S, t)$, $\alpha = 1, 2$, are the director fields which in the special (restricted) theory of Cosserat rods, are subject to the constraints

$$\|\boldsymbol{t}_\alpha(S,t)\| = 1, \qquad \alpha = 1,2, \qquad \text{and} \qquad \boldsymbol{t}_1(S,t) \cdot \boldsymbol{t}_2(S,t) = 0. \tag{4.2}$$

In addition, the admissible motions are required to satisfy the condition

$$\boldsymbol{t}_3(S,t) \cdot \frac{\partial}{\partial S}\varphi(S,t) > 0, \qquad \text{where} \qquad \boldsymbol{t}_3(S,t) := \boldsymbol{t}_1(S,t) \times \boldsymbol{t}_2(S,t) \neq 0. \tag{4.3}$$

We refer to $\mathcal{S}_t(S) := \{\boldsymbol{x} \in \mathbf{R}^3 | [\boldsymbol{x} - \varphi(S,t)] \cdot \boldsymbol{t}_3(S,t) = 0\}$ as the *placement of a cross-section of the rod* at time t. Condition (4.3) limits the amount of *shearing* experienced by each cross-section.

Consequently, at each time $t \in \mathbf{R}_+$ and $S \in [0, L]$, we have an *orthogonal frame* $\{\boldsymbol{t}_I(S,t)\}_{I=1,2,3}$ referred to as a *director frame* in the sequel. Given any *fixed (inertial)* frame $\{\boldsymbol{E}_I(S,t)\}_{I=1,2,3}$ for example, the standard basis in $\mathbf{R}^3$, there is a *unique orthogonal* transformation $\boldsymbol{\Lambda}_t: [0, L] \to SO(3)$, defined for each time t such that

$$\boldsymbol{t}_I(S,t) = \boldsymbol{\Lambda}(S,t)\boldsymbol{E}_I(S,t), \qquad (I = 1,2,3),\ S \in [0, L], \tag{4.4}$$

where $\boldsymbol{\Lambda}(S,t) = \boldsymbol{\Lambda}_t(S)$ for $t \in \mathbf{R}_+$.

Thus, the rod is described by a curve with values in the special orthogonal group which at each point orient the director frame. Abstractly, this latter view leads to the configuration manifold

$$\boxed{\mathcal{C} := \{\boldsymbol{\Phi} = (\varphi, \boldsymbol{\Lambda}) : [0, L] \to \mathbf{R}^3 \times SO(3)\},} \tag{4.5}$$

suitably restricted by prescribed boundary conditions to be specified below. A motion is a curve of configurations $t \mapsto \boldsymbol{\Phi}_t = (\varphi_t, \boldsymbol{\Lambda}_t) \in \mathcal{C}$. Associated with any such motion, there is a sequence of placements $\mathcal{S}_t \subset \mathbf{R}^3$ of a physical body defined by (4.1).

Associated with the configuration manifold one has the collection of tangent spaces $T_{\boldsymbol{\Phi}}\mathcal{C}$ for $\boldsymbol{\Phi} \in \mathcal{C}$, defined as

$$T_{\boldsymbol{\Phi}}\mathcal{C} := \{\mathbf{v}_{\boldsymbol{\Phi}} = (v_\varphi, \hat{\omega}_{\boldsymbol{\Lambda}}): [0, L] \to \mathbf{R}^3 \times T_{\boldsymbol{\Lambda}}SO(3)\}, \tag{4.6}$$

where, in accordance with the earlier notation, the tangent field $S \in [0, L] \to \hat{\omega}_{\boldsymbol{\Lambda}}(S) \in T_{\boldsymbol{\Lambda}(S)}SO(3)$ admits the following *right* realization:
Given $\hat{\omega}: [0, L] \to so(3)$, set

$$\hat{\omega}_{\boldsymbol{\Lambda}}(S) := (\boldsymbol{\Lambda}(S), \hat{\omega}(S)\boldsymbol{\Lambda}(S)), \qquad \text{for} \quad S \in [0, L]. \tag{4.7}$$

We refer to the right realization as *spatial* representation. Following continuum mechanics conventions we have used lowercase letters for the representation.

The *canonical phase space* for our geometrically exact rod model is the *cotangent bundle* $T^*\mathcal{C} := \{\bigcup_{\Phi \in \mathcal{C}} T^*_{\Phi}\mathcal{C}\}$, where

$$T^*_{\Phi}\mathcal{C} := \{\mathbf{p}_{\Phi} = (p_{\varphi}, \hat{\pi}_{\Lambda})\colon [0, L] \to \mathbf{R}^{3*} \times T^*_{\Lambda}SO(3)\}. \tag{4.8}$$

As above, one also has the *right* representation

$$\hat{\pi}_{\Lambda}(S) := (\Lambda(S), \hat{\pi}(S)\Lambda(S)), \tag{4.9}$$

for all $S \in [0, L]$, $\hat{\pi}\colon [0, L] \to so(3)^*$. Finally, we recall (see SIMO, MARSDEN & KRISHNAPRASAD [1988]) that for *pure displacement boundary conditions* $T_{\Phi}\mathcal{C}$ and $T^*_{\Phi}\mathcal{C}$ are in *duality* through the standard L_2 pairing

$$\langle \mathbf{v}_{\Phi}, \mathbf{p}_{\Phi} \rangle := \int_0^L [p_{\varphi}(S) \cdot v_{\varphi}(S) \langle \pi_{\Lambda}(S), \theta_{\Lambda}(S) \rangle] \, dS. \tag{4.10}$$

where $\langle \pi_{\Lambda}(S), \theta_{\Lambda}(S) \rangle$ is defined analogously to the rigid body case.

More generally, definition (4.10) needs to be modified by appending additional boundary terms to accommodate (natural) stress free boundary conditions.

As a function on the phase space $P = T^*\mathcal{C}$, the kinetic energy $K\colon P \to \mathbf{R}$ takes the form

$$K = \int_0^L [\rho_A^{-1} p \cdot p + \pi \cdot \mathbf{I}^{-1} \pi] \, dS \tag{4.11}$$

$$= \int_0^L [\rho_A^{-1} p \cdot p + \pi \cdot \Lambda \mathbf{J}^{-1} \Lambda \pi] \, dS, \tag{4.12}$$

where p and π are the linear and angular velocity respectively, ρ_A is the mass per unit length of the rod, and

$$\mathbf{I} = \Lambda \mathbf{J} \Lambda^T, \qquad \mathbf{J} = \int_A \xi^{\alpha} \xi^{\beta} \rho_{ref} [\delta_{\alpha\beta} \mathbf{1}_3 - \mathbf{E}_{\alpha} \otimes \mathbf{E}_{\beta}] \, dA, \tag{4.13}$$

where ρ_{ref} is the density in the reference configuration. Here ξ^{α} are the integration variables, such that $\xi^3 \in [0, L]$, and $(\xi^1, \xi^2) \in \Omega$.

The potential energy V as a function on $P = T^*\mathcal{C}$ is expressed in terms of a stored energy function $\psi\colon \mathbf{R}^3 \times \mathbf{R}^3 \to \mathbf{R}$ as

$$V = \int_0^L \psi(\Lambda^T(\varphi' - t_3), \Lambda^T \omega) \, dS, \qquad \hat{\omega} := \Lambda' \Lambda^T, \tag{4.14}$$

where $' = \frac{\partial}{\partial S}$. The spatial strains are $\gamma = \varphi' - t_3$ and ω. When the rod is deformed the potential energy gives rise to internal stress resultants defined as

$$n = \frac{\partial \psi}{\partial \gamma}; \qquad m = \frac{\partial \psi}{\partial \omega}. \tag{4.15}$$

Consequently, the Hamiltonian $H\colon T^*\mathcal{C} \to \mathbf{R}$ is given as $H = K + V$.

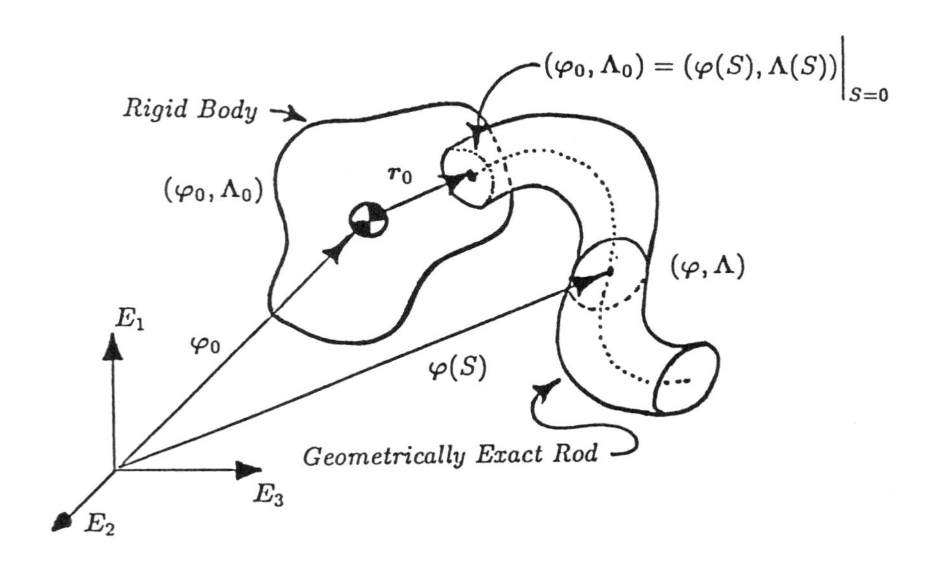

Figure 5.1. Rigid Body With Attached Rod

§5. Coupled Rigid Body – Geometrically Exact Rod

In this section we consider the stability analysis of the relative equilibria for a coupled rigid body and flexible appendage modeled as a fully nonlinear geometrically exact rod. It appears that this development represents one of the first examples of a rigorous nonlinear stability analysis of a realistic, fully nonlinear coupled structural system. The basic configuration is illustrated in figure 5.1. We assume the base of the rod is fixed at a point located a distance $\boldsymbol{r}_0$ from the center of mass of the rigid body. We model this configuration by imposing suitable boundary conditions on the rigid body and geometrically exact rod.

§5A. *Boundary Conditions*

For the rigid body with the attached, flexible rod we require that

$$\varphi_0(t) + \boldsymbol{r}_0(t) = \varphi(S,t)\Big|_{S=0} ; \qquad \Lambda_0(t) = \Lambda(S,t)\Big|_{S=0} \tag{5.1}$$

For a rigid body clamped to the base of the rod we also require that the linear and angular velocities match; *i.e.*,

$$\dot{\varphi} + \boldsymbol{w}_0 \times \boldsymbol{r}_0 = \dot{\varphi}\Big|_{S=0} ; \qquad \hat{\boldsymbol{w}}_0 = \hat{\boldsymbol{w}}\Big|_{S=0} \tag{5.2}$$

These boundary conditions impose further conditions on the admissible variations:

Lemma 5.1. *For the rigid body coupled to a geometrically exact rod, let* $(\delta\varphi_0, \delta\theta_0) \in T_{\Phi_0}P_0$ *and* $(\delta\varphi, \delta\theta) \in T_{\Phi}P$. *Then, the variations of the configuration satisfy the constraints*

$$\delta\varphi_0(t) + \delta\theta_0(t) \times r_0(t) = \delta\varphi(S,t)\Big|_{S=0}; \qquad \delta\theta_0(t) = \delta\theta(S,t)\Big|_{S=0}, \tag{5.3}$$

and the variations in angular momentum satisfy the constraint

$$\mathbf{I}_B^{-1}(\delta\pi_0 + \pi_0 \times \delta\theta_0) = \mathbf{I}^{-1}(\delta\pi + \pi \times \delta\theta)\Big|_{S=0}. \tag{5.4}$$

Proof: Conditions (5.3) follow by direct computation while conditions (5.4) follow by direct computation and application of the Legendre transformation (5.2). ■

We make the assumption that the tip of the rod is stress free. However, at the base of the rod

$$n(S,t)\Big|_{S=0} = n_0; \qquad m(S,t)\Big|_{S=0} = m_0, \tag{5.5}$$

corresponding to the force and moment balance at the boundary.

§5B. *Tangent and Cotangent Spaces: Rigid Body and Rod*

We now outline the appropriate configuration space and the associated tangent bundle and phase space for the problem of a rigid body fixed to one end of a geometrically exact rod and free to move in space.

We set $\Phi_0 = (\varphi_0, \Lambda_0)$, $\Phi = (\varphi, \Lambda)$ and define the configuration manifold of the rigid body with attached rod as

$$\boxed{Q = \left\{q := (\Phi_0, \Phi) \in \mathcal{C}_0 \times \mathcal{C} \mid \beta(\Phi_0) = \Phi(S)\Big|_{S=0}\right\}.} \tag{5.6}$$

where $\beta(\Phi_0) = (\varphi_0 + r_0, \Lambda_0)$. Furthermore, we let $\dot{q} := (\dot{\Phi}_0, \dot{\Phi})$ where $\dot{\Phi}_0 = (\dot{\varphi}_0, \hat{w}_0\Lambda_0)$, and $\dot{\Phi} = (\dot{\varphi}, \hat{w}\Lambda)$.

In view of lemma 5.1 , the tangent bundle associated with this configuration space is given by

$$TQ = \left\{(q; \dot{q}) \mid q \in Q, \dot{\beta}(\Phi_0) = \dot{\Phi}\Big|_{S=0}\right\} \tag{5.7}$$

where $\dot{\beta}(\Phi_0) = (\dot{\varphi}_0 + w_0 \times r_0, \hat{w}_0\Lambda_0)$.

Application of the Legendre transformation to (5.7) with $p = (\mathbf{P}_0, \mathbf{P})$ then yields the phase space

$$\boxed{P := T^*Q = \left\{(q; p) \mid q \in Q,\, p \in \mathsf{FL}TQ\right\}.} \tag{5.8}$$

where $\mathsf{FL}: TQ \to T^*Q$ denotes the Legendre transformation. Elements of the phase space will be denoted as $z := (q, p) \in P$.

§5C. Hamiltonian and Momentum Maps

As a function on P, the kinetic energy in material representation $K: P \to \mathbf{R}$ takes the form

$$K = \tfrac{1}{2}M_B^{-1}p_0 \cdot p_0 + \tfrac{1}{2}\mathbf{I}_B^{-1}\pi_0 \cdot \pi_0 + \int_0^L [\rho_A^{-1}p \cdot p + \pi \cdot \mathbf{I}^{-1}\pi]\, dS. \tag{5.9}$$

Here K is the sum of the kinetic energy of the rigid body, and that of the geometrically exact rod.

The potential energy as a function on $P = T^*Q$ is the same as for the rod and is expressed in terms of a stored energy function $\psi: \mathbf{R}^3 \times \mathbf{R}^3 \to \mathbf{R}$ as

$$V = \int_0^L \psi(\Lambda^T(\varphi' - t_3), \Lambda^T\omega)\, dS, \qquad \hat{\omega} := \Lambda'\Lambda^T. \tag{5.10}$$

Note that the stored energy function depends on the configuration of the rod alone. More generally it may also depend on the configuration of the rigid body. This would be the case, for example, were the rigid body attached to the rod by an elastic hinge. Again, the Hamiltonian is given as $H = K + V$.

Next we turn our attention to the invariance properties of H under group actions. We first compute the *momentum maps* corresponding to the group of rotations and translations.

We have the following group invariance properties.

i. *Left* $SO(3)$*-Invariance.* We note that the reduced expression for the stored energy function ψ appearing in (5.10) is *constructed precisely* to satisfy this invariance property. Thus $G = SO(3)$, Q as defined in (5.6), and $\mathcal{G} = so(3)$, so that

$$\Psi_1(A, (\varphi_0, \Lambda_0, \varphi, \Lambda)) := (A\varphi_0, A\Lambda_0, A\varphi, A\Lambda), \qquad \text{for all } A \in SO(3), \tag{5.11}$$

is the action induced by $SO(3)$ on Q. Given any $\hat{\xi} \in so(3)$, the infinitesimal generator is

$$\begin{aligned} \hat{\xi}_Q((\varphi_0, \Lambda_0, \varphi, \Lambda)) &= \frac{d}{dt}\bigg|_{t=0} (\exp(t\hat{\xi})\varphi_0, \exp(t\hat{\xi})\Lambda_0, \exp(t\hat{\xi})\varphi, \exp(t\hat{\xi})\Lambda) \\ &= (\hat{\xi}\varphi_0, \hat{\xi}\Lambda_0, \hat{\xi}\varphi, \hat{\xi}\Lambda). \end{aligned} \tag{5.12}$$

From the relation $\langle J_1(\alpha_q), \xi\rangle = J_1(\xi)(\alpha_q)$ there follows

$$\boxed{J_1(p_{\varphi_0}, \pi_{\Lambda_0}, p_\varphi, \pi_\Lambda) := \pi_0 + \varphi_0 \times p_0 + \int_0^L (\pi + \varphi \times p)\, dS.} \tag{5.13}$$

ii. $\mathbb{R}^3$-*translational invariance.* The reduced expressions for the stored energy function ψ in (4.14) is also *invariant* under $\mathbb{R}^3$-translations. Consequently, $G = \mathbb{R}^3$, $\mathcal{G} \cong \mathbb{R}^3$, and Q is as defined in (5.6) so that

$$\Psi_2(c, (\varphi_0, \Lambda_0, \varphi, \Lambda)) := (\varphi_0 + c, \Lambda_0, \varphi + c, \Lambda), \qquad \text{for all } \mathbf{c} \in \mathbb{R}^3, \tag{5.14}$$

is the action induced by $\mathbb{R}^3$ on Q. Given $\xi \in \mathbb{R}^3$, the infinitesimal generator is

$$\xi_Q(\varphi_0, \Lambda_0, \varphi, \Lambda) = \frac{d}{dt}\bigg|_{t=0} (\varphi_0 + t\xi, \Lambda_0, \varphi + t\xi, \Lambda) = (\xi, 0, \xi, 0). \tag{5.15}$$

Again, we use $\langle \boldsymbol{J}_2(\boldsymbol{\alpha}_q), \boldsymbol{\xi} \rangle = J_2(\xi)(\boldsymbol{\alpha}_q)$ the corresponding momentum map $\boldsymbol{J}_2 : P \to \mathbb{R}^3$ as

$$\boxed{\boldsymbol{J}_2(\boldsymbol{p}_{\varphi_0}, \boldsymbol{\pi}_{\Lambda_0}, \boldsymbol{p}_{\varphi}, \boldsymbol{\pi}_{\Lambda}) := \boldsymbol{p}_0 + \int_0^L \boldsymbol{p}\, dS.} \tag{5.16}$$

The first step of the energy-momentum method then requires the construction of the energy-momentum functional which in the present context takes the form

$$\begin{aligned} H_{\xi,u} &= K + V - \langle \boldsymbol{J}_1 - \boldsymbol{\mu}_e, \boldsymbol{\xi} \rangle - \langle \boldsymbol{J}_2 - \boldsymbol{v}_e, \boldsymbol{u} \rangle \\ &= V_{\xi,u} + K_{\xi,u} + \langle \boldsymbol{\mu}_e, \boldsymbol{\xi} \rangle + \langle \boldsymbol{v}_e, \boldsymbol{u} \rangle, \end{aligned} \tag{5.17a}$$

where

$$\begin{aligned} V_{\xi,u} = V - \tfrac{1}{2}\Big\{ & M_B \|\xi \times \varphi_0 + \boldsymbol{u}\|^2 + \mathbf{I}_B \xi \cdot \xi \\ & + \int_0^L \rho_A \|\xi \times \varphi + \boldsymbol{u}\|^2\, dS + \int_0^L \mathbf{I}\xi \cdot \xi\, dS \Big\}, \end{aligned} \tag{5.17b}$$

and

$$\begin{aligned} K_{\xi,u} = \tfrac{1}{2}\Big\{ & \|M_B^{-\frac{1}{2}} \boldsymbol{p}_0 - M_B^{\frac{1}{2}}(\xi \times \varphi_0 + \boldsymbol{u})\|^2 + \|\mathbf{I}_B^{-\frac{1}{2}} \boldsymbol{\pi}_0 - \mathbf{I}_B^{\frac{1}{2}} \xi\|^2 \\ & + \int_0^L \|\rho_A^{-\frac{1}{2}} \boldsymbol{p} - \rho_A^{\frac{1}{2}}(\xi \times \varphi + \boldsymbol{u})\|^2\, dS + \int_0^L \|\mathbf{I}^{-\frac{1}{2}} \boldsymbol{\pi} - \mathbf{I}^{\frac{1}{2}} \xi\|^2\, dS \Big\}. \end{aligned} \tag{5.17c}$$

§5D. Energy-momentum functional: First Variation

As pointed out in §2, the computation of the relative equilibria exploits the crucial fact that relative equilibria are critical points of the energy-momentum functional. We have the following result

Proposition 5.2. *(First variation of $H_{\xi,u}$). The Euler-Lagrange equations associated with the critical points of $H_{\xi,u}$ are*

$$\left.\begin{array}{ll} \text{Rigid Body:} & \text{Flexible Appendage:} \\ M_B^{-1}\boldsymbol{p}_{0,e} = \boldsymbol{\xi}\times\boldsymbol{\varphi}_{0,e}+\boldsymbol{u}, & \rho_A^{-1}\boldsymbol{p}_e = \boldsymbol{\xi}\times\boldsymbol{\varphi}_e+\boldsymbol{u}, \\ \mathbf{I}_{B,e}^{-1}\boldsymbol{\pi}_{0,e} = \boldsymbol{\xi}, & \mathbf{I}_e^{-1}\boldsymbol{\pi}_e = \boldsymbol{\xi}, \\ \boldsymbol{\xi}\times\boldsymbol{p}_{0,e} = \boldsymbol{n}_0, & \boldsymbol{\xi}\times\boldsymbol{p}_e = \boldsymbol{n}'_e, \\ \mathbf{I}_{B,e}^{-1}\boldsymbol{\pi}_{0,e}\times\boldsymbol{\pi}_{0,e} = \boldsymbol{m}_{0,e}+\boldsymbol{\varphi}_{0,e}\times\boldsymbol{n}_{0,e}. & \mathbf{I}_e^{-1}\boldsymbol{\pi}_e\times\boldsymbol{\pi}_e = \boldsymbol{m}'_e+\boldsymbol{\varphi}'_e\times\boldsymbol{n}_e \end{array}\right\} \tag{5.18}$$

where $\boldsymbol{n}_0$, and $\boldsymbol{m}_0$ are as defined in (5.5), and $\boldsymbol{n}$, and $\boldsymbol{m}$ are as defined in (4.15).

Proof: Follows by a direct computation of

$$\delta H_{\xi,u} = \delta V_{\xi,u} + \delta K_{\xi,u}. \tag{5.19a}$$

From (5.17b) we find

$$\begin{aligned} \delta V_{\xi,u} &= \delta V - M_B(\boldsymbol{\xi}\times\boldsymbol{\varphi}_0+\boldsymbol{u})\cdot\boldsymbol{\xi}\times\delta\boldsymbol{\varphi}_0 - \mathbf{I}_B\boldsymbol{\xi}\times\boldsymbol{\xi}\cdot\delta\boldsymbol{\theta}_0 \\ &\quad - \int_0^L \rho_A(\boldsymbol{\xi}\times\boldsymbol{\varphi}+\boldsymbol{u})\cdot\boldsymbol{\xi}\times\delta\boldsymbol{\varphi}\,dS - \int_0^L \mathbf{I}\boldsymbol{\xi}\times\boldsymbol{\xi}\cdot\delta\boldsymbol{\theta}\,dS; \end{aligned} \tag{5.19b}$$

whereas from (5.17c) we obtain

$$\begin{aligned} \delta K_{\xi,u} &= M_B^{-1}(\boldsymbol{p}_0 - M_B(\boldsymbol{\xi}\times\boldsymbol{\varphi}_0+\boldsymbol{u}))\cdot(\delta\boldsymbol{p}_0 - M_B(\boldsymbol{\xi}\times\delta\boldsymbol{\varphi}_0)) \\ &\quad + (\mathbf{I}_B^{-1}\boldsymbol{\pi}_0\times\boldsymbol{\pi}_0 + \mathbf{I}_B\boldsymbol{\xi}\times\boldsymbol{\xi})\cdot\delta\boldsymbol{\theta}_0 + (\mathbf{I}_B^{-1}\boldsymbol{\pi}_0 - \boldsymbol{\xi})\cdot\delta\boldsymbol{\pi}_0 \\ &\quad + \int_0^L \rho_A^{-1}(\boldsymbol{p} - \rho_A(\boldsymbol{\xi}\times\boldsymbol{\varphi}+\boldsymbol{u}))\cdot(\delta\boldsymbol{p} - \rho_A(\boldsymbol{\xi}\times\delta\boldsymbol{\varphi}))\,dS \\ &\quad + \int_0^L (\mathbf{I}^{-1}\boldsymbol{\pi}\times\boldsymbol{\pi} + \mathbf{I}\boldsymbol{\xi}\times\boldsymbol{\xi})\cdot\delta\boldsymbol{\theta} + (\mathbf{I}^{-1}\boldsymbol{\pi} - \boldsymbol{\xi})\cdot\delta\boldsymbol{\pi}\,dS. \end{aligned} \tag{5.19c}$$

Furthermore, for δV after integration by parts and using (5.3) we obtain the result that

$$\begin{aligned} \delta V &= -\delta\boldsymbol{\theta}_0\cdot(\boldsymbol{m}_0 + \boldsymbol{r}_0\times\boldsymbol{n}_0) - \delta\boldsymbol{\varphi}_0\cdot\boldsymbol{n}_0 \\ &\quad + \int_0^L \left[-(\boldsymbol{m}' + \boldsymbol{\varphi}'\times\boldsymbol{n})\cdot\delta\boldsymbol{\theta} - \boldsymbol{n}'\cdot\delta\boldsymbol{\varphi}\right] dS. \end{aligned} \tag{5.20}$$

By standard arguments in the calculus of variations we conclude that the stationarity of the first variation requires conditions (5.18) to hold. ■

We note that the relative equilibrium conditions (5.18) are precisely those one would expect to obtain by a 'bare hands' computation using the definition of a relative equilibrium and the the laws of motion. These results are of course in complete agreement with those obtained by the relative equilibrium theorem.

The following theorem describes several properties of a relative equilibrium of the rod.

Theorem 5.3. *At a relative equilibrium configuration of the rigid body with attached, geometrically exact rod, the following conditions hold.*

i. *The spatial angular velocity field is constant for all $S \in [0, L]$; i.e.,*

$$\hat{w}_e := [\dot{\Lambda}_e \Lambda_e^T] \equiv \text{constant}, \tag{5.21}$$

so that

$$\Lambda_e(S,t) = \exp[t\hat{w}_e]\Lambda_0(S). \tag{5.22}$$

ii. *The center of mass of the rod, defined by $t \mapsto r_e^0 \in \mathbb{R}^3$ where*

$$\left.\begin{aligned} r_e^0 &:= \frac{1}{M}\Big(M_B \varphi_{0,e} + \int_0^L \rho_A(S)\varphi_e(S,t)\, dS\Big), \\ M &:= M_B + \int_0^L \rho_A(S)\, dS, \end{aligned}\right\} \tag{5.23}$$

undergoes uniform motion with constant velocity

$$v_e^0 := \frac{1}{M}\Big(p_0 + \int_0^L p(S)\, dS\Big) = u + \xi \times r_e^0. \tag{5.24}$$

iii. *The fixed spatial rotation axis $\xi \in \mathbb{R}^3$ is an eigenvector of the extended inertia dyadic relative to the center of mass; i.e.,*

$$\mathbb{I}_\infty^0 \xi = \lambda \xi, \tag{5.25}$$

where we define

$$\left.\begin{aligned} \mathbb{I}_\infty^0 &:= \mathbb{I}_\infty + M[\|r_e^0\|^2 \mathbf{1}_3 - r_e^0 \otimes r_e^0] \\ \mathbb{I}_\infty &:= \mathbf{I}_B + M_B[\|\varphi_0\|^2 \mathbf{1}_3 - \varphi_0 \otimes \varphi_0]) \\ &\quad + \int_0^L (\mathbf{I} + \rho_A[\|\varphi_0\|^2 \mathbf{1}_3 - \varphi_0 \otimes \varphi_0])\, dS. \end{aligned}\right\} \tag{5.26}$$

iv. *The total linear and angular momentum at a relative equilibrium satisfy the condition*

$$\xi \times \mu_e + u \times \ell_e = 0. \tag{5.27}$$

Proof: see SIMO, POSBERGH & MARSDEN [1989]. ■

Recall the first variation (5.20). If we set $\boldsymbol{u} = \mathbf{0}$ (corresponding to a reduction to center of mass) then the computation of the second variation is straight forward. Thus, for $\delta^2 H_\xi(z_e)$ on TP at equilibrium we have

$$
\begin{aligned}
\delta^2 H_\xi(z_e)\cdot(\delta z, \Delta z) &= \delta^2 V_\xi(q_e)\cdot(\delta \boldsymbol{q}, \Delta \boldsymbol{q}) + \delta^2 K_\xi(z_e)\cdot(\delta z, \Delta z)\\
&= (\delta\boldsymbol{\pi}_0 + \boldsymbol{\pi}_{0,e}\times\delta\boldsymbol{\theta}_0)\cdot \mathbf{I}_{B,e}^{-1}(\Delta\boldsymbol{\pi}_0 + \boldsymbol{\pi}_{0,e}\times\Delta\boldsymbol{\theta}_0)\\
&\quad + \delta\boldsymbol{\theta}_0\cdot[\boldsymbol{\xi}\times\Delta\boldsymbol{\pi}_0 + \boldsymbol{\pi}_{0,e}\times(\boldsymbol{\xi}\times\Delta\boldsymbol{\theta}_0)] - \delta\boldsymbol{\pi}_0\cdot(\boldsymbol{\xi}\times\Delta\boldsymbol{\theta}_0)\\
&\quad - \delta\boldsymbol{\varphi}_0\cdot\Delta\boldsymbol{p}_0\times\boldsymbol{\xi} + \delta\boldsymbol{p}\cdot[M_B^{-1}\Delta\boldsymbol{p}_0 - \boldsymbol{\xi}\times\Delta\boldsymbol{\varphi}_0]\\
&\quad + \int_0^L (\delta\boldsymbol{\pi} + \boldsymbol{\pi}_e\times\delta\boldsymbol{\theta})\cdot\mathbf{I}_e^{-1}(\Delta\boldsymbol{\pi} + \boldsymbol{\pi}_e\times\Delta\boldsymbol{\theta})\\
&\quad + \delta\boldsymbol{\theta}\cdot[\boldsymbol{\xi}\times\Delta\boldsymbol{\pi} + \boldsymbol{\pi}\times(\boldsymbol{\xi}\times\Delta\boldsymbol{\theta})] - \delta\boldsymbol{\pi}\cdot(\boldsymbol{\xi}\times\Delta\boldsymbol{\theta})\\
&\quad - \delta\boldsymbol{\varphi}\cdot\Delta\boldsymbol{p}\times\boldsymbol{\xi} + \delta\boldsymbol{p}\cdot[\rho_A^{-1}\Delta\boldsymbol{p} - \boldsymbol{\xi}\times\Delta\boldsymbol{\varphi}]\, dS + \delta^2 V(q_e)
\end{aligned}
\tag{5.28}
$$

Subsequently, we will set $\Delta z \in \mathcal{S}_{RIG}$. Elements in $\mathcal{S}_{RIG}$ are of the form $\Delta z = (\Delta\boldsymbol{q}, \widetilde{\Delta\boldsymbol{p}})$, where $\Delta\boldsymbol{q} = (\Delta\boldsymbol{\varphi}_0, \Delta\boldsymbol{\theta}_0, \Delta\boldsymbol{\varphi}, \Delta\boldsymbol{\theta})$ is given by

$$
\left.\begin{aligned}
\Delta\boldsymbol{\varphi}_0 &= \boldsymbol{\eta}\times\boldsymbol{\varphi}_0; & \Delta\boldsymbol{\varphi} &= \boldsymbol{\eta}\times\boldsymbol{\varphi};\\
\Delta\boldsymbol{\theta}_0 &= \boldsymbol{\eta}, & \Delta\boldsymbol{\theta} &= \boldsymbol{\eta},
\end{aligned}\right\}
\tag{5.29a}
$$

and $\widetilde{\Delta\boldsymbol{p}} = (\Delta\boldsymbol{p}_0, \Delta\boldsymbol{\pi}_0, \Delta\boldsymbol{p}, \Delta\boldsymbol{\pi})$ is given by

$$
\left.\begin{aligned}
\Delta\boldsymbol{p}_0 &= M_B\boldsymbol{\zeta}\times\boldsymbol{\varphi} + \boldsymbol{\eta}\times\boldsymbol{p}_0; & \Delta\boldsymbol{p} &= \rho_A\boldsymbol{\zeta}\times\boldsymbol{\varphi} + \boldsymbol{\eta}\times\boldsymbol{p};\\
\Delta\boldsymbol{\pi}_0 &= \mathbf{I}_B\boldsymbol{\zeta} + \boldsymbol{\eta}\times\boldsymbol{\pi}_0, & \Delta\boldsymbol{\pi} &= \mathbf{I}\boldsymbol{\zeta} + \boldsymbol{\eta}\times\boldsymbol{\pi},
\end{aligned}\right\}
\tag{5.29b}
$$

for $\boldsymbol{\eta}, \boldsymbol{\zeta} \in \mathbf{R}^3$

To perform the second variation test we will need the following result.

Lemma 5.4. *Let $\Delta\boldsymbol{q} \in \mathcal{V}_{RIG}$, and let $\boldsymbol{q}_e \in Q$ be a configuration in relative equilibrium. Then, in general $\delta^2 V(\boldsymbol{q}_e) \neq 0$; in fact for $\delta^2 V(\boldsymbol{q}_e) : \mathcal{V}_{RIG}\times TQ \to \mathbf{R}$ we find*

$$
\begin{aligned}
\delta^2 V(\boldsymbol{q}_e)(\Delta\boldsymbol{q}, \delta\boldsymbol{q}) &= (\boldsymbol{\xi}\times\boldsymbol{p}_{0,e})\times\delta\boldsymbol{\varphi}_0\cdot\boldsymbol{\eta} + (\boldsymbol{\xi}\times\boldsymbol{\pi}_{0,e})\times\delta\boldsymbol{\theta}_0\cdot\boldsymbol{\eta}\\
&\quad - \int_0^L [(\boldsymbol{\xi}\times\boldsymbol{p}_e)\times\delta\boldsymbol{\varphi}\cdot\boldsymbol{\eta} + (\boldsymbol{\xi}\times\boldsymbol{\pi}_e)\times\delta\boldsymbol{\theta}\cdot\boldsymbol{\eta}]\, dS.
\end{aligned}
\tag{5.30}
$$

Proof: Follows by a direct calculation involving integration by parts. For details see SIMO, POSBERGH & MARSDEN [1989]. ■

§5E. The second variation of the energy-momentum functional

For $\delta^2 H_\xi(z_e) : TP \times \mathcal{S}_{RIG} \to \mathbf{R}$ from (5.28), (5.29) and (5.30) we have

$$\begin{aligned}
\delta^2 H_\xi(z_e) = \zeta \cdot \Big\{ & \delta\pi_0 + \pi_{0,e} \times \delta\theta_0 + \varphi_{0,e} \times \delta p_0 + p_{0,e} \times \delta\varphi_0 \\
& + \mathbf{I}_B(\delta\theta_0 \times \xi) + \xi \times \mathbf{I}_B \delta\theta_0 \\
& + \int_0^L [\delta\pi + \pi_e \times \delta\theta + \varphi_e \times \delta p + p_e \times \delta\varphi \\
& + \mathbf{I}_e(\delta\theta \times \xi) + \xi \times \mathbf{I}_e \delta\theta] \, dS \Big\} \\
& - (\zeta \times \xi) \cdot \Big\{ \mathbf{I}_B \delta\theta_0 + M_B \varphi_{0,e} \times \delta\varphi_0 + \int_0^L [\mathbf{I}_e \delta\theta + \rho_A \varphi_e \times \delta\varphi] \, dS \Big\} \\
& + (\eta \times \xi) \cdot \Big\{ \delta\pi_0 + \delta\varphi_{0,e} \times p_{0,e} + \varphi_{0,e} \times \delta p_{0,e} \\
& + \int_0^L [\delta\pi + \delta\varphi \times p_e + \varphi_e \times \delta p] \, dS \Big\}
\end{aligned} \tag{5.31}$$

We can now state the following Corollary in the case of a geometrically exact rod attached to a rigid body.

Corollary 5.5. *At a relative equilibrium $z_e \in P$, for any $\Delta z \in \mathcal{S}_{RIG}$ and $\delta z \in \mathcal{S}_{INT}$ we have*

$$\boxed{\delta^2 H_\xi(z_e)(\Delta z, \delta z) = 0.} \tag{5.32}$$

so that, restricted to $\mathcal{S} \subset T_{z_e}P$ the second variation of the energy-momentum functional can be written in the **uncoupled** *form*

$$\boxed{\delta^2 H_\xi(z_e)\Big|_{\mathcal{S}} = \delta^2 H_\xi(z_e)\Big|_{\mathcal{S}_{RIG}} + \delta^2 H_\xi(z_e)\Big|_{\mathcal{S}_{INT}} .} \tag{5.33}$$

Proof: Equation (5.32) follows from expression (5.31) and the definition of $\mathcal{V}_{INT}$. On the other hand, (5.33) follows immediately from the bilinearity of $\delta^2 H_\xi: \mathcal{S} \times \mathcal{S} \to \mathbf{R}$, and the equilibrium conditions. ■

Proposition 5.6. *Let $\Delta z \in \mathcal{S}_{RIG}$, and let $z_e \in P$ be a (relative) equilibrium configuration. Let the second variation be regarded as a function $\delta^2 H_\xi: \mathcal{S}_{RIG} \times \mathcal{S}_{RIG} \to \mathbf{R}$. Then, at an equilibrium, $\delta^2 H_\xi(q_e)$ on $\mathcal{S}_{RIG} \times \mathcal{S}_{RIG}$ can be written as the following quadratic form*

$$\boxed{\delta^2 H_\xi(z_e) = (\mu_e \times \eta) \cdot [\mathbb{I}_\infty^{-1} - \lambda \mathbf{1}_3](\mu_e \times \eta)} \tag{5.34}$$

Proof: See SIMO, POSBERGH & MARSDEN [1989]. ■

We note that on $\mathcal{S}_{INT}$ the situation is the same as for the rod alone. This is because *all the flexibility* is assigned to the rod (*i.e.*, V is a depends only on Φ_e, the configuration of the rod). This would not be the case were the rod attached to the rigid body by an elastic hinge.

§5F. *Stability Conditions associated with $\mathcal{V}_{INT}$.*

We complete our stability analysis by deriving conditions which guarantee the definiteness of the component on $\mathcal{V}_{INT}$. According to our outline in §2B, the space $\mathcal{V} \subset T_{q_e}Q$ is given by

$$\mathcal{V} := \{\delta q \in T_{q_e}Q \mid \boldsymbol{\xi} \cdot \int_0^L [\mathbf{I}_e\delta\boldsymbol{\theta} + \rho_A\boldsymbol{\varphi}_e \times \delta\boldsymbol{\varphi}]\, dS = 0\ \}. \tag{5.35}$$

One can show that the evaluation of the Lie derivative condition in (2.26) yields the following explicit expression

$$\begin{aligned}\langle \delta q, \mathcal{L}_{\eta_Q(q_e)}\frac{\delta L_{\xi}}{\delta q}(q_e)\rangle_g &= \boldsymbol{\zeta} \cdot \int_0^L [\boldsymbol{\pi}_e \times \delta\boldsymbol{\theta} + 2\mathbf{p}_e \times \delta\boldsymbol{\varphi} + \mathbf{I}_e(\delta\boldsymbol{\theta} \times \boldsymbol{\xi}) + \boldsymbol{\xi} \times \mathbf{I}_e\delta\boldsymbol{\theta}]\, dS \\ &\quad - (\boldsymbol{\zeta} \times \boldsymbol{\xi}) \cdot \int_0^L [\mathbf{I}_e\delta\boldsymbol{\theta} + \rho_A\boldsymbol{\varphi}_e \times \delta\boldsymbol{\varphi}]\, dS \\ &= 0\end{aligned} \tag{5.36}$$

for all $\delta\boldsymbol{\Phi} \in \mathcal{V}_{INT}$, and all $\boldsymbol{\zeta} \in \mathbf{R}^3$ such that $\boldsymbol{\zeta} \cdot \boldsymbol{\xi} = 0$. This condition defines $\mathcal{V}_{INT} \subset \mathcal{V}$.

As a quadratic form, the second variation of the effective potential evaluated at an equilibrium configuration can now be written as

$$\delta^2 V_{\xi}\Big|_e = \delta^2 V(\boldsymbol{\Phi}_e) - \int_0^L \Big\{\rho_A \|\boldsymbol{\xi} \times \delta\boldsymbol{\varphi}\|^2 + \delta\boldsymbol{\theta} \cdot \hat{\boldsymbol{\xi}}(\hat{\boldsymbol{\pi}}_e - \mathbf{I}_e\hat{\boldsymbol{\xi}}_e)\delta\boldsymbol{\theta}\Big\}\, dS \tag{5.37}$$

where

$$\delta^2 V(\boldsymbol{\Phi}_e) = \int_0^L \mathsf{A}(\boldsymbol{\Phi}_e)(\delta\boldsymbol{\Phi}, \delta\boldsymbol{\Phi})\, dS \tag{5.38}$$

and $\mathsf{A}(\boldsymbol{\Phi}_e)(\delta\boldsymbol{\Phi}, \delta\boldsymbol{\Phi})$ is referred to as the *total elasticity tensor*.

To bound (5.37) from below we introduce the following auxiliary eigenvalue problem.

Eigenvalue Problem. Let $\delta^2 V(\boldsymbol{\Phi}_e)$ be as above. Then, from (5.38) we have

$$\boxed{\delta^2 V(\boldsymbol{\Phi}_e)(\delta\boldsymbol{\Phi}, \delta\boldsymbol{\Phi}) = \langle \delta\boldsymbol{\Phi}, \mathsf{A}(\boldsymbol{\Phi}_e)\delta\boldsymbol{\Phi}\rangle} \tag{5.39}$$

Let $M := \begin{bmatrix} \rho_A 1_3 & 0 \\ 0 & I \end{bmatrix}$ be the matrix form of the kinetic energy Riemannian metric. Consider the symmetric variational eigenvalue problem:

Find $(\lambda, \eta) \in \mathbf{R} \times \mathcal{V}_{INT}$ such that, for all $\delta\Phi \in \mathcal{V}_{INT}$;

$$\boxed{\langle \delta\Phi, \mathsf{A}(\Phi_e)\eta \rangle = \lambda \langle \delta\Phi, M\eta \rangle} . \tag{5.40}$$

We note the following:

1. Suppose that Φ_e is the *reference configuration,* so that the geometric term $G(\Phi_e) \equiv 0$. Then, since $\ker[\Xi(\Phi_e)] = \mathcal{V}_{RIG}$ so that $\eta \notin \mathcal{V}_{RIG}$ by construction, it follows that (5.40) has *positive eigenvalues* since M is always positive definite and $C(\Phi)|_{\Phi=\text{identity}}$ is positive definite. The solutions of (5.40) are, in fact, the *natural frequencies* of the *system* at configuration Φ_e.

2. In general, of course, $\mathsf{A}(\Phi_e)$ *cannot* be *positive definite* for *all* Φ_e (restricted to $\mathcal{V}_{INT}$) since this would imply *convexity* of the stored energy function and hence uniqueness; clearly, an unacceptable restriction. See MARSDEN & HUGHES [1983, Chapter 6], OGDEN [1984] and CIARLET [1988] for a detailed discussion of some appropriate restrictions on the stored energy functions.

3. If $A(\Phi_e)$ *is indeed positive definite at* $\Phi_e \in C$, then (5.40) has *positive* eigenvalues. The *lowest* one is given by the Rayleigh quotient

$$\boxed{\lambda_0(\Phi_e) = \inf_{\delta\Phi \in \mathcal{V}_{INT}} \frac{\langle \delta\Phi, \mathsf{A}(\Phi_e)\delta\Phi \rangle}{\langle \delta\Phi, M\delta\Phi \rangle},} \tag{5.41}$$

so that, trivially, one has the inequality

$$\langle \delta\Phi, \mathsf{A}(\Phi_e)\delta\Phi \rangle \quad \geq \quad \lambda_0(\Phi_e)\langle \delta\Phi, M\delta\Phi \rangle > 0, \tag{5.42}$$

for all $\delta\Phi \in \mathcal{V}_{INT}$.

With these observations in mind we have

Theorem 5.7. *Let the relative equilibrium* $\Phi_e \in C$ *be such that* $\mathsf{A}(\Phi_e)$, *as defined by (5.38) is positive definite, and let* $\lambda_0(\Phi_e)$ *be the lowest eigenvalue of problem (5.40) as given by (5.41), so that (5.42) holds. Then, stability on* $\mathcal{V}_{INT}$ *requires that*

$$\boxed{\delta^2 H_\xi \Big|_e \geq \lambda_0(\Phi_e)\langle \delta\Phi, M\delta\Phi \rangle - \int_0^L \left\{ \rho_A \|\xi \times \delta\varphi\|^2 + \delta\theta \cdot \hat{\xi}[\bar{\pi}_e - \mathbf{I}_e\hat{\xi}_e]\delta\theta \right\} dS.} \tag{5.43}$$

Remark. A sharp estimate of the constant $\lambda_0(\Phi_e) > 0$ in (5.42) can be obtained by solving explicitly the local form of the eigenvalue problem (5.40). Integration by parts of (5.40) and use of standard results in the calculus of variations yields the second order linear eigenvalue problem

$$\Xi^*(\Phi_e)[C(\Phi_e)\Xi(\Phi_e)\eta] = \lambda M\eta \quad \text{in} \quad [0, L], \tag{5.44}$$

along with the (linearized) boundary conditions

$$\Xi(\Phi_e)\eta\Big|_{S=0,L} = 0, \tag{5.45}$$

and the constraint conditions

$$\int_O^L M\delta\Phi \, dS = 0. \tag{5.46}$$

Here, $\Xi^*(\Phi_e)$ is given by the expression

$$(\delta n, \delta m) \mapsto \Xi^*(\Phi_e)(\delta n, \delta m) = \begin{Bmatrix} -\delta n' \\ -\delta m' - \varphi_e' \times \delta n \end{Bmatrix}. \tag{5.47}$$

At a straight reference configuration $\Phi_e = (\varphi_0, 1)$ where $\varphi_0 = Se_3$, for $S \in [0, L]$, and $e_3 = (0, 0, 1)$, equation (5.42) reduces to the classical problem for a Timoshenko beam model if $C(\Phi_e)$ is assumed to be constant and diagonal form of the

$$C(\Phi_e) = \operatorname{diag}[GA_1, GA_2, EA, EI_1, EI_2, GI]. \tag{5.48}$$

Explicitly, setting $\eta = (\eta_1, \eta_2, \eta_3, \theta_1, \theta_2, \theta_3)$, we have

$$-\begin{Bmatrix} GA_1(\eta_1'' - \theta_2') \\ GA_2(\eta_2'' + \theta_1') \\ EA\theta_3'' \\ EI_1\theta_1'' - GA_2(\eta_2' + \theta_1) \\ EI_2\theta_2'' + GA_1(\eta_1' - \theta_2) \\ GJ\theta_3'' \end{Bmatrix} = \lambda_0 \begin{Bmatrix} \rho_A\eta_1 \\ \rho_A\eta_2 \\ \rho_A\eta_3 \\ I_1\theta_1 \\ I_2\theta_2 \\ J\theta_3 \end{Bmatrix}, \tag{5.49}$$

along with the stress free boundary conditions which now take the form

$$\left.\begin{aligned} [\eta_1' - \theta_2]\Big|_{S=0,L} &= [\eta_2' + \theta_1]\Big|_{S=0,L} = \eta_3'\Big|_{S=0,L} = 0 \\ \theta_1'\Big|_{S=0,L} &= \theta_2'\Big|_{S=0,L} = \theta_3'\Big|_{S=0,L} = 0 \end{aligned}\right\} \tag{5.50}$$

and the constraint condition (5.28) which *eliminates* rigid body modes. An explicit solution to this problem can be found in elementary text books, *i.e.* GRAFF [1975].

§6. Conclusions

In this paper we used the energy-momentum method to investigate the stability of the relative equilibria of a uniformly rotating rigid body with an attached flexible appendage. A fundamental decomposition of the space of admissible variations was employed, which decoupled the problem into a 'rigid body' stability problem and an 'internal vibration' problem. The stability conditions on the first of two subspaces, denoted by $\mathcal{S}_{RIG}$, corresponded to that of a classical rigid body. On the complement to this space, denoted by $\mathcal{S}_{INT}$, the stability conditions corresponded to requiring the rate of rotation to be below that of the lowest frequency of excitation of an associated eigenvalue.

Acknowledgements

We thank P. S. Krishnaprasad, Debbie Lewis, John Maddocks and Tudor Ratiu for their input during innumerable discussions about this work.

References

R. Abraham & J. E. Marsden [1978], *Foundations of Mechanics,* Second Edition, Addison-Wesley, Reading.

V. I. Arnold [1978], *Mathematical Methods of Classical Mechanics,* Springer, Berlin.

V.I. Arnold, [1966], "An a priori estimate in the theory of hydrodynamic stability," *Izv. Vyssh. Uchebn. Zaved. Matematicka,* **54**, 3-5, (Russian).

P. G. Ciarlet [1988], *Mathematical Elasticity,* North-Holland, Amsterdam.

D. D. Holm, J.E. Marsden, T. Ratiu, & A Weinstein [1985], "Nonlinear stability of fluid and plasma equilibria," *Physics Reports* , **123**, 1-116.

K. F. Graff [1975], *Wave Motion in Elastic Solids,* Ohio State University press, Columbus.

P. S. Krishnaprasad & J.E. Marsden [1987], "Hamiltonian structure and stability for rigid bodies with flexible attachments," *Arch. Rational Mech. Anal.,* **98**, 71-93.

D. G. Luenberger [1984], *Linear and Nonlinear Programming,* Second Edition, Addison-Wesley, Reading.

J. E. Marsden & T. J. R. Hughes [1983], *Mathematical Foundations of Elasticity,* Prentice-Hall, Englewood Cliffs.

J. E. Marsden, J. C. Simo, D. R. Lewis & T. A. Posbergh [1989], "A Block Diagonalization Theorem in the Energy-Momentum Method," these proceedings.

J. E. MARSDEN & A. WEINSTEIN [1974], "Reduction of Symplectic Manifolds with Symmetry," *Rep. Math. Phys.*, **5**, 121-130.

R. W. OGDEN [1984], *Non-linear Elastic Deformations*, Ellis Horwood series in mathematics and its applications, E. Horwood, Chichester.

T. A. POSBERGH, P. S. KRISHNAPRASAD & J. E. MARSDEN [1987], Stability Analysis of a Rigid Body with a Flexible Attachement Using the Energy-Casimir Method," *Contemp. Math.*, AMS, **68**, 253-273.

T. A. POSBERGH [1988], *Modeling and Control of Mixed and Flexible Structures*, Systems Research Center, TR88-58, U. of Maryland, College Park.

T. A. POSBERGH & J. C. SIMO [1988], "Nonlinear Dynamics of Flexible Structures: Geometrically Exact Formulation and Stability," *Proc. of the 27th Conf. Decision and Contr.*, Austin, 1732-1737.

J. C. SIMO [1985], "A finite strain beam formulation. The three dimensional dynamic problem. Part I," *Comp. Meth. Appl. Mech. Engng.*, **49**, 55-70.

J. C. SIMO & L. VU-QUOC [1986], "A 3-Dimensional Finite Strain Rod Model. Part II: Geometric and Computational Aspects," *Comp. Meth. Appl. Mech. Engng.*, **58**, 79-116.

J. C. SIMO & L. VU-QUOC [1988a], "On the Dynamics in Space of Rods Undergoing Large Overall Motions -A Geometrically Exact Approach," *Comp. Meth. Appl. Mech. Engng.*, **66**, 125-161.

J. C. SIMO & L. VU-QUOC [1988b], "The Role of Nonlinear Theories in the Dynamics of Fast Rotating Flexible Structures," *J. Sound Vibration*, **119**, 487-508.

J. C. SIMO, J.E. MARSDEN & P.S. KRISHNAPRASAD [1988], "The Hamiltonian Structure of Elasticity. The Material and Convective Representation of Solids, Rods and Plates," *Arch. Rational Mech. Anal.*, **104**, 125-183.

J. C. SIMO & L. VU-QUOC [1988c], "A Geometrically Exact Rod Model Incorporating Shear and Torsion-Warping Deformation," *J. Solids and Structures*, submitted for publication.

J. C. SIMO, T. A. POSBERGH & J.E. MARSDEN [1989], "Stability of Coupled Rigid Body and Geometrically Exact Rods: Block Diagonalization and the Energy-Momentum Method ," (preprint).

L. VU-QUOC & J.C. SIMO [1988], "A Novel Approach to the Dynamics of Multicomponent Flexible Satellites," *J. Guidance, Dyn. and Contr.*, **10**, 549-558.

Contemporary Mathematics
Volume **97**, 1989

Controllability of Poisson Control Systems with Symmetries

Gloria Sánchez de Alvarez

Dedicated to Roger Brockett on the occasion of his 50th birthday

Abstract

A mathematical framework for Poisson control systems by using geometric methods of Hamiltonian mechanics is formulated. Controllability criteria are given by relating controllability on the phase space and that on the reduced phase space. Application to a spacecraft with internal rotors is given.

§1 *Introduction*

This paper formalizes a theoretical framework for Poisson control systems and shows that the presence of symmetries determines conditions for the notion of controllability. Brockett [1977] introduced the notion of Hamiltonian control systems. A further development has been initiated by van der Schaft [1983] by formulating Hamiltonian systems for which the state space is a symplectic manifold.

To formulate the concept of Poisson control systems as well as formulating and solving the problem of controllability, we apply the geometric methods of symplectic and Poisson manifolds and reduction (see Marsden and Weinstein [1974]; Weinstein [1983]; Marsden, Ratiu and Weinstein [1984], Marsden and Ratiu [1986] and Sánchez de Alvarez [1986]). Using these methods, we solve the (local weak) controllability problem of these systems by showing controllability in a reduced space. This is generally simpler than a direct approach on the original space state, since the computation of the dimensionality of the Lie algebra generated by the vector fields of the original system is not always easy.

To illustrate our techniques, in Section **2** we study a model for the spacecraft system, namely, a rigid body with rotors - free spinning rotors or driven rotors - which can be described as a Poisson control system whose state space is $\mathfrak{so}(3)^*$ (the dual of the Lie algebra of $\mathbf{SO}(3)$) with the Lie-Poisson bracket by means of reduction. This system has been studied previously in several papers by using different approaches:

1 A differential geometric method is used by Crouch [1984] to study the attitude control problem of the spacecraft with momentum exchange actuators.

0271-4132/89 $1.00 + $.25 per page

2 Bonnard [1981], [1984] uses techniques of Riemannian geometry and Lie groups to set the general problem of the control of the dynamic system evolving in a general Lie group G by describing the motion on a geodesic of a left invariant Riemannian structure on G. In case of G is compact (e.g., $\mathbf{SO}(3)$), he obtains global results for this problem.

However, none of these works take into account information concerning the underlying symmetries and reduced phase space of the system.

3 Krishnaprasad [1985] shows that the system is Hamiltonian relative to a certain Lie-Poisson structure and uses this structure to study its stability. In section **2** and in Sánchez de Alvarez [1986], we derive his structure by reduction, and realize it as a Poisson control system. This puts this system into the framework of reduction so that techniques from this area may be applied to study structure properties or stability behavior (see Krishnaprasad [1985], Krishnaprasad and Marsden [1987]).

The Lie-Poisson structure of the free spinning rotor case (no internal torques) and the driven rotor case (an internal torque is exerted by a motor on the platform) are derived in Section **4** by applying the reduction procedure since there are two symmetries given by the action of $\mathbf{SO}(3)$ and S^1 on the coupled motion.

In the last section, we introduce the concept of local weak controllabiltity relative to symmetries in the original space, and we show its equivalence with local weak controllabiltiy of the reduced system. The local weak controllability of the reduced system is shown to be a necessary and sufficient condition for local weak controllability of the unreduced system by using Poisson maps.

Acknowledgements I wish to thank Jerry Marsden, Eduardo Sontag, and Sudeshna Sengupta for their helpful comments on this work.

§2 *Hamiltonian Structure of the Spacecraft*

The following example motivates the mathematical formulation of Poisson control systems. Figure 1 illustrates a system of two rigid bodies which are coupled so that there is one degree of freedom in relative motion α which corresponds to the angle of rotation of the body **R** (rotor) with respect to a main body **P** (platform). **R** rotates relative to **P** with (in general) a variable angular speed $\dot{\alpha}$ and **P** precesses and spins in a manner similar to that of a single symmetric rigid body. Thus there is a symmetry group, the rotation group $\mathbf{SO}(3)$, acting on the system so that the total angular momentum remains constant. Under some assumptions on the coupled motion, there exist two others integrals: P_α the component of the relative angular momentum of the rotor about its axis of rotation (principal axis) and a type of "energy" integral. These facts suggest the presence of an S^1 symmetry for the coupled motion.

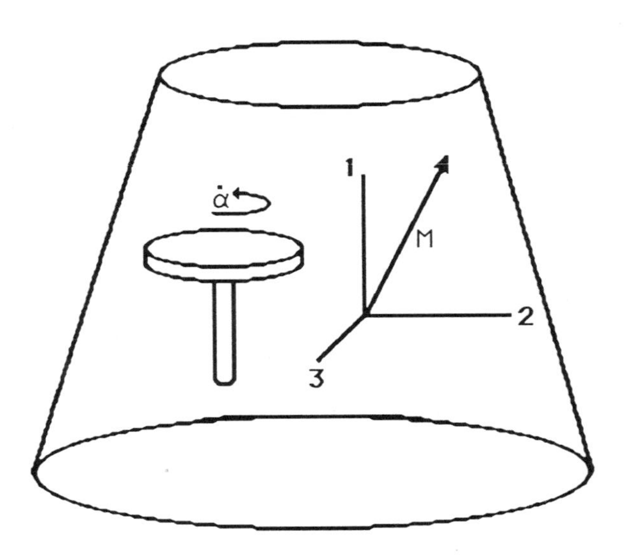

Fig. 1 Spacecraft configuration.

We will distinguish two cases. First, when there are no external or internal torques on the system (free spinning rotor), and second, when we assume internal torques (driven rotor).

Let J_1, J_2, J_3 be the moments of inertia of **P** and $I_1, I_2 = I_3 = I$ the moments of inertia of **R**. If $(\omega_1, \omega_2, \omega_3)$ denotes the angular velocity of the system, the kinetic energy for the free spinning rotor is $\mathbf{E}_s = \frac{1}{2}[J_1\omega_1^2 + J_2\omega_2^2 + J_3\omega_3^2 + I_1(\dot{\alpha} + \omega_1)^2 + I\omega_2^2 + I\omega_3^2]$; for the driven rotor the kinetic energy is $\mathbf{E}_d = \frac{1}{2}[(J_1 + I_1)\omega_1^2 + (J_2 + I)\omega_2^2 + (J_3 + I)\omega_3^2] + \frac{1}{2}I_1\dot{\alpha}^2 + I_1\omega_1\dot{\alpha}$.

Thus, if M_i denotes the i-th coordinate of the total angular momentum in body coordinates, we have $M_1 = J_1\omega_1 + I_1\omega_1 + I_1\dot{\alpha}$, $M_2 = J_2\omega_2 + I\omega_2$, $M_3 = J_3\omega_3 + I\omega_3$, and the equations of motion for the free system are:

$$\dot{M}_1 = -\frac{M_2M_3}{(J_2 + I)} + \frac{M_2M_3}{(J_3 + I)}$$

$$\dot{M}_2 = -\frac{M_1M_3}{(J_3 + I)} + \frac{M_3M_1}{J_1} - \frac{M_3P_\alpha}{J_1}$$

$$\dot{M}_3 = -\frac{M_1M_2}{J_1} + \frac{M_1M_2}{(J_2 + I)} + \frac{M_2P_\alpha}{J_1}$$

The equations of motion for the driven system are:

$$\dot{M}_1 = -\frac{M_2M_3}{(J_2 + I)} + \frac{M_2M_3}{(J_3 + I)}$$

$$\dot{M}_2 = -\frac{M_1M_3}{(J_3 + I)} + \frac{M_3M_1}{(J_1 + I_1)} - \frac{M_3P_\alpha}{(J_1 + I_1)}$$

$$\dot{M}_3 = -\frac{M_1M_2}{(J_1 + I_1)} + \frac{M_1M_2}{(J_2 + I)} + \frac{M_2P_\alpha}{(J_1 + I_1)}$$

where $P_\alpha = I_1(\omega_1 + \dot{\alpha})$ for the free case represents the angular momentum of **R** along its axis of symmetry, and $P_\alpha = I_1\dot{\alpha}$ represents the relative angular momentum of **R** along its axis of symmetry for the driven rotor. These equations are Hamiltonian relative to a Lie-Poisson bracket on the dual of a Lie algebra, as given by Krishnaprasad [1985]. In Sánchez de Alvarez [1986] the brackets are derived by reduction from canonical brackets.

§3 *Mathematical Formulation of Poisson Control Systems*

In what follows, we will present briefly the mathematical notions and refer to the methods needed for the introduction of Poisson control systems which generalize the concept of Hamiltonian systems introduced by Brockett [1977] and reformulated later by van der Schaft [1981], [1983].

Preliminaries Let P be a manifold of dimension n and $C^\infty(P)$ the set of all real infinitely differentiable functions defined on P.

A ***Poisson manifold*** $(P, \{\ ,\ \})$ is a manifold with a real bilinear map $\{\ ,\ \}$ on $C^\infty(P) \times C^\infty(P) \rightarrow C^\infty(P)$ which is skew-symmetric, a derivation in each factor and satisfies Jacobi's identity (see Weinstein [1983]). A well-known example is $\mathbf{so}(3)^* \cong \mathbb{R}^3$, the dual of the Lie algebra of the Lie group $\mathbf{SO}(3)$ with the Lie-Poisson structure $\{f, g\}(m) = -\langle m, \nabla f \times \nabla g\rangle$; see Marsden, et al. [1984].

The derivation property assures the existence of ***Hamiltonian vector fields*** X_f for each function $f \in C^\infty(P)$.

Poisson vector fields are those whose corresponding flows f_t are ***Poisson automorphisms***, i.e., $\{f \circ f_t, g \circ f_t\} = \{f, g\} \circ f_t$. Locally Hamiltonian vector fields are Poisson vector fields, but there are Poison vector fields that are not locally Hamiltonian; see Weinstein [1987a], [1987b].

Let TP be the tangent bundle of a Poisson manifold P and let $\tau_P : TP \rightarrow P$ be the projection. Given a (smooth) function f on P, let $\dot{f}$ be the function on TP defined by $\dot{f}(v_p) = \mathbf{d}f(p) \cdot v_p$, where $p = \tau_P(v_p)$, and let $\tilde{f} = f \circ \tau_P$.

Theorem 3.1 (Sánchez de Alvarez [1986], [1987]). TP *has a unique Poisson structure satisfying the bracket relations* $\{\tilde{f}, \tilde{g}\} = 0$, $\{f, g\}^{\cdot} = \{\dot{f}, \dot{g}\}$, *and* $\{\tilde{f}, \dot{g}\} = \{\dot{f}, \tilde{g}\} = \{f, g\}^{\sim}$.

A submanifold L of a Poisson manifold P is ***Lagrangian*** iff for all $f, g \in C^\infty(P)$ satisfying $X_g \in TL$, we have $\{f, g\}|_L = 0$ iff $X_f \in TL$. For example, one checks that in $P = \mathbf{so}(3)^* \cong (\mathbb{R}^3, \times)$, any plane through the origin is a Lagrangian submanifold. The following proposition shows that many Lagrangian submanifolds can be obtained as graphs of Poisson vector fields.

Proposition 3.2 (Sánchez de Alvarez [1986], [1987]). *Let* P *be a Poisson manifold and let* TP *be its tangent bundle. Then a vector field* X *on* P *is Poisson iff its graph* S *in* TP *is Lagrangian.*

Poisson Control Systems A ***Poisson control system*** is a nonlinear control system $\Sigma(E, P, \mathbf{f})$

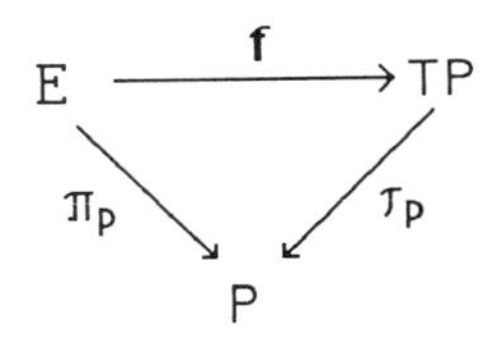

such that the graph of $\mathbf{f}$, $\Gamma_{\mathbf{f}}$, is a Lagrangian submanifold in TP. By choosing coordinates x for P and (x, u) and E, we write the control system as $\dot{x} = \mathbf{f}(x, u)$, where $\mathbf{f}(x, u)$ is a Poisson vector field for each constant u.

We will consider the case in which on bundle charts, for every constant u, $\mathbf{f}(x, u)$ is a Hamiltonian vector field on P, such that the input-state evolution equations take the Hamiltonian form $\dot{x} = \{x, H^u\}_P$ where $H^u \in C^\infty(P)$.

Let $\Sigma(E, P, \mathbf{f})$ be a Poisson control system and let θ and φ be actions of a Lie group G on E and P, respectively. These actions are called ***Poisson actions*** for Σ if the function $\mathbf{f}$ is equivariant with respect to these actions, that is, if for all $g \in G$ the following diagram commutes

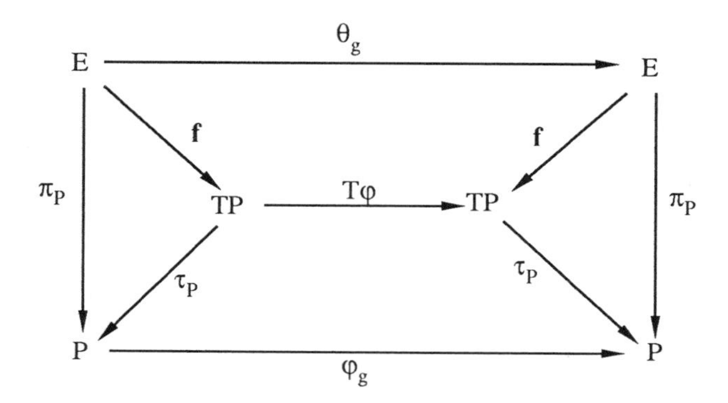

and $\{f_1, f_2\}_P \circ \varphi_g = \{f_1 \circ \varphi_g, f_2 \circ \varphi_g\}_P$ for all $f_1, f_2 \in C^\infty(P)$.

§4 *Reduction of Poisson Control Systems*

In this section, we apply the reduction procedure (Marsden and Weinstein [1974]) to a given Poisson control system for which the vector fields are Hamiltonian in order to get a new system with a reduced state space, using symmetries.

Consider a fiber bundle (E, P, π_P) and a Lie group G acting freely and properly on E, such that E/G is a manifold. In addition, assume that the action θ of G on E leaves the fibers invariant. Then the submersion

$$\tau : E \to E/G$$

is a morphism which takes fibers of E onto fibers of E/G. Thus τ induces a morphism $\tilde{\tau}$ such that $\tilde{\tau} \circ \pi_P = \pi_{P/G} \circ \tau$; in other words, the following diagram commutes:

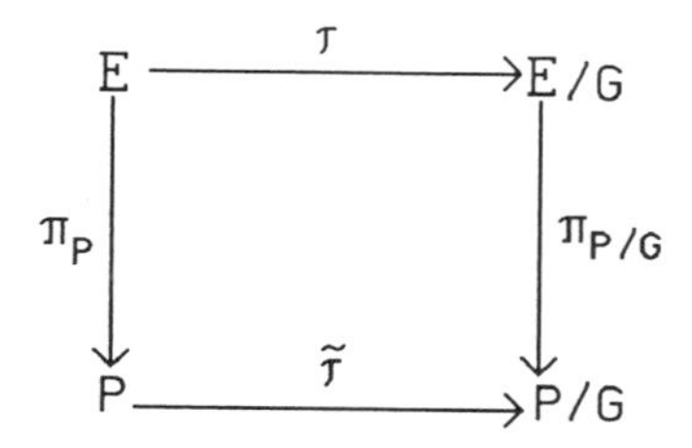

where if P is identified with the zero section on E, and likewise P/G is identified with the zero section on E/G, then $\tilde{\tau}$ may be identified with the restriction $\tau|_P : P \to P/G$ since τ is the identity map on the fibers.

4.1 Theorem (Sánchez de Alvarez, [1986]) *Let* $\Sigma(E, P, \mathbf{f})$ *be a Poisson control system and* G *a Lie group acting freely and properly on* E *and* P *by Poisson maps. Assume that* G *leaves the fibers of* E *invariant. Then the control system* $\tilde{\Sigma}(E/G, P/G, \tilde{\mathbf{f}})$ *is a Poisson control system where* $\tilde{\mathbf{f}}$ *is defined by* $\mathbf{f} = \tilde{\mathbf{f}} \circ t$, *i.e., the following diagram commutes*

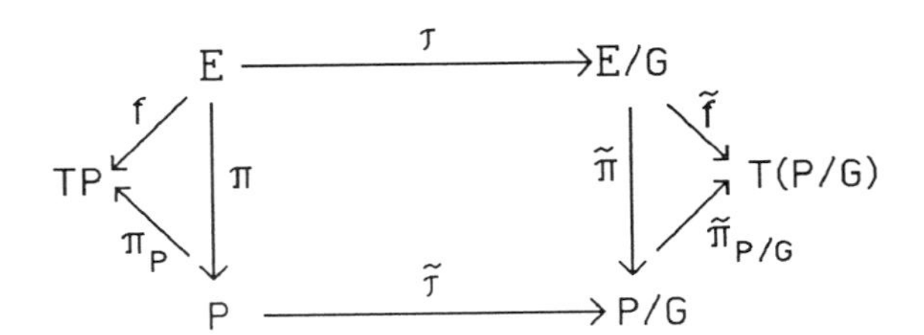

In the spacecraft system, the configuration space is $SO(3) \times S^1$ with coordinates $\varphi, \psi, \theta, \alpha$, where φ, ψ, θ, denote the Euler angles of the platform and α the angle of rotation of the rotor with respect to the platform; see Krishnaprasad [1985]. Thus the phase space is $P = T^*SO(3) \times T^*S^1$.

Groups acting on the system S^1 acts on $SO(3)$ by the trivial action, and acts on S^1 by the left action given by $(\alpha_0, (\varphi, \psi, \theta, \alpha)) \mapsto (\varphi, \psi, \theta, \alpha + \alpha_0)$. By lifting this action to $T^*(SO(3) \times S^1)$ we get a Poisson action. The corresponding momentum map $\mathbf{J}$ from $T^*SO(3) \times T^*S^1$ to $\mathbb{R}$, the Lie algebra of S^1, is given by $\mathbf{J}(x) = P_\alpha$; see Abraham and Marsden [1978], page 276. By restricting to the level set P_α = constant, and modding out by S^1 we reduce to $T^*SO(3)$.

Next, the group $G = SO(3)$ also acts on the configuration space by left translation on $SO(3)$ and trivially on S^1. The momentum map for this action is the total angular momentum; see Abraham and Marsden [1978], page 276. Thus by the Poisson reduction theorem we get $T^*SO(3)/SO(3) \cong so(3)^*_-$ with the (–) Lie-Poisson bracket:

$$\{F, G\}(M) = -\langle M, \nabla F \times \nabla G \rangle$$

(see Marsden et al. [1984]).

In the case of the free spinning rotor, there are no internal torques, i.e., P_α = constant, and the reduced Hamiltonian is the total kinetic energy of the assembly:

$$\mathbf{H}_s = \frac{1}{2}\frac{(M_1 - P_\alpha)^2}{J_1} + \frac{M_2^2}{(J_2 + I)} + \frac{M_3^2}{(J_3 + I)} .$$

For the case of the driven rotor, there are internal torques, so we can assume there is a motor fixed in the spacecraft that supplies a torque to the rotor. This torque is given by $\frac{d}{dt}(I_1\omega_1 + I_1\dot{\alpha})$, the derivative of the component of angular momentum of the rotor in the direction of its axis of rotation, where ω_1 is the component of the angular velocity of the platform along the axis of rotation. The motor is designed such that its exerted torque is equal to $I_1\dot{\omega}_1$; this is equivalent to setting $P_\alpha = I_1\dot{\alpha}$ = constant which represents the relative angular momentum of the rotor. The reduced Hamiltonian is

$$\mathbf{H}_d = \frac{1}{2}\frac{(M_1 - P_\alpha)^2}{(J_1 + I_1)} + \frac{M_2^2}{(J_2 + I)} + \frac{M_3^2}{(J_3 + I)} .$$

§5 *Controllability of Poisson Control Systems*

We now explore controllability of Poisson control systems in the presence of symmetries. It will be shown that symmetries constitute an advantage for control, for it may be possible to reduce the dimensionality of the Lie algebra $\mathcal{L}$ generated by the given family $\mathcal{F}$ of Hamiltonian vector fields.

The following definitions are given by Hermann and Krener [1977] for nonlinear control systems : Given an open subset $U \subset P$, $x_1 \in E$ is called U-***accessible*** from $x_0 \in U$, written $x_1\mathcal{A}_U x_0$, if there exists $u(\cdot) \in \mathcal{U}$, the space of control functions, such that the corresponding solution $x(t)$, $t \in [t_0, t_1]$ satisfies $x(t_0) = x_0$, $x(t_1) = x_1$ and $x(t) \in U$ for all $t \in [t_0, t_1]$. We denote by $\mathcal{A}_U(x_0)$ the set of all points in P that are U-accessible from x_0. We write $\mathcal{A}(x_0) = \mathcal{A}_P(x_0)$.

The system Σ is called ***controllable*** at x_0 if $\mathcal{A}(x_0) = P$, and ***controllable*** if $\mathcal{A}(x) = P$ for every $x \in P$. Similarly, Σ is called ***locally controllable at*** x_0 if for every neighborhood V of x_0, $\mathcal{A}_V(x_0)$ is also a neighborhood of x_0, and ***locally controllable*** if this holds at each $x \in P$.

Let U be an open subset of P and let $x', x'' \in P$; then x' is U-***weakly accessible*** from x'', written $x' w\mathcal{A}_U x''$, when there exists $x_0, \ldots, x_k$ such that $x' = x_0$, $x_k = x''$ and if either $x_i\mathcal{A}_U x_{i-1}$ or $x_{i-1}\mathcal{A}_U x_i$ for $i = 1, \ldots, k$. We denote by $w\mathcal{A}_U(x_0)$ the set of all points in P that are U-weakly accessible from x_0. We write $w\mathcal{A}(x_0) = w\mathcal{A}_P(x_0)$.

Σ is ***weakly controllable*** at x_0 if $w\mathcal{A}(x_0) = P$; and if $w\mathcal{A}(x) = P$ for every $x \in P$, then Σ is ***weakly controllable.***

Σ is ***locally weakly controllable*** at x_0 if for every neighborhood V of x_0, $w\mathcal{A}_V(x_0)$ is a neighborhood of x_0. Σ is ***locally weakly controllable*** if it is locally weakly controllable at every $x \in P$.

For Poisson control systems $\Sigma(E, P, \mathbf{f})$ under the action of a Lie group, a new notion of controllability is introduced by using not only the dynamics of the system, but also the symmetries. Let G be a Lie group acting on Σ by Poisson maps. Given U, an open subset of P and given $x_0 \in U$, we say that $x_1 \in P$ is *U-**accessible relative to symmetries*** from x_0, written $x_1 \mathcal{A}^r{}_U x_0$, if either there exists an admissible control $u(\cdot) \in \mathcal{U}$ such that the corresponding solution $x(t)$ of Σ satisfies $x(t_0) = x_0$, $x(t_1) = x_1$ and $x(t) \in U$ for all $t \in [t_0, t_1]$ *or* $x_1 = f_g(x_0)$ for some $g \in G$. If in the definition of local weak controllability we replace the notion of "accessibility" by "accessibility relative to symmetries", then Σ is said to be ***locally weakly controllable relative to symmetries.***

The following proposition shows that local weak controllability relative to symmetries of a Poisson control system is equivalent to local weak controllability of the reduced control system.

Proposition 5.1 *Let Σ be a Poisson control system and G a Lie group acting freely and properly on E and P by Poisson maps. Then Σ is locally weakly controllable relative to symmetries iff the reduced system $\tilde{\Sigma}$ is locally weakly controllable.*

Proof Assume that Σ is locally weakly controllable relative to symmetries. Let $y_0 \in P/G$, then there exists $x_0 \in P$ such that $y_0 = p(x_0)$. Given a neighborhood V of y_0 and $y_1 \in V \subset P/G$ of y_0, there exists a neighborhood $U = p^{-1}(V)$ of x_0 and $x_1 \in U$ such that $y_1 = p(x_1)$. Since P is locally weakly controllable relative to symmetries, there exist trajectories $x^{u_i}(t) \in U$ $i = 1,\ldots, k$, corresponding to Hamiltonian vector fields $\mathbf{X}^{u_i} \in S$ such that $x_0 = x^{u_1}(t_0)$, $x_1 = x^{u_k}(t_1)$, $x^{u_i}(t) = f_{g(t)}(d^i(t))$ for curves $d^i(t) \subset P$, $g(t) \subset G$, $p(x^{u_i}(t)) = p(d^i(t))$ and $x^{u_1} \circ \ldots \circ x^{u_k} \in U$ (see Abraham and Marsden [1978], page 305). Let $y^i(t) = p(x^{u_i}(t))$ be the induced integral curves on P/G. Then $y^1 \circ \ldots \circ y^k \in V$ and $y^1(t_0) = y_0$, $y^k(t_1) = y_1$, therefore, $\tilde{\Sigma}$ is locally weakly controllable.

Conversely, let $x_0 \in P$ and denote by $[x_0]$ the corresponding orbit of x_0 in P/G. Given any neighborhood $U \subset P$ of x_0 and $x_1 \in U$, then either $[x_0] = [x_1]$ or $[x_0] \neq [x_1]$. If $[x_0] = [x_1]$, then there exists $g \in G$ such that $x_1 = f_g x_0$. If $[x_0] \neq [x_1]$ and $[x_1] \in \tilde{V} \subset P/G$, a neighborhood of $[x_0]$, for $\tilde{\Sigma}$ is locally weakly controllable there exist trajectories $[x^i(t)] \in P/G$ corresponding to the vector fields $\tilde{Y}^{u_i} \in \tilde{\Sigma}$ on P/G, $i = 1,\ldots, r$ such that $[x^1] \circ \ldots \circ [x^r] \in \tilde{V}$, $[x_0] = [x^1(t_0)]$ and $[x_1] = [x^r(t_1)]$. Let $d^i(t)$ be a smooth curve in P such that $[d^i(t)] = [x^i(t)]$, then $d^1(t_0) = x_0$, $d^r(t_1) = x_1$ and $d^i(t) = f_{g(t)} x^i(t)$. Thus, Σ is locally weakly controllable relative to symmetries. ■

Let $\mathcal{L}$ be the smallest subalgebra of vector fields on P which contains $\mathcal{F}$, the set of the given Hamiltonian vector fields. An element of $\mathcal{L}$ is a finite linear combination of elements of $\mathcal{F}$ which have the form $[X^1, [X^2,..., [X^{k-1}, X^k],...]]$ where $X^i = g(., u^i)$ for some $u^i \in U$, and $[X^i, X^j]$ is the Jacobi-Lie bracket defined by $[X^i, X^j]_\alpha = X^j_\beta \frac{\partial X^i_\alpha}{\partial x_\beta} - X^i_\beta \frac{\partial X^j_\alpha}{\partial x_\beta}$. The space of tangent vectors spanned by the vector fields of $\mathcal{L}$ at x, denoted by $\mathcal{L}(x) \subset T_xP$ is a Lie algebra of dimension $k \leq n$ = dimension of P. Σ is said to satisy the ***controllability rank condition*** at x_0 if $k \doteq n$, and Σ is said to satisfy the ***controllability rank condition*** if this holds for every $x \in P$.

Hermann and Krener [1977] proved the following result which reduces the notion of local weak controllability to an algebraic criteria.

Theorem 5.2 *If Σ satisfies the controllability rank condition at x_0 then Σ is locally weakly controllable at x_0.*

The reduction theorem **4.1** for Poisson control systems provides a technique which reduces the dimension of the state space. Therefore, the rank of controllability condition of the Lie algebra $\tilde{\mathcal{L}}$ of the reduced system is smaller than the rank controllability condition of the original system. It would be desirable to know if local weak controllability of the reduced system $\tilde{\Sigma}$ gives local weak controllability of the original system Σ. That is what, generically, the next theorem says for the case of $P = T^*G$, a generalization of a result given by Crouch [1984].

Theorem 5.3 *Let G be a Lie group acting by Poisson maps on a Poisson control system $\Sigma(E, T^*G, \mathbf{f})$. If the reduced system $\tilde{\Sigma}(\tilde{E}, \mathfrak{g}^*, \tilde{\mathbf{f}})$ satisfies the controllability rank condition, then Σ also satisfies the controllability rank condition.*

Proof Let $\tilde{\mathcal{F}}$ be the given family of Hamiltonian vector fields on $\tilde{\Sigma}$ and denote by $\tilde{\mathbf{X}}_u \in \tilde{\mathcal{F}}$ the vector fields $\mathbf{f}(x, u)$ for $u(.) \in U$, and denote by $\tilde{\mathcal{L}}$ the Lie algebra generated by $\tilde{\mathcal{F}}$. Then the tangent vector $\tilde{\mathbf{X}}_u(x) \in T_x\mathfrak{g}^* \cong \mathfrak{g}^*$ can be imbedded into a vector $\mathbf{X}_u$ tangent to $G \times \mathfrak{g}^* \cong T^*G$ at the point (g, x) by means of the map:

$$\chi : T_x\mathfrak{g}^* \rightarrow T_{(g,x)}(G \times \mathfrak{g}^*)$$

defined by:

$$\chi(\tilde{\mathbf{X}}_u(x)) = (T_gL_g^{-1}\, \tilde{\mathbf{X}}_u(x), \tilde{\mathbf{X}}_u(x)) := \mathbf{X}_u(g, x).$$

Let $\mathcal{L}$ be the Lie algebra generated by the new family $\mathcal{F}$ of Hamiltonian vector fields $\mathbf{X}_u$ on $G \times \mathfrak{g}^*$. Consider $G \times \mathfrak{g}^*$ with the Poisson bracket introduced by Krishnaprasad and Marsden [1987]:

$$\{F, G\} = \{F, G\}_{\mathfrak{g}^*_-} + \{F, G\}_G - d_x F \cdot \left(\frac{dG}{dg}\right)_G + d_g G \cdot \left(\frac{dF}{dx}\right)_G$$

where we endow G with the trivial bracket. Thus, by using the fact that $[\mathbf{X}_u, \mathbf{X}_v] = -\mathbf{X}_{\{F,K\}}$ where F and K are the corresponding Hamiltonian functions for $\mathbf{X}_u$ and $\mathbf{X}_v$, respectively, we find that

$$[\mathbf{X}_u, \mathbf{X}_v] = \left[d_x T_g L_g^{-1} \tilde{X}_u(x) \cdot \tilde{X}_v(x) - d_x T_g L_g^{-1} \tilde{X}_v(x) \cdot \tilde{X}_u,\ [\tilde{X}_u, \tilde{X}_v]\right].$$

Since $\tilde{X}_u(x)$ generates $T_x \mathfrak{g}^*$, the first terms in the above bracket span the tangent space $T_g G$, and hence, the vectors $\mathbf{X}_u(g,x)$ and $[\mathbf{X}_u, \mathbf{X}_v](g, x)$ span the tangent space $T_{(g,x)} G \times \mathfrak{g}^*$ for all $(g, x) \in G \times \mathfrak{g}^*$. Thus the system Σ satisfies the controllability rank condition at (g, x). ■

References

R. Abraham and J. Marsden, *Foundations of Mechanics,* 2nd Edition, Benjamin/Cummings, Reading, Mass., [1978].

B. Bonnard, Contrôlabilité des systemès nonlinéaires, *C.R. Acad. Sci. Paris, Série I* **292**, [1981] 535-537.

__________, Contrôlabilité de systémès mécaniques sur les groupes de Lie, *SIAM J. Control Optim.* **22**, [1984] 711-722.

R.W. Brockett, Control theory and analytical mechanics, in C. Martin and R. Hermann, *Lie groups: History, Frontiers and Applications,* **Vol. VII**, Math. Sci. Press, Brookline, Mass. [1977] 1-46.

P.E. Crouch, Spacecraft attitude control and stabilization: Applications of geometric control theory to rigid body models, *IEEE Trans. Automat. Control* **29**, [1984] 321-331.

R. Hermann and A.J. Krener, Nonlinear controllability and observability, *IEEE Trans. Automat. Control* **22**, [1977] 728-740.

P. S. Krishnaprasad, Lie-Poisson structures, dual-spin spacecraft and asymptotic stability, *Nonlinear Anal., Theoretical Methods and Applications* **9**, [1985] 1011-1035.

P.S. Krishnaprasad and J. Marsden, The Hamiltonian structure and stability of rigid bodies with flexible attatchments,, *Arch. Rat. Mech. An.* **98**, [1986] 71-93 .

J. Marsden, T. Ratiu and A. Weinstein, Semidirect products and reduction in mechanics, *Trans. Amer. Math. Soc.* **281**, [1984]147-177.

J. Marsden and A. Weinstein, Reduction of symplectic manifolds with symmetry, *Rep. Mathematical Phys.* **5**, [1974] 121-130.

J. Marsden and T. Ratiu, Reduction of Poisson manifolds, to appear in *Lett. and Math. Phys.*, [1986].

G. Sánchez de Alvarez, *Geometric methods of classical mechanics applied to control theory,* Ph. D. Thesis, University of California, Berkeley, [1986].

____________________, Dinámica analítica de las variedades de Poisson, preprint, Universidad Los Andes, Merida, Venzuela [1987].

A. van der Schaft, Symmetries and conservation laws for Hamiltonian systems with inputs and outputs: A generalization of Noether's theorem, *Syst. Contr. Letters* **1**, [1981], 108-115.

____________________, System theoretic descriptions of physical systems, *Mathematics Center Amsterdam,* [1983].

A. Weinstein, The local structure of Poisson manifolds, *J. Diff. Geom.* **18**, [1983], 523-557.

____________________, Symplectic groupoids and Poisson manifolds, *Bull. Amer. Math. Soc.* **16**, [1987], 101-104.

____________________, Poisson geometry of the principal series and nonlinearizable structures, *J. Diff. Geom.* **25**, [1987], 55-73.

Departamento de Matemáticas
Facultad de Ciencias
Universidad Los Andes
Mérida, Venezuela.

Contemporary Mathematics
Volume 97, 1989

CHAOS IN A RAPIDLY FORCED PENDULUM EQUATION

Jürgen Scheurle

ABSTRACT. We consider a rapidly forced pendulum equation and discuss the questions of splitting and transversal intersections of separatrices. It turns out that the Melnikov function predicts these phenomena correctly, provided that the amplitude of the forcing function is of sufficiently high algebraic order with respect to the forcing period, as the latter tends to zero. This result is by no means standard, since both phenomena are exponentially small with respect to the forcing period.

1. INTRODUCTION. In a couple of papers [5] and [6] by Holmes, Marsden and Scheurle, both upper and lower estimates have been established for the separatrix splitting of rapidly forced systems with a homoclinic orbit. One example to which the general theory has been applied, is the forced pendulum equation

$$\ddot{\phi} + \sin\phi = \delta \sin(t/\varepsilon). \tag{1.1}$$

Here t is the time variable, and dots denote derivatives with respect to t. There are two types of results.

First, fix $\eta > 0$ and let $0 < \varepsilon \leqq 1$ and $0 \leqq \delta \leqq \delta_\circ$, where $\delta_\circ$ is sufficiently small. Consider the first order system corresponding to (1.1) in the (ϕ, v)-plane with $v = d\phi/dt$. For $\delta = 0$, this has homoclinic orbits $\Gamma^\pm$ given by

$$\begin{aligned} \overline{\phi}(t) &= \pm\ 2 \arctan(\sinh(t)) \\ \overline{v}(t) &= \pm\ 2\ \mathrm{sech}(t) \end{aligned} \quad , \quad -\infty < t < \infty , \tag{1.2}$$

corresponding to the equilibrium points at $\phi = \pm\pi$, $v = 0$. If the separatrices split for $\delta > 0$, they do so by an amount that is no more than $\delta\, C \exp(-\frac{1}{\varepsilon}(\frac{\pi}{2} - \eta))$, where $C = C(\delta_\circ)$ is a constant depending on $\delta_\circ$ but is uniform in ε and δ. So the local stable

1980 Mathematics Subject Classification (1985 Revision).58,34.

and unstable invariant manifolds corresponding to the perturbed equilibrium points are exponentially close with respect to ε, in a full neighborhood of the origin in the (ε,δ)-plane.

Second, if we replace δ by $\varepsilon^p\delta$, $p > 8$, then we have the sharper estimate

$$(1.3) \qquad \varepsilon^p \delta C_1 e^{-\pi/2\varepsilon} \leq \text{splitting distance} \leq \varepsilon^p \delta C_2 e^{-\pi/2\varepsilon},$$

where C_1 and C_2 are positive constants. So in this second case, the separatrices really split for $\delta > 0$. Moreover, Melnikov's criterion applies to predict transversal intersections for the perturbed invariant manifolds. For the original equation (1.1), this assures the splitting and transversal intersection of the separatrices within the wedge-like region $0 < \delta \leq \delta_0 \varepsilon^p$. As far as I know, this is the first result of that type for equations like (1.1) and specific values of δ which are not exponentially small with respect to ε. The standard proofs of Melnikov's criterion only work for $\delta \leq C \exp(-\pi/2\varepsilon)$, where C is some constant (cf. [4] , [9], [12]). The analyticity argument of Cushman [2] and Koslov [7] could be used to prove transversal intersections of the perturbed invariant manifolds for most parameter values. For results on exponentially small upper bounds for the splitting distance see also Lazutkin et al. [8], Neisthadt [10], and Fontich and Simó [3].

In many applications the splitting of separatrices is described by equations of the form (1.1), with δ being a prescribed function of ε (cf. [6]). This shows that one needs to have a theory for specific values of δ. Transversal intersections of stable and unstable invariant manifolds are of interest, because they obstruct the integrability and establish chaos in the behaviour of the modeled systems [1], [4]. Typically the forcing amplitude δ is proportional to ε, though. So, in general only results of the first type mentioned above apply, but not of the second one. Nevertheless, we feel that also the second result is one step forward towards a better understanding of the phenomenon of exponentially small splitting of separatrices. Actually, already Poincaré [11] has faced this problem and recognized its difficulty, when he tried to prove nonintegrability for the three-body problem.

In this note we shall outline a proof of the second result. The basic ideas have already been described in [5] and [6]. But we fill in additional details. In particular we shall explain, where the restriction $p > 8$ comes from.

2. LOWER ESTIMATE OF THE SPLITTING DISTANCE. Consider the system

$$(2.1)\qquad \begin{aligned} \dot{\phi} &= v \\ \dot{v} &= -\sin\phi + \varepsilon^p \delta \sin(t/\varepsilon) \end{aligned}$$

which is equivalent to the forced pendulum equation (1.1) with δ replaced by $\varepsilon^p \delta$. First let $\delta = 0$. Then there are saddle type equilibrium points at $\phi = \pm\pi$, $v = 0$ which can be identified, because of the 2π-periodicity of the angular variable ϕ. The corresponding stable and unstable invariant manifolds coincide and form the homoclinic orbits $\Gamma^\pm$ given by (1.2). For small $\delta > 0$, the saddle point is perturbed into a nearby $2\pi/\varepsilon$-periodic solution. This is again of saddle type, and its local stable and unstable invariant manifolds are close to $\Gamma^\pm$. For simplicity we only consider those pieces of the invariant manifolds which are close to Γ^+ and call them $W^+_{\delta,\varepsilon}$ and $W^-_{\delta,\varepsilon}$, respectively. We want to estimate their distance from below.

For each time $t = t_\circ$, the distance between $W^+_{\delta,\varepsilon}$ and $W^-_{\delta,\varepsilon}$ is measured along a certain one-dimensional cross section which is transversal to Γ^+. The <u>splitting distance</u> is defined to be the maximal distance with respect to all $t_\circ$. The parameter δ is considered to be small, say $0 < \delta \leqq \delta_\circ$, but fixed, whereas ε is supposed to vary in the interval $0 < \varepsilon \leqq 1$. In particular, we are interested in the asymptotic behaviour of the splitting distance as $\varepsilon \to 0$.

In the standard Melnikov theory, first order perturbations of Γ^+ with respect to δ are used to approximate $W^\pm_{\delta,\varepsilon}$. Based on this approximation, the distance between $W^+_{\delta,\varepsilon}$ and $W^-_{\delta,\varepsilon}$ is measured by the socalled <u>Melnikov function</u> $M_\varepsilon(t_\circ)$ times the factor δ. In our case the Melnikov function

$$(2.2)\qquad M_\varepsilon(t_\circ) = 2\pi\, \varepsilon^p \operatorname{sech}(\pi/2\varepsilon) \sin(t_\circ/\varepsilon)$$

is exponentially small, whereas the higher order corrections of the δ-perturbations of $\Gamma^+$are of order $O(\varepsilon^{2p})$ as $\varepsilon \to 0$. So it is not clear a-priori whether $\delta M_\varepsilon(t_\circ)$ is a good measure for the distance between $W^+_{\delta,\varepsilon}$ and $W^-_{\delta,\varepsilon}$ for any ε. Rather we have to show that the difference $\Delta_{\delta,\varepsilon}(t_\circ)$ between $\delta M_\varepsilon(t_\circ)$ and the exact value for the distance is of the same exponential order with respect to ε as $M_\varepsilon(t_\circ)$. Then, of course, we can use $M_\varepsilon(t_\circ)$ to estimate the splitting distance from above and below.

In order to estimate $\Delta_{\delta,\varepsilon}(t_\circ)$, we construct the terms of order greater than or equal to two in the δ-expansion of $W^\pm_{\delta,\varepsilon}$ iteratively. Thus we obtain a sequence of functions $\Delta^n_{\delta,\varepsilon}(t_\circ)$ which converge to $\Delta_{\delta,\varepsilon}(t_\circ)$, as $n\to\infty$, and can be estimated.

The following representations of $W^\pm_{\delta,\varepsilon}$ are used:

$$
\begin{aligned}
W^\pm_{\delta,\varepsilon}: \quad u &= u^\pm(t,t_\circ,\varepsilon,\delta) \\
&= \bar{u}(t-t_\circ) + \delta\,\varepsilon^{p-3} w_1^\pm(t-t_\circ,t/\varepsilon,\varepsilon,\delta) \\
&\quad + \delta^2 \varepsilon^{p-2} w_2^\pm(t-t_\circ,t/\varepsilon,\varepsilon,\delta),
\end{aligned}
\tag{2.3}
$$

where

$$
u = \begin{pmatrix} \phi \\ v \end{pmatrix},
$$

and $\bar{u}(t)$ has components given by (1.2) with the + sign and represents Γ^+. The functions $w_1^\pm$ and $w_2^\pm$ are determined as follows (cf. [5] and [6]):

Set

$$
s = t - t_\circ \quad \text{and} \quad \theta = t/\varepsilon . \tag{2.4}
$$

Then

$$
w_1^\pm(s,\theta) = \varepsilon^3 L^\pm_\varepsilon \begin{pmatrix} 0 \\ \sin\theta \end{pmatrix} \tag{2.5}
$$

and $w_2^\pm(s,\theta)$ are the limits as $n\to\infty$ of the functions ${}^{(n)}w_2^\pm(s,\theta)$ defined iteratively by

$$
\begin{aligned}
{}^{(n+1)}w_2^\pm(s,\theta) &= L^\pm_\varepsilon F({}^{(n)}w_2^\pm(s,\theta),s,\theta), \quad n=0,1,2,\ldots \\
{}^{(0)}w_2^\pm(s,\theta) &\equiv 0 .
\end{aligned}
\tag{2.6}
$$

These functions converge in the topology of certain spaces $X^\pm_\varepsilon$ of vector valued functions $f(s,\theta)$ which are defined and analytic in the half strips

$$
S^\pm_\varepsilon = \{ s \in \mathbb{C} \,/\, |\mathrm{Im}\, s| \leq \tfrac{\pi}{2} - \varepsilon \quad \text{and} \quad \mathrm{Re}\, s \gtrless 0 \}
$$

of the complex s-plane, and 2π-periodic with respect to the real variable θ. The topology is induced by the norm $\|\cdot\|$ composed out of the sup norm over $S^\pm_\varepsilon$ in the variable s and the Sobolev H^1 norm in the variable θ. The linear operators $L^\pm_\varepsilon$ are defined in terms of Fourier series. For any function

$$(2.7) \qquad f(s,\theta) = \sum_{k=-\infty}^{\infty} f_k(s)\, e^{ik\theta}, \qquad f_k = \begin{pmatrix} f_{k,1} \\ f_{k,2} \end{pmatrix},$$

where $s \in S_\varepsilon^\pm$ and $\theta \in \mathbb{R}$, we set

$$(2.8) \qquad L^\pm f(s,\theta) = \sum_{k=-\infty}^{\infty} a_k^\pm(s)\, e^{ik\theta},$$

where

$$a_o^+(s) = \int_0^s \Phi_1(s,\sigma)\, f_o(\sigma)\, d\sigma - \int_s^\infty \Phi_2(s,\sigma)\, f_o(\sigma)\, d\sigma$$

$$a_o^-(s) = \int_0^s \Phi_1(s,\sigma)\, f_o(\sigma)\, d\sigma + \int_{-\infty}^s \Phi_2(s,\sigma)\, f_o(\sigma)\, d\sigma$$

$$(2.9) \qquad \begin{aligned} a_k^+(s) &= \int_{\alpha_\varepsilon}^s \Phi_1(s,\sigma)\, e^{ik(\sigma-s)/\varepsilon} f_k(\sigma)\, d\sigma \\ &\quad - \int_s^\infty \Phi_2(s,\sigma)\, e^{ik(\sigma-s)/\varepsilon} f_k(\sigma)\, d\sigma \\ a_k^-(s) &= \int_{\alpha_\varepsilon}^s \Phi_1(s,\sigma)\, e^{ik(\sigma-s)/\varepsilon} f_k(\sigma)\, d\sigma \\ &\quad + \int_{-\infty}^s \Phi_2(s,\sigma)\, e^{ik(\sigma-s)/\varepsilon} f_k(\sigma)\, d\sigma \end{aligned}$$

with $\alpha_\varepsilon = \pm i(\frac{\pi}{2} - \varepsilon)$ for $k \gtrless 0$, and

$$(2.10) \qquad \begin{aligned} \Phi_1(s,\sigma)\, f_k(\sigma) &= \tfrac{1}{2}\big[\{\cosh\sigma + \operatorname{sech}\sigma - \sigma \operatorname{sech}\sigma \tanh\sigma\} f_{k,1}(\sigma) \\ &\quad - \{\sinh\sigma + \sigma \operatorname{sech}\sigma\} f_{k,2}(\sigma)\big] \begin{pmatrix} \operatorname{sech} s \\ -\operatorname{sech} s \tanh s \end{pmatrix} \\ \Phi_2(s,\sigma)\, f_k(\sigma) &= \tfrac{1}{2}\big[\{\operatorname{sech}\sigma \tanh\sigma\} f_{k,1}(\sigma) \\ &\quad - \{\operatorname{sech}\sigma\} f_{k,2}(\sigma)\big] \begin{pmatrix} \sinh s + s \operatorname{sech} s \\ \cosh s + \operatorname{sech} s - s \operatorname{sech} s \tanh s \end{pmatrix} \end{aligned}$$

Note, that in (2.9) we have chosen other values for α_ε than in [5] and [6]. There, $\alpha_\varepsilon = 0$. With the present choice the method works for much more general equations. Finally, the nonlinear operator F is defined by

$$F(w_2^\pm, \cdot, \cdot) = \frac{1}{\delta^2 \varepsilon^{p-2}} \begin{pmatrix} F_1 \\ F_2 \end{pmatrix}$$

$$(2.11) \qquad F_1 = 0$$

$$F_2 = -\sin(\bar{\phi} + \delta \varepsilon^{p-3} \phi_1^\pm + \delta^2 \varepsilon^{p-2} \phi_2^\pm) + \sin(\bar{\phi}) + \cos(\bar{\phi})(\delta \varepsilon^{p-3} \phi_1^\pm + \delta^2 \varepsilon^{p-2} \phi_2^\pm)$$

for any functions

$$w_2^\pm = \begin{pmatrix} \phi_2^\pm \\ v_2^\pm \end{pmatrix}$$

of s and θ. In (2.11), $\phi_1^\pm$ is the first component of $w_1^\pm$ from (2.5).

Concerning the background for these definitions we refer to [5]. For simplicity we have dropped the ε and δ dependence occasionally in the notation. Subsequently, C_i $(i = 3,4,5,\ldots)$ always denote real constants which are independent of ε and δ. They may depend on δ_0, though.

LEMMA 2.1. The operators $L_\varepsilon^\pm$ and F map the spaces $X_\varepsilon^\pm$ into themselves. Moreover, if $8 < p \in \mathbb{N}$, then for any constant C_3 there is a constant C_4, such that $\|w_2^\pm\| \leq C_3$ implies $\|L_\varepsilon^\pm F(w_2^\pm, \cdot, \cdot)\| \leq C_4$ for $w_2^\pm \in X_\varepsilon^\pm$.

Proof: We just outline the basic ideas of a proof and thereby explain where the assumption $p > 8$ is used.

Assume that for any $f \in X_\varepsilon^\pm$, we can estimate $\|f\|$ from above and below in terms of the Fourier coefficients f_k. Then it suffices to estimate a_k in terms of f_k for each k in (2.9), in order to estimate $\|L_\varepsilon^\pm f\|$ in terms of $\|f\|$. If the first component of f vanishes, one obtains

$$(2.12) \qquad \|L_\varepsilon^\pm f\| \leq \frac{C_5}{\varepsilon^3} \|f\| .$$

Here the negative power of ε on the right-hand side stems from the fact that according to (2.10), in (2.9) the integral kernels have first, respectively second order poles at $\sigma = \pm i\pi/2$ and $s = \pm i\pi/2$, which are ε-close to the half strips $S_\varepsilon^\pm$. In particular, for $w_1^\pm$ in (2.5) it follows that $\|w_1^\pm\|$ is uniformly bounded with respect to ε and δ.

To estimate $\|F(w_2^{\pm},\cdot,\cdot)\|$, we expand F into the following Taylor series:

$$(2.13)\qquad F(w_2^{\pm},\cdot,\cdot) = \begin{pmatrix} 0 \\ -\sum_{n=2}^{\infty} \frac{1}{n!}\frac{\sin^{(n)}(\bar{\Phi})}{\delta^2 \varepsilon^{p-2}}(\delta\varepsilon^{p-3}\phi_1^{\pm} + \delta^2\varepsilon^{p-2}\phi_2^{\pm})^n \end{pmatrix}$$

Assuming that the spaces $X_\varepsilon^{\pm}$ are continuously embedded into $C^0(S_\varepsilon^{\pm},\mathbb{C})$ and observing that the functions

$$(2.14)\qquad \sin^{(n)}(\bar{\phi}(s)) = \begin{cases} 2(-1)^{\frac{n}{2}}\dfrac{\sinh(s)}{\cosh^2(s)}, & \text{if } n \text{ is even} \\[2ex] (-1)^{\frac{n-1}{2}}\dfrac{1-\sinh^2(s)}{\cosh^2(s)}, & \text{if } n \text{ is odd} \end{cases}$$

have second order poles at $s = \pm i\pi/2$, we obtain

$$(2.15)\qquad \|F(w_2^{\pm},\cdot,\cdot)\| \leqq C_6 \sum_{n=2}^{\infty} \frac{1}{(n-1)!}\frac{C_7^n}{\delta^2\varepsilon^p}\|\delta\varepsilon^{p-3}w_1^{\pm} + \delta^2\varepsilon^{p-2}w_2^{\pm}\|^n$$
$$\leqq C_8\,\varepsilon^3,$$

provided that $p \geqq 9$. Combining the estimates (2.12) and (2.15), we obtain the estimate in Lemma 2.1.

We finally note that in case of the standard Sobolev space H^1, the assumptions made in the proof follow from Parseval's identity and Sobolev's well known embedding theorem, respectively. But in our case, additional arguments are necessary to verify them.

REMARK. Since in (2.9) we are integrating over σ, the σ-poles actually only lead to a factor $\varepsilon^{-\mu}$, $0 < \mu < 1$, on the right-hand side of the inequality in (2.12). Thus ε^{-3} can be replaced by $\varepsilon^{-(2+\mu)}$ in (2.12). Making use of this fact, one can lower the lower bound for p by 1 or 2 in the assumptions of Lemma 2.1.

It also follows from the above estimates that sufficiently large balls $\|w_2^{\pm}\| \leqq C_3$ in $X_\varepsilon^{\pm}$ are mapped into themselves by $L_\varepsilon^{\pm}F$, if δ is sufficiently small (depending on C_3 but not on ε). Therefore the iterates ${}^{(n)}w_2^{\pm}$ defined in (2.6), are uniformly bounded in $X_\varepsilon^{\pm}$. This is important for the following estimates. If p is too small, this property seems to be missing.

LEMMA 2.2. If $8 < p \in \mathbb{N}$ and δ_0 is sufficiently small, then there is a constant C_9, such that

(2.16) $$\Delta^n_{\delta,\varepsilon}(\varepsilon\theta) := \| {}^{(n)}w_2^+(0,\theta) - {}^{(n)}w_2^-(0,\theta) \| \leq \varepsilon^2 C_9 e^{-\pi/2\varepsilon}$$

holds for all $\theta \in \mathbb{R}$ and $n \in \mathbb{N}$.

This lemma can essentially be proved along the lines of the proof of Fact 2 in [5]. We do not prove it here. We just mention that the factor ε^2 in the estimate (2.16) is gained by the fact that the distance from $s = 0$ to $s = \pm i\pi/2$, where the poles of the kernels in the representation (2.9) of $L^{\pm}_{\varepsilon}$ lie, is of order $O(1)$ as $\varepsilon \to 0$.

Passing to the limit in (2.16), as $n \to \infty$, one finds

(2.17) $$\Delta_{\delta,\varepsilon}(\varepsilon\theta) = \| w_2^+(0,\theta) - w_2^-(0,\theta) \| \leq \varepsilon^2 C_9 e^{-\pi/2\varepsilon}.$$

Moreover, using the method of residues, one obtains

(2.18) $$w_1^+(0,\theta) - w_1^-(0,\theta) = -\frac{M_\varepsilon(\varepsilon\theta)}{\varepsilon^{p-3}} \begin{pmatrix} 0 \\ 1 \end{pmatrix},$$

where M_ε is the Melnikov function defined in (2.2). Thus, for the distance between $W^+_{\delta,\varepsilon}$ and $W^-_{\delta,\varepsilon}$ at time $t = t_0$ we have the estimate

(2.19) $$\begin{aligned} d_{\delta,\varepsilon}(t_0) &:= \| u^+(t_0,t_0,\varepsilon,\delta) - u^-(t_0,t_0,\varepsilon,\delta) \| \\ &\lesseqgtr \delta \, | M_\varepsilon(t_0) | \pm \delta^2 \varepsilon^{p-2} \Delta_{\delta,\varepsilon}(t_0) \\ &= \delta \, | M_\varepsilon(t_0) | \pm O(\delta^2 \varepsilon^p e^{-\pi/2\varepsilon}) , \end{aligned}$$

as ε and δ tend to zero. Consequently, if $M_\varepsilon(t_0) \not\equiv 0$ and δ_0 is sufficiently small (depending on t_0), then there are positive constants C_{10} and C_{11}, such that

(2.20) $$\delta C_{10} |M_\varepsilon(t_0)| \leq d_{\delta,\varepsilon}(t_0) \leq \delta C_{11} |M_\varepsilon(t_0)|$$

holds. Also, under these assumptions the difference vector $u^+(t_0,t_0,\varepsilon,\delta) - u^-(t_0,t_0,\varepsilon,\delta)$ is transversal to Γ^+ and to both $W^{\pm}_{\delta,\varepsilon}$. Therefore, taking the supremum over all t_0 in (2.20) and using (2.2), the estimates in (1.3) follow.

REMARKS. It follows from (2.18), that $W^+_{\delta,\varepsilon}$ and $W^-_{\delta,\varepsilon}$ intersect each other if $M_\varepsilon(t_0)$ changes sign at some value of t_0. To see this, observe that the maps $t_0 \mapsto (u^{\pm}(t_0,t_0,\varepsilon,\delta),t_0)$ define continuous paths on the two-dimensional oriented manifolds $W^{\pm}_{\delta,\varepsilon}$ in the extended phase space with coordinates (ϕ,v,t). These paths lie in a region, where the vector $\begin{pmatrix} 0 \\ 1 \end{pmatrix}$ is transversal to $W^+_{\delta,\varepsilon}$ and $W^-_{\delta,\varepsilon}$.

By the above techniques, one can also prove the estimate

$$(2.21) \qquad \left| \frac{d}{dt_\circ} \left(u^+(t_\circ,t_\circ,\varepsilon,\delta) - u^-(t_\circ,t_\circ,\varepsilon,\delta) + \delta M_\varepsilon(t_\circ)\binom{0}{1}\right)\right| = O\left(\delta^2 \varepsilon^{p-1} e^{-\pi/2\varepsilon}\right)$$

as ε and δ tend to zero. Consequently, for $\delta > 0$, $W^+_{\delta,\varepsilon}$ and $W^-_{\delta,\varepsilon}$ intersect each other transversally, since $M_\varepsilon(t_\circ)$ has a simple zero at $t_\circ = 0$.

BIBLIOGRAPHY

1. V.I. Arnol'd, Mathematical Methods of Classical Mechanics, Springer-Verlag, New York, 1978.

2. R. Cushman, Examples of non-integrable analytic Hamiltonian vector fields with no small divisors, Trans. Am. Math. Soc. 238 (1978), 45-55.

3. E. Fontich and C. Simó, The splitting of separatrices for analytic diffeomorphisms, preprint 1988.

4. J. Guckenheimer and P. Holmes, Nonlinear Oscillations, Dynamical Systems, and Bifurcations of Vector Fields, Springer-Verlag, New York 1983.

5. P. Holmes, J. Marsden and J. Scheurle, Exponentially small splitting of separatrices, to appear in Proc. Nat. Acad..

6. P. Holmes, J. Marsden and J. Scheurle, Exponentially small splitting of separatrices in KAM theory and degenerate bifurcations, to appear in Cont. Math..

7. V. Koslov, Integrability and non-integrability in Hamiltonian mechanics, Russian Math. Surveys 38 (1984), 1 - 76.

8. V. Lazutkin, G. Shachmanski and M. Tabanov, Separatrix splitting for the standard and semi-standard mappings, preprint.

9. V. Melnikov, On the stability of the center for time periodic perturbations, Trans. Moscow Math. Soc. 12 (1963), 1 - 57.

10. A. Neisthadt, The separatrices of motions in systems with rapidly rotating phase, P.M.M. USSR 48 (1984), 133 - 139.

11. H. Poincaré, Sur la probléme des trois corps et les équations de la dynamique, Acta Math. 13 (1890), 1 - 271.

12. J. Sanders, Melnikov's method and averaging, Celestial Mech. 28 (1982), 171 - 181.

INSTITUT FÜR ANGEWANDTE MATHEMATIK
UNIVERSITÄT HAMBURG
BUNDESSTRASSE 55
D-2000 HAMBURG 13
WEST GERMANY

and

DEPARTMENT OF MATHEMATICS
COLORADO STATE UNIVERSITY
FORT COLLINS, CO 80523
USA

Contemporary Mathematics
Volume **97**, 1989

ACCURATE TIME CRITICAL CONTROL OF MANY BODY SYSTEMS

N. Sreenath[1]

Abstract

Rapid large angle reorientation of articulated many body systems necessitate the use of nonlinear control since linear controllers cannot adequately handle model nonlinearities. Results available in the literature for spacecraft with momentum wheels and/or reaction jets discuss the use of approximate nonlinear control laws based on nonlinear optimal control formulation. Such control laws work well for slow maneuvers, but zero terminal error cannot be realized. Accurate nonlinear control can be achieved by finding a suitable nonlinear feedback and a nonlinear transformation to linearize the multibody system, and formulating the proper time-optimal control for the resultant linear system. We consider a planar multibody system connected in the form of a tree and prove that one only needs all joint torques and an external torque on one of the bodies, to realize such a control. The validity of such a controller has been demonstrated for a planar two-body system on the OOPSS system (Object Oriented Planar System Simulator) which simulates as well as animates the system in any desired coordinate frame.

1. Introduction

Rapid large angle maneuvers sometimes referred to as *step stare* maneuvers [19] necessitate the use of sophisticated control techniques to handle inherent nonlinear coupled dynamics of the modern day spacecraft. A number of linear control techniques have been advocated in the past for rotational maneuvers of single axis spacecraft. Breakwell [2], Hefner, Kawauchi, Meltzer and Williamson, [10] suggest linearization

[1]Systems Eng, Case Western Reserve Univ., Cleveland, OH 44106.

around operating points of the nonlinear system for multi axis single body spacecraft control. Quartararo [19] attempts linearization for a planar two-body system. Langer, Hemami, and Yurkovich [18] design a linear controller for a linearized version of a similar system using reduced and unreduced models of the plant. Tardy response of such controllers has led many researchers to look for nonlinear control techniques which not only speed up the response but also promote effective use of the actuators (like momentum wheel motors and gas jets).

Traditionally the nonlinear control approach to such problems problems have been formulated as optimal control problems which can be further classified under approximate and exact methods. We mention a few of each here. Relaxation and continuation techniques have been used by Vadali and Junkins [25] and Junkins and Turner [16] which resulted in suboptimal solutions. Carrington and Junkins [5] and, Dwyer [6] used polynomial truncation to solve the optimal control problem approximately. Approximate methods save computation time and may be adequate for slow maneuvers, but zero terminal conditions cannot be realized [9].

Some exact techniques involve finding a nonlinear feedback and coordinate transformation which linearizes the nonlinear dynamics (known as exact or feedback linearization). Acceptable controllers could then be designed for the linearized system. This technique has been successfully applied for single body spacecraft with gas jets (external torques) by Dwyer [7] or with momentum wheels (internal torques) [8]. Variable structure control has also been applied for spacecraft by Vadali [26] and by Dwyer and Sira-Ramirez [9] (see other references therein).

In this paper we are concerned with rapid large angle reorientation of articulated planar many-body systems via exact linearization procedures. The paper is divided as follows. The problem is formulated in Section 2 as a time optimal control problem using the model discussed in Sreenath [22] and Sreenath, Oh, Krishnaprasad and Marsden [23]. A controllability result given in Section 3 is used to linearize the system using feedback and coordinate transformation. The problem is reposed using the linearized dynamics, in Section 4 and solved. The time optimal control strategy is presented in the form of an algorithm. Simulation results are discussed in Section 5. The Appendix contains the model description and a Lemma used for feedback linearization.

2. Time Optimal Problem Formulation

We consider a planar multibody system of N bodies in space with a tree connected topology (no loops). The bodies are modeled as rigid bodies and are interconnected by means of friction free pin (one degree of freedom) joints. The system is assumed to be conservative with no external or internal disturbance torques or forces acting on the system, modulo input torques described later. For ease of computation the bodies are labeled using increasing integers along a path starting from body 1 (chosen arbitrarily). A joint connecting body i and another body with lesser body label is labeled $(i-1)$ (see Figure 1 for an example).* A reference coordinate frame is fixed at the system center of mass, and local coordinate frames are fixed at the center of mass of each body. The angle the local frame of reference of body i makes with the reference coordinate frame is referred to as the body angle θ_i.

The dynamics of such a system can be described by means of Newton-Euler, Lagrange's or Hamiltonian equations. We chose here to describe the evolution of motion on the cotangent bundle using Hamilton's principle. We state without proof that there is a rotational and a translational symmetry in the system and the system could be reduced by these symmetries, see Sreenath, Oh, Krishnaprasad, Marsden [23], Sreenath [22] and Abraham and Masden [1] for proof. The reduced Hamiltonian

$$\mathbf{H} = \frac{1}{2}\underline{\mu}^T \mathbf{J}^{-1} \underline{\mu}, \tag{1}$$

and the flow on the reduced manifold along with an output $\underline{y}$, defined for convenience, is given by

$$\left.\begin{aligned} \dot{\underline{x}} &= f(\underline{x}) + \underline{g}\underline{u} \\ \underline{y} &= h(\underline{x}) = \underline{\omega} \end{aligned}\right\} \tag{2}$$

where $f(\cdot)$ and $\underline{g}$ are given in the Appendix along with certain inherent properties. Also

$$\begin{aligned} \underline{x} &= [\mu_1, \mu_2, \ldots, \mu_N, \theta_{2,1}, \ldots, \theta_{i,j}, \ldots, \theta_{N,N-1}]^T, \\ \underline{u} &= [u_1, u_2, \ldots, u_{N-1}, u_N]^T \\ &= \left[T_1, T_2, \ldots, T_{N-1}, T_1^{ext}\right]^T, \\ \underline{\omega} &= [\omega_1, \omega_2, \ldots, \omega_N]^T, \\ &= \mathbf{J}^{-1}\underline{\mu}. \end{aligned}$$

Here ω_i is the angular velocity of body i in the reference coordinate frame and

*See pages 434-436 for figures and tables.

$\theta_{i,j} = (\theta_i - \theta_j)$ is the relative angle between adjacent bodies i and j with $i > j$. T_i is the input torque at joint i for $i = 1, \ldots, N-1$. This torque produced by a joint motor is also known as an internal torque and changes the energy of the multi-body system while limiting the flow to the symplectic leaf. The symplectic leaf here represents a constant angular momentum surface. The input (external) torque T_1^{ext} acting on body 1 changes both energy and the angular momentum in the system. **J** is a symmetric pseudo-inertia matrix.

The formulation of the *time optimal* control problem is given next. Here we are interested in formulating controls to take the system from a set of initial states, i.e., set of orientation, and angular velocities (so conjugate momenta) to another such desired set in *minimum time.*
The problem is stated as follows:

$$\textbf{Problem 0} \qquad \max_{\underline{u}} \; -\int_0^T dt, \tag{3}$$

subject to (2) and

$$\left.\begin{array}{rcll} \underline{x}(0) & = & \underline{x}_0 & \text{(Initial condition)} \\ \underline{x}(T) & = & \underline{x}_T & \text{(Final condition)} \\ |u_i| & \leq & u_{max_i} & \text{(Control constraint)} \end{array}\right\}, \tag{4}$$

where T is the *minimum time.*

It is customary to set up an *optimal control* problem either as a *calculus of variations* problem using *maximum principle* or, as a *dynamic programming* problem using the well known *Hamilton-Jacobi-Bellman* (HJB) equation [4]. The former technique is used here. Problem 0 could be recast using *maximum principle* by defining a Hamiltonian for the optimal control problem as

$$\mathbf{H}_{opt}(\underline{x}, \underline{p}, \underline{u}) = \underline{p}^T \left[f(\underline{x}) + \underline{g}\underline{u} \right], \tag{5}$$

where

$$\underline{p} = \left[\underline{p}_1^T, \underline{p}_2^T \right]^T, \tag{6}$$

is known as a vector of *auxiliary* or *costate* variables.

The necessary conditions for the optimal control problem using *maximum principle* are :

$$\textbf{Problem 1} \qquad \mathbf{H}^*_{opt}(\underline{x}^*, \underline{p}^*, \underline{u}^*) = \max_{u_i} \underline{p}^{*T} \left[f(\underline{x}^*) + \underline{g}\underline{u}^* \right], \tag{7}$$

subject to

$$\left.\begin{array}{rcl} \dot{\underline{x}}^* & = & f(\underline{x}^*) + \underline{g}\underline{u}^* \\ \dot{\underline{p}}^* & = & -\frac{\partial H}{\partial \underline{x}} \\ & = & -\frac{\partial f(\underline{x}^*)}{\partial \underline{x}^*}.\underline{p}^* \end{array}\right\}, \tag{8}$$

and the initial, final and control constraint conditions are as in (4) evaluated at $\left[\underline{x} = \underline{x}^*, \underline{p} = \underline{p}^*, \underline{u} = \underline{u}^*\right]$, where $\underline{x}^*$, $\underline{p}^*$ and $\underline{u}^*$ are the values taken by the state, costate and the control respectively along the optimal trajectory.

In addition, the following conditions should be satisfied

$$\frac{\partial \mathbf{H}^*_{opt}}{\partial \underline{u}^*} = \underline{0}, \tag{9}$$

$$\underline{p}^{*T}(T).f(\underline{x}^*)(T) = 1, \tag{10}$$

where $\underline{0}$ is a N column vector of 0's. The last equation is termed as the *terminal condition.*

Remark 2.1 : Simple calculation shows that Problem 1 is *singular*, i.e., $\frac{\partial H^*_{opt}}{\partial \underline{u}^*}$ is not a function of $\underline{u}$. Higher derivatives of (9) are needed to use the above conditions to solve for the optimal control $\underline{u}^*$.

Problem 1 is difficult to solve analytically. However, numerical techniques exist to solve the above problem by posing it as a two-point boundary value (TBVP) problem. This is numerically intensive and difficult to solve in general. Our interest here is to solve the optimal control problem analytically. The following strategy is adopted to this end.

We first prove by means of simple arguments that the system (2) is completely *controllable*. This result is used to formulate an input of the form

$$\underline{u} = \alpha(\underline{x}) + \beta(\underline{x}).\underline{v}, \tag{11}$$

$\beta(\cdot)$ invertible, which *Input/Output (I/O) linearizes* the system (2) by *feedback equivalence*, see [3], [11], [12] [13], [14] and [15]. By reformulating Problem 0 with (11) and the new input $\underline{v}$ we will show that an analytical solution is possible.

3. Exact Linearization

We present the controllability result next.

Theorem 3.1 : The multibody system given by (2) is controllable.

Proof : Differentiating the output in (2) yields

$$\begin{aligned} \dot{\underline{y}} &= \frac{\partial h}{\partial \underline{x}}\dot{\underline{x}} \\ &= \frac{\partial h}{\partial \underline{x}} f(\underline{x}) + \frac{\partial h}{\partial \underline{x}} g\underline{u}. \end{aligned} \tag{12}$$

Again from (2) we obtain,

$$\begin{aligned} \frac{\partial h}{\partial \underline{x}} g &= \left[\mathbf{J}^{-1} \middle| \begin{bmatrix} -\underline{\mu}^T \mathbf{J}^{-1} \frac{\partial \mathbf{J}}{\partial \theta_{2,1}} \mathbf{J}^{-1} \\ \cdots \\ -\underline{\mu}^T \mathbf{J}^{-1} \frac{\partial \mathbf{J}}{\partial \theta_{i,j}} \mathbf{J}^{-1} \\ \cdots \\ -\underline{\mu}^T \mathbf{J}^{-1} \frac{\partial \mathbf{J}}{\partial \theta_{N,N-1}} \mathbf{J}^{-1} \end{bmatrix}^T \right] \cdot \begin{bmatrix} \underline{Q} \mid \underline{w} \\ \underline{O}_{N-1,N} \end{bmatrix}, \\ &= \mathbf{J}^{-1} \left[\underline{Q} \mid \underline{w} \right]. \end{aligned} \tag{13}$$

We know that $\underline{Q}$ is full rank from Lemma A.1, so $\left[\underline{Q} \mid \underline{w} \right]$ is full rank and

$$\frac{\partial h}{\partial \underline{x}} g = \text{full rank},$$

this means that from (12) we can choose

$$\underline{u} = \left[\frac{\partial h}{\partial \underline{x}} g \right]^{-1} \left[-\frac{\partial h}{\partial \underline{x}} f(\underline{x}) + \underline{v} \right], \tag{14}$$

where $\underline{v}$ is the new input. The block diagram (Figure 2) illustrates a realization by feedback. The dynamics with the input as in (14) reduces (13) to

$$\dot{y} = \underline{v}, \tag{15}$$

i.e., from (2), the above equation becomes

$$\dot{\underline{\omega}} = \underline{v}. \tag{16}$$

Further, let

$$\underline{\Theta} = [\theta_1, \theta_2, \cdots, \theta_N]^T, \tag{17}$$

where θ_i is the angle made by the local coordinate system of body i with the reference coordinate system, i.e.,

$$\dot{\underline{\Theta}} = \underline{\omega}. \tag{18}$$

Use (16) and (18) to get

$$\left.\begin{aligned} \begin{bmatrix} \dot{\underline{\Theta}} \\ \dot{\underline{\omega}} \end{bmatrix} &= \begin{bmatrix} \underline{O}_{N,N} & \mathbf{I_N} \\ \underline{O}_{N,N} & \underline{O}_{N,N} \end{bmatrix} \begin{bmatrix} \underline{\Theta} \\ \underline{\omega} \end{bmatrix} + \begin{bmatrix} O \\ \mathbf{I_N} \end{bmatrix} \underline{v} \\ y &= \underline{\omega} \end{aligned}\right\}. \tag{19}$$

Choice of a suitable nonlinear control (14) has reduced the system in (2) to a I/O linearized system given by (19), which can be easily verified to be controllable (using the rank condition arguments of the Grammian [17]). Since the controllability property of a system is invariant under feedback and transformation, system (2) is *controllable.* Also Compare (11) and (14) to conclude that they are in the same form.

Q.E.D.

Remark 3.2: It is clear from (19) that I/O linearization *decouples* the dynamics of all the bodies in the multibody system.

4. Time Optimal Control

The optimal control problem Problem 0 reformulated using feedback equivalence is given next. Choosing the control as in (14) for the system (2) results in the flow and the output represented by (19), and Problem 0 can be rewritten as

$$\textbf{Problem 2} \qquad \max_{\underline{u}} \; -\int_0^T dt \tag{20}$$

subject to

$$\dot{\underline{z}} = \underline{Az} + \underline{Bv}. \tag{21}$$

with $\underline{z} = \left[\underline{\Theta}^T, \underline{\omega}^T\right]^T$; the initial and final conditions (4) can be appropriately rewritten as follows along with the control constraint,

$$\left.\begin{aligned} \underline{z}(0) &= \underline{z}_0 = \left[\underline{\Theta}^T(0), \underline{\omega}^T(0)\right]^T \\ \underline{z}(T) &= \underline{z}_T = \left[\underline{\Theta}^T(T), \underline{\omega}^T(T)\right]^T \\ |v_i| &\le v_{max_i}. \end{aligned}\right\} \tag{22}$$

where $\underline{v} = [v_1, \ldots, v_N]$.

Remark 4.1 : The constraint on the new controls $\underline{v}$ (i.e., the saturation of v_i) can be interpreted as the constraint on angular acceleration of the individual bodies in the system. One could estimate this from practical considerations. On the other hand, if the control constraints (4), and the range of the angular velocities and the angles are known, the following method can be used to estimate v_{max_i}.

Firstly, rewrite $\underline{v}$ using (14) as

$$\begin{aligned} \underline{v} &= \left[\frac{\partial h}{\partial \underline{x}} g\right] \underline{u} + \left[\frac{\partial h}{\partial \underline{x}} f(\underline{x})\right], \\ &= [E_1(\underline{z}, \underline{u}), \ldots, E_N(\underline{z}, \underline{u})]^T \end{aligned} \tag{23}$$

v_{max_i} and v_{min_i} which are also the extremum values of the angular acceleration of body i can be found as follows:

$$\begin{aligned} v_{max_i} &= \max_{\underline{z}, \underline{u}} E_i(\underline{z}, \underline{u}) \\ v_{min_i} &= \min_{\underline{z}, \underline{u}} E_i(\underline{z}, \underline{u}) \end{aligned}$$

Note that in (22) we have assumed that $|v_{min_i}| = v_{max_i} > 0$ for all i, in general this need not be the case.

4.1 Solution by Maximum Principle

We focus our attention on solving Problem 2. Using *maximum principle*, the Hamiltonian for the optimal control problem along the optimal trajectory $\left[\underline{z}^*, \underline{p}^*, \underline{v}^*\right]$ is

$$\mathbf{H}^*_{opt}(\underline{z}^*, \underline{p}^*, \underline{v}^*) = \max_{\underline{v}} \left\{\underline{p}_1^{*T} \underline{\omega}^* + \underline{p}_2^{*T} \underline{v}\right\}, \tag{24}$$

The other necessary conditions for the optimal control problem Problem 2 are given by the initial, final and control constraint conditions (22) at $\left[\underline{z} = \underline{z}^*, \underline{p} = \underline{z}^*, \underline{v} = \underline{z}^*\right]$. The evolution of the auxiliary variables are given by

$$\left.\begin{aligned} \dot{\underline{p}}_1^* &= \underline{0} \\ \dot{\underline{p}}_2^* &= \underline{p}_1^* \end{aligned}\right\} \tag{25}$$

or

$$\left.\begin{aligned} \underline{p}_1^*(t) &= \underline{p}_1(0) \\ \underline{p}_2^*(t) &= \underline{p}_1^* \, t + \underline{p}_2^*(0) \end{aligned}\right\}, \tag{26}$$

for $t \in [0, T]$ and $\underline{p}^* = \left[\underline{p}_1^*, \underline{p}_2^*\right]$. The corresponding terminal condition is

$$\underline{p}_1^{*T}(0).\underline{\omega}^*(T) + \underline{p}_2^{*T}(T).\underline{v}^*(T) = 1. \tag{27}$$

The facts listed below follow from the *maximum principle* :

(a) The Hamiltonian given by (24) is a constant over the optimal trajectory and greater than or equal to zero.

(b) The problem is singular since $\frac{\partial \mathbf{H}}{\partial \underline{v}}$ is not a function of $\underline{v}$.

(c) Optimal control is extremal, i.e., bang-bang control. So from (24), if $p^*_{2i}(t) > 0$ then $v^*_i = v_{max_i}$ or if $p^*_{2i}(t) < 0$ then $v^*_i = -v_{max_i}$, where $\underline{p}^*_2 = [p^*_{21}, \ldots, p^*_{2N}]$.

(e) It is easily verified from (26) that the optimal input can have at most one switching from $+v_{max_i}$ to $-v_{max_i}$ or from $-v_{max_i}$ to v_{max_i}.

Integrating (21) for the switching cases, after some manipulation, solving for t_{fi} – the minimum time necessary for body i to reach the desired state, t_{si} – and the switching time,

$$\begin{aligned} t_{si} &= -\frac{\omega_i(0)}{I_i v_{max_i}} \\ &\quad \pm \frac{1}{2v_{max_i}} \sqrt{2\left(\omega_i(T)^2 + \omega_i(0)^2\right) + 4I_i\left[\theta_i(T) - \theta_i(0)\right] v^2_{max_i}} \qquad (28) \\ t_{fi} &= 2\, t_{si} - \frac{\omega_i(T) - \omega_i(0)}{I_i\, v_{max_i}}. \qquad (29) \end{aligned}$$

For the no switching case to hold

$$\omega_i(T)^2 - \omega_i(0)^2 = 2I_i v_{max_i}\left[\theta_i(T) - \theta_i(0)\right], \qquad (30)$$

should be satisfied, then

$$t_{fi} = \frac{\omega_i(T) - \omega_i(0)}{I_i v_{max_i}}. \qquad (31)$$

Here

$$I_i = \begin{cases} 1 & if\ p_2(0) > 0 \\ -1 & if\ p_2(0) < 0 \end{cases} \qquad (32)$$

Thus we have a maximum of four values for the switching case *or* a maximum of two values for the no switching case of the pair (t^j_{si}, t^j_{fi}) (see (28) – (32)). The index j indicates the real and feasible values of the pair (t^j_{si}, t^j_{fi}) with feasibility being defined as $(0 < t^j_{si} < t^j_{fi})$. Since each body in the system is decoupled, we take the minimum of t^j_{fi} for each body i and take the maximum of these minimum times for all bodies. This is defined as the *minimum time* T, i.e.,

$$T = \max_i \left\{ \min_j t^j_{fi} \right\}. \qquad (33)$$

This means that for some body k

$$t_{fk} = \min_j t^j_{fk} = T, \tag{34}$$

i.e., body k takes the maximum time among all the bodies in the multibody system to reach the desired final angle and angular velocity starting from the initial values. The time optimal control for the switching case of body k is shown in Figure 3.

For all other bodies $i \neq k$ the optimal input (see Figure 4) has one switching from $\pm v_{max_i}$ to $\mp v_{max_i}$ at $t = t_{si}$ and another switching from $\mp v_{max_i}$ to 0 at $t = t_{zi}$. The value of the pair (t_{si}, t_{zi}) such that $0 < t_{si} < t_{zi} \leq T$ are found by solving the following quadratic equations and selecting the feasible values :

$$\begin{aligned} \theta_i(T) &= \theta_0 + \omega_i(0)\, T + \frac{I_i\, v_{max_i}}{2}\left[4t_{si}t_{zi} - 2t_{si}^2 - t_{zi}^2\right] \\ &\quad + \omega_i(T)\,(T - t_{zi}), \end{aligned} \tag{35}$$

$$t_{si} = \frac{1}{2}\left[t_{zi} + \frac{\omega_i(T) - \omega(0)}{I_i v_{max_i}}\right]. \tag{36}$$

4.2 Algorithm

The *time optimal control* strategy given above could be summed up as follows given the setup of Problem 0:

(i) Choose a feedback as in (14) and linearize the system.

(ii) Translate all initial, final and control constraint conditions as in (4) to (22).

(iii) Find T, k, from (28) - (34).

(iv) Find feasible (t_{zi}, t_{si}), for all $i \neq k$, $i = 1, \cdots, N$ using (35) and (36).

(v) Formulate *optimal control* for each axis as in Figures 3 and Figure 4.

5. Simulation using OOPSS

The optimal control algorithm above was validated on the **OOPSS** [2] software system [21]. OOPSS is capable of constructing Hamiltonian models of tree connected

[2]Object Oriented Planar System Simulator

planar multibody systems and automatically generate the computer code to simulate the dynamics. Control strategies like proportional-derivative (PD) control and time optimal control could be chosen. On request by the user, OOPSS will also run the simulation through a well designed user interface (Figure 5) allowing the user to interactively change various parameters and animate the evolution of motion. It is implemented on a LISP[3] machine using a combination of MACSYMA (for symbolic model generation and automatic code generation), FORTRAN (for numerical simulation) and LISP implementation of FLAVORS (for animation). Planar two and three body systems can be animated in the present versions of OOPSS.

A planar two body example was chosen with a joint motor (internal) torque and a gas jet (external torque) on body 1 as control inputs. Figure 5 shows an OOPSS animation window for this system. The biggest *pane* of this window displays the animation in the reference coordinate frame which is assumed to be at the center of mass of the two-body system or in the *inertial frame*. Each body is represented by a stick figure with a big filled-in-circle at one end (representing the center of mass of the body) and a small circle at another indicating the joint connecting the other body. The center of mass of the system which is a fixed point in the reference frame is the small circle in the center of the pane. The two small panes below this pane display the motion of the system in the joint frame and body 1 local frame of reference respectively. The bottom right pane is a *mouse sensitive command pane*, for user interaction. Above this are the plot and the data display panes. For more details regarding the OOPSS the reader is referred to Sreenath and Krishnaprasad [21].

The physical parameters used for the simulation is given in Table 1. Table 2 gives the minimum time required for the maneuvers along with the various switching times involved. A *rest-to-constant-spin* maneuver was simulated. The body angles and angular velocities as a function of time is shown in the plot pane of Figure 5. *Rest-to-rest* maneuvers were also simulated with different saturation values of the maximum angular accelerations v_{max_i}. For the same intial and final conditions it was found that the optimal time reduced by a factor of four with a magnitude increase in v_{max_i}. Graphs of the various angles, angular velocities with time are given in Figure 6 and Figure 7.

[3]versions exist on a TI Explorer color LISP machine as well as on Symbolics 3600 series machines.

6. Summary

The problem of accurate rapid reorientation of articulated multibody systems in space is difficult to solve because of tight coupling of the dynamics between the bodies when the bodies are modeled as rigid bodies. This problem was formulated as a time optimal control problem and solved explicitly for the case of planar system interconnected in the form of a tree. We first linearize a Hamiltonian model of the system using a combination of nonlinear transformation and feedback and then solve for the optimal control using maximum principle. An algorithm for accurate time optimal control was presented and validated using OOPSS software system for the case of a planar multibody system.

Appendix

The reduced Hamiltonian dynamics of a planar multibody system connected in the form of a tree structure with joint (internal) torques $T_k,\ k = 1, \ldots, N-1$ and an external torque T_1^{ext} on body 1 is given by [22] Chapter 3, Theorem 3.6.1 to be

$$\dot{\mu}_1 = \sum_{\forall j,\ s.t.\ J(j)=1} \frac{\partial \mathbf{H}}{\partial \theta_{j,1}} \tag{37}$$

$$\vdots \qquad \vdots$$

$$\dot{\mu}_i = \sum_{\forall j,\ s.t.\ J(j)=i} \frac{\partial H}{\partial \theta_{j,i}} - \frac{\partial H}{\partial \theta_{i,J(i)}} \tag{38}$$

$$\text{for } i = 2, \ldots, N-1,$$

$$\vdots \qquad \vdots$$

$$\dot{\mu}_N = -\frac{\partial \mathbf{H}}{\partial \theta_{N,J(N)}} - T_{N-1}, \tag{39}$$

$$\dot{\theta}_{i,J(i)} = \frac{\partial \mathbf{H}}{\partial \mu_i} - \frac{\partial \mathbf{H}}{\partial \mu_{J(i)}} \quad \text{for } i = 2, \ldots, N. \tag{40}$$

here, $J(i)$ is the body label of the body connected to body i with $J(i) < i$. The reduced Hamiltonian $\mathbf{H}$ is given by

$$\mathbf{H} = \frac{1}{2} \underline{\mu}^T \mathbf{J}^{-1} \underline{\mu} \tag{41}$$

$\mathbf{J}$ is an $N \times N$ nonsingular, symmetric pseudo-inertia matrix whose elements are functions of $\underline{\theta}$. The μ_i's are known as *conjugate momenta*; also,

$$\mu_s = \sum_{i=1}^{N} \mu_i \tag{42}$$

where μ_s is the system angular momentum.

The reduced dynamical equations with the output $\underline{y}$ defined as a vector of body angular velocities could be also written as,

$$\left.\begin{array}{rcl} \dot{\underline{x}} & = & f(\underline{x}) + \underline{g}\underline{u} \\ \underline{y} & = & h(\underline{x}) \end{array}\right\}, \tag{43}$$

where the state vector $\underline{x}$ and control (input) vector $\underline{u}$ are given by

$$\underline{x} = \left[\mu_1, \ldots, \mu_N, \theta_{k,J(k)},\ k = 2, \ldots, N\right], \tag{44}$$

$$\underline{u} = \left[T_1, \ldots, T_{N-1}, T_1^{ext}\right]^T, \tag{45}$$

and

$$f(\underline{x}) = \left[\begin{array}{c} \sum_{\forall j,\ s.t.\ J(j)=1} \frac{\partial H}{\partial \theta_{j,1}} \\ \sum_{\forall j,\ s.t.\ J(j)=i} \frac{\partial H}{\partial \theta_{j,i}} - \frac{\partial H}{\partial \theta_{i,J(i)}} \\ \text{i=2, \ldots, N-1} \\ -\frac{\partial H}{\partial \theta_{N,J(N)}} \\ \frac{\partial H}{\partial \mu_k} - \frac{\partial H}{\partial \mu_{J(k)}}, \quad k = 2, \ldots, N \end{array}\right], \tag{46}$$

$$\underline{g} = \left[\begin{array}{c} \underline{Q} \mid \underline{w} \\ \underline{O}_{N-1,N} \end{array}\right], \tag{47}$$

and $\underline{Q}$ is a $N \times N-1$ matrix whose elements are either 0, +1, or −1, $\underline{w} = [1, 0, \ldots, 0]^T$, and $\underline{O}_{N-1\times N}$ is the $N-1\times N$ *null matrix.*

The output $\underline{y}$ is the vector of inertial angular velocities $\underline{\omega}$ of the system and is given by :

$$\underline{y} = h(\underline{x}) = \underline{\omega} = \mathbf{J}^{-1}\underline{\mu}. \tag{48}$$

Lemma A.1 : The matrix $\underline{Q}$ which represents a *directional vertex matrix* of a *connected graph*, is full rank.

Proof : Some of the properties of the matrix $\underline{Q}$ are listed below:

(a) $\underline{Q}$ is a $N \times N-1$ matrix.

(b) Elements of any column of $\underline{Q}$ except for two nonzero elements, a (+1) and a (-1), are all zeros. Addition of all the rows of $\underline{Q}$ thus results in a *zero row vector*. Addition of any $r < N$ rows of $\underline{Q}$ results in a nonzero row vector (since $\underline{Q}$ represents a *directional vertex matrix* of a *connected graph.* See Seshu and Reed [20] (Lemma 5-1(b) and Theorem 5-1).

(c) The non-zero elements of the first row of Q are always 1.

A linear combination of $(\underline{Q}_1, \dots, \underline{Q}_N)$, the rows of $\underline{Q}$, equation

$$\sum_{k=1}^{N} c_k \underline{Q}_k = \underline{O}_{1,N-1},$$

has the only non-trivial solution ,

$$c_1 = c_2 = \cdots = c_N = 1, \tag{49}$$

modulo a scale facto,r from property (b).

This proves that the matrix $\underline{Q}$ has *rank* $(N-1)$. since only one independent relation exists among the rows of $\underline{Q}$.

Q.E.D.

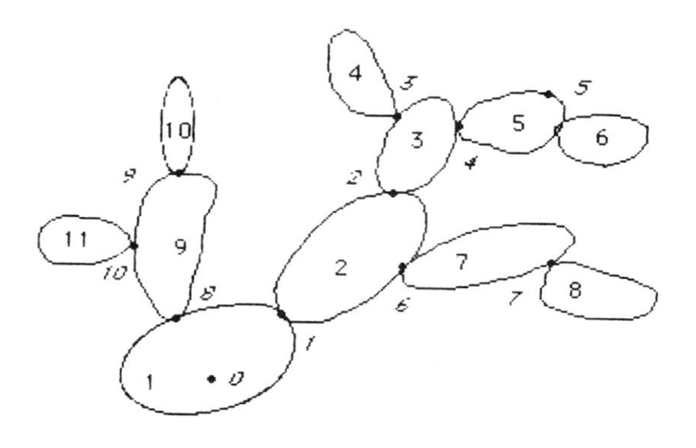

Figure 1 : Planar body system example with 11 bodies

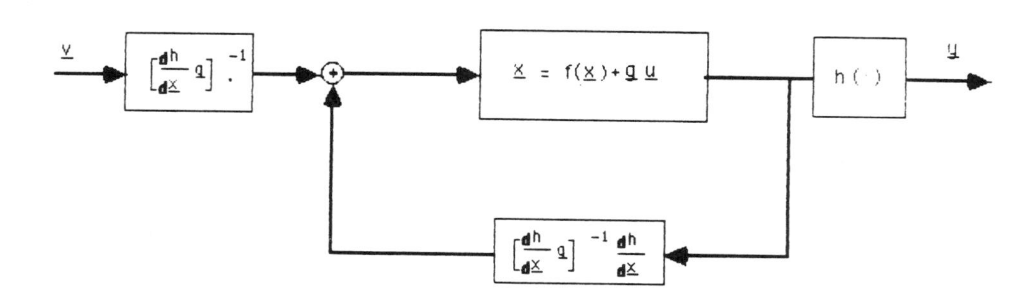

Figure 2 : Input/Output feedback linearization

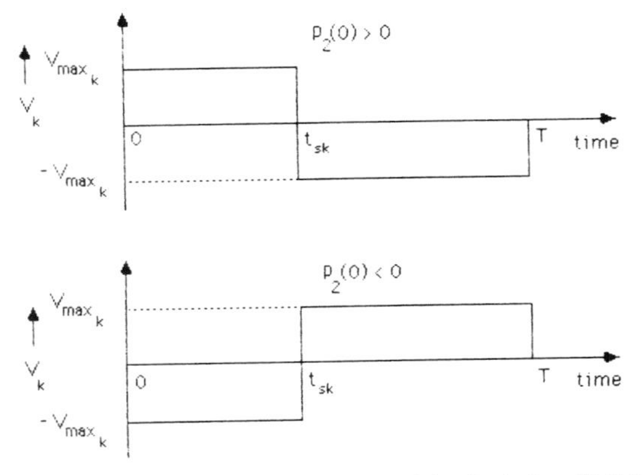

Figure 3 : Optimal control for the body with $T=t_{fk}$

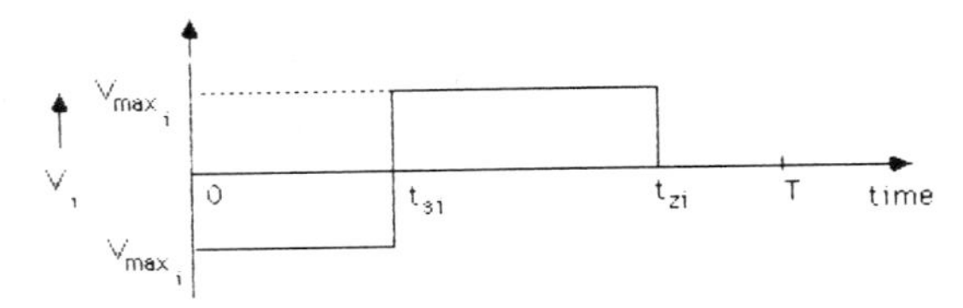

Figure 4 : Optimal control for bodies with $T \neq t^{i}_{fi}$

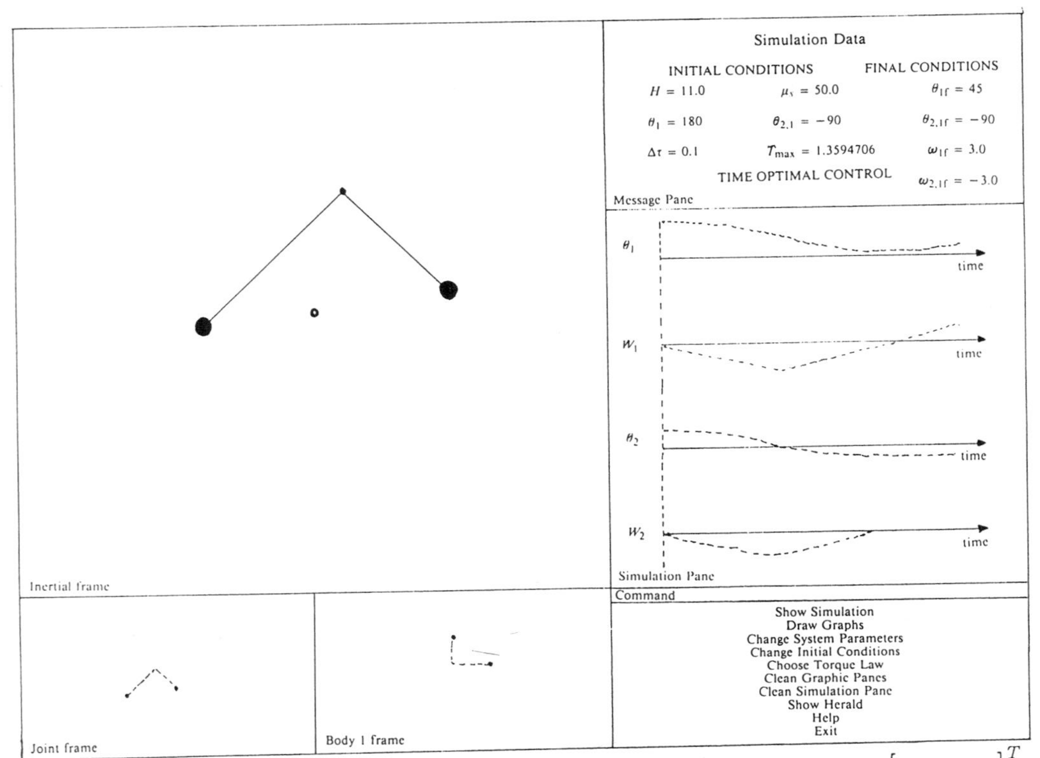

Figure 5 : OOPSS window with a *rest-to-spin* maneuver, $\underline{z}(0) = \left[\pi, \frac{\pi}{2}, 0, 0\right]^T$ and $\underline{z}(T) = \left[\frac{\pi}{4}, -\frac{\pi}{4}, 3.0, 0\right]^T$.

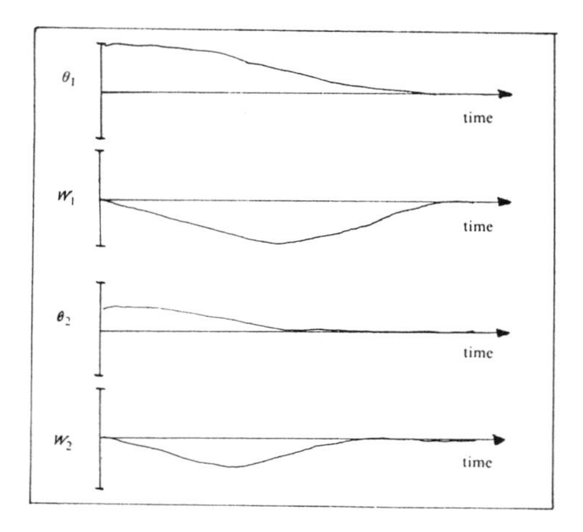

Figure 6 : *Rest-to-rest* maneuver with $v_{max} = v_{max_1} = v_{max_2} = 10.0$ rad/sec^2, $\underline{z}(0) = \left[\pi, \frac{\pi}{2}, 0, 0\right]^T$ and $\underline{z}(T) = [0,0,0,0]^T$.

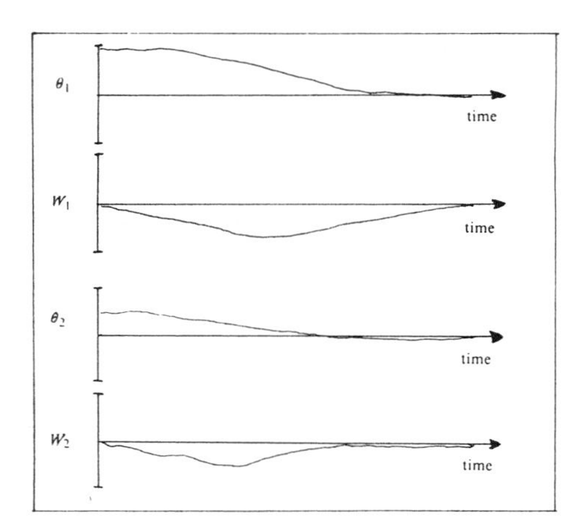

Figure 7 : *Rest-to-rest* maneuver with $v_{max} = v_{max_1} = v_{max_2} = 100.0$ rad/sec^2, $\underline{z}(0) = \left[\pi, \frac{\pi}{2}, 0, 0\right]^T$ and $\underline{z}(T) = [0,0,0,0]^T$.

Table 1

	Mass	Inertia	Dist. to center of mass from joint
Body 1	125 kg.	70 kg-m^2	0.8 m
Body 2	100 kg.	50 kg-m^2	0.6 m

Table 2

	Maneuver	V_{max}	k	T (seconds)	t_{s1} (seconds)	t_{s2} (seconds)	t_{z2} (seconds)
Figure 5	Rest-to-spin	10.0	1	1.3595	0.4854	0.4854	0.9708
Figure 6	Rest-to-rest	10.0	1	1.121	0.5605	0.3963	0.7927
Figure 7	Rest-to-rest	100.0	1	0.35 45	0.1173	0.1253	0.2507

Bibiliography

[1] Abraham, R., and Marsden, J. E., *Foundations of Mechanics*, Benjamin /Cummings, Reading, Mass., 1978.

[2] Breakwell, J. A., "Optimal feedback slewing of flexible spacecraft", *J of Guidance, Control and Dyn.*, Vol 4, pp. 472-479, Sept-Oct 1981.

[3] Brockett, R. W., "Feedback invariants of nonlinear systems", *IFAC Congress*, Helsinki, pp. 1115, 1978.

[4] Bryson, A. and Ho, *Applied Optimal Control*, Addison-Wesley, 1978.

[5] Carrington, C. K., and Junkins, J. L., "Nonlinear feedback control of spacefraft slew manwuvers", *J Astro. Sci.* Vol 32, pp. 29-45, Jan-March 1984.

[6] Dwyer, T. A. W. III, and Sena, R. P.,"Control of spacecraft slewing maneuvers",*Proc. 21st IEEE Conf. on Deci. and Control*, pp. 1142-1144, New York, 1982.

[7] Dwyer, T. A. W. III, "Exact nonlinear control of large angle rotational maneuvers", *IEEE Trans. on Automatic Control*, Vol. AC-29, pp. 769-784, Sept. 1984.

[8] Dwyer, T. A. W. III, "Exact nonlinear control of spacecraft slewing maneuvers with internal momentum transfer", *Journal of Guidance, Control, and Dynamics*, Vol 9, No.2, pp. 240-247, March-April, 1986.

[9] Dwyer, T. A. W. III, and Sira-Ramirez, H., "Variable Structure control of spacecraft arritude maneuvers", *J Guidance, Control, and Dyn.*, Vol 11, No. 3, 262-270, May-June 1988.

[10] Hefner, R. D., Kawauchi, B., Meltzer, S., and Williamson, R., "A terminal controller fot the pointing of flexible spacecraft", *Proc. AIAA Guidance and Control Conf.*, AIAA, New York, 1980.

[11] Hunt L. R., Su, R., and Meyer, G., "Global transfer of nonlinear systems". *IEEE Transactions on Automatic Control,* Vol. AC-29, pp. 24-31, Jan. 1983.

[12] Hunt L. R., Su, R., and Meyer, G., "Design of multi-input nonlinear systems", *Differential Geometric Control Theory*, Ed. R. Brockett, R. Millman,

H. Sussman, Stuttgart: Birkhauser, pp. 268-298, 1983.

[13] Isidori, A., "Non-Linear Control Systems : An Introduction", *Lecture Notes in Control and Information Science*, Vol 72, Springer-Verlag, New York, 1985.

[14] Isidori, A., and Krener, A. J., "On feedback equivalence of nonlinear systems", *System & Control Letters*, Vol 2, No. 2, August 1982.

[15] Jakubczyk, V. J., and Respondek, W., "On linearization of control systems", *Bull. Acad, Pol, Sci, Ser. Sci. Math,* 28, pp. 517-522, 1980.

[16] Junkins, J. L., and Turner, J.D., "Optimal Spacecraft Rotational Maneuvers", *Elsevier Science Publishers*, Amsterdam, The Netherlands, 1986.

[17] Kailath, T., "Linear Systems", *Prentice Hall*, Englewood Cliffs, N.J., 1980.

[18] Langer, F. D., Hemami, H., and Yurkovich, S., "Control and simulation of constrained dynamic two-link system", *Int. J. Cont.* Vol 47, No. 5, pp 1341-1354, 1988.

[19] Quartararo, R., "Two-body control for rapid attitude maneuvers", *Advances in Astronautical Sciences*, L.A. Morine ed., Vol 42, pp. 397-422,

[20] Seshu, S., and Reed, M. B., "Linear Graphs and Electrical Networks", *Addison-Wesley Publishing Company, Inc.*, Reading, Mass., 1961.

[21] Sreenath, N., and Krishnaprasad, P.S., "Multibody Simulation in an object-oriented programming environment", *Advances in Symbolic Computation*, Ed. R. Grossman, SIAM Publ., 1988.

[22] Sreenath, N., *Modeling and Control of Multibody Systems*, Ph.D. Thesis, Univ. of Maryland, College Park, 1987.

[23] Sreenath, N., Oh, Y.G., Krishnaprasad, P. S., Marsden, J. E., "Dynamics of coupled planar rigid bodies, Part I : modeling, reduction and stability",*J. Dynamics and Stability of Systems*, Vol 2, No. 0, 1988.

[24] Vadali, S. R., and Junkins, J. L., "Spacecraft large angle rotational maneuvers with optimal momentum transfer", AIAA paper 82-1469, Aug 1882.

[25] Vadali, S.R., and Junkins, J. L., "Optimal open loop and stable feedback control of rigid spacecraft attitude maneuvers", *J. of Astronautical Sciences*, Vol. 32, Jan-March pp. 105-122, 1984.

[26] Vadali, S. R., "Variable structure control of spacecraft large angle maneuver", *J Guidance, Control and Dynamics*, Vol 9, pp235-239, March-April 1986.

Contemporary Mathematics
Volume 97, 1989

HAMILTONIAN CONTROL SYSTEMS: DECOMPOSITION AND CLAMPED DYNAMICS

A.J. van der Schaft

ABSTRACT. After defining Hamiltonian and Poisson control systems it is shown how a Hamiltonian control system admits a local canonical decomposition reflecting its controllability and observability properties. Under extra assumptions, the constrained or clamped dynamics of a Hamiltonian system is shown to be Hamiltonian, and a conjecture is formulated for the general case.

0. INTRODUCTION. In this paper, which is partly of a survey nature, we summarize and motivate the definition of Hamiltonian and Poisson control systems. In section 2 we show, under some regularity assumptions, how a Hamiltonian control system can be locally decomposed into a set of subsystems reflecting the controllability and observability properties of the system. This will define a natural Poisson mapping from the system onto a Poisson control system of lower dimension, but having the same input-output behavior. In the third section we show how under extra assumptions the clamped dynamics of a Hamiltonian control system is given by a Hamiltonian vectorfield on a symplectic submanifold. Alternatively it can be represented as a Hamiltonian vectorfield on the full state space manifold endowed with the Poisson structure given by a certain Dirac bracket (see also [MB,BM]). A conjecture is formulated for the general case.

1. DEFINITION OF A HAMILTONIAN CONTROL SYSTEM. We first summarize some theory about Poisson brackets, cf. [W], see also [LM,SA]. Let M be a manifold, and let $C^\infty(M)$ denote the smooth real-valued functions on M.
A *Poisson structure* on M is a bilinear map from $C^\infty(M) \times C^\infty(M)$ into $C^\infty(M)$,

---

1980 *Mathematics Subject Classification* (1985 *Revision*).93C10, 58F05, 93B10.

called the *Poisson bracket*, and denoted as

$$(1) \qquad (F,G) \rightarrow \{F,G\} \qquad F,G \in C^{\infty}(M)$$

which satisfies for any $F,G,H \in C^{\infty}(M)$ the following properties

$$(2a) \qquad \{F,G\} = -\{G,F\} \qquad \text{(skew-symmetry)}$$

$$(2b) \qquad \{F,\{G,H\}\} + \{G,\{H,F\}\} + \{H,\{F,G\}\} = 0 \qquad \text{(Jacobi)}$$

$$(2c) \qquad \{F,GH\} = \{F,G\}H + G\{F,H\} \qquad \text{(Leibnitz)}$$

M together with a Poisson structure is called a *Poisson* manifold.
It follows from (2c) that every $F \in C^{\infty}(M)$ defines a unique vector field, denoted as X_F, satisfying

$$(3) \qquad X_F(G) = \{F,G\} \qquad G \in C^{\infty}(M).$$

X_F is called the *Hamiltonian vectorfield* corresponding to the Hamiltonian F. The Jacobi-identity (2b) immediately implies the following relation between Poisson brackets and Lie brackets of Hamiltonian vectorfields, namely

$$(4) \qquad [X_F,X_G] = X_{\{F,G\}} \qquad F,G \in C^{\infty}(M).$$

Since by (2a) and (3)

$$(5) \qquad dG(X_F) = X_F(G) = \{F,G\} = -\{G,F\} = -X_G(F) = -dF(X_G),$$

it follows that the Poisson bracket $\{F,G\}$ depends on F and G only through their derivatives dF and dG. Hence in local coordinates $x_1,..,x_r$ for M we can write

$$(6) \qquad \{F,G\}(x) = \sum_{i,j=1}^{r} w_{ij}(x) \frac{\partial F}{\partial x_i}(x) \frac{\partial G}{\partial x_j}(x)$$

for the smooth functions $w_{ij}(x) = \{x_i,x_j\}$, $i,j = 1,\dots,r$. The *rank* of the Poisson structure in a point $x \in M$ is now defined as the rank of the $r \times r$ matrix $W(x)$ with (i,j)-th element $w_{ij}(x)$. By skew-symmetry of $W(x)$ the rank of any Poisson structure is even. The Poisson structure is said to be *non-degenerate* if its rank is everywhere equal to dim M. A non-degenerate Poisson structure defines a *symplectic form* ω on M by setting

$$(7) \qquad \omega(X_F,X_G) = \{F,G\}$$

M together with a non-degenerate Poisson structure is called a *symplectic manifold.*

THEOREM 1. (Lie, Darboux, Weinstein [W]). *Let M be a Poisson manifold, and let the rank of the Poisson structure in* $x_0 \in M$ *be equal to 2n. Then there exist local coordinates* $(q_1,\dots,q_n,\ p_1,\dots,p_n,\ y_1,\dots,y_s)$ *centered around*

x_0, *such that*

(8a) $\{p_i, q_j\} = \delta_{ij}$

(8b) $\{q_i, q_j\} = \{p_i, p_j\} = \{q_i, y_j\} = \{p_i, y_j\} = 0$

(8c) $\{y_i, y_j\} = v_{ij}(y)$ with $v_{ij} = 0$ at x_0

Thus the submanifold $y_1 = \ldots = y_s = 0$ *is a symplectic manifold which furthermore can be extended to a unique maximal symplectic submanifold, called the* symplectic leaf *through* x_0.

In particular if M *is a symplectic manifold then there exist local coordinates* $(q_1, \ldots, q_n, p_1, \ldots, p_n)$, *called* canonical, *such that*

(9) $\{p_i, q_j\} = \delta_{ij}, \ \{q_i, q_j\} = \{p_i, p_j\} = 0. \qquad i,j = 1, \ldots, n$

It is easily seen that in canonical coordinates (9) a Hamiltonian vectorfield X_H takes the familiar form

(10) $\dot{q}_i = \dfrac{\partial H}{\partial p_i}, \qquad \dot{p}_i = -\dfrac{\partial H}{\partial q_i} \qquad i = 1, \ldots, n$

A vectorfield X on M is called a *Poisson* vectorfield if it is an infinitesimal automorphism of the Poisson structure, i.e.

(11) $X(\{F,G\}) = \{X(F), G\} + \{F, X(G)\} \qquad F, G \in C^\infty(M)$

It immediately follows from (2b) and (3) that a Hamiltonian vectorfield is necessarily a Poisson vectorfield.

Conversely, if X is a Poisson vectorfield for a *non-degenerate* Poisson structure then it can be easily seen (for instance in canonical coordinates) that for any $x_0 \in M$ there exists a neighborhood V of x_0 and a function $H \in C^\infty(V)$ such that $X = X_H$ on V, i.e. X is a *locally* Hamiltonian vectorfield. Unfortunately this is not true for a general Poisson structure ([W]). We are now ready to give (see [B],[S1],[S2],[SA]):

DEFINITION 2. *Let* M *be a Poisson manifold. A nonlinear control system on* M *is called a* Poisson control system *if it is of the form*

(12a) $\dot{x} = X_0(x) - \sum_{j=1}^{m} u_j X_{H_j}(x) \qquad x \in M, \qquad x(0) = x_0$

(12b) $y_j = H_j(x) \qquad j = 1, \ldots, m$

with X_0 *a Poisson vectorfield, and* $H_1, \ldots, H_m \in C^\infty(M)$. *The scalars* $u_1, \ldots, u_m$ *are called the controls or inputs, and* $y_1, \ldots, y_m$ *the observations or outputs.*

If M *is a symplectic manifold then (12) is called a* Hamiltonian control system, *and locally there will exist a function* H_0, *called the internal*

Hamiltonian, such that $X_0 = X_{H_0}$.

REMARK. This definition can be generalized by considering a smooth output map from M to an m-dimensional *output manifold* Y. In local coordinates $(y_1,\dots,y_m)$ for Y we then write $C = (H_1,\dots,H_m)$ and we recover (12b).
This makes clear that $(y_1,\dots,y_m, u_1,\dots,u_m)$ can be regarded as natural coordinates for the cotangent bundle T^*Y, which has the physical interpretation of the controls being co-vectors (forces).
More geometrically a Poisson or Hamiltonian control system can be defined as a Lagrangian submanifold of $TM \times T^*Y$, where T^*Y is endowed with its natural Poisson bracket, and the Poisson structure on TM is the canonical prolongation of the Poisson structure on M. For the Hamiltonian case this was done in [S1,S2], and for the Poisson case in [SA]. This last definition also suggests a more general form of the local equations of a Hamiltonian control system, namely in canonical coordinates (q,p) for M and (y,u) for T^*Y

$$
(13)\qquad
\begin{aligned}
\dot{q}_i &= \frac{\partial H}{\partial p_i}(q,p,u) \\
\dot{p}_i &= -\frac{\partial H}{\partial q_i}(q,p,u) \\
\dot{y}_j &= -\frac{\partial H}{\partial u_j}(q,p,u)
\end{aligned}
\qquad
\begin{aligned}
&i = 1,\dots,n \\
\\
&j = 1,\dots,m
\end{aligned}
$$

see [B], [S1,S2]. Notice that if H(q,p,u) is *affine* in u, i.e. of the form $H_0 - \sum_j u_j H_j$, then (13) reduces to (12).

The particular choice of the outputs (12b) corresponds to what in the control literature of flexible structures is sometimes called the case of *collocated* sensors and actuators. Every pair (y_j,u_j), $j = 1,\dots,m$, is a pair of dual variables, i.e. if y_j is a particular generalized configuration coordinate, then u_j will be the corresponding generalized external force. The relation between inputs and outputs in (12) can be seen as the dynamic generalization of a static *reciprocal* relation $u_j = -\frac{\partial V}{\partial y_j}(y_1,\dots,y_m)$, $j = 1,\dots,m$, see [AM p. 413] for references. (This is most evident in the more general form (13)). The definition is also similar to the definition of a driving-point impedance in electrical network theory, or of a transfer matrix in the theory of elastic structures.
Finally we remark that in the case of a Hamiltonian control system equation (12a) can be alternatively seen as a *time-varying* Hamiltonian differential equation with time-dependent Hamiltonian (H_0 locally defined)

(14) $$H_0 - \sum_{j=1}^{m} u_j(t)\, H_j$$

where the input functions $u_1(t),\dots,u_m(t)$ are, say, piecewise continuous. It immediately follows that we have an energy-balance

(15) $$\frac{dH_0}{dt} = \sum_{j=1}^{m} u_j\, \dot{y}_j$$

which expresses the fact that the increase of the internal Hamiltonian (energy) equals the external power (think of $\dot{y}_j$ as a generalized velocity, and u_j as the corresponding generalized external force).

EXAMPLE. A typical simple example of a Hamiltonian control system is provided by a two-link rigid frictionless robot manipulator.

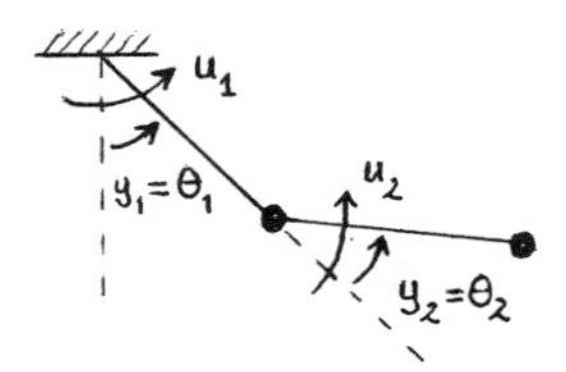

Fig. 1.

In this case the inputs u_1 and u_2 can be taken as the external torques at the two joints (delivered by actuators), and the corresponding outputs y_1 and y_2 are the (relative) angles of both links. □

In the sequel we will restrict ourselves to Hamiltonian control systems, although in the analysis of these Hamiltonian systems Poisson control systems (with degenerate Poisson bracket) will naturally show up. For Poisson control systems per se we refer to [K,KM,SA].

First we show that the *controllability* and *observability* properties of a Hamiltonian control system are to a large extent characterized by the following Lie subalgebra of $C^\infty(M)$, viewed as a Lie algebra under the Poisson bracket.

DEFINITION 3. *Consider a Hamiltonian control system (12). Define* $\mathcal{H}$ *as the smallest subalgebra of* $(C^\infty(M),\{\,,\})$ *satisfying*

(16a) $$H_1,\dots,H_m \in \mathcal{H}$$

(16b) $$\textit{if } F \in \mathcal{H},\ \textit{then } X_0(F) \in \mathcal{H}$$

Define the codistribution $d\mathcal{H}$ *as* $d\mathcal{H}(x) = \mathrm{span}_{\mathbb{R}}\{dF(x) \mid F \in \mathcal{H}\}$, $x \in M$.

A nonlinear control system (12) is called *locally strongly accessible* from $x_0 \in M$ if for any neighborhood V of x_0 and any time $T > 0$, sufficiently small, the set of points reachable from x_0 in exactly time T, using piecewise constant inputs such that the resulting state space trajectories $x(t)$, $t \in [0,T]$, remain in V, has a non-empty interior. If this holds for any $x_0 \in M$ then the system is called locally strongly accessible. A sufficient (and almost necessary) condition for local strong accessibility can be given in terms of the Lie brackets of the vectorfields $X_0, X_{H_1}, \ldots, X_{H_m}$ as follows (see [SJ]). Define the Lie algebra $\mathcal{C}$ as the smallest Lie subalgebra of the Lie algebra of vectorfields on M (under the Lie bracket) satisfying

(17a) $\quad X_{H_1}, \ldots, X_{H_m} \in \mathcal{C}$

(17b) $\quad$ if $X \in \mathcal{C}$, then $[X_0, X] \in \mathcal{C}$

The Lie algebra $\mathcal{C}$ defines a distribution C by $C(x) = \text{span}_{\mathbb{R}}\{X(x) \mid X \in \mathcal{C}\}$. Now the system is locally strongly accessible from x_0 if $\dim C(x_0) = \dim M$.
On the other hand, in the Hamiltonian case it immediately follows from (4) that

(18) $\quad \mathcal{C} = \{X_F \mid F \in \mathcal{H}\}$

and thus, since the Poisson structure is non-degenerate, $\dim d\mathcal{H}(x_0) = \dim M$ implies local strong accessibility from x_0.
A system is said to be *locally observable* at x_0 if there exists a neighborhood W of x_0 such that for every neighborhood $V \subset W$ of x_0 the following holds. If $x_1 \in V$ is such that for any piecewise constant input function $u\colon [0,T] \to \mathbb{R}^m$, with the property that the corresponding state space trajectories $x(t)$, $t \in [0,T]$, for $x(0) = x_0$ and for $x(0) = x_1$ both remain in V, the output functions for initial states x_0 and x_1 coincide on $[0,T]$, then necessarily $x_0 = x_1$. If this holds for any $x_0 \in M$ then the system is called *locally observable*, and if we may take in the above definition $W = M$ for every x_0, then the system is *observable*.
Now in order to distinguish nearby points x_0 and x_1 on the basis of their output functions we may clearly use the functions $H_1, \ldots, H_m$, but also all (repeated) Lie derivatives of $H_1, \ldots, H_m$ along the vectorfields of the system, i.e. all functions of the form

(19) $\quad L_{X_1} L_{X_2} \cdots L_{X_k} H_j, \qquad j = 1, \ldots, m$

with X_i in the set $\{X_0, X_{H_1}, \ldots, X_{H_m}\}$. Then, using (3), it can be readily concluded that the condition $\dim d\mathcal{H}(x_0) = \dim M$ will imply local observability at x_0. Summarizing (see [S2]):

THEOREM 4. *Suppose* $\dim d\mathcal{H}(x_0) = \dim M$, *then the Hamiltonian control system is locally strongly accessible from* x_0 *and locally observable at* x_0. *Furthermore if* $\dim d\mathcal{H}(x) =$ *constant then the system is locally strongly accessible if and only if it is locally observable.*

REMARK. The concept of local strong accessibility is clearly not a very satisfactory notion of controllability. It is an open problem how one can use the Hamiltonian structure to define a stronger controllability notion for which sufficient conditions can de derived similar to the one in Theorem 4.

2. LOCAL DECOMPOSITIONS. Now let assume that $\dim d\mathcal{H}(x) =$ constant and *less* than $\dim M$, so that by Theorem 4 the Hamiltonian control system is not locally observable nor locally strongly accessible. We shall show how we can give a canonical decomposition of the Hamiltonian control system with regard to its "controllable" and "observable" parts. In the linear case this decomposition reduces to the well-known Kalman- decomposition of a linear control system (cf. [Ka]).
First let us recall the notion of a function group (see [W], [S5]).

DEFINITION 5. *Let* M *be a symplectic manifold. A collection* $\mathcal{F}$ *of smooth functions on* M *is called a* function group *if*

(20a) $\quad \mathcal{F}$ *is a Lie algebra under the Poisson bracket,*

(20b) $\quad$ *if* $F_1,\ldots,F_s \in \mathcal{F}$ *and* $G:\mathbb{R}^s \to \mathbb{R}$ *smooth, then* $G(F_1,\ldots,F_s) \in \mathcal{F}$.

Let F_i, $i \in I$ (I index set) be functions on M, then we will denote by span $(F_i;\ i \in I)$ the smallest function group containing these functions. For any function group $\mathcal{F}$ we define its polar group $\mathcal{F}^\perp$ as

$$\mathcal{F}^\perp = \{G \in C^\infty(M) \mid \{G,F\} = 0,\ \forall F \in \mathcal{F}\} \tag{21}$$

Indeed it can be easily checked [W] that $\mathcal{F}^\perp$ is again a function group.
Finally a function group $\mathcal{F}$ defines a codistribution $d\mathcal{F}$ by setting

$$d\mathcal{F}(x) = \operatorname{span}_{\mathbb{R}}\{dF(x) \mid F \in \mathcal{F}\} \tag{22}$$

THEOREM 6 *Let* M *be a symplectic manifold, and let* $\mathcal{F}$ *be a function group on* M. *Assume that* $\dim d\mathcal{F}(x) =$ *constant, and* $\dim d(\mathcal{F} \cap \mathcal{F}^\perp)(x) =$ *constant. Let* $\dim d\mathcal{F} = k$, *and* $\dim d(\mathcal{F} \cap \mathcal{F}^\perp) = r$. *Then locally there exist canonical coordinates* $(q_1,\ldots,q_n,\ p_1,\ldots,p_n)$ for M such that

(23) $\quad \mathcal{F} = \text{span}\,(q_1,\dots,q_\ell,\ q_{\ell+1},\dots,q_{\ell+r},\ p_1,\dots,p_\ell)$

where $2\ell + r = k$.

Proof (see also [W],[S5]). Since $\dim d\mathcal{F} = k$ we can locally find k independent functions $F_1,..,F_k \in \mathcal{F}$ such that $\mathcal{F} = \text{span}\,(F_1,\dots,F_k)$. Since $\mathcal{F}$ is a function group it follows that

(24) $\quad \{F_i,F_j\} = w_{ij}(F_1,\dots,F_k), \qquad i,j = 1,\dots,k,$

for certain smooth functions $w_{ij}\colon \mathbb{R}^k \to \mathbb{R}$. It is immediately checked that these functions w_{ij} define a Poisson structure on $\mathbb{R}^k$, as in (6). Furthermore by the assumption that $\dim d(\mathcal{F} \cap \mathcal{F}^\perp)(x) = \text{constant} = r$ it follows that this Poisson structure on $\mathbb{R}^k$ has constant rank 2ℓ, where $2\ell = k-r$. Thus by Theorem 1 there exist local coordinates $q_1',\dots,q_\ell',\ q_{\ell+1}',\dots,q_{\ell+r}',\ p_1',\dots,p_\ell'$ on $\mathbb{R}^k$ satisfying

(25) $\quad \{p_i',q_j'\} = \delta_{ij},\quad \{q_i',q_j'\} = \{p_i',p_j'\} = 0$

Now define, with $F := (F_1,\dots,F_k)$,

(26) $\quad q_i = q_i' \circ F, \quad i = 1,\dots,\ell+r$

$\quad p_i = p_i' \circ F, \quad i = 1,\dots,\ell$

By (25) $(q_1,..,q_{\ell+r},\ p_1,\dots,p_\ell)$ form a set of partial canonical coordinates on M, which can be extended to a full set of canonical coordinates $(q_1,..,q_n,\ p_1,\dots,p_n)$. Then by construction (23) holds. □

Now let us return to the Lie algebra $\mathcal{H}$ (Definition 3), characterizing controllability and observability (Theorem 4). Let $\bar{\mathcal{H}} = \text{span}\,\mathcal{H}$ be the smallest function group containing $\mathcal{H}$. Since $\mathcal{H}$ is a Lie subalgebra of $C^\infty(M)$ it follows that

(27) $\quad d\bar{\mathcal{H}}(x) = d\mathcal{H}(x), \qquad x \in M$

THEOREM 7. *Consider a Hamiltonian control system (12). Assume that* $\dim d\mathcal{H}(x) = \text{constant} = k$, *and that* $\dim d(\bar{\mathcal{H}} \cap \bar{\mathcal{H}}^\perp)(x) = \text{constant} = r$. *Then locally there exist canonical coordinates*

(28) $\quad (q_1,..,q_\ell,q_{\ell+1},..,q_{\ell+r},q_{\ell+r+1},..,q_n,p_1,..,p_\ell,p_{\ell+1},..,p_{\ell+r},\ p_{\ell+r+1},\dots,p_n)$

$\quad =: (q^1,q^2,q^3,p^1,p^2,p^3)$

where $k = 2\ell + r$, *such that*

(29) $\quad \bar{\mathcal{H}} = \text{span}\,(q^1,q^2,p^1),\ \bar{\mathcal{H}}^\perp = \text{span}\,(q^2,q^3,p^3)$

Let H_0 *be a (locally defined) function such that* $X_0 = X_{H_0}$, *then* H_0 *(the internal Hamiltonian) satisfies*

(30) $\{H_0,\bar{\mathcal{H}}\} \subset \bar{\mathcal{H}}, \qquad \{H_0,\bar{\mathcal{H}}^{\perp}\} \subset \bar{\mathcal{H}}^{\perp}$

which implies that H_0 *is locally of the form*

(31) $H_0(q,p) = H_0^1(q^1,p^1,q^2) + H_0^3(q^3,p^3,q^2) + (p^2)^T f(q^2)$

for a smooth function $f:\mathbb{R}^r \to \mathbb{R}^r$. *Furthermore* $H_i \in \mathcal{H}$, $i = 1,\dots,m$, *only depend on* q^1,q^2,p^1. *Therefore the Hamiltonian control system is locally of the form*

$$(32a)\qquad \begin{cases} \dot{q}^1 = \dfrac{\partial H_0^1}{\partial p^1}(q^1,p^1,q^2) - \displaystyle\sum_{j=1}^m u_j \dfrac{\partial H_j}{\partial p^1}(q^1,p^1,q^2) \\ \dot{p}^1 = -\dfrac{\partial H_0^1}{\partial q^1}(q^1,p^1,q^2) + \displaystyle\sum_{j=1}^m u_j \dfrac{\partial H_j}{\partial q^1}(q^1,p^1,q^2) \end{cases}$$

$$(32b)\qquad \dot{q}^2 = f(q^2)$$

$$(32c)\qquad \dot{p}^2 = -\left(\frac{\partial f}{\partial q^2}(q^2)\right)^T p^2 - \frac{\partial H_0^1}{\partial q^2}(q^1,p^1,q^2) - \frac{\partial H_0^3}{\partial q^2}(q^3,p^3,q^2) + \sum_{j=1}^m u_j \frac{\partial H_j}{\partial q^2}(q^1,p^1,q^2)$$

$$(32d)\qquad \begin{cases} \dot{q}^3 = \dfrac{\partial H_0^3}{\partial p^3}(q^3,p^3,q^2) \\ \dot{p}^3 = -\dfrac{\partial H_0^3}{\partial q^3}(q^3,p^3,q^2) \end{cases}$$

$$(32e)\qquad y_j = H_j(q^1,p^1,q^2) \qquad j = 1,\dots,m$$

Proof. Application of Theorem 6 yields (29). By definition $\{H_0,F\} = X_0(F) \in \bar{\mathcal{H}}$ for any $F \in \bar{\mathcal{H}}$, i.e. $\{H_0,\bar{\mathcal{H}}\} \subset \bar{\mathcal{H}}$. By the Jacobi-identity (2b) we have for any $F \in \bar{\mathcal{H}}$ and $F^{\perp} \in \bar{\mathcal{H}}^{\perp}$

(33) $\{\{H_0,F^{\perp}\},F\} = -\{\{F^{\perp},F\},H_0\} - \{\{F,H_0\},F^{\perp}\} = 0$

since $\{F,H_0\} \in \bar{\mathcal{H}}$. Thus $\{H_0,F^{\perp}\} \in (\bar{\mathcal{H}}^{\perp})^{\perp}$. For a function group $\bar{\mathcal{H}}$ with dim $d\bar{\mathcal{H}}(x)$ = constant it easily follows [S5] that $(\bar{\mathcal{H}}^{\perp})^{\perp} = \bar{\mathcal{H}}$, and thus (30) follows. Writing out (30) in the coordinates (28) yields

$$\{H_0,q^1\} = \frac{\partial H_0}{\partial p^1} \in \bar{\mathcal{H}}, \quad \{H_0,p^1\} = -\frac{\partial H_0}{\partial q^1} \in \bar{\mathcal{H}}$$

$$\{H_0,q^2\} = \frac{\partial H_0}{\partial p^2} \in \bar{\mathcal{H}} \cap \bar{\mathcal{H}}^{\perp} \tag{34}$$

$$\{H_0,q^3\} = \frac{\partial H_0}{\partial p^3} \in \bar{\mathcal{H}}^{\perp}, \ \{H_0,p^3\} = -\frac{\partial H_0}{\partial q^3} \in \bar{\mathcal{H}}$$

and (31) and (32) follow. □

REMARK. It follows (see Theorem 4) that the sub-system (32a,b,e) is *locally observable*. Furthermore for fixed q_2 the sub-systems (1a,c) as well as (1a) are *locally strongly accessible*.

Notice that the input-output behavior of (32) is fully determined by the sub system (32 a,b,e). This sub-system is itself a *Poisson control* system if we endow (q^1,p^1,q^2) with the natural bracket $\{p_i,q_j\} = \delta_{ij}$, $\{q_i,q_j\} = 0 = \{p_i,p_j\}$. (Notice that $(q^1,q^2,q^3,p^1,p^2,p^3) \to (q^1,q^2,p^2)$ is then a *Poisson mapping* [W]). Alternatively, since (1b) yields for any initial state a unique time-function $q_2(t)$, the input-output behavior of (32) is described by the *time-varying Hamiltonian control* system (32a,e). In particular, if the initial state $(\bar{q}^1,\bar{p}^1,\bar{q}^2)$ satisfies $f(\bar{q}^2) = 0$ then the input-output behavior is given by the Hamiltonian control system (32a,e) (with $q^2 = \bar{q}^2$ constant).

In the above theorem we assumed that dim $d\mathcal{H}(x)$ and dim $d(\bar{\mathcal{H}} \cap \bar{\mathcal{H}}^{\perp})(x)$ are *constant* (at least locally). A particular case where we can do *without* these assumptions is if the Lie algebra $\mathcal{H}$ is *finite*-dimensional (see [BG,S6]). Then it is well-known that $\mathcal{H}^*$ (the dual of $\mathcal{H}$) has a naturally defined Poisson structure (cf.[W],[LM]). Indeed let $\varphi_1,\ldots,\varphi_N$ be a basis of $\mathcal{H}$. Since $\mathcal{H}$ is a Lie algebra there exist constants $c_{ijk} \in \mathbb{R}$ such that

$$\{\varphi_i,\varphi_j\} = \sum_{k=1}^{N} c_{ijk}\,\varphi_k \tag{35}$$

Viewing $\varphi_1,\ldots,\varphi_N$ as linear coordinate functions $z_1,\ldots,z_N$ on $\mathcal{H}^*$, we define the following Poisson bracket on $\mathcal{H}^*$

$$\{z_i,z_j\}_{\mathcal{H}^*} = w_{ij}(z_1,\ldots,z_N) := \sum_{k=1}^{N} c_{ijk}\,z_k \tag{36}$$

The time-evolution of $\varphi_1,\ldots,\varphi_N$ along the Hamiltonian system is described as

$$\frac{d\varphi_i}{dt} = X_0(\varphi_i) - \sum_{j=1}^{m} u_j\,\{H_j,\varphi_i\} = \sum_{k=1}^{N} a_{ik}\,\varphi_k - \sum_{j=1}^{m} u_j \sum_{k=1}^{N} b^j_{ik}\,\varphi_k \tag{37}$$

for certain constants a_{ik}, $b^j_{ik} \in \mathbb{R}$. Furthermore, since $H_1,\ldots,H_m \in \mathcal{H}$ it follows that y_j is of the form

$$(38) \qquad y_j = \sum_{k=1}^{N} h_k^j \varphi_k \qquad j = 1,\dots,m$$

Interpreting φ_i, $i = 1,\dots,N$, again as linear coordinate functions z_i on $\mathcal{H}^*$ this yields the following *bilinear* system on $\mathcal{H}^*$

$$(39) \qquad \begin{aligned} \dot{z}_i &= \sum_{k=1}^{N} a_{ik} z_k - \sum_{j=1}^{m} u_j \sum_{k=1}^{N} b_{ik}^j z_k, \qquad & i = 1,\dots,N \\ y_j &= \sum_{k=1}^{N} h_k^j z_k & j = 1,\dots,m \end{aligned}$$

which has the same input-output behavior as the original Hamiltonian control system (cf.[FK]). Since the map $x \xrightarrow{\varphi} (\varphi_1(x),\dots,\varphi_N(x))$ is a Poisson mapping [W] from M to $\mathbb{R}^N$ it immediately follows that (39) is a *Poisson control system*, which since $\varphi_1,\dots,\varphi_N \in \mathcal{H}$, is automatically *observable*. Furthermore the maximal integral manifold of the distribution defined by the strong accessibility algebra $\mathcal{C}$ of (39) (cf.(17)) through a point $z_0 \in \mathcal{H}^*$ is exactly the symplectic leaf of the Poisson structure (see Theorem 1), which in this case is the co-adjoint orbit through z_0 of $\mathcal{H}$ acting on $\mathcal{H}^*$ via its co-adjoint representation. We thus obtain the following specialization of Theorem 7 (compare [S6]).

PROPOSITION 8. *Let (12) be a Hamiltonian control system with finite-dimensional Lie algebra $\mathcal{H}$. Then (39) is a Poisson control system on $\mathcal{H}^*$ with the same input-output behavior. Furthermore assume that the Poisson vectorfield $\dot{z}_i = \sum_{k=1}^{N} a_{ik} z_k$, $i = 1,\dots,N$, is tangent to the co-adjoint orbit $\mathcal{L}(z_0)$ through $z_0 = \varphi(x_0)$ (for example if the Poisson vectorfield is locally Hamiltonian). Then the system (39) can be restricted to an observable and locally strongly accessible Hamiltonian control system on $\mathcal{L}(z_0)$.*

3. CLAMPED DYNAMICS. It is well-known that two sorts of dynamics are of primary importance for control systems. First of all there is the *free dynamics* (or unconstrained modes), which is simply the dynamics for $u = (u_1,\dots,u_m) = 0$. For a Hamiltonian control system (12) this is just the locally Hamiltonian vectorfield $\dot{x} = X_0(x)$. Secondly there is the *clamped dynamics* (or constrained modes), which is dually defined as the systems dynamics *compatible* with the *constraints* $y = (y_1,\dots,y_m) = 0$ (or equal to a constant). (The importance of this last dynamics for control purposes can be already understood by considering the system together with a simple output feedback $u_j = k_j y_j$, $k_j \in \mathbb{R}$, $j = 1,\dots,m$.) Under some extra conditions we will show that for a Hamiltonian control system the clamped dynamics is also

Hamiltonian, and alternatively can be described as a Hamiltonian vectorfield on the state space manifold endowed with a modified Poisson structure.
For simplicity of exposition we will throughout assume that the locally Hamiltonian vectorfield X_0 in (12) is globally Hamiltonian, i.e. $X_0 = X_{H_0}$ for $H_0 \colon M \longrightarrow \mathbb{R}$.
Define inductively for $F, G \in C^\infty(M)$

$$\text{(40)} \qquad ad_F^0\, G = G,\ ad_F^k G = \{F, ad_f^{k-1} G\}, \qquad k \geq 1$$

We define for the Hamiltonian control system integers ρ_i, $i = 1,\dots,m$, as follows. Let ρ_i be the smallest integer ≥ 0 such that

$$\text{(41)} \qquad \exists\ j \in \{1,\dots,m\} \text{ for which } \{H_j,\ ad_{H_0}^{\rho_i} H_i\} \neq 0$$

We assume throughout that $\rho_i < \infty$, $i = 1,\dots,m$. Define the $m \times m$ matrix $A(x)$, $x \in M$, with (i,j)-th element $\{H_j,\ ad_{H_0}^{\rho_i} H_i\}(x)$. Recall that a submanifold N of a symplectic manifold M with symplectic form ω (see (7)) is called a *symplectic submanifold* if $\omega|_N$ is non-degenerate.

THEOREM 9 [S7,S3] *Assume that* rank $A(x) = m$ *for every point* x *in*

$$\text{(42)} \qquad N^* = \{x \in M \mid H_i(x) = ad_{H_0} H_i(x) = \dots = ad_{H_0}^{\rho_i} H_i(x) = 0,\ i = 1,\dots,m\}$$

Then the functions $H_i, \dots, ad_{H_0}^{\rho_i} H_i$, $i = 1,\dots,m$, *are independent on* N^*. *Assume that* N^* *is non-empty. Then* N^* *is a symplectic submanifold of* M, *with codimension* $\sum_{i=1}^{m} (\rho_i + 1)$, *and the clamped dynamics is given as*

$$\text{(43)} \qquad \dot{\bar{x}} = X_{\bar{H}_0}(\bar{x}),\ \bar{x} \in N^*$$

where $X_{\bar{H}_0}$ *is the Hamiltonian vectorfield on* N^* *with regard to the symplectic form* $\bar{\omega} := \omega|_{N^*}$ *and Hamiltonian* $\bar{H}_0 := H|_{N^*}$.

In the terminology of Bergmann and Dirac (cf. [D]) the functions H_i, $i = 1,\dots,m$, are the *primary* constraints and the functions $ad_{H_0}^k H_i$, $k = 1,\dots,\rho_i$, $i = 1,\dots,m$, the *secondary* constraints. The key fact used in the proof of the above theorem is that by definition of ρ_i and the Jacobi-identity (2b)

$$\text{(44)} \qquad \{ad_{H_0}^k H_j,\ ad_{H_0}^{\rho_i - k} H_j\} = (-1)^k \{H_j,\ ad_{H_0}^{\rho_i} H_i\},\ k \leq \rho_i,$$

which implies [S7] that there are no first-class (primary or secondary) constraints. (A constraint is said to be first-class if its Poisson bracket with every other constraint function is zero on N^*.)
The feedback controls $u_j = \alpha_j(x)$, $j = 1,\dots,m$, which have to be applied in order that the system (12) remains to evolve on the constraint manifold N^*

are uniquely given as

$$\begin{pmatrix} \alpha_1 \\ \vdots \\ \alpha_m \end{pmatrix}(x) = A^{-1}(x) \begin{pmatrix} ad_{H_0}^{(\rho_1+1)} H_1 \\ \vdots \\ ad_{H_0}^{(\rho_m+1)} H_m \end{pmatrix}(x) \tag{45}$$

Notice the close similarities of our considerations with the theory of Hamiltonian differential equations with constraints. In this last case the required external forces which are needed in order for the constraints to remain satisfied are interpreted as *reaction forces*, passively originating from physically imposed constraints. In our case the constraints are *actively* imposed using the feedback controls $u_j = \alpha_j(x)$ as given in (45). For a theory of Hamiltonian control systems with additional physical constraints we refer to [Bl,BM,MB]. An important subclass of Hamiltonian control systems for which the assumptions of Theorem 9 are automatically satisfied is formed by what are called, in analogy with [AM], the *simple* Hamiltonian control systems. In this case the state space manifold M is given as a cotangent bundle T^*Q, with Q the configuration manifold, and in natural coordinates $(q,p) = (q_1,..,q_n,p_1,..,p_n)$ for T^*Q (q coordinates for Q) the functions $H_0,H_1,..,H_m$ are of the form

$$\begin{aligned} H_0(q,p) &= \frac{1}{2} \sum_{i,j=1}^{n} g^{ij}(q)p_i p_j \\ H_j(q,p) &= H_j(q) \qquad j = 1,..,m \end{aligned} \tag{46}$$

with the matrix G(q) with (i,j)-th element $g^{ij}(q)$ positive definite for all q. If we assume for simplicity that $H_1,..,H_m$ are independent, then locally we may take $H_j = q_j$, $j = 1,..,m$. It is now easily seen that $\rho_j = 1$, $j = 1,..,m$, and that $-A(x)$ equals the leading m×m submatrix of G(x), which by definition is positive definite.

EXAMPLE. Consider the Hamiltonian control system given in the example above Definition 3.

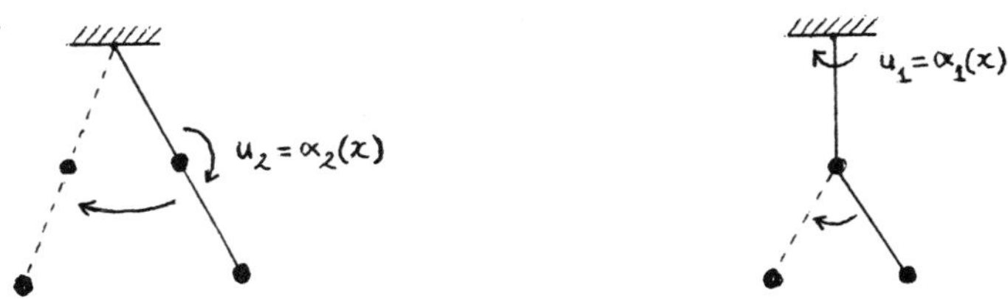

Fig. 2.

If we wish to maintain the constraint $\theta_2 = 0$ with the torque u_2, then the clamped dynamics is as in Fig. 2. Alternatively if we want to maintain the constraint $\theta_1 = 0$ using the torque u_1 then we obtain the second clamped dynamics of Fig. 2. □

An interesting open problem is the relation between the free dynamics $\dot{x} = X_{H_0}(x)$, and the clamped dynamics $\dot{\bar{x}} = X_{\bar{H}_0}(\bar{x})$. For instance for a single-input simple Hamiltonian control system with an internal energy H_0 which is positive around an equilibrium point (as in the above example) it can be readily seen that the eigenvalues of the linearized free and clamped dynamics are both located on the imaginary axis and *intertwining*.

Next question which needs to be addressed is what one can say about the clamped dynamics if the matrix $A(x)$ does *not* have full rank. Under some constant rank assumptions it is still possible to compute a maximal constraint submanifold $N \subset M$ given as the zero set of all primary and secondary constraints. However in general N will not be symplectic, and the feedback $u = \alpha(x)$ which makes N invariant will not be unique anymore.

CONJECTURE. Let $N \subset M$ be the maximal constraint submanifold (if existing). Suppose the set of maps $u = \bar{\alpha}(x)$ such that the vectorfield $X_{H_0}(x) - \sum_{j=1}^{m} \bar{\alpha}_j(x) X_{H_j}(x)$ is tangent to N can be written as a family $u = \alpha(x) + \beta(x)v$ parametrized by a vector $v \in \mathbb{R}^{m'}$, $m' \leq m$, and $\beta(x)$ an $m \times m'$ matrix. Consider the control system on N

$$(47) \qquad \dot{\bar{x}} = \left(X_{H_0}(\bar{x}) - \sum_{j=1}^{m} \alpha_j(\bar{x}) X_{H_j}(\bar{x})\right) - \sum_{j=1}^{m'} \Big(\sum_{i=1}^{m} X_{H_i}(\bar{x}) \beta_{ij}(\bar{x}) \Big) v_j \quad , \quad \bar{x} \in N,$$

with controls $v_1, \ldots, v_{m'}$, and with strong accessibility algebra $\bar{\mathcal{C}}$, cf. (17). Suppose that the distribution $\bar{C}$ generated by $\bar{\mathcal{C}}$ has constant dimension, so that at least locally we can factor out by this distribution to obtain a new manifold $\tilde{N}$, with the projection $\pi : N \to \tilde{N}$ satisfying $\ker \pi_* = \bar{C}$, and the control system (47) projects under π to a single vectorfield $\dot{\tilde{x}} = \tilde{f}(\tilde{x})$ on $\tilde{N}$. Then $\tilde{N}$ will be a symplectic manifold with symplectic form $\tilde{\omega}$ satisfying $\pi^* \tilde{\omega} = \omega|_N$, and $\dot{\tilde{x}} = \tilde{f}(\tilde{x})$ will be a Hamiltonian vectorfield on $(\tilde{N}, \tilde{\omega})$ with Hamiltonian $\tilde{H}$ satisfying $\tilde{H}(\pi(\bar{x})) = H_0(\bar{x}) - \sum_{j=1}^{m} \alpha_j(\bar{x}) H_j(\bar{x})$, $\bar{x} \in N$.

REMARK 1 For *linear* Hamiltonian systems this conjecture was proved in [S3].

REMARK 2 Notice that if the above conjecture is true then the existence of first class (primary or secondary) constraints has to imply the existence of first class *primary* constraints.

Now let us return to the case that the assumptions of Theorem 9 are satisfied. We will even strengthen the assumptions by requiring that $A(x)$ has rank m *everywhere*. The clamped dynamics (43) can now alternatively be formulated using the *Dirac bracket* (compare with [MB]). Denote the constraint functions H_i, $ad_{H_0}H_i, \ldots, ad_{H_0}^{\rho_i}H_i$, $i = 1,\ldots,m$ (which by Theorem 9 are everywhere independent) as $\psi_j(x)$, $j = 1,\ldots,m^* := \sum_{i=1}^{m}(\rho_i+1)$. It follows from Theorem 9 that the $m^*\times m^*$ matrix $C(x)$ with (i,j)-th element $c_{ij}(x) = \{\psi_i,\psi_j\}(x)$ has rank m^*. Hence we can define on M the Dirac bracket (see for instance [DLT])

$$(48) \qquad \{F,G\}^* = \{F,G\} - \sum_{i,j=1}^{m}\{F,\psi_i\}c^{ij}\{\psi_j,G\} \qquad F,G \in C^\infty(M)$$

where $c^{ij}(x)$ denotes the (i,j)-th element of the inverse matrix $C^{-1}(x)$. It can be proved that the Dirac bracket defines a modified Poisson structure on M (i.e. satisfies (2)). Notice that the Hamiltonian vectorfields $X^*_{\psi_k}$ with respect to the Dirac bracket for a constraint function ψ_k are automatically zero. Indeed for any $G \in C^\infty(M)$

$$(49) \qquad \{\psi_k,G\}^* = \{\psi_k,G\} - \sum_{i,j}\{\psi_k,\psi_i\}c^{ij}\{\psi_j,G\} =$$

$$= \{\psi_k,G\} - \sum_{i,j}c_{ki}c^{ij}\{\psi_j,G\} = \{\psi_k,G\} - \{\psi_k,G\} = 0$$

In particular the vectorfields $X^*_{H_j}$, $j = 1,\ldots,m$, are zero. It follows that the rank of the Dirac bracket is everywhere $2n-m^*$, and that the symplectic leaf through any point of N^* is N^* itself.
If we now define the Hamiltonian vectorfield (with respect to the Dirac bracket)

$$(50) \qquad \dot{x} = X^*_{H_0}(x) \quad , \quad x \in M$$

which by construction is tangent to N^*, then restricted to N^* this vectorfield coincides with the clamped dynamics (43).

EXAMPLE Consider a simple Hamiltonian system (see (46)), with $H_j(q) = q_j$, $j= 1,\ldots,m$. Then $\rho_i = 1$, $i = 1,\ldots,m$, and $N^* = \{(q,p) \mid q_1 = \ldots = q_m = \sum_{j=1}^{n}g^{1j}p_j = \ldots = \sum_{j=1}^{m}g^{mj}p_j = 0\}$. It follows that

$$(51)\qquad C(q,p) = \begin{bmatrix} 0 & -G_{11}(q) \\ G_{11}(q) & S(q,p) \end{bmatrix}$$

with $G_{11}(q)$ the m×m matrix with (i,j)-th element $g^{ij}(q)$, and S the m×m matrix with (i,j)-th element

$$(52)\qquad \sum_k \Big(\sum_s \Big(g^{is} \frac{\partial g^{jk}}{\partial q_s} - g^{js} \frac{\partial g^{ik}}{\partial q_s}\Big) p_k\Big)$$

Clearly

$$(53)\qquad C^{-1}(q,p) = \begin{bmatrix} G_{11}^{-1} S G_{11}^{-1} & G_{11}^{-1} \\ -G_{11}^{-1} & 0 \end{bmatrix}$$

and the expression for the Dirac bracket follows. □

Let us finally consider the problem of *invertibility* for a Hamiltonian control system (12), i.e. we wish to reconstruct from the observed output functions $y_1(t),..,y_m(t)$, $t \in [0,T]$, $T > 0$, the input functions $u_1(t),..,u_m(t)$, $t \in [0,T]$ which give rise to these outputs (for some fixed initial state), or alternatively we want to compute input functions $u_1(\cdot),..,u_m(\cdot)$, such that certain desired output functions $y_1(\cdot),..,y_m(\cdot)$ result. Using the definition of ρ_i it can be easily checked that the k-th time-derivative of $y_j = H_j(x)$ along the Hamiltonian control system for $k \le \rho_j$ equals

$$(54)\qquad y_j^{(k)} = ad_{H_0}^k H_j(x) \quad , \quad k = 0,1,..,\rho_j \quad , \quad j = 1,..,m$$

while

$$(55)\qquad \begin{bmatrix} y_1^{(\rho_1+1)} \\ \vdots \\ y_m^{(\rho_m+1)} \end{bmatrix} = \begin{bmatrix} ad_{H_0}^{(\rho_1+1)} H_1 \\ \vdots \\ ad_{H_0}^{(\rho_m+1)} H_m \end{bmatrix} (x) - A(x)u$$

Furthermore by Theorem 9 the constraint functions $z_{jk} = ad_{H_0}^k H_j(x)$, $k = 0,1,..,\rho_j$, $j = 1,..,m$, are all independent. By Theorem 9 we can take canonical coordinates (q^*,p^*) for N^* such that (q^*,p^*,z), with $z = (z_{jk}\ ;\ k = 0,1,..,\rho_j,\ j = 1,..,m)$, are local coordinates for M. Then the Hamiltonian vectorfield with respect to the Dirac bracket and the time-dependent Hamiltonian $H_0(q^*,p^*,z) - \sum_{j=1}^m u_j(t)H_j(q^*,p^*,z)$ is simply given as

$$
(56) \qquad \begin{aligned}
\dot{z} &= 0 \\
\dot{q}^* &= \frac{\partial H_0}{\partial p^*}(q^*,p^*,z) \\
\dot{p}^* &= -\frac{\partial H_0}{\partial q^*}(q^*,p^*,z)
\end{aligned}
$$

By invertibility of $A(x)$ it follows from (47) that (compare (45))

$$
(57) \qquad \begin{bmatrix} u_1 \\ \vdots \\ u_m \end{bmatrix} = A^{-1}(q^*,p^*,z) \begin{bmatrix} ad_{H_0}^{(\rho_1+1)}H_1 \\ \vdots \\ ad_{H_0}^{(\rho_m+1)}H_m \end{bmatrix} (q^*,p^*,z) - A^{-1}(q^*,p^*,z) \begin{bmatrix} y_1^{(\rho_1+1)} \\ \vdots \\ y_m^{(\rho_m+1)} \end{bmatrix}
$$

Since by (54) the components of z consist of the time-derivatives of the outputs, it follows from (57) that the inputs can be determined on the basis of the output functions $y_1,..,y_m$, together with the knowledge of the dynamics (56), called the *inverse system*. Notice that we can interpret (56) also as a Hamiltonian control system (in general form (13)) living on N^*, with *inputs* z, which for $z = 0$ reduces to the clamped dynamics (43). Similar considerations were used in [HS] for the investigation of input-output decoupling with stability for Hamiltonian control systems, and in [S4] for the problem of singular optimal control. Finally we remark that the notion of clamped dynamics naturally arises in *interconnecting* Hamiltonian control systems (see also [MB]).

REFERENCES

[AM] R.A. Abraham & J.E. Marsden, Foundations of Mechanics (2nd edition), Benjamin/Cummings, Reading Mass., 1978.

[B] R.W. Brockett, "Control theory and analytical mechanics", in Geometric Control Theory (eds. C. Martin & R. Hermann), Vol VII of Lie Groups: History, Frontiers and Applications, Math Sci Press, Brookline, 1977, 1-46.

[BG] J. Basto Goncalves, "Realization theory for Hamiltonian systems", SIAM J. Control and Opt., 25, 1987, 63-73.

[Bl] A.M. Bloch, "Stabilizability of nonholonomic control systems", preprint 1988.

[BM] A.M. Bloch & N.H. McClamroch, "Stabilization of Hamiltonian systems with constraints", in Analysis and Control of Nonlinear Systems (eds. C.I. Byrnes, C.F. Martin, R.E. Saeks), North-Holland, Amsterdam, 1988, 385-392.

[D] P. Dirac, Lectures on Quantum Mechanics, Belfer Graduate School of Science Monographs No. 3, Yeshiva University, 1964.

[DLT] P. Deift, F. Lund, E. Trubowitz, "Nonlinear wave equations and constrained harmonic motion", Commun. Math. Phys. 74, 1980, 141-188.

[FK] M. Fliess & I. Kupka, "A finiteness criterion for nonlinear input-output differential systems", SIAM J. Contr. and Opt. 21, 1983, 721-728.

[HS] H.J.C. Huijberts & A.J. van der Schaft, "Input-output decoupling with stability for Hamiltonian systems", University of Twente, Dept. Applied Math., Memo 722, 1988.

[K] P.S. Krishnaprasad, "Lie-Poisson structures, dual-spin spacecraft and asymptotic stability", Nonlinear Anal. Theor. Meth. and Appl., 7, 1984, 1011-1035.

[Ka] R.E. Kalman, "Mathematical description of linear dynamical system", SIAM J. Control, 1, 1963, 152-192.

[KM] P.S. Krishnaprasad & J.E. Marsden, "Hamiltonian structures and stability for rigid bodies with flexible attachments", Arch. Rat. Mech. & Anal., 98, 1987, 71-93.

[LM] P. Libermann & C.-M. Marle, Symplectic Geometry and Analytical Mechanics, D. Reidel, Dordrecht, 1987.

[MB] N.H. McClamroch & A.M. Bloch, "Control of constrained Hamiltonian systems and applications to control of constrained robots", in Dynamical Systems Approaches to Nonlinear Problems in Systems and Circuits (eds. F.M.A. Salam, M.L. Levi), SIAM, 1988, 394-403.

[S1] A.J. van der Schaft, "Hamiltonian dynamics with external forces and observations", Mathematical Systems Theory, 15, 1982, 145-168.

[S2] A.J. van der Schaft, System theoretic descriptions of physical systems, CWI Tract No. 3, Centrum voor Wiskunde en Informatica, Amsterdam, 1984.

[S3] A.J. van der Schaft, "On feedback control of Hamiltonian systems" in Theory and Applications of Nonlinear Control Systems (eds. C.I. Byrnes, A. Lindquist), North-Holland, Amsterdam, 1986, 273-290.

[S4] A.J. van der Schaft, "Optimal control and Hamiltonian input-output systems", in Algebraic and Geometric Methods in Nonlinear Control Theory (eds. M. Fliess, M. Hazewinkel), D. Reidel, Dordrecht, 1986, 389-407.

[S5] A.J. van der Schaft, "Controlled invariance for Hamiltonian systems", Math. Systems Theory, 18, 1985, 257-291.

[S6] A.J. van der Schaft, "Hamiltonian and quantum mechanical control systems", in Information Complexity and Control in Quantum Physics (eds. A. Blaquière, S. Diner, G. Lochak), CISM Courses and Lect. No. 294, Springer, Wien, 1987, 277-296.

[S7] A.J. van der Schaft, "Equations of motion for Hamiltonian systems with constraints", J. Phys. A : Math. Gen, 20, 1987, 3271-3277.

[SA] G. Sanchez de Alvarez, Geometric Methods of Classical Mechanics applied to Control Theory, Ph.D. Thesis, Dept. Mathematics, Univ. of California, Berkeley, 1986.

[SJ] H.J. Sussmann & V. Jurdjevic, "Controllability of nonlinear systems", J. Diff. Eqns., 12, 1972, 95-116.

[W] A. Weinstein, "The local structure of Poisson manifolds", J. Differential Geometry, 18, 1983, 523-557.

DEPARTMENT OF APPLIED MATHEMATICS
UNIVERSITY OF TWENTE
P.O. BOX 217, 7500 AE ENSCHEDE
THE NETHERLANDS

Contemporary Mathematics
Volume 97, 1989

GRAPH-THEORETICAL METHODS IN MULTIBODY DYNAMICS

Jens Wittenburg

ABSTRACT: The incidence matrix and the path matrix of a directed system graph are used for establishing various expressions in the kinematics and dynamics of linear or nonlinear multibody systems. An application to vibration theory yields a method for eliminating a zero eigenvalue from differential equations of motion. Loop closure conditions are formulated for systems with closed kinematical loops, and the notion of reduced mass is generalized from two-body systems to n-body systems.

1. INTRODUCTION

A multibody system is an assembly of rigid or deformable bodies labeled 1 to n which is moving relative to a reference body 0. The motion of body 0 is assumed to be given as function of time. Body 0 is either a material body or a reference frame. Bodies 0 to n are interconnected by two types of elements called joints and force elements. Joints are purely kinematical devices which restrict the number of degrees of freedom of one body relative to its contiguous body by means of kinematical constraints. Forces in joints are constraint forces which have no contribution to virtual work. All kinematical constraints existing between two bodies are counted as one single joint. The term joint has therefore a very general meaning. To give an example: A complicated spatial mechanism interconnecting two bodies represents a single joint if the mechanism is considered as massless.

In contrast to joints force elements do not influence the number of degrees of freedom. They produce forces which are determined either by the location and velocity of the system (spring and damper forces, control forces) or as functions of time. In contrast to joints there can be any number of force elements between two bodies.

The formulation of various kinematical quantities for multibody systems as well as of nonlinear differential equations for large motions is of great importance for industrial applications. It is interesting also as a subject of theoretical

0271-4132/89 $1.00 + $.25 per page

mechanics. For both aspects the reader is referred to [1,2,3]. In the present paper some new relationships and methods are described.

2. PARAMETERS OF SYSTEM TOPOLOGY

In classical mechanics either a single rigid body or very special systems of two or three bodies were investigated. It was therefore unnecessary to introduce mathematical parameters which describe the topology, i.e. the interconnection structure of a system. In the mechanics of general multibody systems such parameters play a dominant role. These parameters are graph-theoretical quantities. A system of bodies 0 to n with interconnections 1 to m is represented by a graph. The bodies are represented by the vertices and the interconnections by the arcs. For each multibody system there is one graph associated with the interconnection by joints and another with the interconnection by force elements. The definitions and relationships discussed in this section are valid for both graphs.

The order in which the vertices and the arcs are labeled is arbitrary. The arcs are given a sense of direction which is indicated by an arrow (Fig.1). The labels of the two vertices interconnected by (or incident with) arc j (j=1...m) are called $i^+(j)$ and $i^-(j)$, respectively, such that the arrow is pointing away from vertex $i^+(j)$. The integer functions $i^+(j)$ and $i^-(j)$ for j=1...m are used

Fig.1

for defining the (n×m) incidence matrix $\underline{S}$ with elements

$$S_{ij}=\begin{cases} +1 & \text{if } i=i^+(j) \\ -1 & \text{if } i=i^-(j) \\ 0 & \text{else} \end{cases} \qquad i=1\ldots n,\ j=1\ldots m. \tag{1}$$

The joint graph has tree structure if m=n. In the case m>n it has closed loops. In this case a "spanning tree" is constructed from the graph by deleting m-n arcs (Fig.2). There exists more than one spanning tree for a given graph. One of them is chosen arbitrarily. Without loss of generality it is assumed that the deleted arcs have the labels n+1...m. For the graph of Fig.2 the incidence matrix $\underline{S}$ reads

Fig.2: Directed system graph with closed loops (arcs 1 to 9) and a spanning tree (arcs 1 to 5)

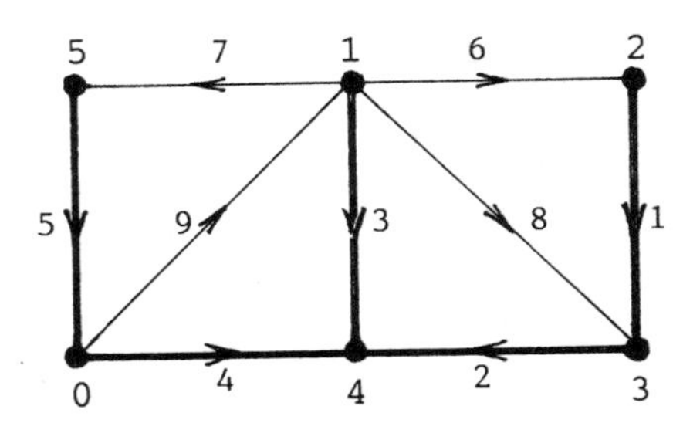

$$\underline{S} = \begin{array}{c} \\ 1 \\ 2 \\ 3 \\ 4 \\ 5 \end{array} \begin{array}{c} \begin{array}{ccccc|cccc} 1 & 2 & 3 & 4 & 5 & 6 & 7 & 8 & 9 \end{array} \\ \left[\begin{array}{ccccc|cccc} 0 & 0 & 1 & 0 & 0 & 1 & 1 & 1 & -1 \\ 1 & 0 & 0 & 0 & 0 & -1 & 0 & 0 & 0 \\ -1 & 1 & 0 & 0 & 0 & 0 & 0 & -1 & 0 \\ 0 & -1 & -1 & -1 & 0 & 0 & 0 & 0 & 0 \\ 0 & 0 & 0 & 0 & 1 & 0 & -1 & 0 & 0 \end{array}\right] \end{array} \qquad \underline{T} = \begin{array}{c} \\ 1 \\ 2 \\ 3 \\ 4 \\ 5 \end{array} \begin{array}{c} \begin{array}{ccccc} 1 & 2 & 3 & 4 & 5 \end{array} \\ \left[\begin{array}{ccccc} 0 & 1 & 0 & 0 & 0 \\ 0 & 1 & 1 & 0 & 0 \\ 1 & 0 & 0 & 0 & 0 \\ -1 & -1 & -1 & -1 & 0 \\ 0 & 0 & 0 & 0 & 1 \end{array}\right] \end{array}$$

The incidence matrix for the spanning tree is the submatrix of columns 1 to 5. The submatrix of $\underline{S}$ for a spanning tree will be called $\underline{S}^*$. For the spanning tree the (n×n) path matrix $\underline{T}$ is defined. Its elements are

$$T_{ji} = \begin{cases} +1 & \text{arc } j \text{ is on the path between vertices 0 and } i \text{ and directed toward 0} \\ -1 & \text{arc } j \text{ is on the path between vertices 0 and } i \text{ and directed toward } i \\ 0 & \text{else} \qquad i,j=1\ldots n. \end{cases} \tag{2}$$

The path matrix for the spanning tree in Fig.2 is shown above to the right of the matrix $\underline{S}$.

THEOREM 1: $\underline{TS}^*$ is the unit matrix.

THEOREM 2: $\underline{TS}$ is the so-called cutset matrix of the spanning tree.
It has the following properties. By theorem 1 the columns 1...n form the unit matrix. With each column k=n+1...m one closed loop is associated. It is formed by restoring the previously deleted arc k. Only for the arcs j of this loop the elements $(\underline{TS})_{jk}$ are nonzero. They are -1 if arc j and arc k have the same sense of direction in the loop and +1 if arc j and arc k have opposite directions.
The proofs of theorems 1 and 2 are simple (see [5,1]). For Fig.2 one finds

$$\underline{TS} = \begin{array}{c} \\ 1 \\ 2 \\ 3 \\ 4 \\ 5 \end{array} \begin{array}{c} \begin{array}{ccccc|cccc} 1 & 2 & 3 & 4 & 5 & 6 & 7 & 8 & 9 \end{array} \\ \left[\begin{array}{c|cccc} & -1 & 0 & 0 & 0 \\ & -1 & 0 & -1 & 0 \\ \text{unit matrix} & 1 & 1 & 1 & -1 \\ & 0 & -1 & 0 & 1 \\ & 0 & -1 & 0 & 0 \end{array}\right] \end{array}$$

3. THE MATRICES $\underline{S}$, $\underline{S}^*$ AND $\underline{T}$ IN VIBRATION THEORY

In Fig.3 body 0 is inertial space. Bodies 0...n are interconnected by springs 1...m with spring constants k_i($i=1...m$). The bodies 1...n move along the horizontal. For body i ($i=1...n$) the location x_i in inertial space is used as variable with $x_i=0$ ($i=1...n$) in the position when all springs are unstressed. All coordinates x_i have the same sense of direction. Define $x_o=0$.

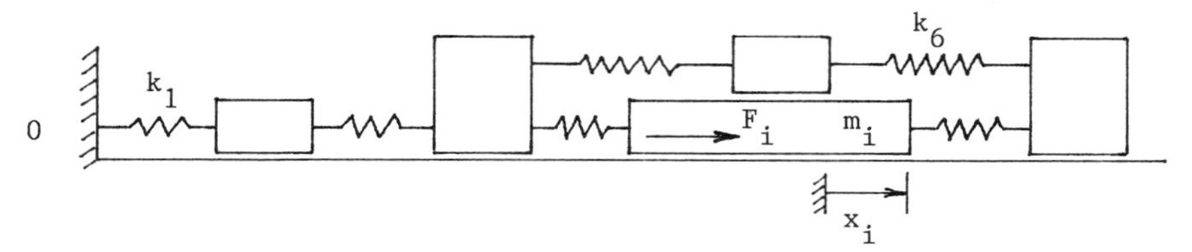

Fig.3: Vibration system

Let $\underline{S}$ be the incidence matrix associated with the graph of spring interconnections. The force exerted by spring j is $K_j=k_j(x_{i^-(j)}-x_{i^+(j)})$. The resultant of all spring forces acting on body i is

$$\sum_{j=1}^{m} S_{ij}K_j = \sum_{j=1}^{m} S_{ij}k_j(x_{i^-(j)}-x_{i^+(j)}) \qquad i=1\ldots n. \tag{3}$$

Because of (1) the relationship holds

$$x_{i^-(j)}-x_{i^+(j)} = -\sum_{k=1}^{n} S_{kj}x_k \qquad j=1\ldots m. \tag{4}$$

With this Newton's law for body i reads

$$m_i\ddot{x}_i = F_i - \sum_{j=1}^{m}\sum_{k=1}^{n} S_{ij}k_jS_{kj}x_k \qquad i=1\ldots n. \tag{5}$$

In matrix form all n equations are combined as

$$\underline{m}\ddot{\underline{x}}+\underline{SkS}^T\underline{x}=\underline{F} \tag{6}$$

with diagonal matrices $\underline{m}$ and $\underline{k}$. The exponent T indicates transposition. These equations are used if there are springs connecting the system to body 0. If there are no such springs then the characteristic equation of (6) has an eigenvalue zero which expresses the fact that for $\underline{F}=\underline{0}$ the whole system can move with constant velocities $\dot{x}_i \equiv v=\text{const}$ ($i=1...n$). In order to construct a system of differential equations without eigenvalue zero we introduce as variables n independent relative displacements, i.e. spring elongations

$$z_j=x_{i^-(j)}-x_{i^+(j)} \qquad j=1\ldots n. \tag{7}$$

In a system with n vertices and with tree structure there are just n such variables. If the system has closed loops as is the case in Fig.3 we select a spanning tree and use the variables $z_1 \ldots z_n$ associated with its arcs as independent variables. Let $\underline{S}^*$ and $\underline{T}$ be the incidence matrix and the path matrix, respectively, of the spanning tree. (7) can then be written

$$z_j = -\sum_{k=1}^{n} S^*_{kj} x_k \qquad j=1\ldots n \quad \text{or in matrix form} \quad \underline{z} = -\underline{S}^{*T}\underline{x}. \tag{8}$$

From theorem 1 follows $\underline{x} = -\underline{T}^T\underline{z}$. (9)

With (8) and (9) the differential equations (6) take the form

$$\underline{mT}^T\ddot{\underline{z}} + \underline{Sk}(\underline{TS})^T z = -\underline{F}$$

and after premultiplication with $\underline{m}^{-1}$ and with $\underline{S}^{*T}$

$$\ddot{\underline{z}} + \underline{S}^{*T}\underline{m}^{-1}\underline{Sk}(\underline{TS})^T\underline{z} = -\underline{S}^{*T}\underline{m}^{-1}\underline{F}. \tag{10}$$

The product $\underline{TS}$ is the cutset matrix appearing in theorem 2. In the special case of a system without closed loops $\underline{T}$ is the inverse of $\underline{S}$ and, furthermore, $\underline{S}^*$ is identical with $\underline{S}$ so that (10) is reduced to

$$\ddot{\underline{z}} + \underline{S}^T\underline{m}^{-1}\underline{Skz} = -\underline{S}^T\underline{m}^{-1}\underline{F}. \tag{11}$$

Independent of whether the system has or does not have closed loops the equations contain a single equation which is uncoupled from the remaining n-1 equations and these latter ones have no eigenvalue zero if the bodies 1...n are all interconnected by springs. In the case of Fig.3 with $k_1=0$, for example, the column matrix $\underline{k}(\underline{TS})^T\underline{z}$ does not contain z_1 so that the equation for z_1 is uncoupled from the remaining equations. For the special case of a system of two masses m_1 and m_2 with one spring interconnecting the two the single remaining equation reads $\ddot{z}_2 + k_2(m_1+m_2)/(m_1m_2)z_2 = F_2/m_2 - F_1/m_1$. The expression $m_1m_2/(m_1+m_2)$ is known as the reduced mass of the two- body system. In Sec.5 a matrix of reduced masses will be developed for n-body systems.

4. ARBITRARY RIGID-BODY SYSTEMS.

For a multibody system with n rigid bodies i=1...n the principle of virtual power reads (see [1,2])

$$\sum_{i=1}^{n}\left[\delta\dot{\vec{r}}_i\cdot(m_i\ddot{\vec{r}}_i - \vec{F}_i) + \delta\vec{\omega}_i\cdot(J_i\cdot\dot{\vec{\omega}}_i - \vec{M}^*_i)\right] = 0 \tag{12}$$

(mass m_i, inertia tensor J_i, absolute translational acceleration $\ddot{\vec{r}}_i$ of the body center of mass, absolute angular acceleration $\dot{\vec{\omega}}_i$, resultant force $\vec{F}_i$ on body i, $\vec{M}^*_i = \vec{M}_i - \vec{\omega}_i \times J_i\cdot\vec{\omega}_i$, resultant torque $\vec{M}_i$ about the body i center of mass; $\vec{F}_i$ and $\vec{M}_i$

contain external as well as internal forces resulting from force elements).

For systems not kinematically constrained to inertial space the radius vector $\vec{r}_o$ of the composite system center of mass and the radius vector $\vec{R}_i$ from there to the body i center of mass is introduced, so that $\vec{r}_i=\vec{r}_o+\vec{R}_i$ for i=1...n. With this (12) yields Newton's law for the composite system center of mass, $\ddot{\vec{r}}_o=(1/m_{total})\sum_{j=1}^{n} F_j$, and in addition

$$\sum_{i=1}^{n}\left[\delta\dot{\vec{R}}_i\cdot(m_i\ddot{\vec{R}}_i-\vec{F}_i)+\delta\vec{\omega}_i\cdot(J_i\cdot\dot{\vec{\omega}}_i-\vec{M}_i^*)\right]=0. \tag{13}$$

(12) and (13) are written in the matrix forms

$$\delta\dot{\underline{\vec{r}}}^T\cdot(\underline{m}\ddot{\underline{\vec{r}}}-\underline{\vec{F}})+\delta\underline{\vec{\omega}}^T\cdot(\underline{J}\cdot\dot{\underline{\vec{\omega}}}-\underline{\vec{M}}^*)=0 \quad \text{and} \quad \delta\dot{\underline{\vec{R}}}^T\cdot(\underline{m}\ddot{\underline{\vec{R}}}-\underline{\vec{F}})+\delta\underline{\vec{\omega}}^T\cdot(\underline{J}\cdot\dot{\underline{\vec{\omega}}}-\underline{\vec{M}}^*)=0 \tag{14a,b}$$

with obvious definitions of the matrices (see [1,2]). Between $\underline{\vec{r}}$ and $\underline{\vec{R}}$ there exists the relationship

$$\underline{\vec{R}}=\underline{\mu}^T\underline{\vec{r}} \tag{15}$$

with the dimensionless (n×n) matrix $\underline{\mu}$ with elements $\mu_{ij}=\delta_{ij}-m_i/m_{total}$ (i,j=1...n; Kronecker δ_{ij}).

In [1,2] it was shown how $\delta\dot{\underline{\vec{r}}}$, $\ddot{\underline{\vec{r}}}$, $\delta\underline{\vec{\omega}}$ and $\dot{\underline{\vec{\omega}}}$ can be expressed in terms of a minimal set of generalized joint coordinates $\underline{q}$:

$$\delta\dot{\underline{\vec{r}}}=\underline{\vec{a}}_1\delta\dot{\underline{q}}, \quad \ddot{\underline{\vec{r}}}=\underline{\vec{a}}_1\ddot{\underline{q}}+\underline{\vec{b}}_1, \quad \delta\underline{\vec{\omega}}=\underline{\vec{a}}_2\delta\dot{\underline{q}}, \quad \dot{\underline{\vec{\omega}}}=\underline{\vec{a}}_2\ddot{\underline{q}}+\underline{\vec{b}}_2. \tag{16}$$

For the matrices $\underline{\vec{a}}_1$, $\underline{\vec{b}}_1$, $\underline{\vec{a}}_2$ and $\underline{\vec{b}}_2$ explicit expressions are available if the joint graph for the multibody system has tree structure. For the present discussion it suffices to explain the role of the path matrix $\underline{T}$ in the expressions for $\underline{\vec{a}}_1$ and $\underline{\vec{a}}_2$. Let $\vec{\Omega}_j$ be the angular velocity of body $i^-(j)$ relative to body $i^+(j)$. $\vec{\Omega}_j$ is a linear combination of the first time derivatives of generalized coordinates chosen for joint j. The absolute velocity $\vec{\omega}_i$ of body i (i=1...n) is the sum (or difference depending on the sense of direction of the arcs in the system graph) of the $\vec{\Omega}_j$ for all arcs j along the path from vertex 0 to vertex i. Hence with (2)

$$\vec{\omega}_i=-\sum_{j=1}^{n}T_{ji}\vec{\Omega}_j \quad i=1...n \quad \text{or in matrix form} \quad \underline{\vec{\omega}}=-\underline{T}^T\underline{\vec{\Omega}}. \tag{17}$$

With each $\vec{\Omega}_j$ being a linear combination of generalized joint velocities we can write $\underline{\vec{\Omega}}=\underline{\vec{p}}^T\dot{\underline{q}}$ where the matrix $\underline{\vec{p}}^T$ is composed of the coefficient vectors (for example unit vectors along rotation axes if rotation angles are used as joint variables). With this (17) becomes

$$\underline{\vec{\omega}}=-(\underline{\vec{p}T})^T\dot{\underline{q}} \tag{18}$$

whence follows for (16)

$$\underline{\vec{a}}_2=-(\underline{\vec{p}T})^T. \tag{19}$$

An analogous formula is found for the matrix $\underline{\vec{a}}_1$. For the sake of simplicity we consider a multibody system in which none of the bodies rotates. In analogy to $\vec{\Omega}_j$ in the previous discussion let $\vec{v}_j$ be the velocity of body $i^-(j)$ relative to body $i^+(j)$. The absolute velocity of body i is

$$\dot{\vec{r}}_i=-\sum_{j=1}^{n} T_{ji}\vec{v}_j \qquad i=1...n. \tag{20}$$

This corresponds to (17). It follows that

$$\underline{\vec{a}}_1=-(\underline{\vec{k}T})^T \tag{21}$$

where $\underline{\vec{k}}$ is a matrix of coefficient vectors of joint velocities (for example axial unit vectors if cartesian coordinates are used as joint variables).

The expressions of $\underline{\vec{a}}_1$ for more general cases and of $\underline{\vec{b}}_1$ and $\underline{\vec{b}}_2$ are similarly simple. Each matrix is a sum of products of other matrices whose elements have simple physical interpretations. On the basis of such formulations it is simple to write a general-purpose computer program for the kinematics and dynamics of arbitrary multibody systems. MESA VERDE (MEchanism, SAtellite, VEhicle and Robot Dynamics Equations) which was developed at Karlsruhe is such a program. It is extensively used in various branches of industry.

Substitution of (16) into (14a) yields the minimal set of differential equations for the variables $\underline{q}$ in the form

$$\underline{A}\ddot{\underline{q}}=\underline{B} \qquad \text{with} \qquad \underline{A}=\underline{\vec{a}}_1^T\cdot\underline{m}\underline{\vec{a}}_1+\underline{\vec{a}}_2^T\cdot\underline{J}\cdot\underline{\vec{a}}_2, \qquad \underline{B}=\underline{\vec{a}}_1^T\cdot(\underline{\vec{F}}-\underline{m}\underline{\vec{b}}_1)+\underline{\vec{a}}_2^T\cdot(\underline{\vec{M}}^*-\underline{J}\cdot\underline{\vec{b}}_2). \tag{22}$$

These equations result from

$$\delta\dot{\underline{q}}^T(\underline{A}\ddot{\underline{q}}-\underline{B})=0. \tag{23}$$

If a system has closed kinematical loops then there exists an independent subset $\ddot{\underline{q}}^*$ of $\ddot{\underline{q}}$ such that

$$\ddot{\underline{q}}=\underline{G}\ddot{\underline{q}}^*+\underline{H}, \qquad \dot{\underline{q}}=\underline{G}\dot{\underline{q}}^*, \qquad \delta\dot{\underline{q}}=\underline{G}\delta\dot{\underline{q}}^* \tag{24}$$

with scalar matrices $\underline{G}$ and $\underline{H}$. Substitution into (23) yields the minimal set of differential equations

$$\underline{A}^*\ddot{\underline{q}}^*=\underline{B}^* \quad \text{with} \quad \underline{A}^*=\underline{G}^T\underline{A}\underline{G}, \quad \underline{B}^*=\underline{G}^T(\underline{B}-\underline{A}\underline{H}). \tag{25}$$

In contrast to the matrices $\underline{A}$ and $\underline{B}$ no simple closed-form expressions exist for the matrices $\underline{G}$ and $\underline{H}$. Industrial users of a general-purpose computer program are not willing to develop expressions for $\underline{G}$ and $\underline{H}$ for every closed loop individually. The following automatic procedure solves this problem in a way which does not require any kinematics analysis of closed loops. In each closed kinematical loop one arbitrarily chosen body is replaced by two identical twin bodies of half of the original mass. The twin bodies are connected to the system as is shown in Fig.4. For the resulting tree-structured system the matrices $\underline{\vec{a}}_1$, $\underline{\vec{b}}_1$, $\underline{\vec{a}}_2$ and $\underline{\vec{b}}_2$ of (16) are constructed. If i and j are the labels of two twin bodies then the loop closure conditions require that $\ddot{\vec{r}}_i\equiv\ddot{\vec{r}}_j$ and $\dot{\vec{\omega}}_i\equiv\dot{\vec{\omega}}_j$. In view of (16) this represents six linear equations for $\ddot{\underline{q}}$ for each closed loop. At each step in the course of numerical integration the set of all linear equations is numerically solved for a set of dependent accelerations in terms of the remaining set $\ddot{\underline{q}}^*$ of independent accelerations. This results in numerical values for the matrices $\underline{G}$ and $\underline{H}$ of (25). For the elements of $\underline{q}$ themselves implicit nonlinear constraint equations are autimatically constructed. They express the fact that (i) a sum of body-fixed vectors around each loop is zero and (ii) the ordered product of the direction cosine matrices of all joints in a closed loop is the unit matrix. For spatial closed loops (i) and (ii) together represent 12 scalar equations and for planar closed loops three scalar equations. These equations are automatically constructed from kinematical quantities for the tree-structured system. At each integration step the equations are iteratively solved. Only few iterations are required since the solution valid for the previous integration step is a good starting point.

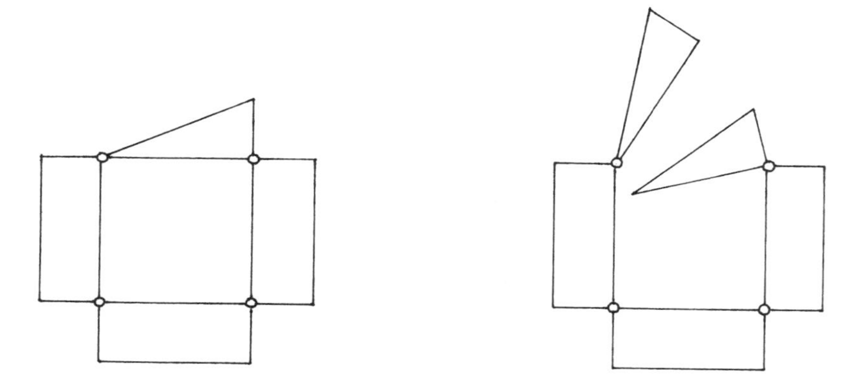

Fig.4: A closed kinematical loop formed by four bodies (a) and the tree-structured system after replacing one body by two twin bodies (b)

5. GENERALIZED REDUCED MASSES

We consider a tree-structured system of n bodies in purely translatory motion for which (16) and (21) result in the equations

$$\delta\underline{\vec{r}}=-(\underline{\vec{k}T})^T\delta\underline{\dot{q}}, \quad \ddot{\underline{\vec{r}}}=-(\underline{\vec{k}T})^T\underline{\ddot{q}}. \tag{26}$$

This is the case if in each joint j the translation of body $i^-(j)$ relative to body $i^+(j)$ has constant direction. As an example consider a system of triangular bodies in planar motion such that neighboring bodies slide along one another. If the system has no other constraints (14b) applies in the special form $\delta\dot{\underline{\vec{R}}}^T\cdot(\underline{m}\ddot{\underline{\vec{R}}}-\underline{\vec{F}})=0$. Together with (15) and (26) this yields the equations of motion

$$\underline{\vec{k}}\cdot\underline{T\mu m}(\underline{T\mu})^T\underline{\vec{k}}\underline{\ddot{q}}=-\underline{\vec{k}T\mu}\cdot\underline{\vec{F}}. \tag{27}$$

These equations are particularly simple if all vectors in the matrix $\underline{k}$ are collinear as is the case for the system of Fig.3, for example. The equations then read $\underline{T\mu m}(\underline{T\mu})^T\underline{\ddot{q}}=-\underline{T\mu F}$. In this case as well as in (27) the coefficient matrix of $\underline{q}$ is a constant matrix with the dimension of mass. It is easy to show that

$$[\underline{T\mu m}(\underline{T\mu})^T]_{ij}=\begin{cases} m_{ii}(m_{total}-m_{ii})/m_{total} & i=j \\ m_{ij}m_{ji}/m_{total} & i\neq j \end{cases} \qquad i,j=1\ldots n. \tag{28}$$

where m_{ii} and m_{ij} are defined as follows (see Fig.3 without the spring k_6). Cut the connections i and j. If i=j then the system is split into two parts. The masses of these two parts are called m_{ii} and $m_{total}-m_{ii}$. In the particular case i=j=1 one of the two masses is zero. For i≠j the cuts split the system in three parts, namely one part "between the connections i and j" and two "outer parts". The masses of the outer parts represent m_{ij} and m_{ji}. If either i=1 or j=1 one of the two masses is zero. (28) defines a matrix of reduced masses with zero elements in the first row and in the first column (if the system has the labeling of Fig.3). For a two-body system the only nonzero element of the matrix is the familiar expression $m_1m_2/(m_1+m_2)$.

6. SUMMARY

The paper establishes some results of theoretical interest and some methods for practical applications in dynamics of multibody systems. The incidence matrix of a system graph and the path matrix of a spanning tree of the graph are used for reducing a system of differential equations with eigenvalue zero to a smaller system without eigenvalue zero. The matrices are also used for an automatic formulation of loop closure constraints for multibody systems with closed kinematical loops. Finally, the path matrix is used for a generalisation of the classical term reduced mass to systems with more than two bodies.

LITERATURE

1. Wittenburg, J., Dynamics of Systems of Rigid Bodies; ser. LAMM vol.33, Teubner 1977

2. Wittenburg, J., Analytical Methods in Mechanical System Dynamics; in Computer Aided Analysis and Optimization of Mechanical System Dynamics (Editor Haug, E.G.), ser.F: Computer and Systems Sciences, vol.9, Springer 1984

3. Wittenburg, J., Analytical Methods in the Dynamics of Multibody Systems; Proc. IUTAM-ISIMM Symp. on Modern Developments in Analytical Mechanics, Turin 1982

4. Busacker, R.G., Saaty, T.L., Endliche Graphen und Netzwerke, Oldenbourg 1968 in English: Finite Graphs and Networks. An Introduction With Aplications, New York 1965

5. Branin, F.H., The Relation Between Kron's Method and the Classical Methods of Network Analysis, Matrix and Tensor Quat., vol.12, 1962, 69-105